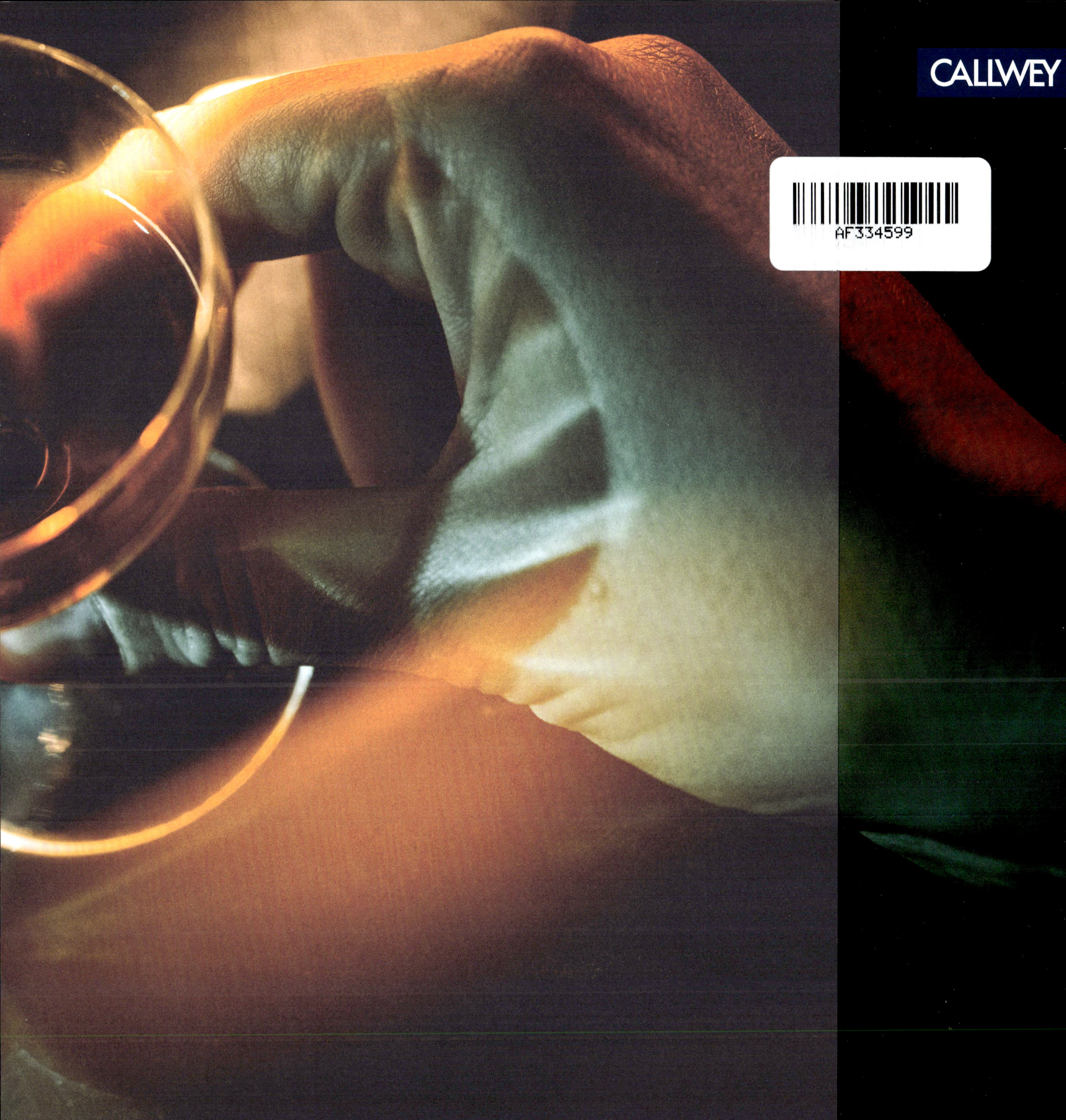
CALLWEY
AF334599

THE BEST NIGHTS OF YOUR LIFE

THE ORIGINAL Jägermeister BOOK

ANJA DELASTIK

Contents

OUR MEISTERS

Kerry King — 12
Bernd Frank — 16
Deva — 20
Paul Breitner — 24
Alfred Kogler — 28
Dizzy Reed — 30
Hans-Joachim Stuck — 34
The Jägermeister Tattoo Clique — 38
Evil Jared — 40
Wolfgang Spurek — 42
Holger Stonjek — 48
Raz Degan — 52
Chefboss — 56
Greg Koch — 60

OUR HISTORY

The origin story — 66
A family since 1878 — 78
In the fast lane — 86
Eintracht Braunschweig — 98
Jägermeister Open Air — 108
Halali – the happy hunter's call — 128
A star is born — 138
Music to everyone's ears — 146
Product and lifestyle in harmony — 156

OUR TASTE

OUR BEST NIGHTS

Familiar diversity _______ 168

The Shot _______ 178

International inspiration for the perfect drink _______ 188

Brand ambassadors

— Nils Boese: The Flavour Wizard _______ 194

— Florian Beuren: The Showman _______ 196

— Willy Shine: The Brand Meister _______ 198

— Sabrina Traubner: The Master-Class Student _______ 200

— Lukáš Čabaj: Young And Wild _______ 202

Wolfenbüttel: for a Jäger at Theo's _______ 208

Berlin: mastering the art of partying _______ 212

New Orleans: liqueur shots on Bourbon Street _______ 216

Dubai: the height of ice-cold pleasure _______ 218

Spain: ultimate house party _______ 220

Buenos Aires: deer crossing _______ 222

Shanghai: "Ye Ge" in Shanghai's underground scene _______ 226

Switzerland: ice-cold heat _______ 230

Czech Republic: ice-cold apparition _______ 234

Netherlands: Rudi and Ralph, the pert Dutch stags _______ 236

Sibiria: the coldest gig of all time _______ 238

Hollywood: shot moments on the big screen _______ 240

South Africa: loud and vibrant _______ 242

India: the 'ice cold tour' _______ 248

Israel: memories for eternity _______ 250

Whenever you see this code, you can enter the Jägermeister world experience in AR. Simply use the app on your smartphone or open your tablet, hold them over the code – and off you go!

DEAR JÄGERM FAN,

What do you associate with Jägermeister? What is the first story you would tell me? Almost every time I ask this question, people's eyes light up and the conversation is filled with emotion and great experiences. Ask a hundred people what they associate with Jägermeister, and you'll receive countless different answers. You'll probably also hear a few stories and anecdotes no Hollywood director could ever have possibly thought up – at least not if they want to make an even remotely realistic film. Jägermeister is much more than just a herbal liqueur or lifestyle brand! Jägermeister is always around. Whether celebrating success, dancing through the night, or enjoying a relaxed après-ski shot at the bar, we all drink Jägermeister: young and old, me and you, our neighbors, people on the other side of the country, and in more than 150 other countries around the globe. The stag-

Jägermeister brand has thus always been very personal for me. Jägermeister has been inextricably linked with my family since day one. The brand continues to shape me and my family, and is omnipresent, whether it be in relation to our family history or as we look to the future together. Jägermeister has become part of our DNA – and that is why the attitudes and values of the Mast family are also palpable within the company and brand – values such as humanity, open-mindedness, tolerance and sustainability, values we want to instil in our children and later even our grand-

EISTER

bearing brand is different – both universal and unique. It embodies perfect harmony through contrast: traditional yet always up with the times, tied to its roots yet at home all over the globe, perfectionist and detail-oriented yet untamed and wild. Jägermeister has always connected people, creating for them – and with them – the best nights of their lives. The brand awakens both the hunter *(Jäger)* and master *(Meister)* in all of us, and has been doing so for generations.

Jägermeister was invented by Curt Mast, my great-grandfather. 'Curt had a highly developed sense of taste,' my grandmother would often say about her father, who spent so much time in his laboratory at the 'Stammhaus' composing spirits. As his great-grandson, the significance of the

children. I myself represent the fifth generation of the Mast-Jägermeister family business, and want to do my bit in ensuring the next generations can live in a world where diversity, joy, community spirit and nature can thrive. For me, this is also part of what the Jägermeister brand stands for.

The product that gave rise to the brand is difficult to describe fully in just a few words, so I'm not even going to try here. All I will say is that in this world-famous, semi-bitter herbal liqueur, apparent opposites are fused into a unique whole through the skilful artistry of our distillers.

Now would be the perfect opportunity to air the secret of this carefully guarded recipe that has remained unchanged since Jägermeister was first invented, known to only a handful of people of

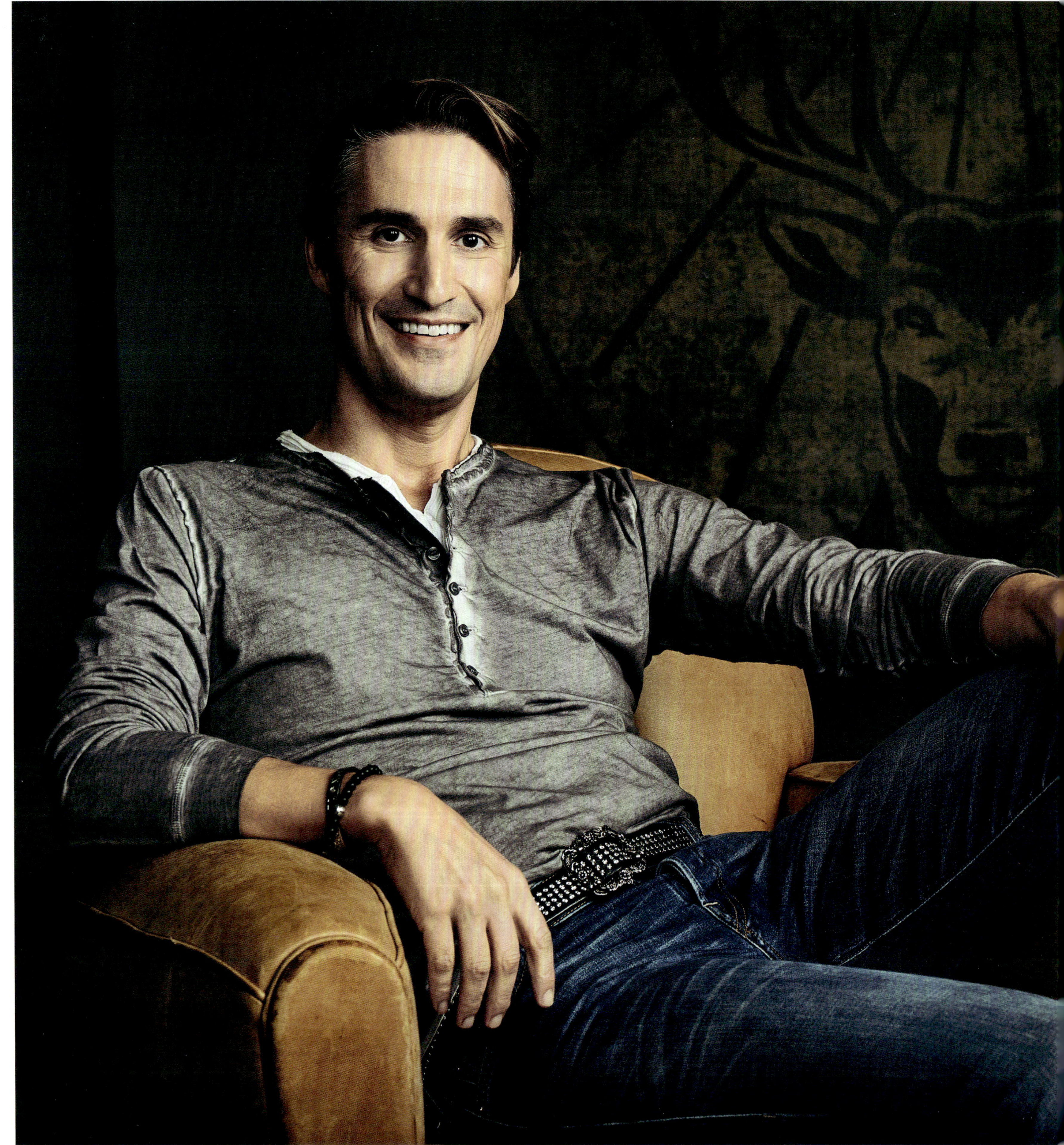

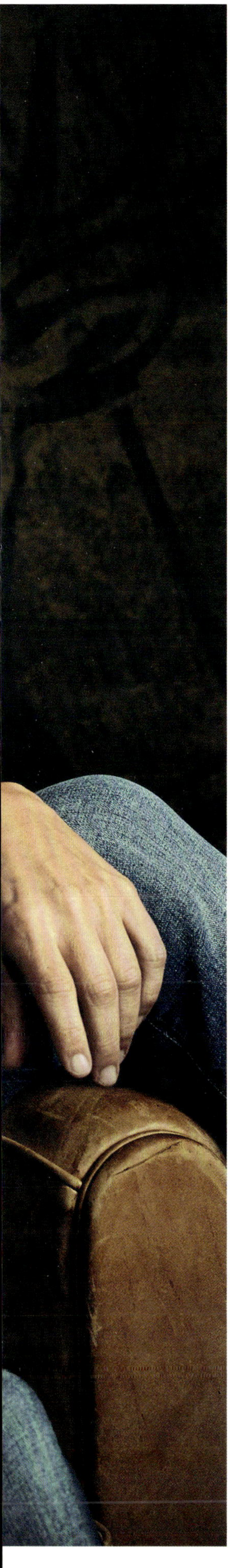

which I am one. However, as this recipe secrecy is also what adds to the Jägermeister legend, I shall resist the temptation to reveal anything. But, to this day, the herb warehouse and production plant are probably my favourite areas of the company. The smell there is amazing! You need only see a small selection of the ingredients to get an idea of the diversity of herbs, roots and blossoms amassed from all over the world. Ever since my great-grandfather Curt Mast invented Jägermeister in the German town of Wolfenbüttel, we have been putting a strong love of detail and great effort into using these ingredients to create a product that, in every corner of the globe, has become synonymous among its fans with an entire category of spirits: herbal liqueur. But very few people know how much emphasis we place on quality, and the fact that every drop of Jägermeister is stored in an oak barrel for around 12 months.

I don't know if my great-grandfather realized at the time how much of a success story the Jägermeister he invented and loved so much would be when he set about applying his perfectionism and love of experimentation to creating a spirit that hunters could use to celebrate their community. That liqueur designed for the hunting community has long since become a legendary shot for the broader community, whether they live in Wolfen-büttel, Berlin, New York, London, Prague, Cape Town, Rio de Janeiro, Shanghai, Moscow or any-where else on the planet. No matter what language we speak, or what our preferences are, we all celebrate together with Jägermeister.

'All for one and one for all' was the motto of the three musketeers, and this was picked up by German band Die Toten Hosen in their 1996 song *Zehn kleine Jägermeister*. Without the diversity of the world and the people growing the ingredients, there would be no Jägermeister. So I find it fitting that this shot, which combines ingredients from around the globe, also brings together people from all corners of the earth – even if it is for just one legendary night.

How is it possible to encapsulate all these aspects of Jägermeister in one single book? That's the question that was raised at the very start of the design process of our first-ever Jägermeister book, which you're holding in your hands right now. The answer soon became clear: it could only be achieved by making a book that celebrates this unique brand – a book providing insights into more than eight decades of Jägermeister, and conveying its diversity. This aim was enough of an incentive for us.

We faced the challenge, collating stories that had never been told in such a way before. The first original Jägermeister brand book is not a chrono-logical history book with a start and finish; it is designed to be browsed and leafed through. I invite you to embark on this journey to discover the world of Jägermeister, and I hope you enjoy it.

Perhaps this book will bring back your own memories of unforgettable nights spent under the sign of the stag. And if there's still space in your games room, rehearsal room or elsewhere in your home, I strongly recommend taking a look at the inside cover. I wish you much joy with this book!

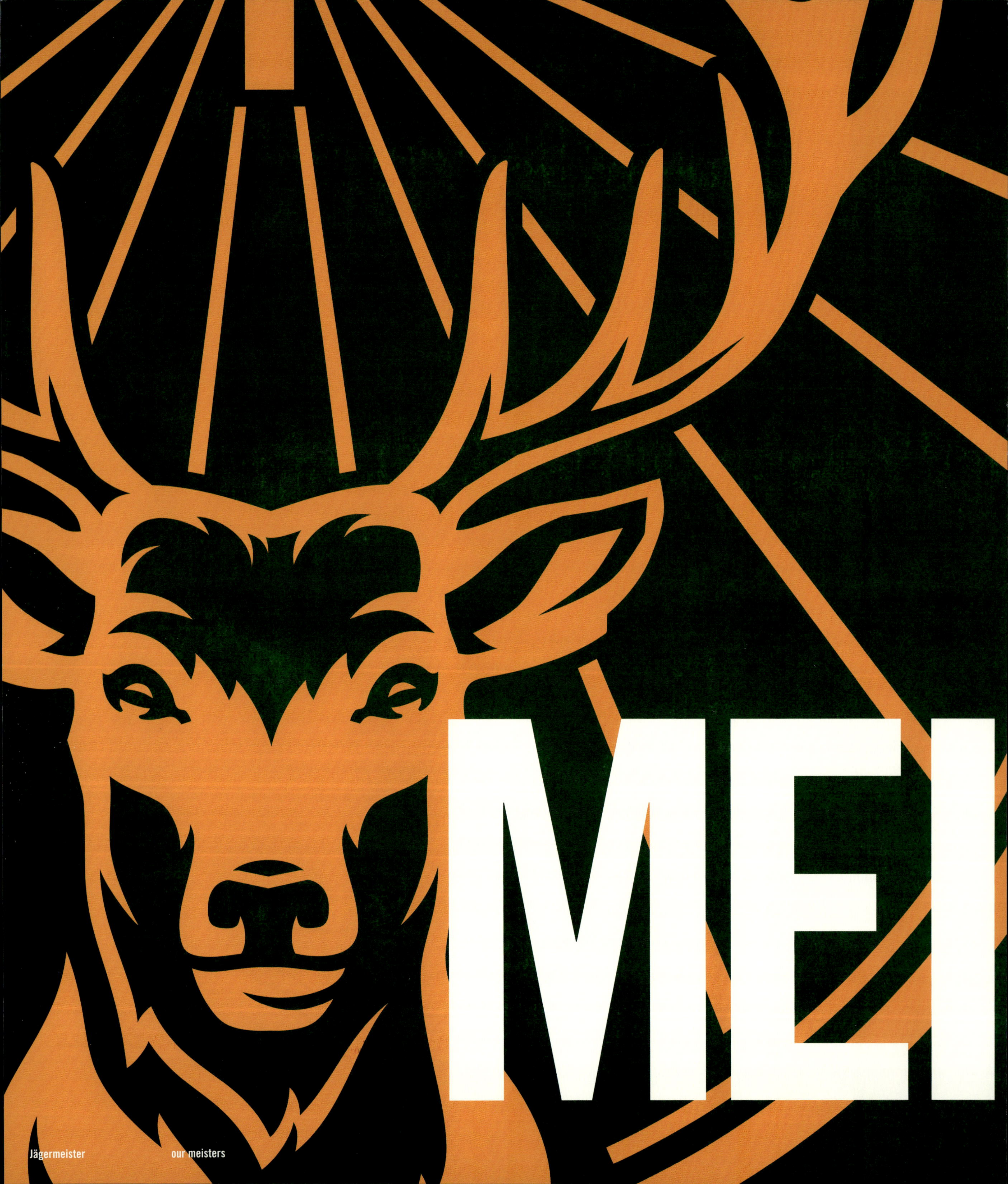

MEI
Jägermeister
our meisters

Kerry King — 12
Bernd Frank — 16
Deva — 20
Paul Breitner — 24
Alfred Kogler — 28
Dizzy Reed — 30
Hans-Joachim Stuck — 34
The Jägermeister Tattoo Clique — 38
Evil Jared — 40
Wolfgang Spurek — 42
Holger Stonjek — 48
Raz Degan — 52
Chefboss — 56
Greg Koch — 60

OUR
STERS

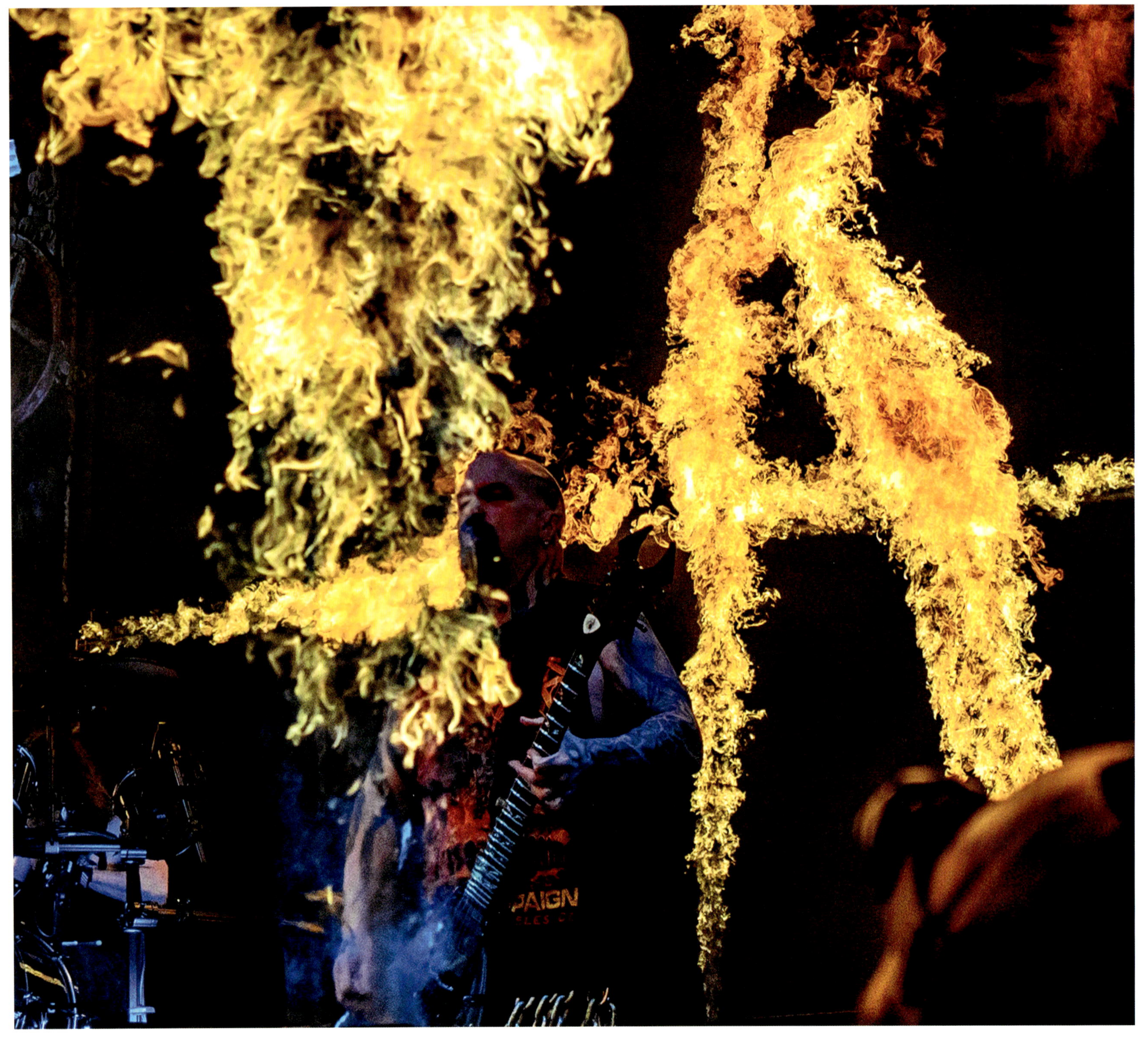

KERRY KING

A conversation with
Slayer's former guitarist

His riffs wrote music history. As the co-founders of a new form of heavy metal known as thrash metal, Slayer and their legendary guitarist Kerry King delighted head-bangers all over the world for nearly four decades. Their album *Reign in Blood* created a paradigm shift in 1986, with fans and critics still rating it as the best thrash record of all time. Slayer played their last-ever concert in Los Angeles in late November 2019, but King is continuing to work on new songs and music projects – and Jägermeister is always on hand, just as it always has been, ideally frozen at minus 18 degrees. But it's much more than just ice-cold drinking pleasure that connects the herbal liqueur and the bald-headed guy with eye-catching tattoos. In 2012, the Californian appeared in the very
first Jägermeister commercial to be broadcast right across the United States, and, with Slayer, he played in more Jägermeister tours than any other band. But that's far from all . . .

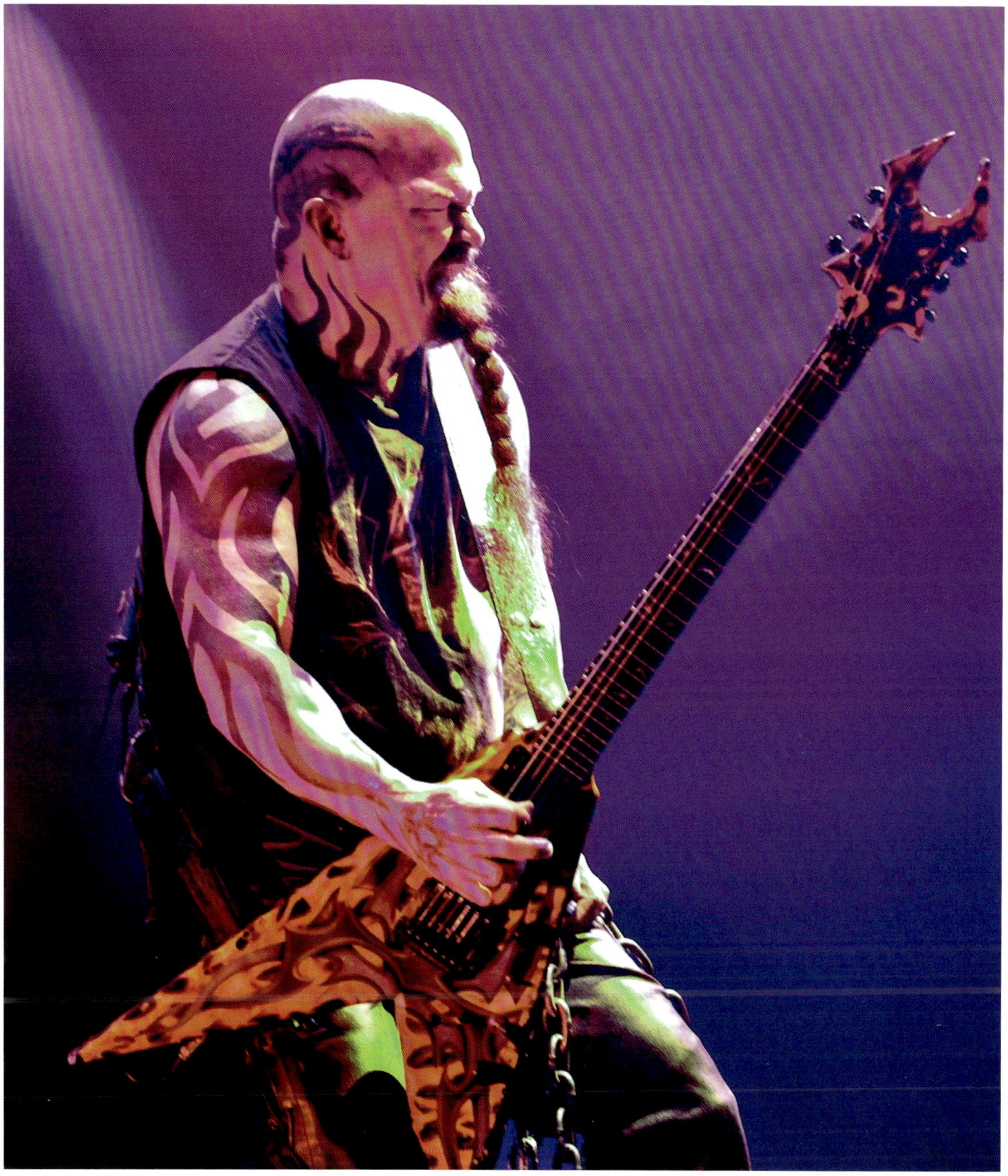

Was there Jägermeister even after the very last Slayer show?

Whenever we drank shots, they would always be Jägermeister, and we kept up this tradition until the very end. If someone wanted to hold a backstage party, they just had to look for my room. It always had ice-cold Jägermeisters.

Slayer did three big Jägermeister tours in 2003, 2004 and 2010. Why?

Jägermeister is unique – there is no other brand that connects me with so much. I like building strong relationships with people who believe in me, and Jägermeister has always supported me and the band. We have a shared past; we're family. And I am still close friends with some of the managers from back in the day. Our first joint tour set entirely new standards, and I can honestly say I'm proud of that.

There have no doubt also been other unique moments in your career to date?

I played in a band for 38 years, and would still be doing it if everyone else had wanted to continue. Whichever way you look at it, we achieved some legendary things. Playing sold-out shows on two consecutive nights at The Forum in Los Angeles, my hometown – that was epic. But so was the very first time we ever performed there, four or five years ago . . . We played to a full house there four times in total. If someone had told me that would happen, I would have asked them if they were smoking crack *(laughs)*! And the super-amazing metal-only festivals we constantly played at – Full Force, Wacken, Hellfest – with all their incredible line-ups. That sort of thing is important to me; that's where our audience is.

You're clearly still a metal fan at heart . . .

Absolutely! Judas Priest, my absolute heroes from pre-Slayer times, have just released their best album in 15 years! And I'm sitting here, without a gig . . . *(laughs)*.

What changes in your life have you seen post-Slayer?

I had a long time to prepare for it. When, six months after the decision to stop, the split was announced, I had already made peace with it. In terms of making music, it's business as usual for me. I have already written a bunch of new songs. I

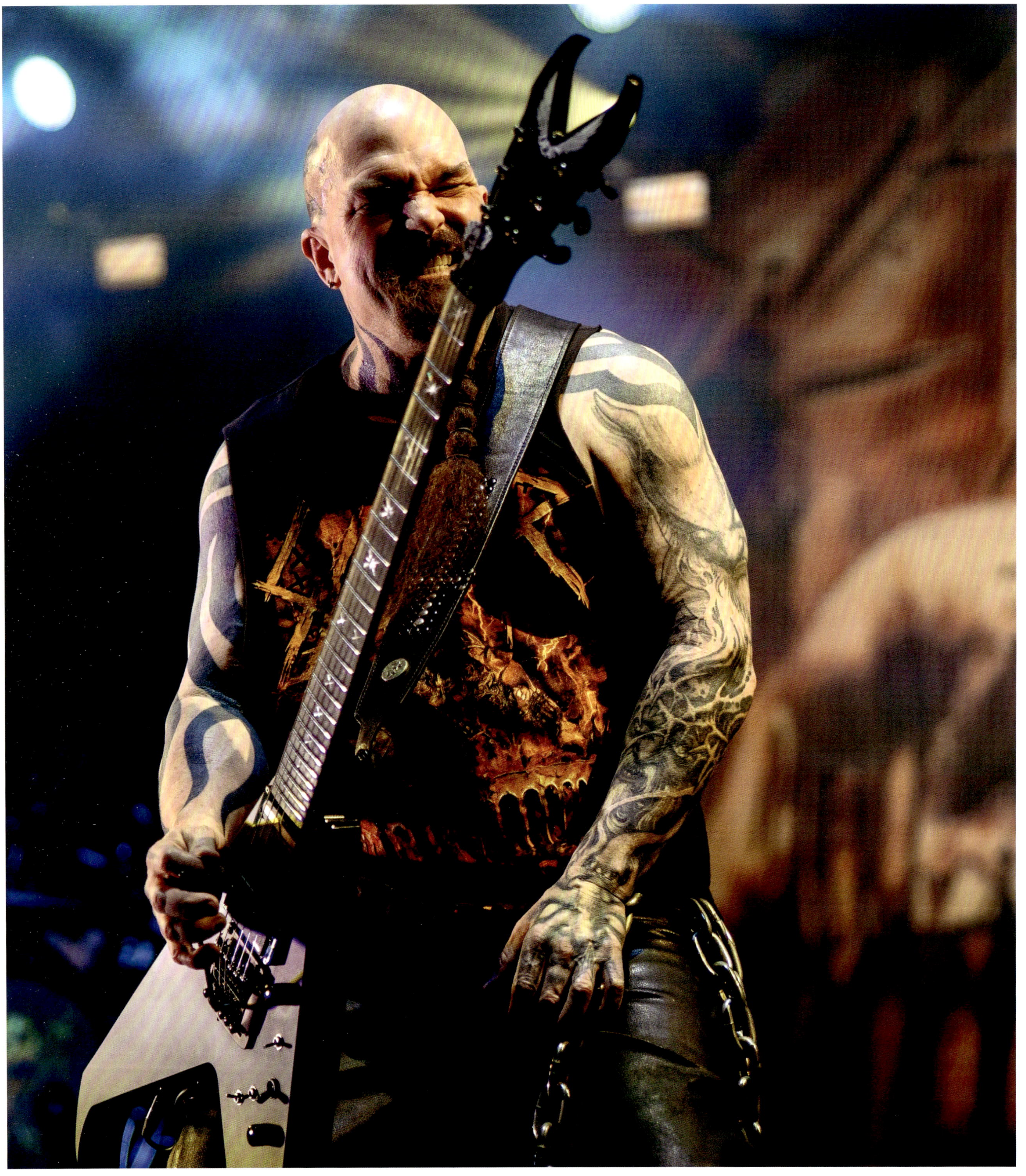

will record the best ten or eleven of them with my musician friends, and release them if the time is right. It's good stuff!

Speaking of good stuff: to celebrate your legacy, Jägermeister brought out a strictly limited edition 1.75 litre bottle as a special Slayer edition in 2019 . . .

Only 500 of these specials were produced at the time, and they sold out within only three days. I feel honoured, and am proud of it – but why only 500? I'm sure they would also have been able to sell 10,000 (laughs)! At least I was able to get hold of a few bottles.

There was also a Slayer Jägermeister Tap Machine . . .

Yes, Jägermeister had one made for us with our logo. We gave it to Jimmy Fallon as a gift the second time he invited us on to *The Tonight Show*, where we were due to perform *Raining Blood*.

Can you think of any other cool Jägermeister moments?

Yes, I can, actually, I still remember one particular concert – I think it was in Canada. At the Jägermeister stand, I saw this super-cool Jägermeister bike, some kind of beach cruiser, matte black with orange rims. So I went over and said, 'Hey, you guys know you now have to get one of these for me, right?!' And now that very bike stands in my garage (laughs).

Do you ride it?

Yes, the area around my new house in Las Vegas is nice and flat, so a chilled bike like that is just perfect for me.

But you still have another Jägermeister accessory, the 'Jäger bling' . . .

Someone gave us these Jägermeister gold chains. Our drummer Paul Bostaph and I were the only ones who didn't lose ours straight away. So we were the cool guys with our chains, drinking Jägermeister after the shows. At one point, Paul asked Jägermeister if they could send us another couple of chains. We were thinking of only a handful, but

they sent us a staggering forty. Whenever someone came backstage to drink a shot of Jägermeister, Paul and I would deck them out with one of these chains (laughs).

Did all the band members join in the drinking?

I could definitely always rely on Paul. In the last five years, we even introduced 'penalty shots', which involved picking up when the other person had played a wrong note during a concert. It wasn't easy, because they would often just be tiny mistakes in very complex passages, which most fans would probably never even notice. As long as you start and stop playing at the same time, it's all good (laughs).

When did you have your last 'penalty shot'?

Probably the last time I played a round of darts. That's one of my hobbies. The rule is that the loser has to buy a round of Jägermeister for everyone else! No discussion!

You call yourself an experienced drinker. So what's your expert take on how to best drink a Jägermeister?

For most Americans – and I assure you I can speak for most of them, as I was the one who familiarized them with the drink – ice-cold is the only real way to drink Jägermeister. And I don't mean frozen like ice, of course. But it really needs to have come out of the freezer. The fridge is okay, better than lukewarm, of course, but not as good as frozen. It took an eternity to get that through to some people. Once I had discovered Jägermeister, I was so happy when we went to tour Germany, because I thought, 'That's the home of Jägermeister – it will be everywhere!' But that didn't in fact go quite as I had expected.

What happened?

I went to the bar at a Hilton hotel and ordered a Jägermeister. But they looked at me like I was actually crazy. The herbal liqueur wasn't at all as culturally established in the country back then as it is today. It's funny, but Germany was pretty late in getting the memo that, 'Hey, it tastes really good ice-cold!' ⚜

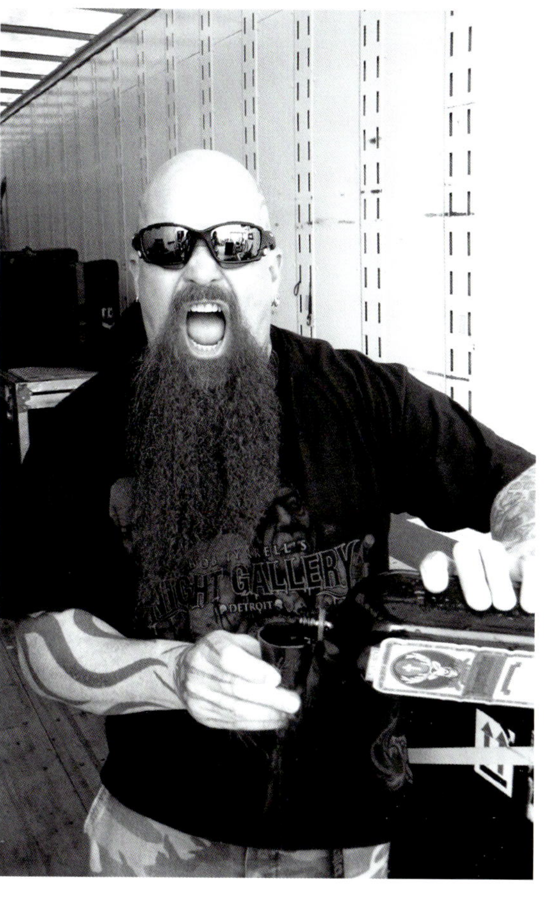

In the last five years, we even introduced "penalty shots", which involved picking up when the other person had played a wrong note during a concert. It wasn't easy, because they would often just be tiny mistakes in very complex passages, which most fans would probably never even notice.

JÄGER(GARDIST) AND COLLECTOR

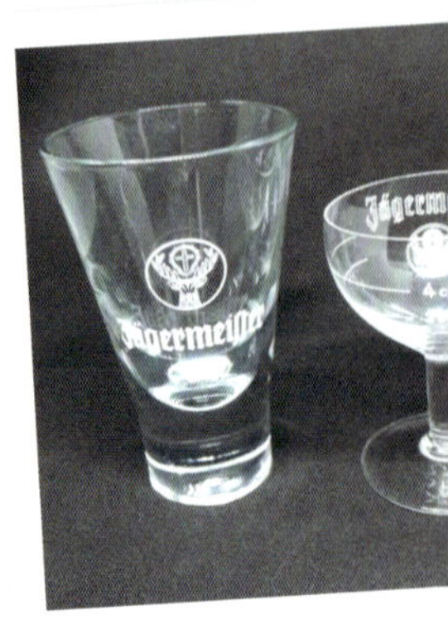

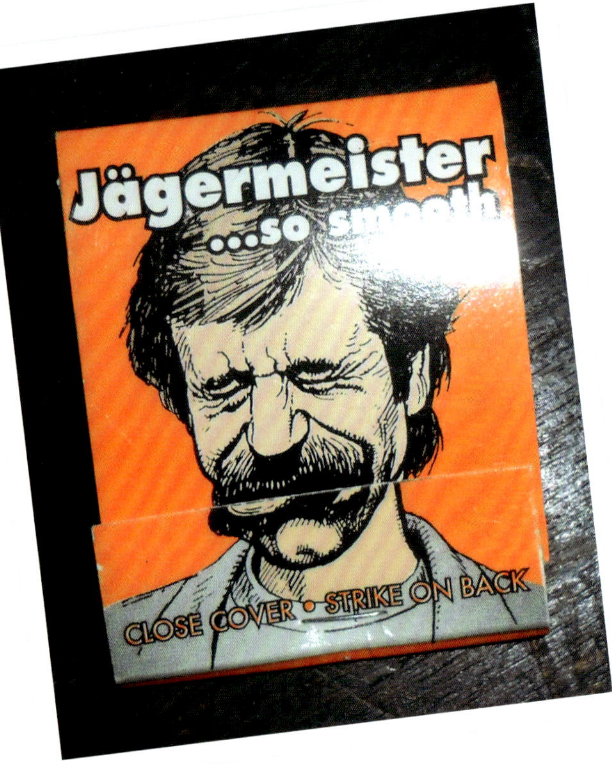

Jägermeister has fans all over the world, but none of them are quite like Bernd Frank. Anyone visiting the designer's home in his adopted town of Mainz will feel as though they have stepped into a museum, for he has one of the most amazing collections of Jägermeister memorabilia anywhere.

In his hallway and his living room various cabinets display Jägermeister garters from Halloween parties, rare glasses from the 1950s and 60s, and a 3-litre bottle of Jägermeister that he is particularly proud of. 'They used to come either as sealed, empty ornamental bottles for display windows or filled,' Frank says. 'Barely any have survived. Before 1954, the bottles were white, not green. Mine is a green display-window bottle.' His greatest wish? 'A full, white bottle in an original wooden box, the way they used to come at the time. That would be my dream! I know of two that exist, but have never seen them for sale anywhere – and even if one were put up for sale, it would probably be too expensive for me anyway.'

He is in contact with other collectors, both through his *Jägermeister Sammlerbörse* (Jägermeister collectors' market) Facebook group, but particularly via his blog, Hochsitz-Cola (www.hochsitz-cola.de) in order to unearth, document and exchange absurd curiosities and valuable unique editions like his window display bottle. This is where Frank shares his expertise, his trophies, and his passion, which began in the late 90s with the legendary 1973 advertising campaign 'I drink Jägermeister because . . .' – 'In my opinion, the campaign is unparalleled in advertising history, just like a lot of other things Jägermeister has done,' says Frank. He started collecting the ads. 'I now have about 1,000 of them; I've never

properly counted them. I had the good fortune of being able to purchase another collection at an auction. The seller, who was a passionate collector, was super-happy when I told him I would be adding them to my collection – he had been afraid somebody would use them to wallpaper their bathroom!' he says. Instead, Frank is now gradually publishing them chronologically on his blog, and also collecting anything and everything else relating to Jägermeister: 'Mainly old promotional items – the older, the better!' One particular highlight was his first visit to the company's Wolfenbüttel headquarters in 2001. 'At the time, a friend and I were given a personal factory tour by the then CEO. Afterwards, we sat for hours with the ad director and were able to go into the advertising archive in the basement,' he enthuses. 'We were even given some rare items, including samples that never went into production.' His favourite object is a simple cardboard stand-up display. On the left is a hunter with binoculars, in the middle is a fawn seen through the binocular lenses, and on the right is a bottle of Jägermeister. 'It's so absurd from a present-day perspective,' laughs Frank. 'This ad wouldn't entice anybody to buy even a drop of Jägermeister. But it was genius at the time, and the latest thing in an era when *Heimat* films idealizing regional German life were still blockbusters at cinemas in this country.'

Frank says, 'I drink Jägermeister because I don't need a hunting licence for my collection.' But it's not just his passion for collecting, nor his height of 2.04 metres that has made this native of Upper Franconia well known right across Mainz. He is field marshal of the *Meenzer Jägergarde* (www.jaegergarde.de [German only]). 'As an adopted Mainz local, I am a huge fan of the *Kneipenfastnacht* (the pub carnival),' he says, 'so it was only logical for me to combine my passion for collecting with the carnival by giving my costumes a Jägermeister theme.' Initially haphazardly thrown together, his costumes have grown more professional each year, to the point where he has become known simply as 'the Jägermeister'. His first 'real' uniform came in 2014, based on that of the fire brigade, when he made the decorations himself, crafting the epaulettes out of Jägermeister coasters. The finished uniform ultimately gave rise to the *Jägergarde*, but it is also special for another reason. 'I'm not actually

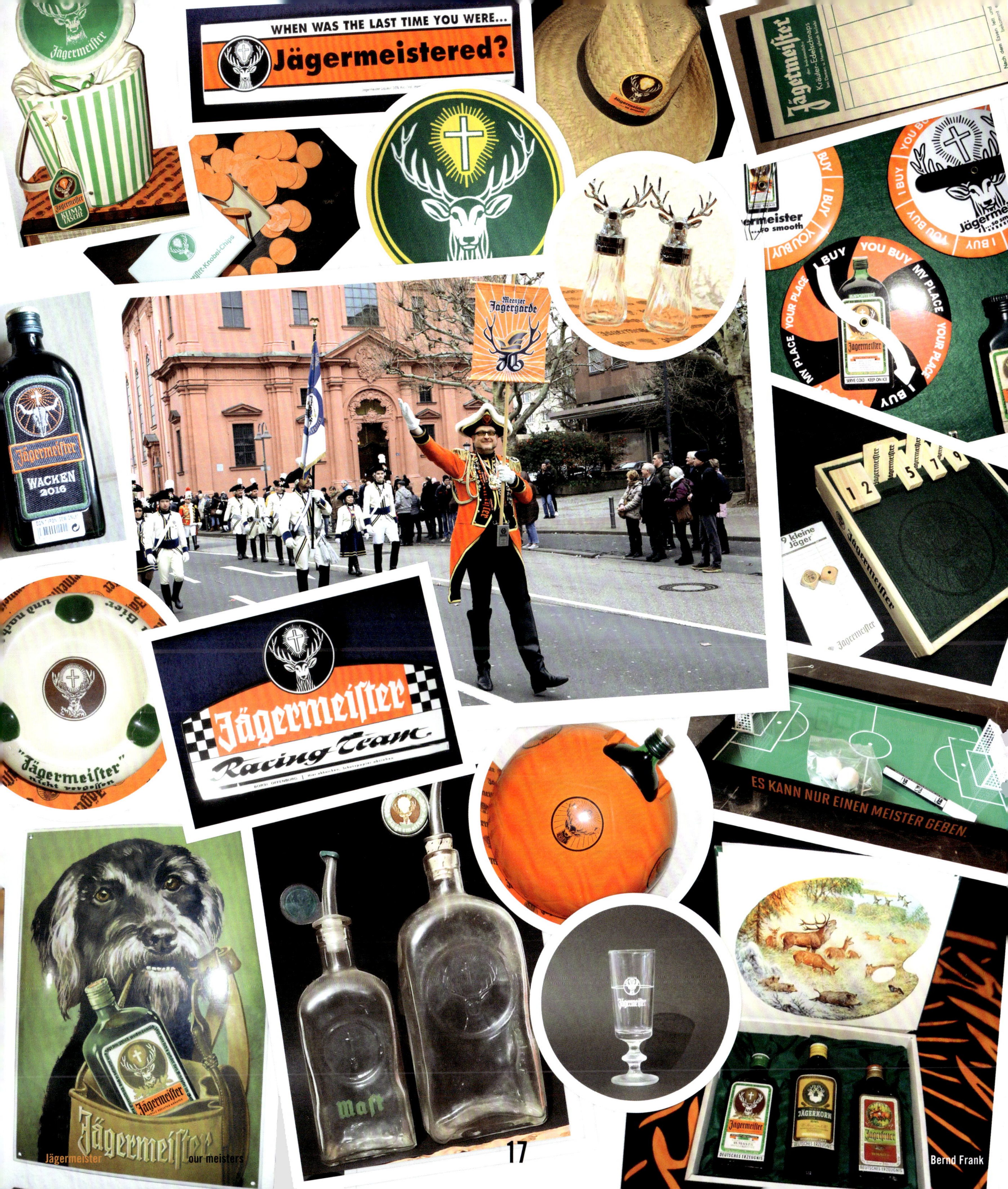

WHEN WAS THE LAST TIME YOU WERE...
Jägermeistered?
WACKEN
2016
Menzer
Jägergarde
Jägermeister
Racing Team
ES KANN NUR EINEN MEISTER GEBEN.
Most
Jägermeister
I BUY
YOU BUY
MY PLACE
YOUR PLACE
JÄGERMEISTER
JÄGERKORN
Jägerfeur

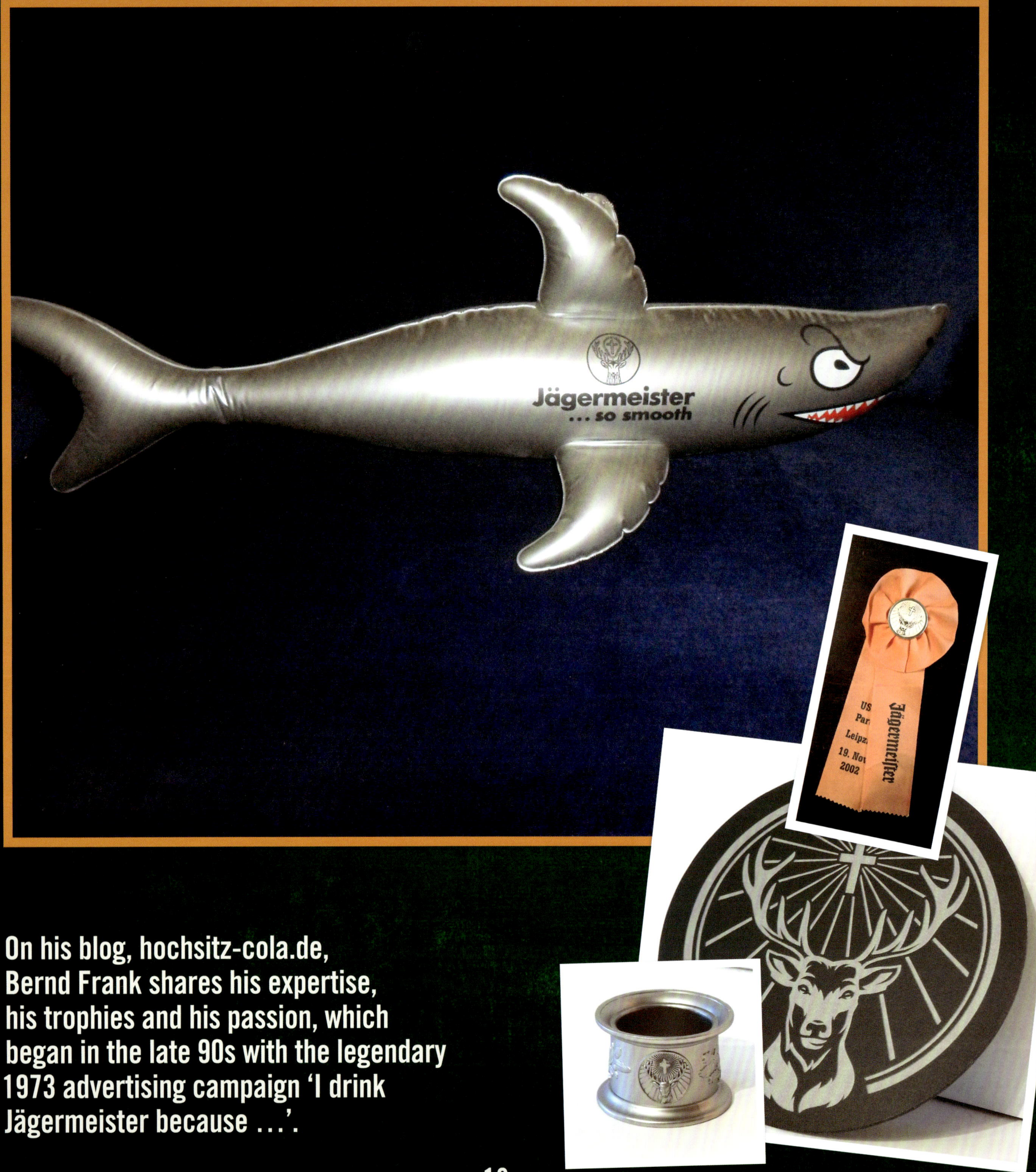

On his blog, hochsitz-cola.de, Bernd Frank shares his expertise, his trophies and his passion, which began in the late 90s with the legendary 1973 advertising campaign 'I drink Jägermeister because …'.

'I drink Jägermeister because I don't need a hunting licence for my collection.'

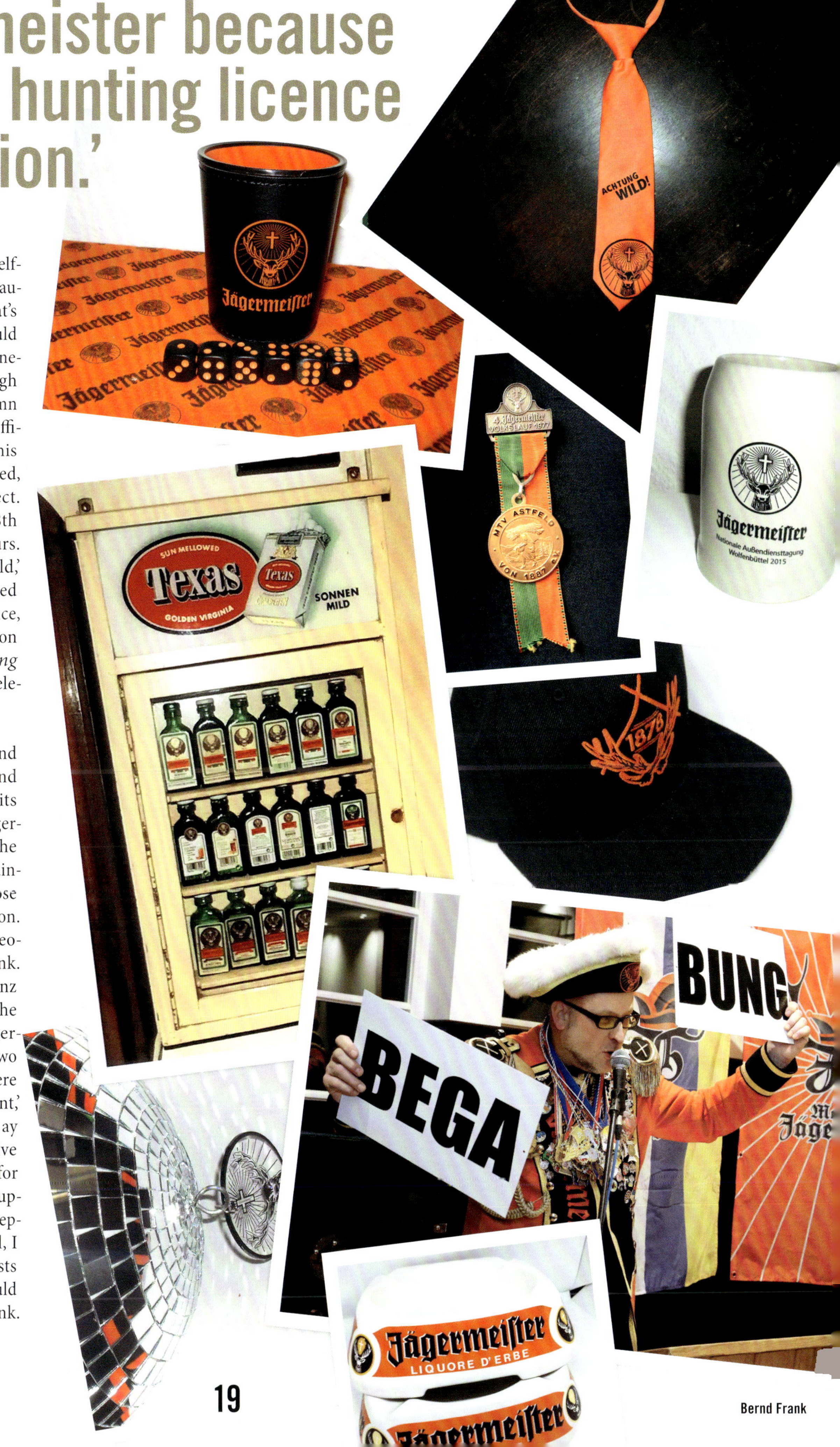

the type of guy who would join a guard,' says the self-confessed exhibitionist. 'I had no interest in the bureaucracy, club cronyism or lack of individuality. That's why it was clear right from the outset that there would only be one member of the Jägergarde.' But as the one-man guard hadn't been able to build a suitably high profile, Frank returned to Wolfenbüttel in the autumn of 2015 to ask for the managers' permission to officially use the Jägermeister lettering and logo for his new, more professional uniform. Jägermeister agreed, and even helped Frank financially with his project. The new uniform drew on the style used in the 18th century, but with Jägermeister lettering and colours. 'For me, orange is just the best colour in the world,' laughs Frank. The 2016 New Year procession marked the Jägergarde's long-awaited first official appearance, with the unusual guardsman attracting the attention of local media, including the *Allgemeine Zeitung Mainz* and SWR (Südwestrundfunk, a German television channel).

This was followed by convoy-escort requests and invitations to receptions held by Mainz guards and associations. In 2018, the Jägergarde organized its own reception for the first time, sponsored by Jägermeister. In 2020, it made its first appearance at the Fastnachtsposse, one of the cornerstones of the Mainzer Fastnacht festival, and participatcd in its first Rose Monday [the day before Shrove Tuesday] procession. 'We are relentlessly pushing on with what some people initially thought was just a bad joke,' says Frank. 'We are now one of the most famous guards in Mainz and well beyond the city's borders.' He also has the Jägergarde to thank for giving him the best Jägermeister nights of his life. 'Our reception, to which two members each from many different associations were invited, has now become something of a cult event,' he explains. 'And, of course, Jägermeister is served ay such an event.' Jägermeister's Mainz representative organized large quantities of the herbal liqueur for one of these receptions, because the supply was supposed to last for another event as well. Plus, the reception was being held on a Wednesday night. 'Well, I thought to myself,' laughs Frank, 'most of the guests will have to work tomorrow so it's unlikely they would want to drink a lot and risk a hangover,' laughs Frank. 'But, in fact, there wasn't a drop left over.'

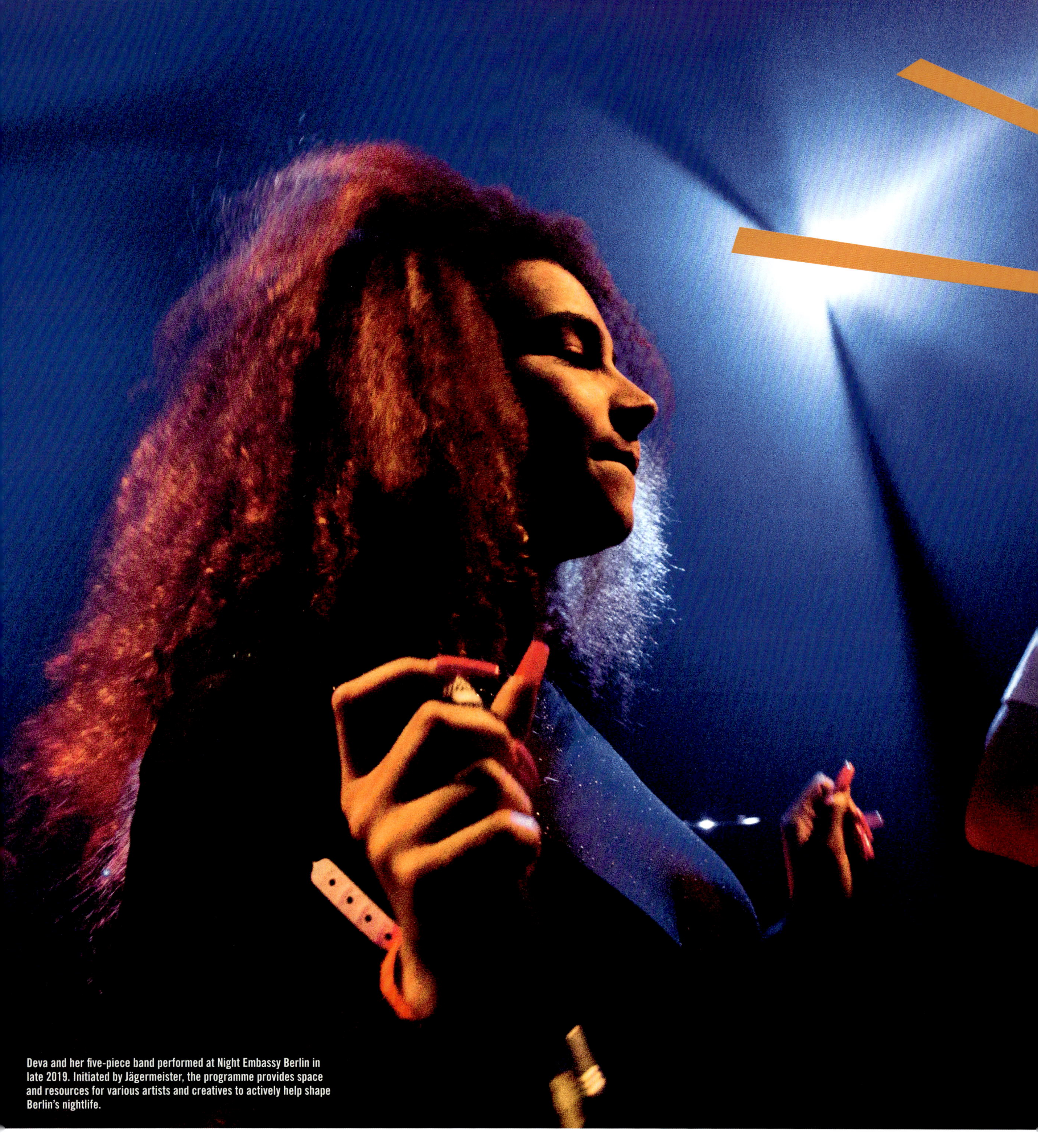

Deva and her five-piece band performed at Night Embassy Berlin in late 2019. Initiated by Jägermeister, the programme provides space and resources for various artists and creatives to actively help shape Berlin's nightlife.

THE MEISTER'S EXAM

To create something unique, you need to break new ground,
take your destiny into your own hands and set your own rules. This is
the inspiring message not only of Jägermeister's innovative 'Be the Meister'
corporate campaign, but also of the message of the song *¿Quien quieres ser tú?* –
'Who do you want to be?' The title tells the story of a young woman trying
to find her place in the world, while also remaining true to herself. And
it is basically also the story of its singer, young Spanish artist Deva.

I was always crazy about music, and wanted to make my own, but never had the opportunity,' says the Cantabrian musician, who has Anglo-Caribbean roots. 'It's a gift that I am able to sing.' So, one day, she started writing songs, eventually recording two of them at a small studio in her hometown of Santander in northern Spain and putting them on YouTube, with no big expectations. Sony Music soon had the 17-year-old on their radar, recognizing her potential and offering her a record deal. Even today, three years later, Deva can hardly believe her luck. 'I would never have thought making music would one day be my career.'

R 'n' B, soul, hip-hop, dancehall, trap – Deva combines influences from all these genres, yet her songs sound so fresh, sassy and modern that they can't be pigeon-holed. The Spanish team of Jägermusic, Jägermeister's music support programme, identified the char-ismatic newcomer's talent early on, offering her assistance. 'The people at Jägermeister have no idea how valuable their support was for me,' says Deva. 'Without it, I wouldn't be where I am today. I had very little experience at the time, particularly when it came to live performances', the artist adds. 'This co-operation was an important step for me to gain experience and grow.'

Her first-ever international performance was also made possible with Jägermeister's support, with Deva and her five-piece band playing at Night Embassy Berlin club in late 2019. Initiated by Jägermeister, the programme provides various artists and creatives with the space and resources to actively shape Berlin's nightlife. 'It was so much fun', Deva enthuses. 'It was our first time in Berlin and it was really exciting – the culture, the appreciation from the people, the food, the beer . . .' Then she laughs. 'We even mixed Jägermeister

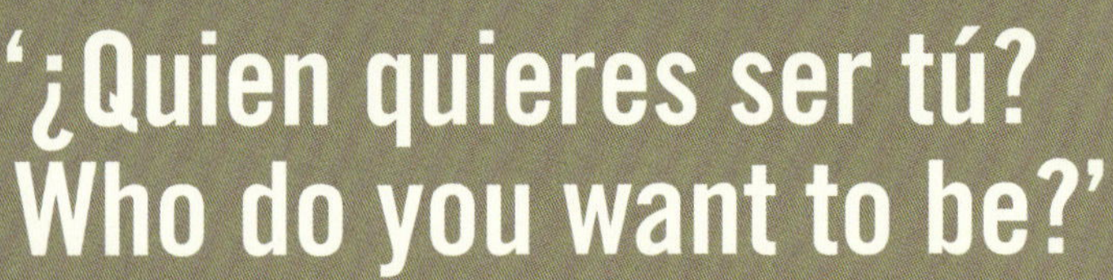

with beer, which was not a good idea!' But the musician does have one special tip she wants to share: 'I drink the herbal liqueur when I have pre-performance jitters, butterflies in my stomach or general collywobbles. One shot and everything's great and you're not nervous anymore.'

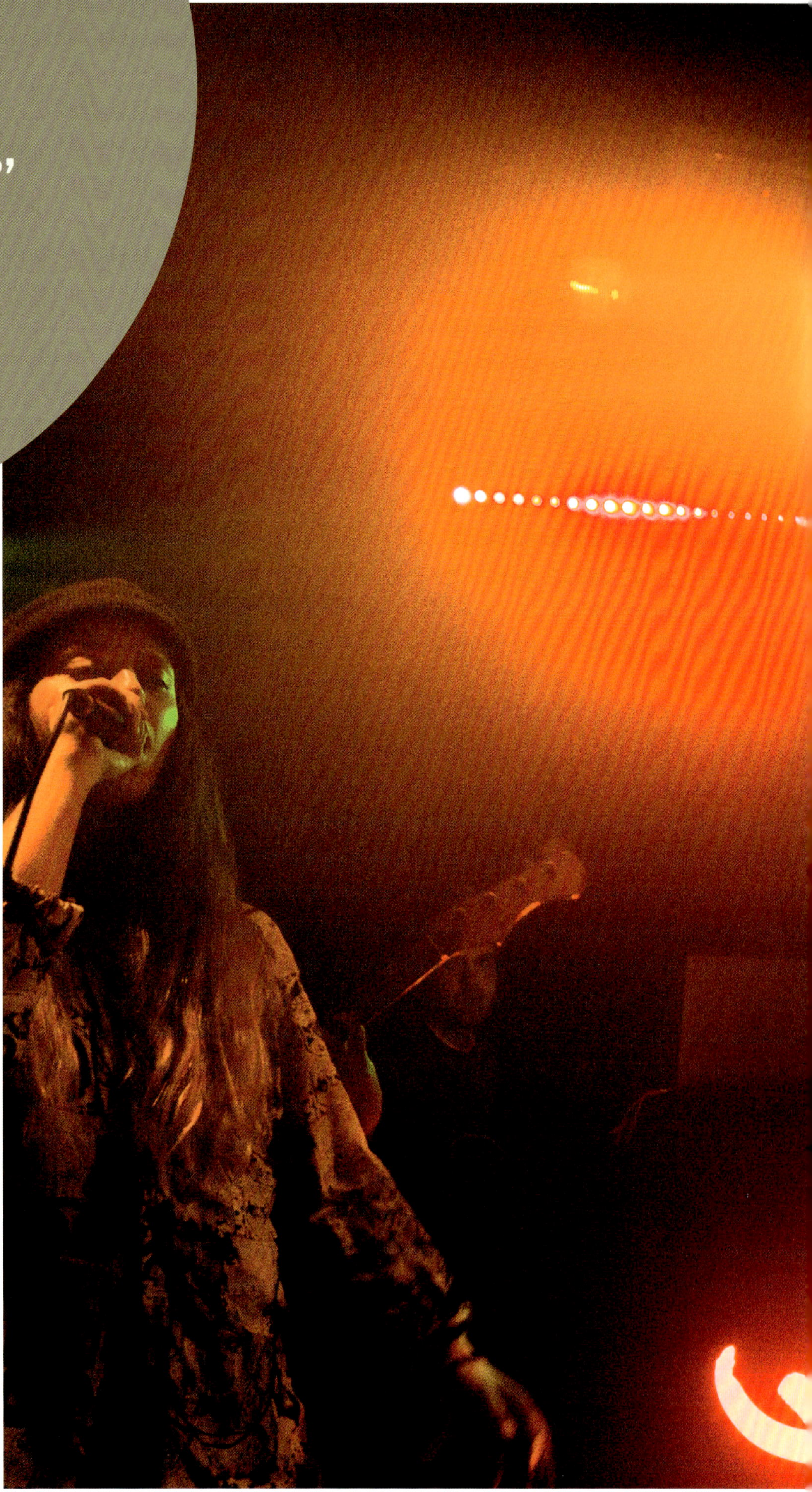

The newcomer had another reason to feel excited in the spring of that same year, when Jägermeister invited her and Rels B, Spain's best-known hip-hop and trap musician, to participate in an extraordinary musical experiment. Under the motto of 'Be the Meister', the two artists were asked to separately create songs and shoot music videos for them in vertical format – a master craftsperson's exam that Deva passed with flying colours. But, she stresses, 'Jägermeister placed their entire trust in me, teamed me up with a much better-known artist, and gave me the resources to make my song as good as possible. It was a mad experience.' The result was just as mad; both songs were mixed together, and the videos placed alongside each other in a split screen. And, lo and behold! Not only did they complement each other, but they also produced something completely new, perfectly encapsulating the campaign's concept: 'Be The Meister means making your own decisions about who you want to be and which life you want to live,' says Deva. *'Rompe las reglas',* she constantly sings, 'break the rules', finally asking, *¿Quien quieres ser tú'* – 'Who do you want to be?'

Deva has long answered this question for herself, which is why she keeps diligently honing her live performances and working on fresh material, breaking old rules and making new ones. Her current style is moving more in the direction of electronic Rhythm and Blues with pop elements, and at the time of writing she was about to release her hotly anticipated debut album. Does she want to keep working with Jägermeister after that, we wondered? 'Ideally for the rest of my life!' ⚜

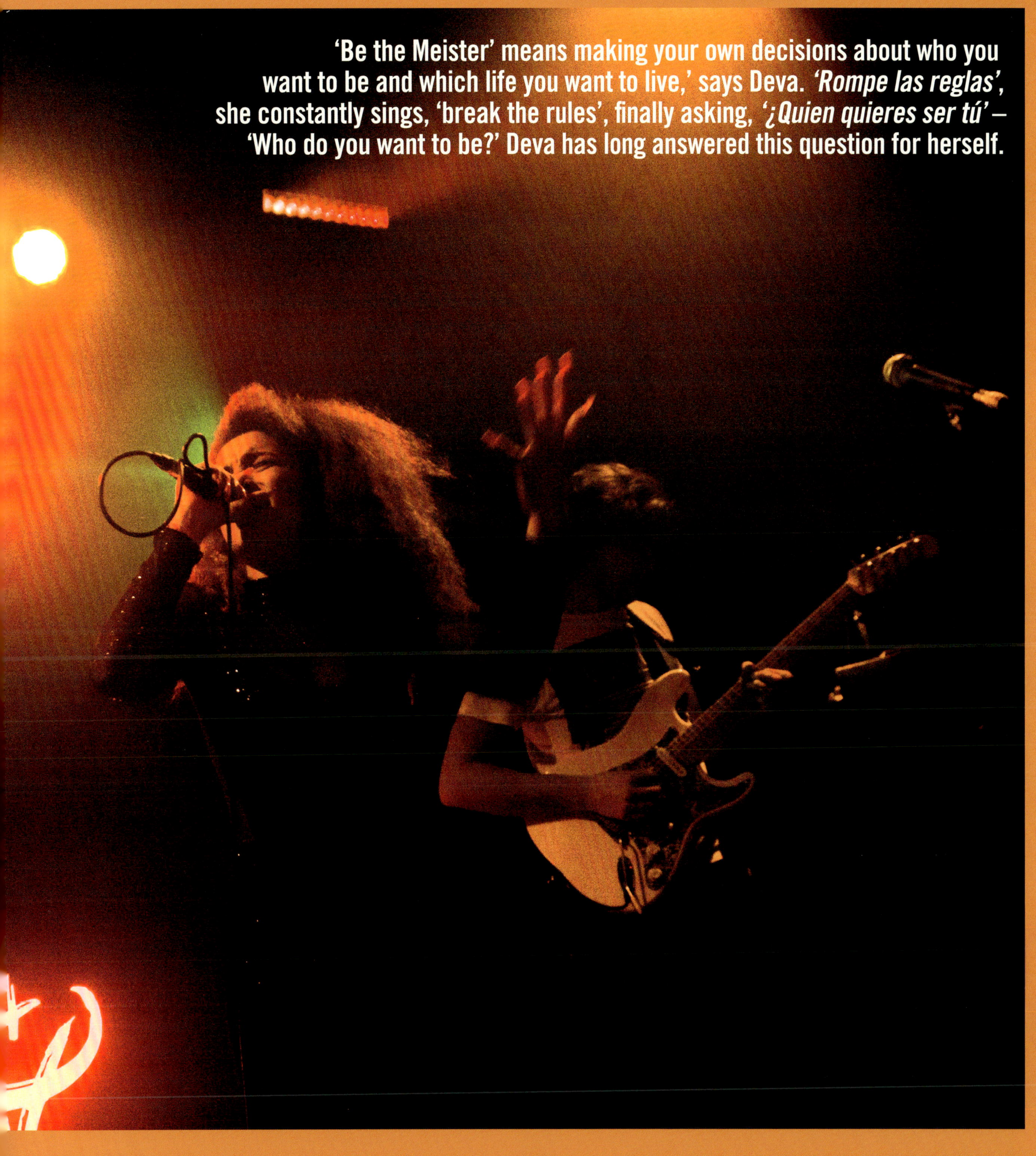
'Be the Meister' means making your own decisions about who you want to be and which life you want to live,' says Deva. 'Rompe las reglas', she constantly sings, 'break the rules', finally asking, '¿Quien quieres ser tú' – 'Who do you want to be?' Deva has long answered this question for herself.

PAUL BREITNER
FRIENDS FOR LIFE

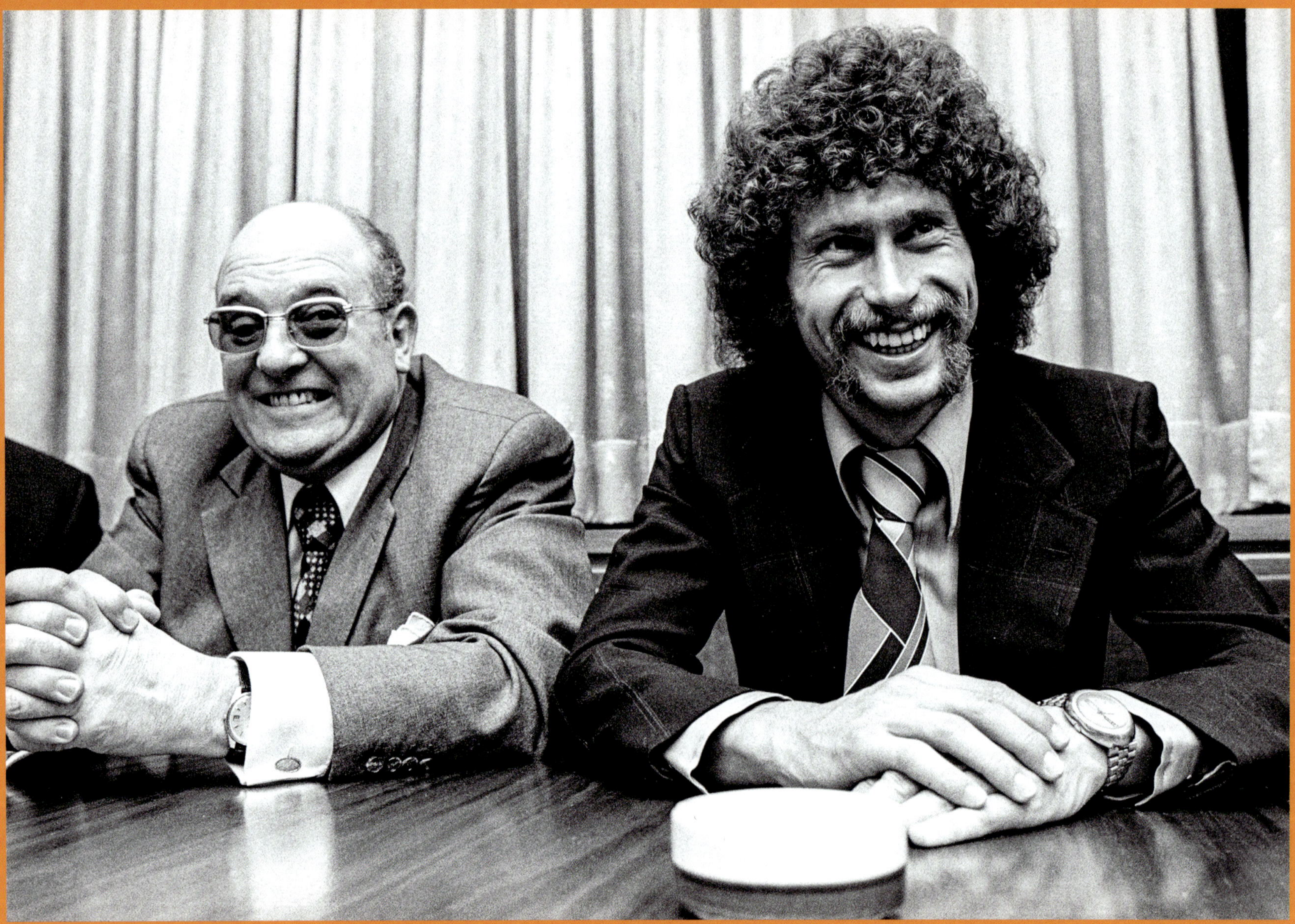

There are nights, experiences and moments that people reminisce about fondly for the rest of their lives. One such example is 7 July 1974, at 5.47 pm, when referee John Taylor blew the final whistle in Germany's World Cup Final against the Netherlands, crowning them World Cup champions for the second time. Football legend Paul Breitner is not the only one who continues to list this moment as one of the best of his entire life. But it's not always just the best moments that leave a lasting memory – the most difficult ones do too. When Jägermeister boss Günter Mast signed global superstar Breitner at Eintracht Braunschweig for a record transfer fee of 1.75 million German marks in April 1977, it was the start of a brief but intense period for Breitner that would shape the rest of his life. Here, the former German national team player talks about the most difficult year of his life, a wonderful friendship, and what happens when someone offers him a Jägermeister today.

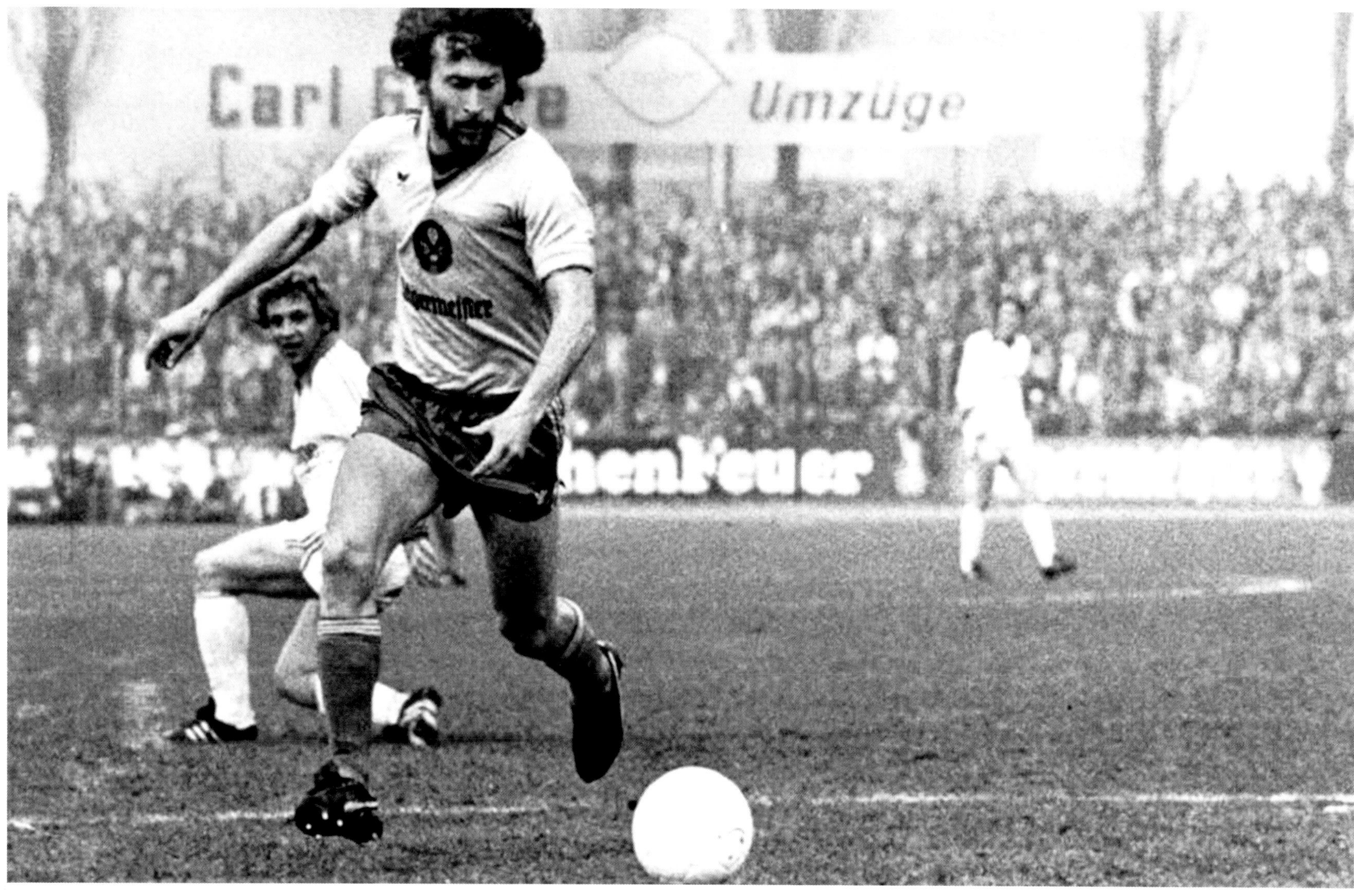

Even before my time in Braunschweig, I loved it when someone offered me a Jägermeister – and today, I love it even more, despite the fact that my year at Eintracht Braunschweig was the worst year of my life. But wherever there are lots of negative experiences and problems, there are also many positives, and relationships grow closer. I made a couple of amazing friends during my toughest year. I also forged what were for me unbelievably positive, solid ties with the Jägermeister brand. And these are not just empty words; I have repeatedly said over the last 42 years, and will continue to say, that Jägermeister is an extremely important part of my life as well as my family's life.

Life in a gilded cage

My family was the main reason I moved from Madrid to Braunschweig in the first place. At that time, Real Madrid was not only a successful club, but also a universe of its own. Even in 1974, it had 600 employees. There were six doctors, who were primarily responsible for the players and their families, and gardeners, carpenters and plumbers on call to attend whenever a player or member of his family needed assistance, for example if his wife called to say, 'Oh, my iron is broken!' Previously, at Bayern Munich, I played around 100 games a year with no more than eight or ten days' holiday. In Madrid, I played 40 games and got six weeks' holiday a year!

We had just won the second championship and the second cup when I received an offer to extend my contract for another four years. The signing was supposed to take place on the last day before our summer holiday, but my gut feeling told me: This is my future and that of the club, and I can't just make a decision on something as important as this overnight. When I was on holiday, I began to ponder and rigorously question what I did and the way we lived. My goodness, we went clubbing three times a week! Nobody cared, the people carried us in their hearts. But I asked myself: do I really want that? Where is the challenge for my brain, my personal progress, my continued learning? As a Real Madrid player, I wasn't allowed to do anything other than play football. And so it gradually became more and more clear to me from day to day that I had to get out of this gilded cage.

When the Jägermeister boss rings twice

It took until nearly Christmas to convince our president, Santiago Bernabéu, to let me go. Within a week, I had offers from clubs in France, England, South America and the USA. My wife wanted to go back to Germany, as our daughters would soon be starting school. She suggested I spend a few months in New York while she stayed home with our daughters and we just visited each other. That was out of the question for me, so I told Uli Hoeness I would be returning to Germany. I was very close friends with him at the time, but going back to Bayern like a prodigal son? I was too proud for that. A few weeks later, I finally reached an agreement with Hamburger SV, but the contract-signing was delayed. The sports editor of

Bild newspaper in Hamburg, who was an acquaintance of mine, knew about the delay and called Günter Mast. 'Mr Mast, you've been saying for years that you want to secure a real star at Eintracht. Well, now you can.' He then called me. 'I've just spoken with Günter Mast, and he would like to talk to you. May he call you?' I said that of course he could, ideally the following day at 2 pm. The next day, my telephone rang at 2 pm on the dot. 'I would like you to be clear with me,' I told Günter Mast. 'If you're looking for a little ballyhoo and PR for your club, I don't mind helping out, but . . .' He replied, 'No, Mr Breitner, you've misunderstood me; I really want to have you in Braunschweig.' We eventually agreed to have another telephone conversation the day after next – once again at 2 pm. When, just like the last time, the telephone rang on the dot of 2 pm that Thursday, the matter was totally clear in my head. Punctuality is among my three or four top criteria for a legitimate partnership.

The following Monday, Günter Mast flew to Madrid with his treasurer and president to negotiate a deal with the club managers. Beforehand, he took me aside and said, 'Mr Breitner, just so we are absolutely clear: I'm not actually very interested in football. Nor do I have any idea about it, but I would like to sign you for three years as my chief attraction to bring great joy to the football fans in the Braunschweig area.' To which I responded, 'Mr Mast, we are partners; we're well matched!'

A tip of the hat to Wolfenbüttel

I was, of course, already aware of the name Günter Mast; I had known of him since at least the quarter-finals of the European championship against England at Wembley Stadium on Saturday 29 April 1972. When we arrived in London on the Thursday before, we found out that ARD had cancelled the live broadcast of the game because advertising boards were going to be displayed along the sidelines. But on the Friday afternoon, we received a message to say that the game would very likely be broadcast, because a Wolfenbüttel-based liqueur manufacturer had bought up all the signboards and would be leaving the spaces blank. This was the greatest coup ever, and it remains unmatched to this day! Thirty-five million people watched the game, and thirty-five million people tipped their hats to Wolfenbüttel. Whenever there are discussions about how football has evolved and who led its developments, I always mention two stories: Günter Mast's coup at Wembley, and how he introduced advertising on shirts in the Bundesliga. And to think that someone like that approaches you and says, 'I don't have any idea about football, but would like you to be my chief attraction.' Yet the only thing that went through my mind at the time was, 'Finally someone who's not a swindler, and who is open – that's the partner for me.'

Tough times in Braunschweig

As soon as I arrived in Braunschweig, I noticed that most of the team cut me off. There was constant jealousy, resentment and envy to the point that, after the second game at the training camp, I said to them: 'You guys are cutting me off, you're not playing with me. Would you rather lose than pass the ball to me? What have I done to you? What's the issue here?' One of them said, 'Paul, surely you understand. We have to beg Adidas whenever we need new shoes, but you have a Puma contract and turn up every week in a new training kit and new flip flops.' I replied, 'Are you for real? I came here to achieve success with you all. You guys were runners-up last year, and with me you will become champions if that's what you want. But obviously you don't. I'll give you six weeks, and then we'll chat again.' But, even six weeks later, the situation had not improved. 'You'll only have to put up with me for this season,' I told them. 'I think I'll be able to convince Günter Mast to let me go.'

This had nothing to do with the fans; they stuck by me. Even our very first training session attracted 5,000 people to the stadium – more than there had been at many official games in the past. Most non-football fans thought, 'Look at this man, coming to Braunschweig and partying or having orgies every day,' when all I really wanted was to have peace and quiet and play good football. The local bus companies would stop outside our door on city tours so people could get out and take photos. But even that wouldn't have bothered me if there hadn't been other negative, dangerous aspects. Like many towns at the time, Braunschweig was still a little provincial. And some people vilified me as a communist, Marxist and Leninist – an image that had stayed with me ever since my *Sturm und Drang* phase between 1970 and 1974. After moving into our house in mid-September, we were kept under police protection 24 hours a day. Every week I would receive at least one or two kidnap or death threats. Someone shot at us through our window, and every few weeks we would wake up to a headless chicken or rabbit on our doorstep.

At the start of the second half of the season, in early January, I finally announced I would be leaving Braunschweig and returning to Bayern. I was no longer too proud. The Braunschweig fans came to me in their droves at training sessions or after games to

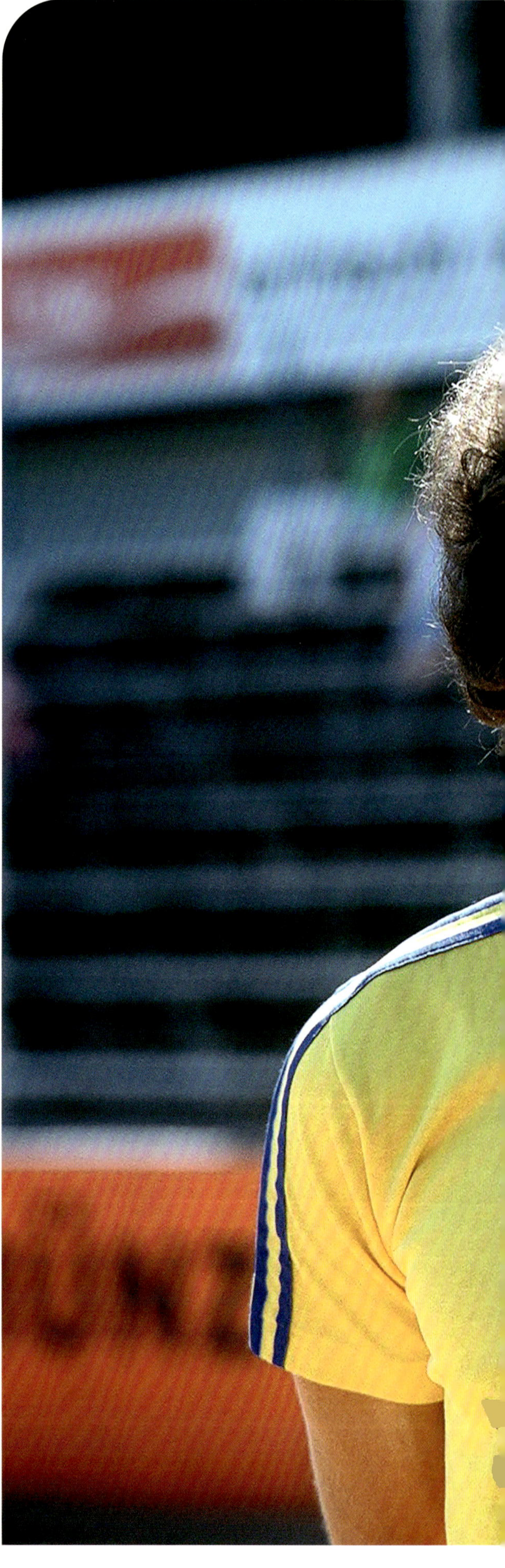

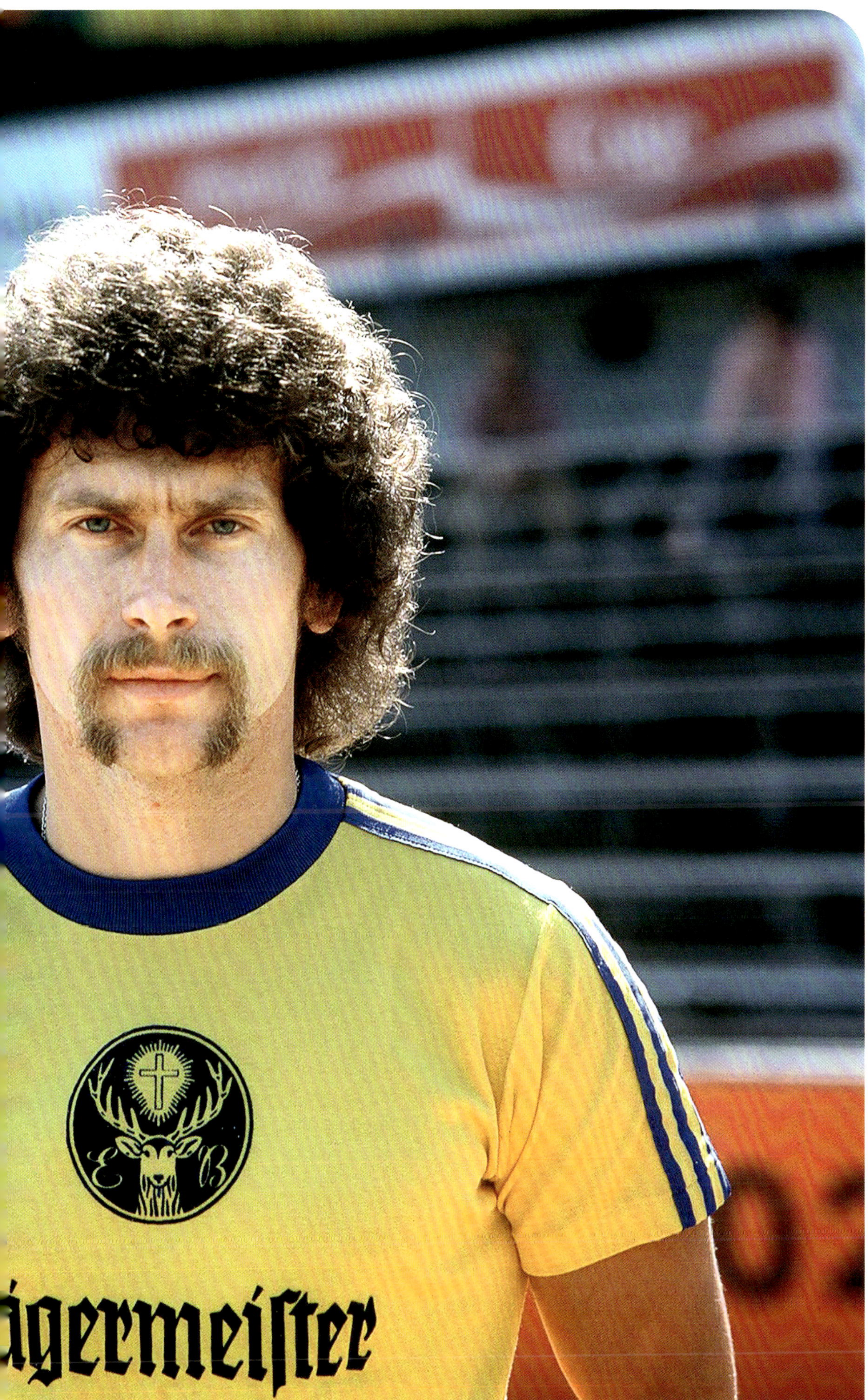

tell me, 'Paul, you're right; you don't belong here. Those morons don't deserve to have you playing with them.' They could see I was giving everything I had right up to the last second, regardless of whether the others wanted that or not. So the people supported me, and they continued to do so whenever I went to Braunschweig.

Kindred spirits

The same was true for Günter Mast. For me, our friendship was above everything that happened that year. We shared many views: always try to keep a consistent line, don't trick people, don't do anything underhand, give of your best, speak openly. And both of us realized that it's difficult to get through life with such a philosophy. It is an incomprehensible path for many. Another crucial factor in his positive attitude towards me was that, soon after arriving in Braunschweig, I had called Günter Mast's long-time secretary, Ingrid Bastian, asking to take a tour of the Jägermeister facility. I just wanted to see where and to whom I belonged. Günter was baffled, and said he was speechless when Ingrid told him. But, of course, he thought it was great, and took me on a tour through the company. I think that was a very important foundation for the wonderful friendship we maintained even after I stopped playing. Apart from my family, of all the people I have met over the years, Günter Mast is one of the very few I would put above the rest. ⩓

Apart from my family, of all the people I have met over the years, Günter Mast is one of the very few I would put above the rest.

ALFRED KOGLER

A LITTLE SIP OF HOME

Anyone who claims you cannot shift an old tree without it dying has obviously never met Alfred Kogler. Upon his retirement, the Hamburg native left his homeland to build a new life for himself on the edge of the prairies of Canada, a country very dear to his heart. But there's one little slice, or rather a little sip, of home that he held on to even when far away: Jägermeister.

Kogler went to Canada for the first time in 1984 to visit his brother-in-law, who was already living there. He travelled through the Rocky Mountains and all the way up to Alaska then, returning in 1988, he explored Vancouver Island. After that, it became clear to him that Canada was where he now needed to live. In 1994, soon after his retirement, the former Deutsche Bundespost (the federal post office) employee and his wife lodged their entry application and they received approval in February 1995. Finally, on 2 July that year, the couple packed their belongings into a shipping container and set off on their Canadian adventure. They settled in Lethbridge, a city 109 kilometres south-east of Calgary in the province of Alberta, where they built a house, quickly became a part of the local community and made new friends – including at the German Canadian Club of Lethbridge, a non-profit association whose members strive to maintain German traditions and foster German culture. And they did so only relatively successfully until, in the year 2000, Alfred Kogler was asked to become the association's chairperson.

The new president introduced a number of innovations to gain new members and inject money into the association's empty coffers. He started a coffee chat every second Sunday, with cakes. 'My wife was famous for her *Eierlikörtorte* [egg-liqueur cake]; people would queue up for it, some even having pre-ordered a piece,' says Kogler, now 87 years old. Once a year, the club organized an Octoberfest, where the women would wear dirndl dresses and the men lederhosen.

They were caught once, but Alfred Kogler managed to convince the customs officer that the herbal liqueur was medication. 'I showed him the label, which said it contained 56 herbs,' says Kogler. 'The customs officer then said, "Okay, go ahead!" and let us through.'

'And I was from Hamburg!' Kogler laughs when he looks back. 'But Bavarian equals German when you're abroad.' The festivities ran for a whole day, but they became so popular during Kogler's presidency that the event was eventually extended and celebrated for the entire weekend.

Hundreds of people would attend the Octoberfest, and there was a raffle for cash and other prizes that had been donated by individuals and companies. The club president was in regular contact with Jägermeister – after all, the herbal liqueur is also part of German tradition – and the club was supplied with advertising, serving trays and decorations from Wolfenbüttel. 'And Jägermeister was, of course, also served at the bar – sometimes until it all ran out, depending on who was in attendance,' Kogler laughs. At the end-of-year Christmas party, the club members would pile up all the Jägermeister bottles they had emptied throughout the year on a table. 'For us,' says Kogler, 'Jägermeister was kind of like a connection to Germany, which is why we loved it so much. It was a drink from home.'

And Alfred Kogler sometimes even smuggled it into the country. Alcohol was expensive in Canada; at the time, a bottle of Jägermeister cost nearly three times as much as it did in Germany, as indeed it still does today. Kogler and his wife would bring four bottles back to Canada after every trip home. 'My wife and I would each carry two,' he says. 'Each person was allowed 1.1 litres, so 2.2. litres in total. But four bottles were 2.8 litres, so, technically, we would be smuggling in 0.6 litres each time.' They were caught once, but Alfred Kogler managed to convince the customs officer that the herbal liqueur was medication. 'I showed him the label, which said it contained 56 herbs,' says Kogler. 'The customs officer then said, "Okay, go ahead!" and let us through.'

Alfred Kogler was president of the city's German Canadian Club (and unofficial Jägermeister fan club) for 13 years until, in 2013, he and his wife decided to

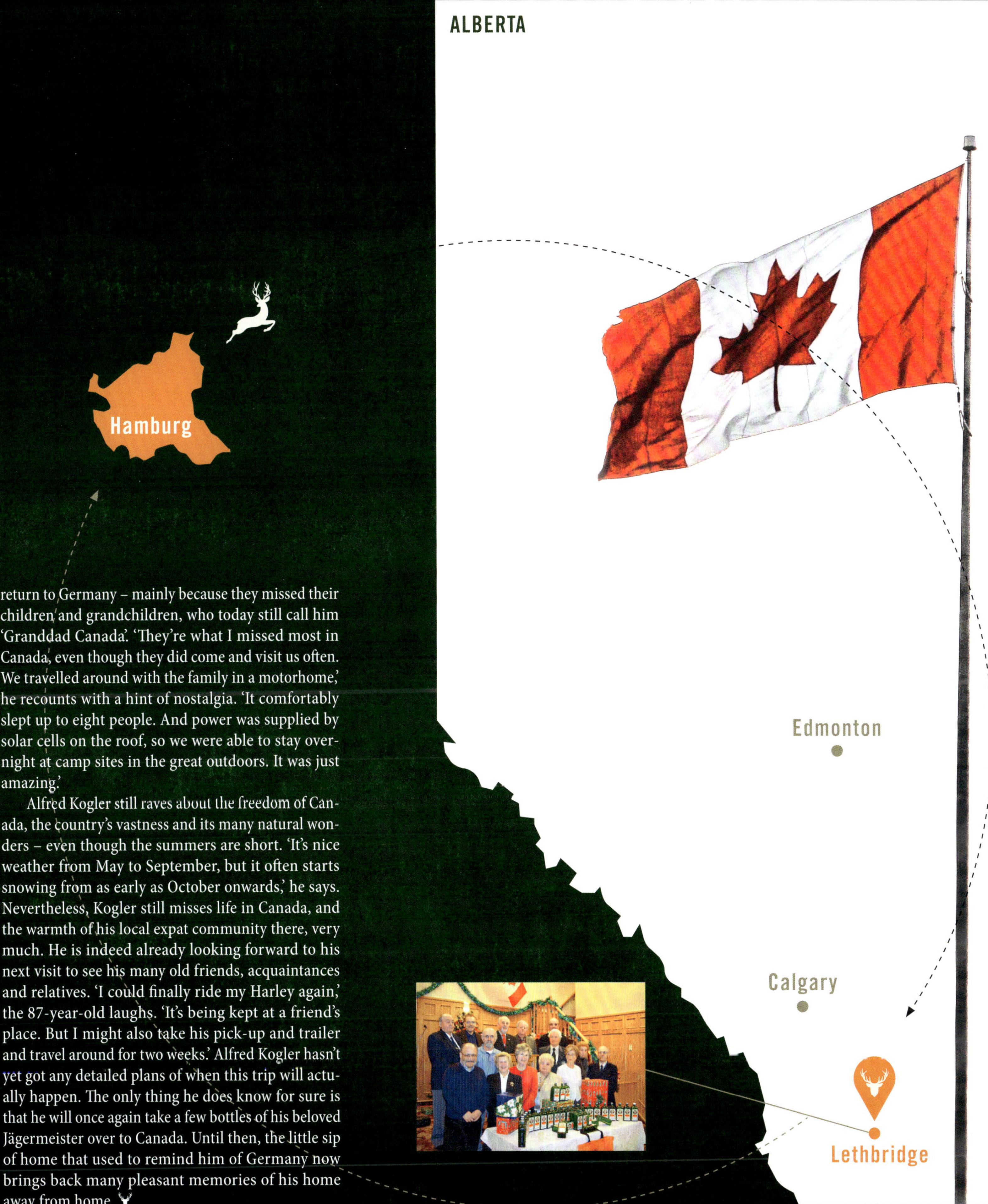

return to Germany – mainly because they missed their children and grandchildren, who today still call him 'Granddad Canada'. 'They're what I missed most in Canada, even though they did come and visit us often. We travelled around with the family in a motorhome,' he recounts with a hint of nostalgia. 'It comfortably slept up to eight people. And power was supplied by solar cells on the roof, so we were able to stay overnight at camp sites in the great outdoors. It was just amazing.'

Alfred Kogler still raves about the freedom of Canada, the country's vastness and its many natural wonders – even though the summers are short. 'It's nice weather from May to September, but it often starts snowing from as early as October onwards,' he says. Nevertheless, Kogler still misses life in Canada, and the warmth of his local expat community there, very much. He is indeed already looking forward to his next visit to see his many old friends, acquaintances and relatives. 'I could finally ride my Harley again,' the 87-year-old laughs. 'It's being kept at a friend's place. But I might also take his pick-up and trailer and travel around for two weeks.' Alfred Kogler hasn't yet got any detailed plans of when this trip will actually happen. The only thing he does know for sure is that he will once again take a few bottles of his beloved Jägermeister over to Canada. Until then, the little sip of home that used to remind him of Germany now brings back many pleasant memories of his home away from home.

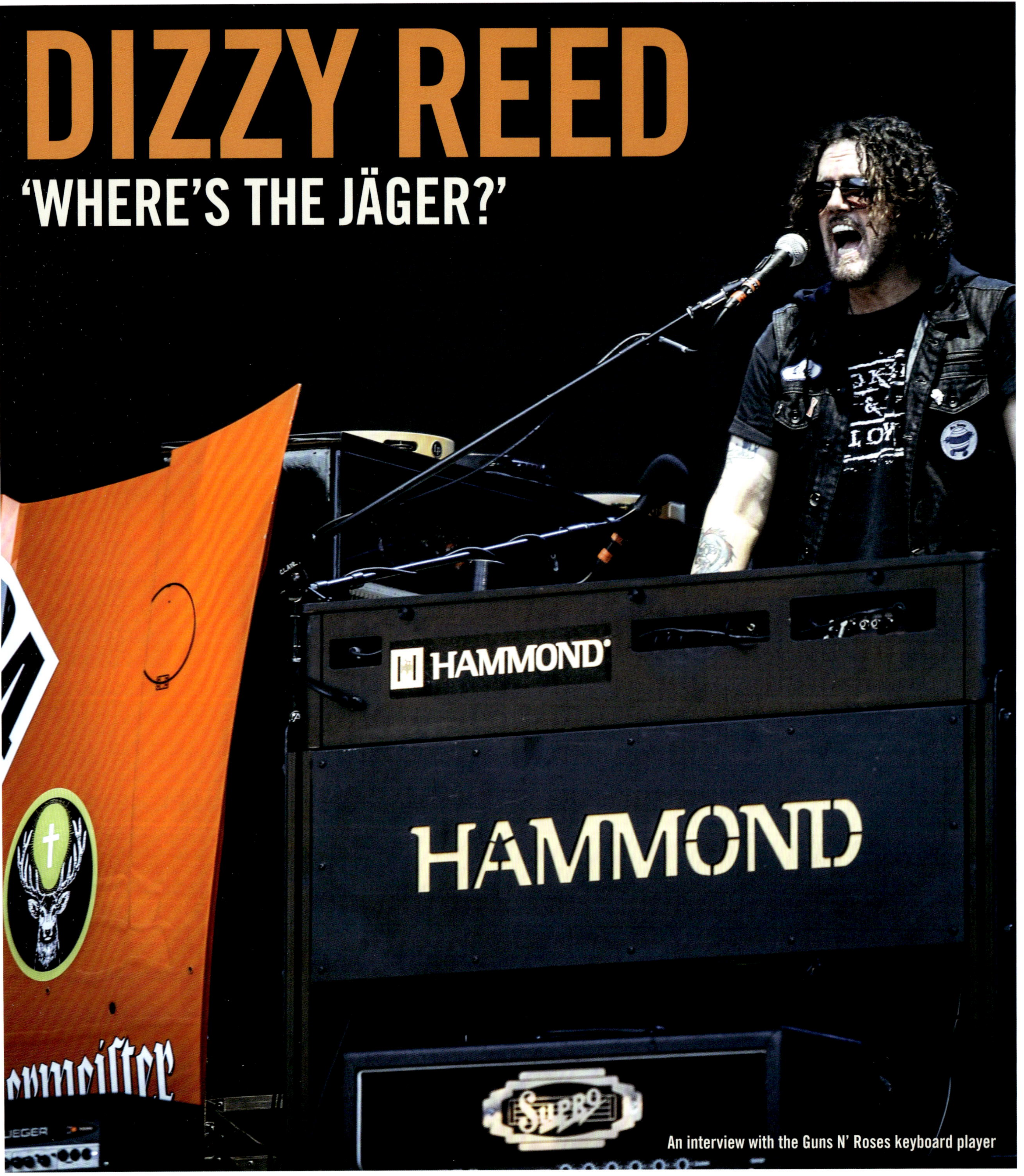

DIZZY REED

'WHERE'S THE JÄGER?'

An interview with the Guns N' Roses keyboard player

It was one of those situations where a nice gulp of Jägermeister would have really helped, but Darren 'Dizzy' Reed didn't know about it at the time. 'I don't normally get nervous, but I couldn't help it in that situation,' says the musician, reminiscing on his very first show with Guns N' Roses. And it wasn't just being held at some club on the Sunset strip, but rather in front of more than 100,000 people at the Maracanã Stadium during the 1991 Rock in Rio. 'It was almost like a dream,' Reed says, 'but it marked the beginning of a wonderful time with a great band.' Until the reunion in April 2016, keyboardist Dizzy Reed had, for two whole decades, been one of the two remaining members from the early 1990s, along with Axl Rose. The Guns N' Roses frontman is incidentally also the reason Reed was able to experience another amazing première just a few months after his epic first performance.

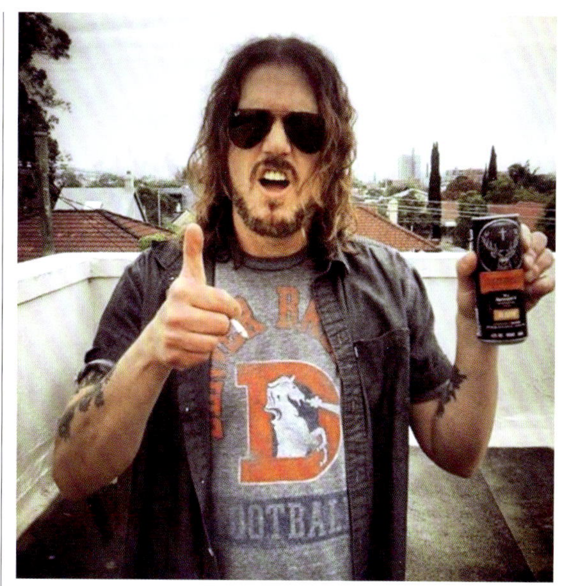

I do remember the exact time I had my first Jägermeister. We'd been in Europe on tour with Guns N' Roses and had just got back to the States. While we were there, *Use Your Illusion 1 & 2* had just come out and we hadn't had an official release party. So we sort of had an impromptu one at a place in Hollywood called On the Rox. I believe it was Axl who had ordered a tray of Jägermeister shots. I guess someone had told him about it. So the tray came by me and I remember trying it and was it was ice cold and it was perfect. And that was the beginning of another long and wonderful relationship. I dove right in and have loved Jägermeister ever since.

Why is it your drink of choice?

For a lot of reasons, but mainly because it stands alone and above everything else. So when I enjoy Jäger, I don't mix it with anything else. I will occasionally put some energy drink in there to have a Jäger Bomb, but other than that, it's best enjoyed on its own. They put so much into it. And whoever came up with the recipe did a great job. It's amazing that they haven't changed it, because everything changes so quickly these days. The care that goes into it really shows how much they appreciate their consumers.

Can you describe what it does to you?

Jägermeister is good for the soul. It makes me feel better about everything. Other spirits don't do that for me. It's a good, mellow feeling, somehow comforting and cosy.

Happy or sad: When do you prefer to drink Jägermeister?

If it's a difficult situation or things aren't going my way, I sometimes change my mindset if I have a little bit of Jägermeister and see the good in things. I'm certainly never into drowning my sorrows or covering up issues that I have. Drinking responsibly is very important to me.

Many other rock musicians have a hard time doing that. How did you learn to draw the line?

I just don't want to let people down who are counting on me. If I've had too much to drink, I can't do my job properly. So I'm not going to do it or I'll wait till later.

Your audiences sometimes bring you shots while you are playing on stage. How exactly did that come about?

I have a few other bands and one of them is Hookers N' Blow, which my lovely wife Nadja is part of. In 2003, shortly after we had started, we were planning some shows on the East Coast. When we were putting the contracts together, they asked me what I wanted on our rider. 'Jägermeister,' I replied. 'That's all I want, that's all I really care about.' But on one of our very first shows, they hadn't filled my rider properly. So when I was on stage, I said, 'Hey, it would be great if anyone could buy me a shot of Jägermeister', appealing to the owner of the club. But I got zero response. So eventually I just started singing a little ditty, that went something like this: 'I love Jägermeister, yes, I do / I love Jägermeister, how about you? / If you buy me a shot, I'll take two / If you're not, well, the show is probably through.' And I kept singing that over and over and eventually there was way more Jägermeister than we could possibly drink. Afterwards the owner of the club was thrilled and said: 'Oh my God, we've made so much money at the bar tonight!' *(laughs).* I didn't intentionally make it a tradition, but that's how it became a thing.

On your recent European run of the Guns N' Roses shows, you used a Jägermeister Racing Porsche hood as part of your stage gear. Please tell me more about that.

Our manager found the hood online and decided he needed to buy it and put it up next to me on stage – as a representation of my love for the Jägermeister.

Is it true that your wife gave you a Jägermeister Christmas gift?

Yes, that's true *(laughs)*. I had this button-up collared Jägermeister shirt and I wore it so often that the collar came off and it looked like it came out of the trash. But I was still wearing it. So she surprised me and bought me a brand new one that is even cooler and I wear it all the time. I just washed it. Also, she made me a Jägermeister lanyard hanger, so I can hang my backstage pass on my belt.

Your wife supports your passion for Jägermeister, it seems.

Well, it is pretty easy to make me happy. And she knows me very well. But I do have a funny story about Nadja and Jägermeister. She is from New Zealand and for the first several years that we were together, she did not have her driver's licence yet. So when I played shows in L.A., I couldn't drink because I was driving, and a lot of times I unfortunately had to leave some shots of Jäger-meister behind on stage. One day, Hookers N' Blow had a gig at The Roxy. The club had just been sold, it was going to be the last show – and Nadja had just gotten her driver's licence. It was the first time we were going to play a show in L.A. and she was going to drive home. I was like: 'Yes, I'm going to go for it tonight, I'm going to have fun and I'm going to drink as much as I want.' So we got to The Roxy and I said: 'Where's the Jäger?' Turns out they were out of Jägermeister! Because they were closing and it was the last show, they hadn't restocked their bar. That one time, after all these years! So I looked at the stage manager and I said: 'Go to the liquor store down the street, please, and buy me a bottle of Jägermeister!' *(laughs)*.

From your almost worst Jägermeister moment to your best: is there one?

Some of the best are those we don't remember *(laughs)*. ⚇

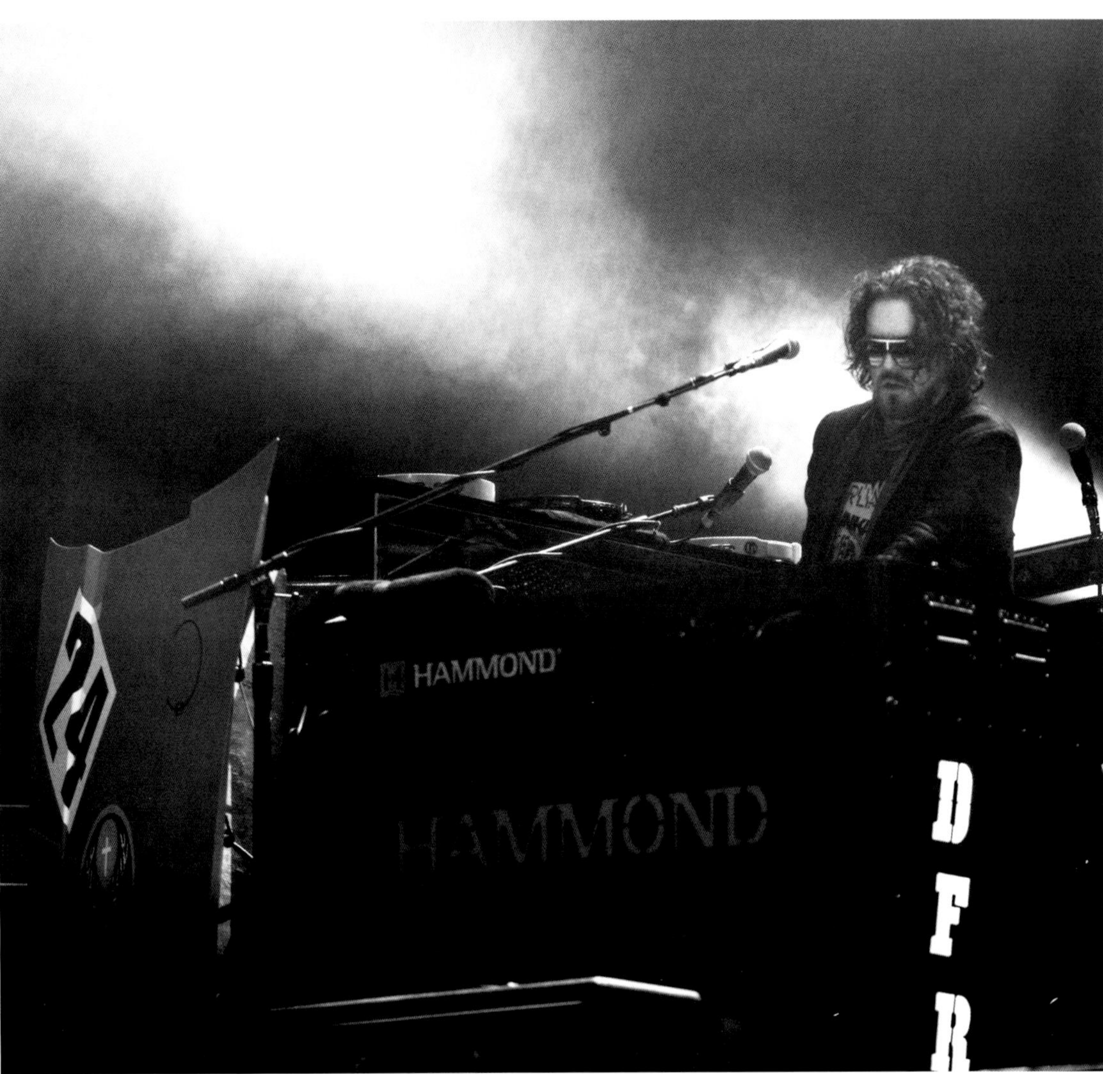

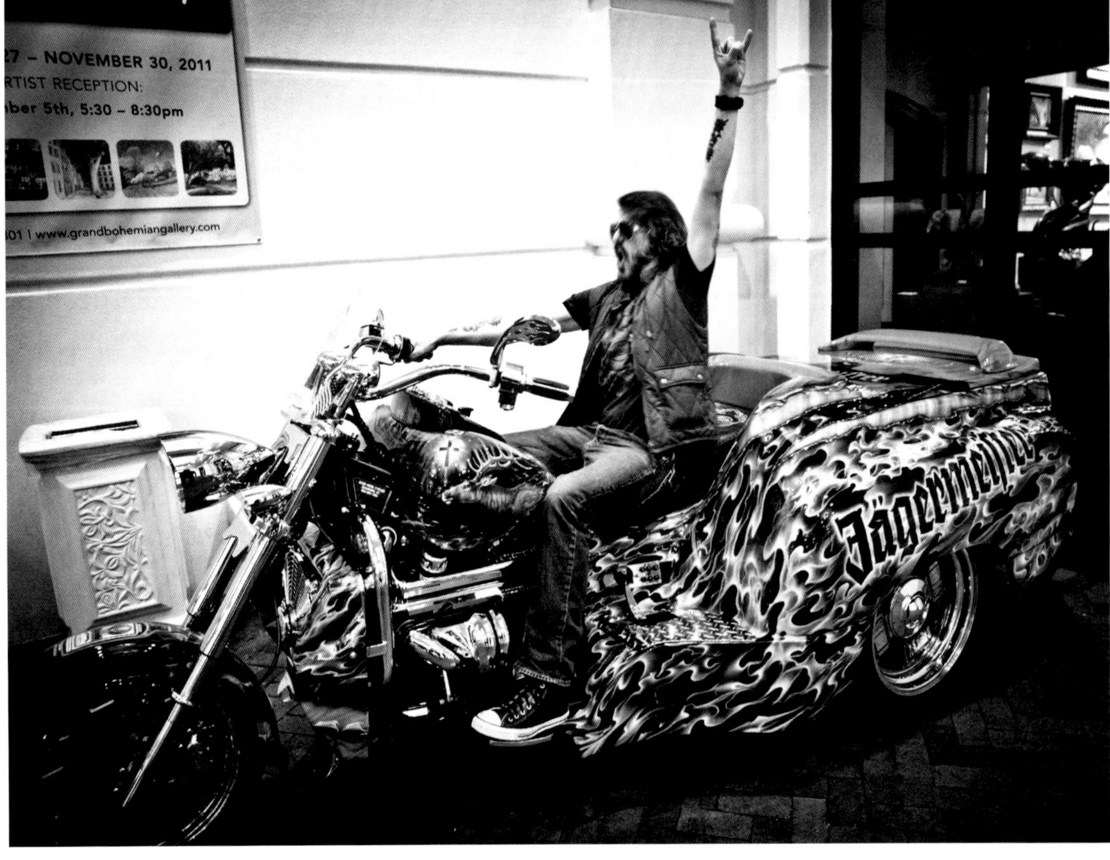

32

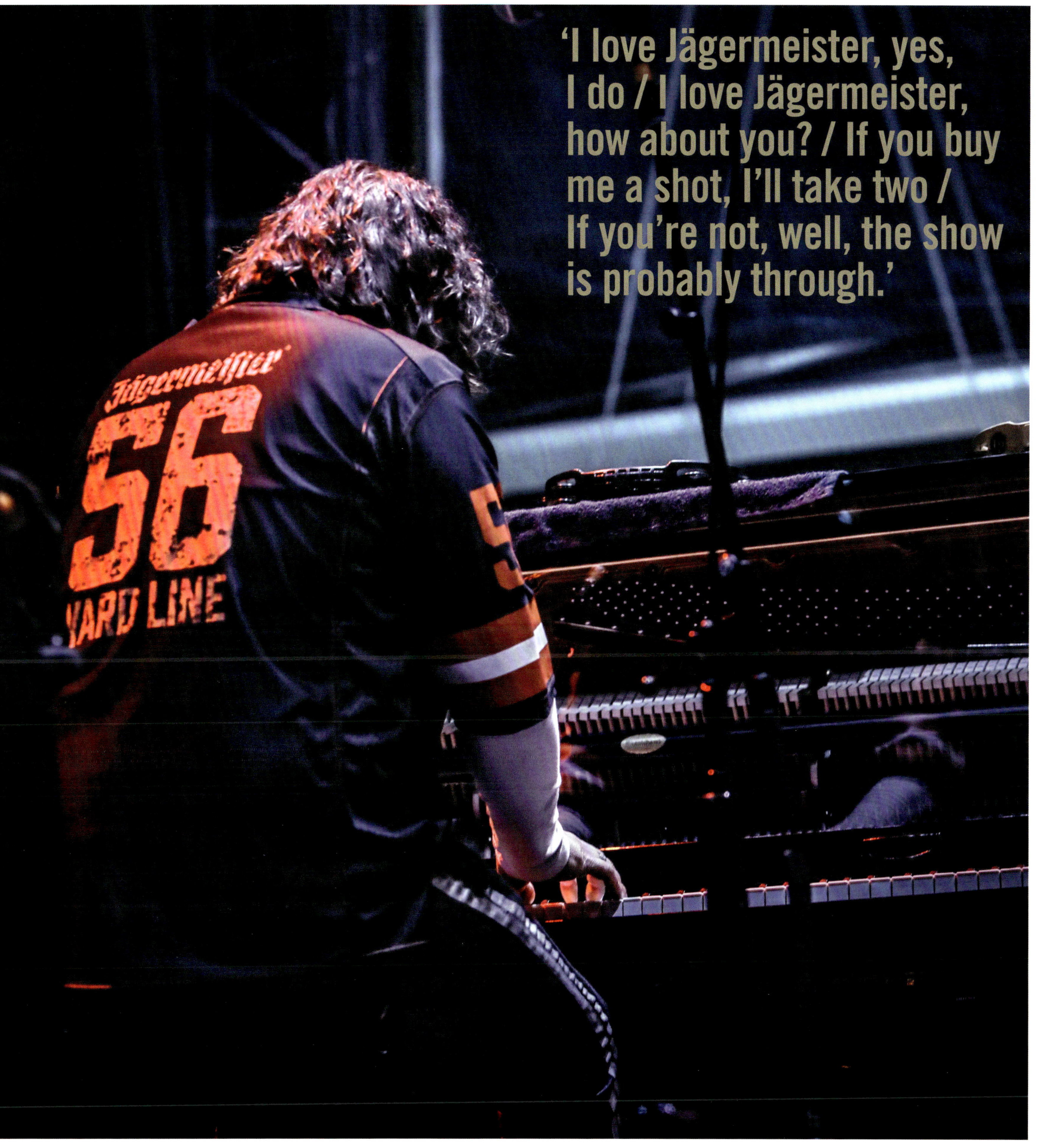
'I love Jägermeister, yes,
I do / I love Jägermeister,
how about you? / If you buy
me a shot, I'll take two /
If you're not, well, the show
is probably through.'

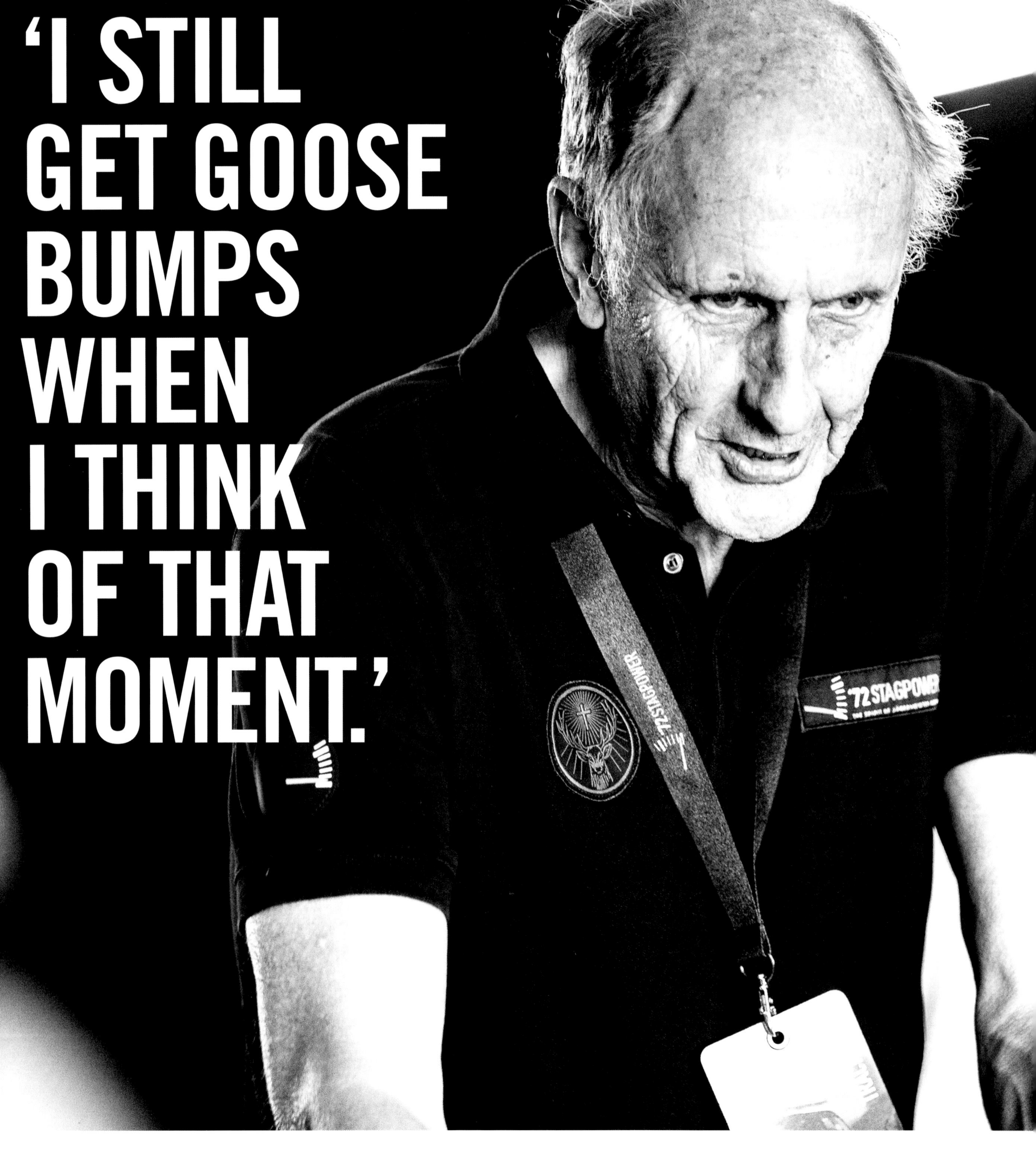

'I STILL GET GOOSE BUMPS WHEN I THINK OF THAT MOMENT.'

HANS-JOACHIM STUCK

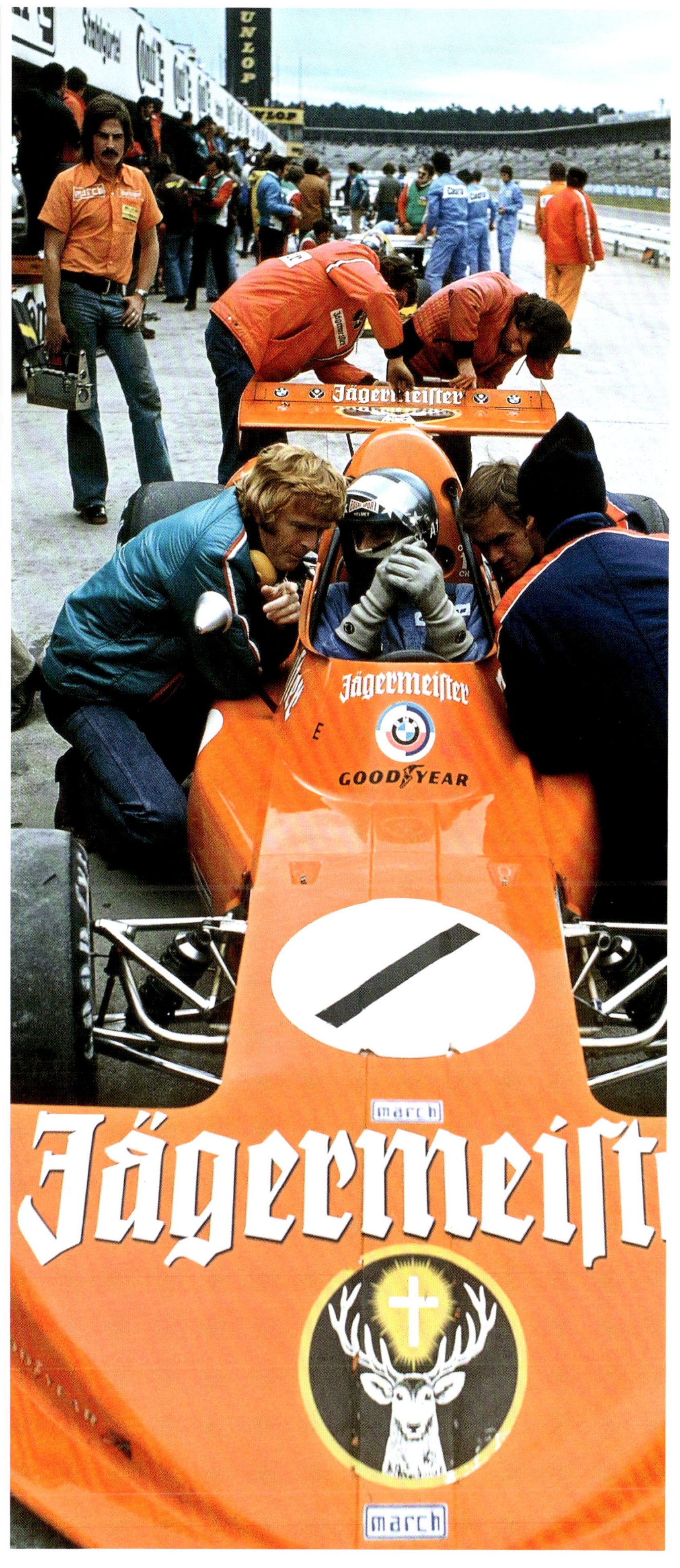

Whether there was a whiff of petrol around at the time is unknown, but one thing is for sure: Hans-Joachim Stuck had motorsport in his veins right from the start. Born in Garmisch in 1951 as the son of the famous "Bergkönig" and Auto Union driver Hans Stuck, Hans-Joachim was still a young boy when he gained his first experiences on the Nürburgring track where his father gave driving lessons. A special permit from the transport minister meant he got his driver's licence at the age of 16, and he started his career in 1969. He ended it 43 years later as a 60-year-old at the 24-hour race at the Nürburgring – the place where his career had first begun in a BMW. 'Strietzel', or 'Stucki' ('Both arc ok, just not Hans,' he says. 'While it's a nice name, it's common as muck') is one of Germany's most famous and successful drivers, competing in more than 1000 races in all kinds of series between 1974 and 1979, including Formula 1. With his iconic star-patterned helmet, the King of Hockenheim (a name gained from his many victories at that track) won numerous titles: endurance champion in 1985, 24 Hours of Le Mans in 1986 and 1987, and DTM champion in 1990. The affable Bavarian spent four years in Formula 2 driving for the Jägermeister-sponsored March-BMW, but also triumphed in other Jägermeister cars, such as the Porsche 956 and 962, and the BMW 320 Group 5. Since 2008, he has worked as the Volkswagen Group's motorsport representative and as one of 245 globally registered FIA stewards.

Hans-Joachim, not only are you considered a motor-racing legend, but you are also rumoured to have enjoyed playing the odd trick on other members of your team . . .

Yes, there is one story *(laughs)*. In 1975, I was travelling in California with my teammate Dieter Quester and Mucki, my girlfriend at the time. Quester was training for the Riverside 6 Hours race, and we were looking on from the corner. He saw us and waved, and the third time he drove past, I hitched up Mucki's T-shirt from behind so that she was standing topless by the racetrack. Quester promptly crashed into the barrier. This was obviously bad news, and we were called to the office of our boss, Jochen Neerpasch. He scolded us, saying it was not a good look for BMW, and was immoral, especially in the United States, of all places. As punishment, we all had to pay our own travel expenses and repair the smashed-up vehicle. But Quester simply said, 'Not to worry; the peepshow was worth it.'

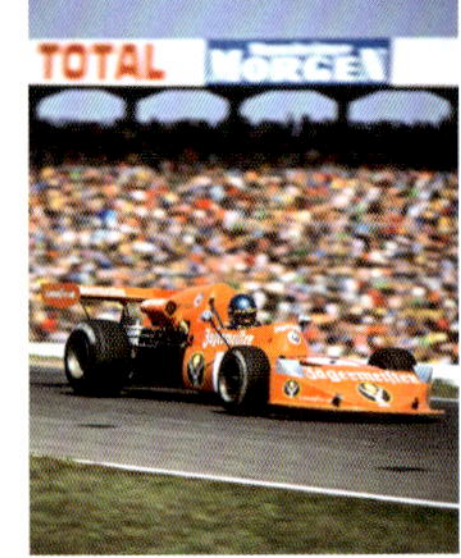

Are racing drivers wired differently to other people?

They're all crazy, that's for sure *(laughs)*! During my Formula-1 days in the 70s I would sit in a car with 100 litres of fuel either side of the cockpit, aluminium bodywork incapable of withstanding anything all around me, and drive round bends at nearly 100 miles an hour – I must have been out of my mind. But at the time we didn't care. Whether it was Niki Lauda, James Hunt or Jochen Maas, our focus was on competing against each other, and being better and faster than each other. Sure, we saw things happen often enough. In my six years in Formula 1, one person on average would die each year. But we pushed that

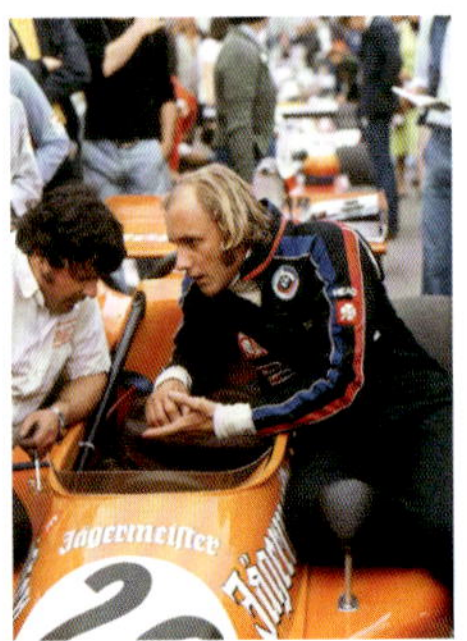

aside, because we wanted to be there again the following year. All high-speed athletes must be bonkers – in a good way *(laughs)*. A ski racer who speeds down the Streif slope on the Hahnenkamm course must be just as crazy.

What other traits do good racing drivers need?

They need to have the right attitude to life in general. Not drinking alcohol, being physically fit – this is hugely important, because condition and concentration are very closely related. There are recurring processes: late braking, steering, accelerating . . . these also involve effort. Then there's the heat inside the car. You have to remember that temperatures can sometimes reach 50 or 60 degrees Celsius in these sorts of cars. If you're not fit on race day, everything is less sharp: you brake earlier, accelerate later and turn in later. Of course you also need to have a certain degree of combat-readiness, along with technical experience and a good seat-of-the-pants feel.

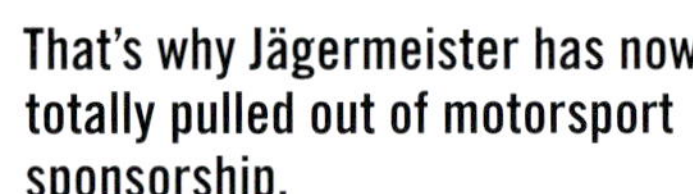

You had all of those, and were therefore very successful. Which victory was your personal highlight?

Definitely the 1985 World Endurance Championship with Porsche. My goal was always to become a Formula 1 world champion, but unfortunately I didn't achieve that. I did, however, enjoy multiple successes at the 24 Hours of Le Mans. The first time I stood on the podium there, I thought to myself, 'Stucki, you've done it!' Along with the Indy 500 and the Monaco Grand Prix, Le Mans is one of the three races any successful race driver needs to have won. Another great milestone was when I was admitted into the FIA Hall of Fame in Paris in 2019.

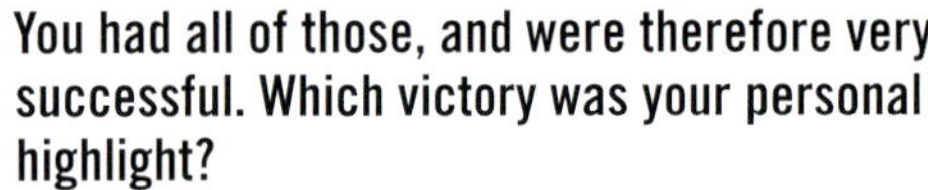

You drove for Jägermeister in Formula 1 and 2, but you also won races in other Jägermeister cars. What are your fondest memories of those racing days?

Most certainly being runner-up in the 1974 European Formula 2 Championship, even though my teammate Depailler beat me by a few points. But that was shortly after I had won the Formula 2 at Hockenheim. When I entered the circuit in the Jägermeister car and tens of thousands of specta-

tors set off skyrockets and firecrackers – I still get goose-bumps when I think of that moment. Without Jägermeister as a sponsor, I would not have been able to drive; Jägermeister was simply a part of me at the time. It was my misfortune to be a teetotaller; in my whole life I have drunk maybe five Jägermeisters in total. It's a bit of a shame actually, because I'm sure we would have got them for free *(laughs)*. But alcohol and motor-racing don't really go together.

That's why Jägermeister has now totally pulled out of motorsport sponsorship.

But it remains a fantastic brand to which I owe a lot. Jägermeister is still an integral part of motorsport. There are a handful of sponsors that people will never forget, and Jägermeister is definitely one of them. When I was able to drive the Jägermeister Formula 1 car two years ago, many people came up to me and said, 'Stucki, how great it was back in the days!'

Looking back now, is there anything you would have done differently in your life?

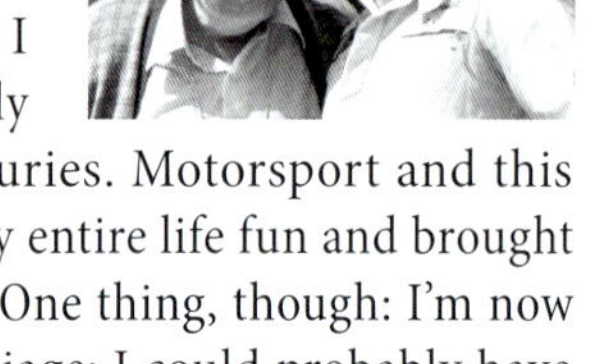

Absolutely nothing. I have survived and only had a few minor injuries. Motorsport and this amazing job made my entire life fun and brought me great enjoyment. One thing, though: I'm now onto my fourth marriage; I could probably have spared myself two of them *(laughs)*!

And what's on your personal wish list for the future?

I have just one wish, but it may sound just as crazy as race-car driving itself. I would really like to go into space and see the earth from there. Hopefully I will be able to! ♥

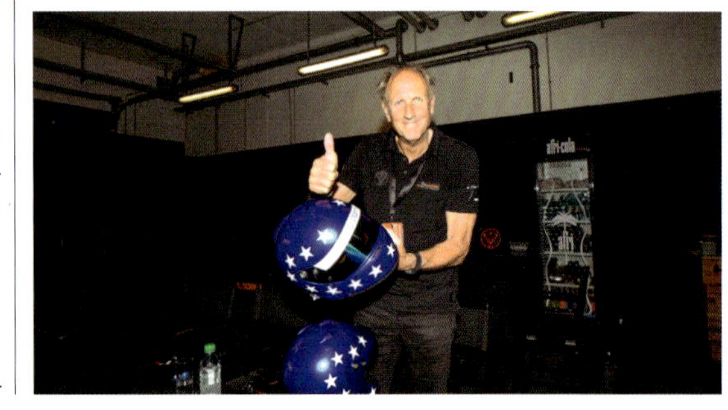

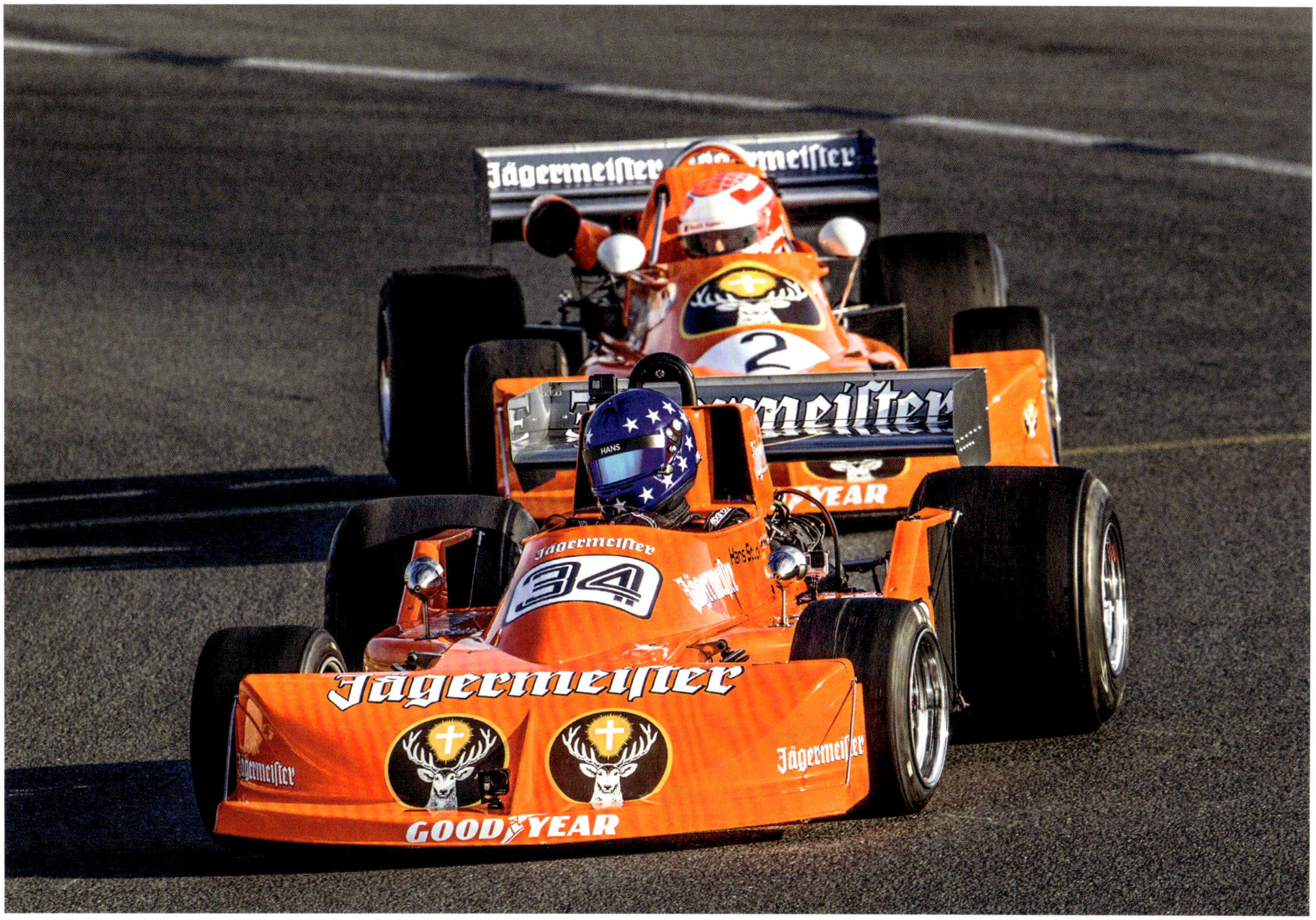

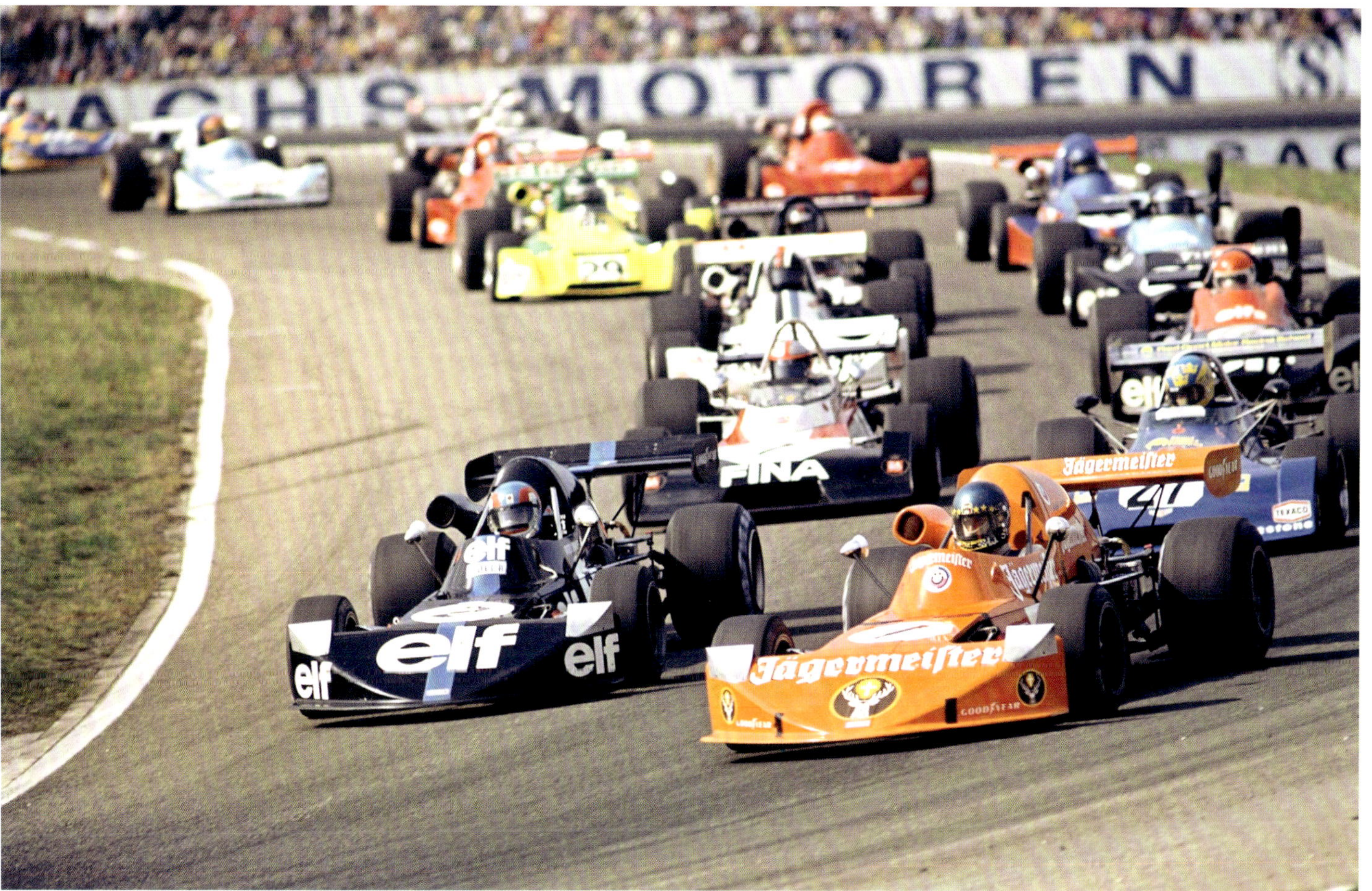

'During my Formula-1 days in the Seventies I would sit in a car with 100 litres of fuel either side of the cockpit, aluminium bodywork incapable of withstanding anything all around me, and drive round bends at nearly 160 kilometres an hour – I must have been out of my mind. But at the time we didn't care.'

'There are a handful of sponsors that people will never forget, and Jägermeister is one of them.'

THE JÄGERMEISTER TATTOO CLIQUE

A penchant that's more than skin-deep

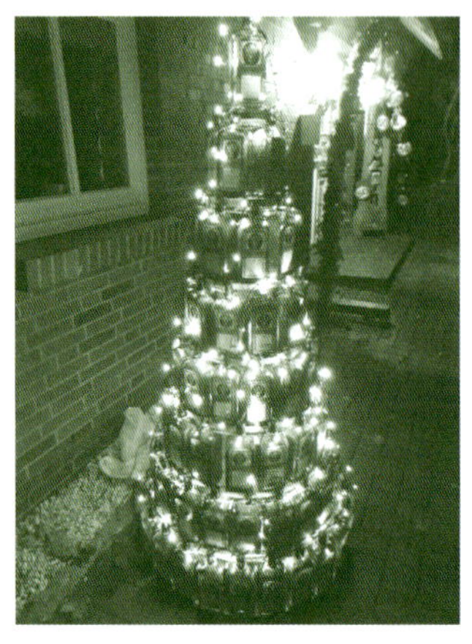

'I'm Björn Rastedt, 35, happily married, with a penchant for Jägermeister.' From this skimpy introduction, you wouldn't guess that Björn's penchant for Jägermeister is a little more than just a surface thing. In fact, it extends 1.2 to 1.5 millimetres into the dermis, to be exact, for he has a tattoo of the Jägermeister stag on the inside of his upper arm.

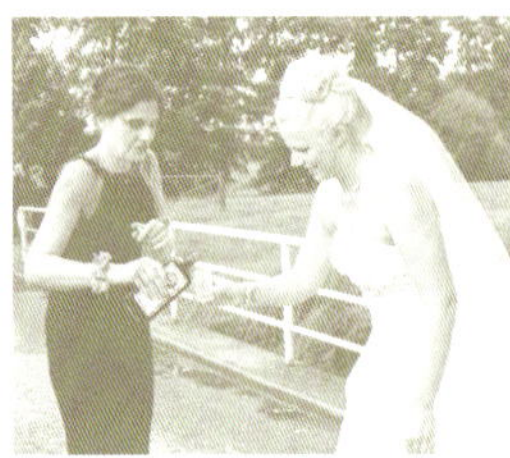

Björn Rastedt is part of a clique of 18 friends from Apen, a small town of 2,432 inhabitants located somewhere between Oldenburg and Leer in Lower Saxony, northern Germany. The group consists of couples, nine men and nine women, all of whom share a strong liking for Jägermeister. 'Otherwise we probably wouldn't be crazy enough to get a tattoo like this,' says Anke Krüger, a founding member of the group. 'We're fully aware that it can never go away now.'

The group started out quite small. 'At some point, those of us who were there in the early days began drinking Jägermeister,' Anke recalls. More and more people joined, and eventually everyone liked it. 'Two of the group members usually have it as a long drink, mixed with Fanta, orange juice, Red Bull or Coke – and others prefer it as a short drink,' says Björn Rastedt. 'It's pretty hardcore in our clique,' adds Anke Krüger. They party together at each others' homes, at concerts and on daytrips, and every year at Whitsun, they spend a long weekend camping. 'Sometimes we'll even have a Jägermeister to wake up in the morning then,' Björn says dryly.

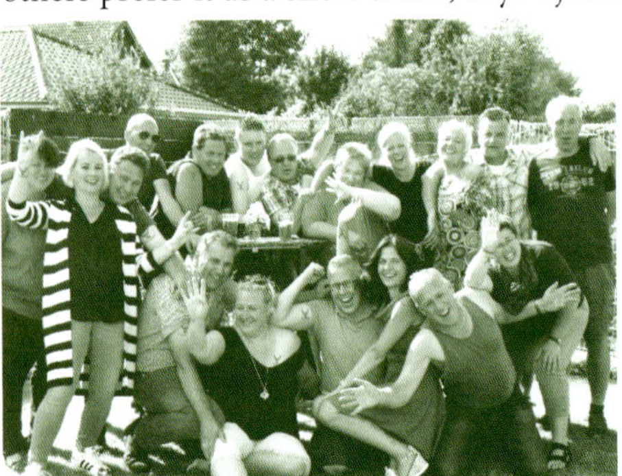

The herbal liqueur even played an important role in the best moment of Björn's life: 'It was an informal wedding with some 200 guests in a garden. I was at the front with my best man, waiting for my bride. That's when the best man whipped out his hip flask and gave me a Jägermeister – and then

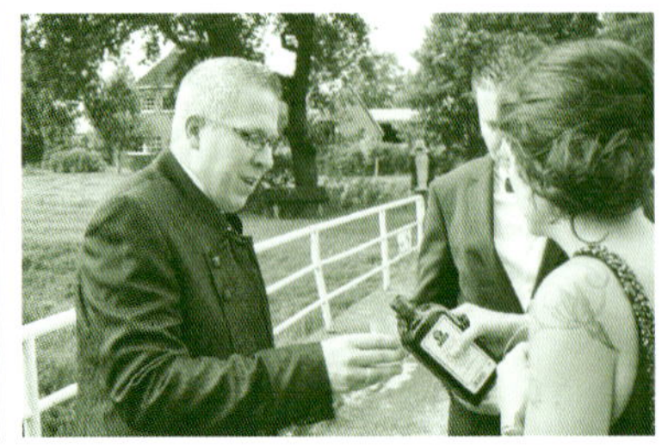

we could proceed.' The herbal liqueur was warmer than might have been ideal on that occasion, but either way Björn Rastedt certainly wouldn't be getting cold feet after his little gulp.

The wonderful but crazy idea of all the members of the clique getting tattoos was born at the end of 2016. 'We made an appointment at the tattoo parlour in May 2017 – five people, with one every hour,' says Björn. The others followed suit later. 'The tattooist initially didn't want to do it, saying more than four people was too much for him,' Anke recalls.

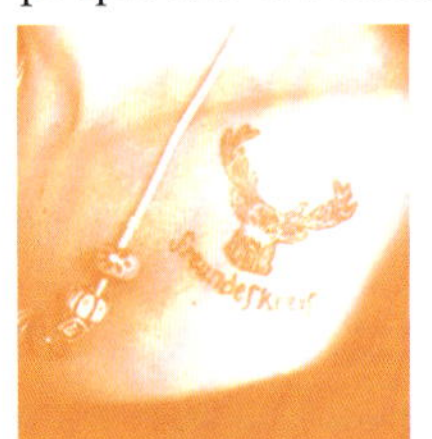

But the tattooist was in fact delighted by the end of the group session. 'He found it really cool, because we are all so different, and he said he would do it again any time.' Others also reacted positively to the Jägermeister tattoos. 'Most people think it's great,' says the mother of two. 'My parents, my aunt .. well, they are now more than used to it.' And at work? The two group members laugh, and Björn says, 'There are, of course, also body parts that are covered up and so aren't immediately visible, as is the case for me.' The clique members' tattoos vary in size and are located on wrists, upper arms, back, torsos and cleavages.

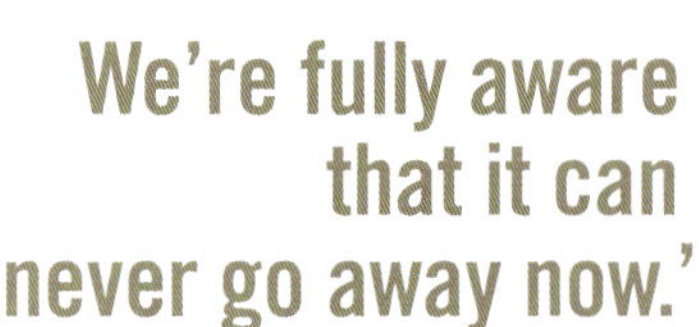

'It's almost like the crowning glory of our love for the herbal liqueur,' enthuses Anke. 'The close-knit group of friends and Jägermeister – for us, that means friendship, family, parties, cohesion, fun, sociability, good vibes, spontaneity, rituals, BBQs, good conversations, endless nights and of course very loud music . . .'

We're fully aware that it can never go away now.'

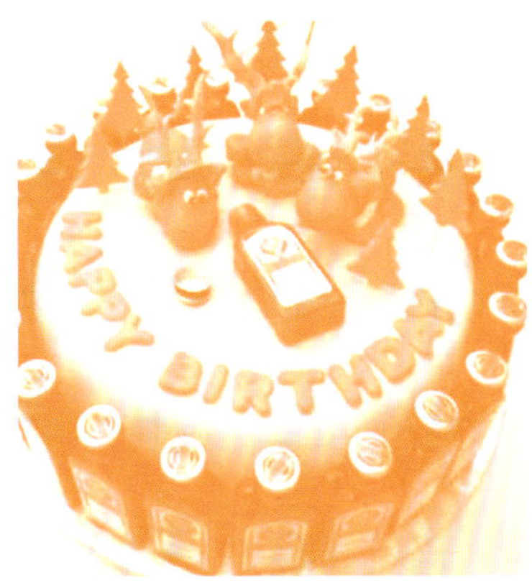

'We love travelling together whenever something's on in the north,' says Björn. This is often the case, because, when it comes to festivals, northern Germany has got it covered – famous festivals up north include the likes of Wacken, Hurricane and Deichbrand. It was at Deichbrand, a music festival near Cuxhaven, that they met the Jägermeister people for the first time. 'Anke and my wife met two women at the camp site washrooms,' recounts Björn. 'All four of us were wearing Jägermeister jumpers and that got us chatting,' Anke remembers. It turned out that the two Jägermeister employees were camping near the group, so they all spent the afternoon together, united by the famous stag.

Jenny, one of the group members, had written to Jägermeister in Wolfenbüttel beforehand to ask if they could get admission wristbands for the Platzhirsch stage, and it was precisely these two women from the Jägermeister crew who had read the letter. The friends had enclosed photos of their tattoos with the letter – certainly a way of showing that your commitment is a permanent one! And so it was that the group gained admission to the Platzhirsch stage that night, with wristbands and tokens for Jägermeister shots, to once again celebrate their penchant for Jägermeister. ⍩

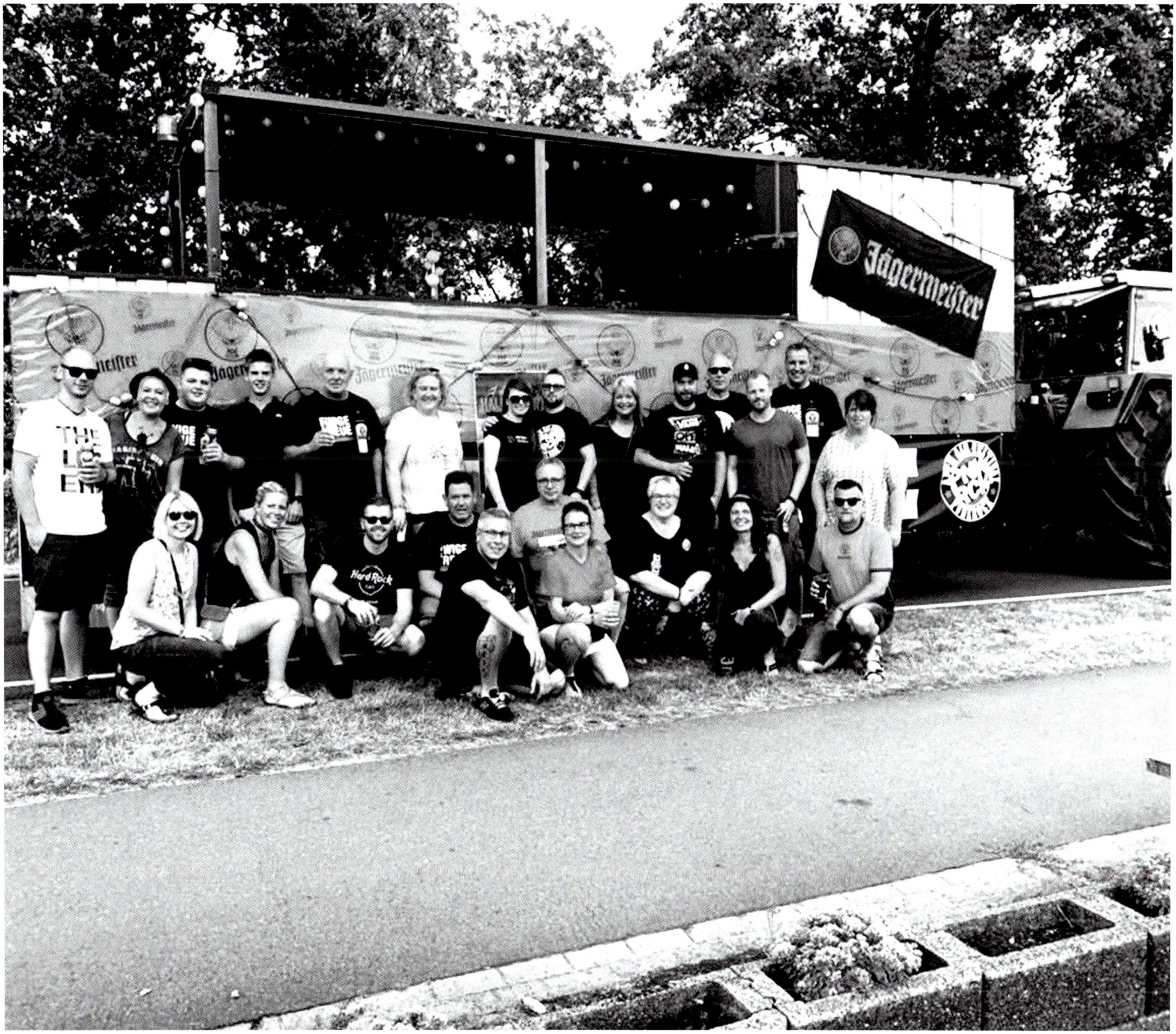

EVIL JARED

From Crocs, the iPod and emos to *American Idol*, GPS and dot-coms, the early 2000s brought a load of new, wild and crazy things. On the big screen, Shrek, Frodo and Jack Sparrow were embarking on adventures, while on the red carpet, glitzy Paris Hilton and Lindsay Lohan were vying for the paparazzi's attention. **And as Jägermeister came out with an animal-centred hit commercial in Germany with its *Achtung Wild!* campaign and the two talking deer Rudi and Ralph, the Bloodhound Gang were taking the charts by storm with trippy pop rock such as *The Bad Touch*, coupled with dubious videos and performances. The most scandalous of them all was bassist and self-confessed Jägermeister fan 'Evil' Jared Hasselhoff, who kept causing a stir with increasingly wild attempts at provocation.**

Memorable moments not only include his sexually explicit appearance on German talk show *TV Total,* but also his shocking interludes during live concerts. The American musician would, for instance, regularly skol a 700ml bottle of Jägermeister on stage through a beer bong – only to throw up on singer Jimmy Pop Ali afterwards. 'Our 2006 show at Rock am Ring was particularly amazing,' Jared recalls. 'It was live on MTV.' Pushing the limit? Definitely. Even he himself can't dispute that: 'If, as a college kid, I had seen an asshole like that skolling a bottle of Jäger on stage, I would have copied him. It was pretty irresponsible. Something like that can easily go wrong, but I didn't think about that at the time.'

But Jägermeister is more than just a tasty means of provocation for the man who now calls Berlin home, even though it was an outrageous rumour that initially sparked his curiosity about the herbal liqueur. 'At college at the time, people were talking about a liqueur from Germany that contained deer's blood and opium. Of course my friends and I had to try it.' And, of course, it turned out that Jägermeister contained nothing of the kind. 'But I still thought it was awesome,' says Jared. So he remained faithful to the drink – and the brand. A few years later, he and the Bloodhound Gang became part of the Jägermusic support programme for American bands in the United States, and thus also part of the corporate communications. 'At the time, there weren't many old hunters left who drank Jägermeister. So the idea was to bring together Jägermeister and young adults who liked rock music,' Jared explains. 'The whole concept was pretty cool, and came at the right time.' It was effectively a contra deal for the Bloodhound Gang: the party combo introduced the brand to their fans, while Jägermeister facilitated larger tours and sponsorship advertising. One of the highlights of the partnership in Germany was ultimately being asked to judge the *Miss Arschgeweih* (Miss Trampstamp) event that kicked off the Jägermeister Rock:Liga. At a large Berlin nightclub in August 2004, the band was responsible for choosing the girl with the best tailbone tattoo out of 50 possible candidates. The second Jägermeister Rock:Liga in 2007 saw the Bloodhound Gang actually participate themselves; three bands played off against each other in different categories over several nights, with a local panel of judges and the audience determining the winner. Though the Bloodhound Gang were just edged out by Deichkind, the event was a big success, and was sold out almost every night.

But 'Evil' Jared considers a different night to be his coolest Jägermeister moment. American radio legend Howard Stern had invited the Bloodhound Gang on to his morning show in New York. Knowing that the host had a bar fitted in his new studio, the guys took him a Jägermeister tap machine and a few bottles of the herbal liqueur as a gift. 'It was maybe five or six o'clock in the morning. We were in the green room waiting for our interview and drinking Jägermeister. Sitting opposite us was an old man, who snarled ferociously at us, like a wolf about to destroy the three little pigs' hut,' Jared recalls. Someone finally whispered to the band that the old man was none other than Henry Hill, the Mafioso and FBI informant on whose story Scorsese's classic *Goodfellas* is based. 'At one point, he came over to us and asked if he could have some of our Jägermeister. I thought to myself, "Dude, you're Henry goddamn Hill! Of course you can get what you want – the most important thing is that you don't knock me off".' They continued drinking together until the band was called into the studio to speak to Howard Stern. When they came back out, they bumped into Henry Hill in the lift again. 'He was a total wreck, and grabbed the bottle from us. But I thought, "Hey, isn't he part of the witness-protection programme?! And now he's going on live radio so everyone will know where he is".' Hill then disappeared. When the band arrived for another live interview the day after, they were asked, 'What the hell did you guys do to Henry Hill?' 'We found out that he hadn't shown up at any of his appointments since meeting us. People had been searching everywhere for him, and finally found him in a steakhouse basement bar. He had 120 dollars on his tab, but only 5 dollars in his pocket, so the bouncers had thrown him out,' says Jared. 'That's definitely my craziest Jägermeister story.'

Looking back, it's clear that the *Achtung Wild!* warning was serious. It didn't just apply to Rudi and Ralph, the two cheeky alpha-deer, but also to Evil Jared and his shameless Bloodhound Gang. Few other bands supported by Jägermeister at that time had such an influence on the *Achtung Wild!* era. And one thing is for sure: 'Without Jägermeister, my life would have been very different,' Jared laughs. 'Much worse.'

CHTUNG WILD! ACHTUNG WILD! ACHTUNG W

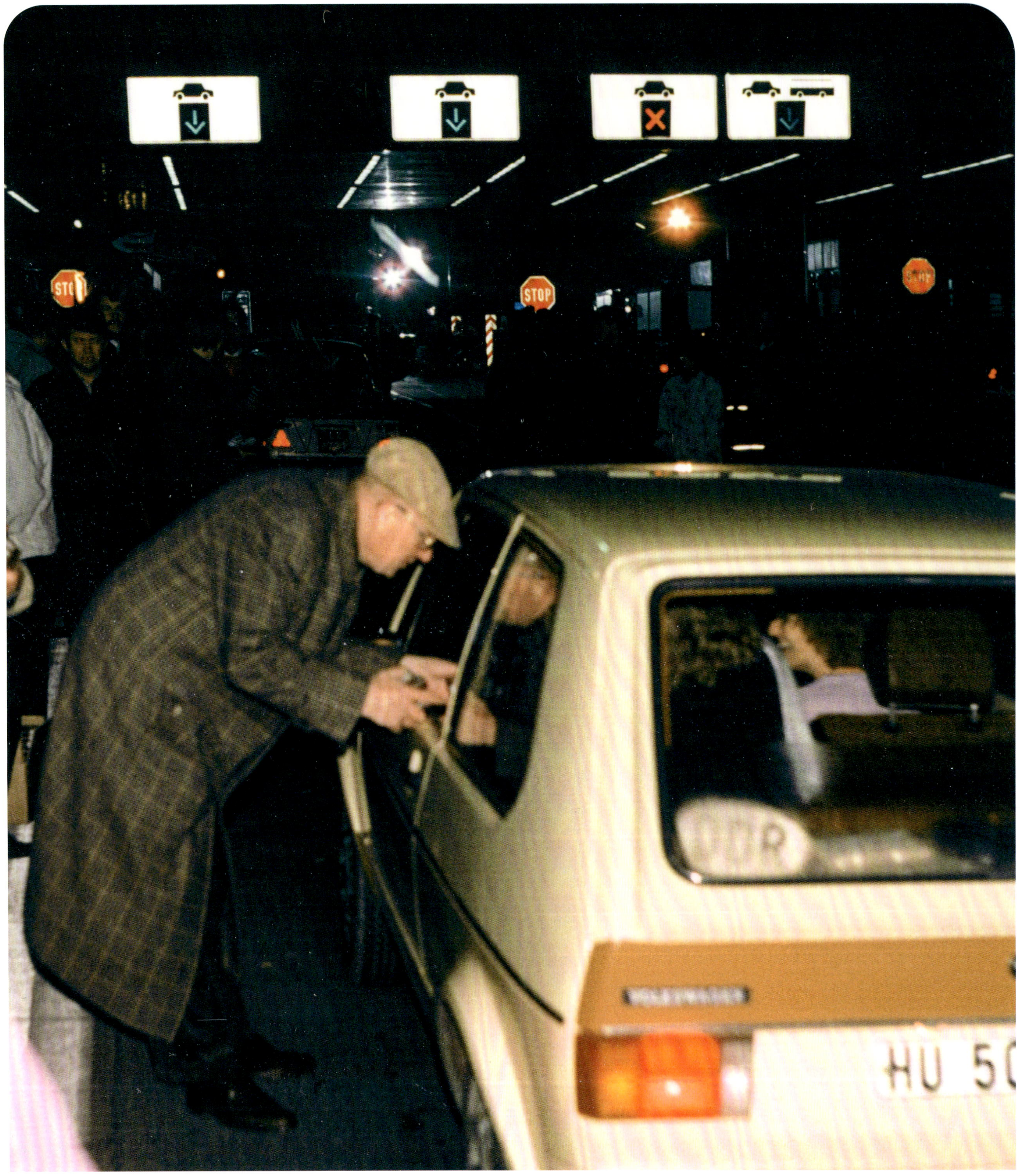

WOLFGANG SPUREK

'JÄGERMEISTER ALWAYS REMINDS ME OF NOVEMBER 1989'

The borders are open! This was the breaking news that spread across divided Germany like wildfire on the night of 9 November 1989. And very soon after, countless East German citizens flocked to the West for the first time. East and West Germans fell into each other's arms at the border crossings, singing, laughing, crying and celebrating together. Perhaps the greatest party ever. Even at the largest border crossing, located halfway between Braunschweig and Magdeburg, the first GDR citizens were allowed to start passing from the Marienborn checkpoint to the Helmstedt checkpoint at around 9.30 pm.

'I also would have loved to cross over that very night of 9 November,' recalls Wolfgang Spurek, who was 40 at the time. A native of East German Magdeburg, he had relatives in the Lower Saxon town of Helmstedt. 'So near – not even half an hour away – and yet inaccessible to us.' His wife was concerned about the prospect of him making the crossing, warning him: 'Are you crazy? They'll lock you up!' So the next day, a Friday, he went to his job as an electrical engineer as usual – but when a colleague told him he had visited his sister in Wolfsburg in the West the night before, he simply could not wait any longer. 'I took the rest of the day off, picked up my daughters from school and made my way to Helmstedt,' Wolfgang says.

Some 20,000 GDR citizens passed through the Helmstedt/Marienborn border crossing within the first two days. By the early hours of 10 November 1989, a traffic jam consisting of thousands of vehicles had built up. Wolfgang Spurek crossed the border into West Germany for the first time at around 7 pm with his daughters, aged 11 and 13,

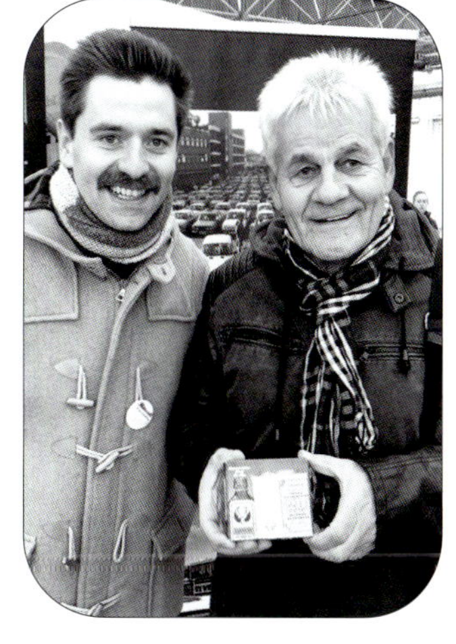

seated in the back of his Polski Fiat. 'It was a massive hubbub,' he recalls. 'Hundreds of thousands of people had gathered on the West German side to welcome us, including Günter Mast, who was handing out Jägermeister.'

The Jägermeister boss was known for his unique flair for extraordinary promotional campaigns, and the Wolfenbüttel headquarters of the company were not even 30 kilometres from the West—East German border. Günter Mast and eight of his employees made their way to the Helmstedt/Marienborn border crossing on the night of 10 November to welcome East German citizens with three-packs of Jägermeister, to which 20-mark notes were stuck. 'We exchanged a few words, and then he handed me his welcome gift through the car window. It was an incredibly nice gesture I will never forget,' Wolfgang Spurek says, still touched by the memory of that unique moment. 'Everything felt so unreal.' His daughters were equally bewildered. 'Papa, we don't understand it,' they said. 'At school, they tell us that the people in the West are basically bad. So why are they giving us so many presents?'

In the weeks that followed, several other spectacular promotions by Jägermeister caused a similar stir, and created a collective festive mood. On 17 November, Jägermeister opened its factory to East German citizens. Over a full three days, there were sausages, cakes and drinks on offer, as well as a Trabant workshop and a competition in which Jägermeister raffled ten VW Golfs to Trabant drivers under the motto 'Swap your Trabi for a Golf'. Those who didn't win slapped a Jägermeister sticker on their numberplate instead. 'You can still see them in our region to this day,' laughs Wolfgang Spurek. 'But now Trabis like that are highly sought-after and worth a lot of money!'

There would be an even greater crush at the Christmas market organized by Jägermeister on 9 December. The flyer had been distributed beforehand at the border, and even the *Magdeburger Volksstimme* had printed an invitation. Wolfgang Spurek was one of the 25,000 people who went to the Wolfenbüttel Christmas market – a record for the city of 50,000, whose citizens actively helped to arrange this collective Christmas party. Just a year later, Jägermeister had become one of the best-known and most popular spirits brands in eastern Germany – and it had become a pan-German company by the time the foundation stone was laid for a new bottling plant in the east Saxon town of Kamenz. Since then, the herbal liqueur has been shipped to the entire world from the two facilities in Wolfenbüttel and Kamenz.

'Whenever I see a Jägermeister anywhere, it always reminds me of that night in November 1989,' says Wolfgang Spurek. 'Nothing like that will ever happen again; I will never experience anything like that again.' Even today, he still sometimes gets goosebumps when he drives down the A2 past the former border. And the young-at-heart 71-year-old still shakes his head in disbelief when he thinks back to that time. Perhaps that's why he has kept his Jägermeister welcome gift, including the 20-mark note – as a souvenir of that unique moment, but also as proof that he really was there.

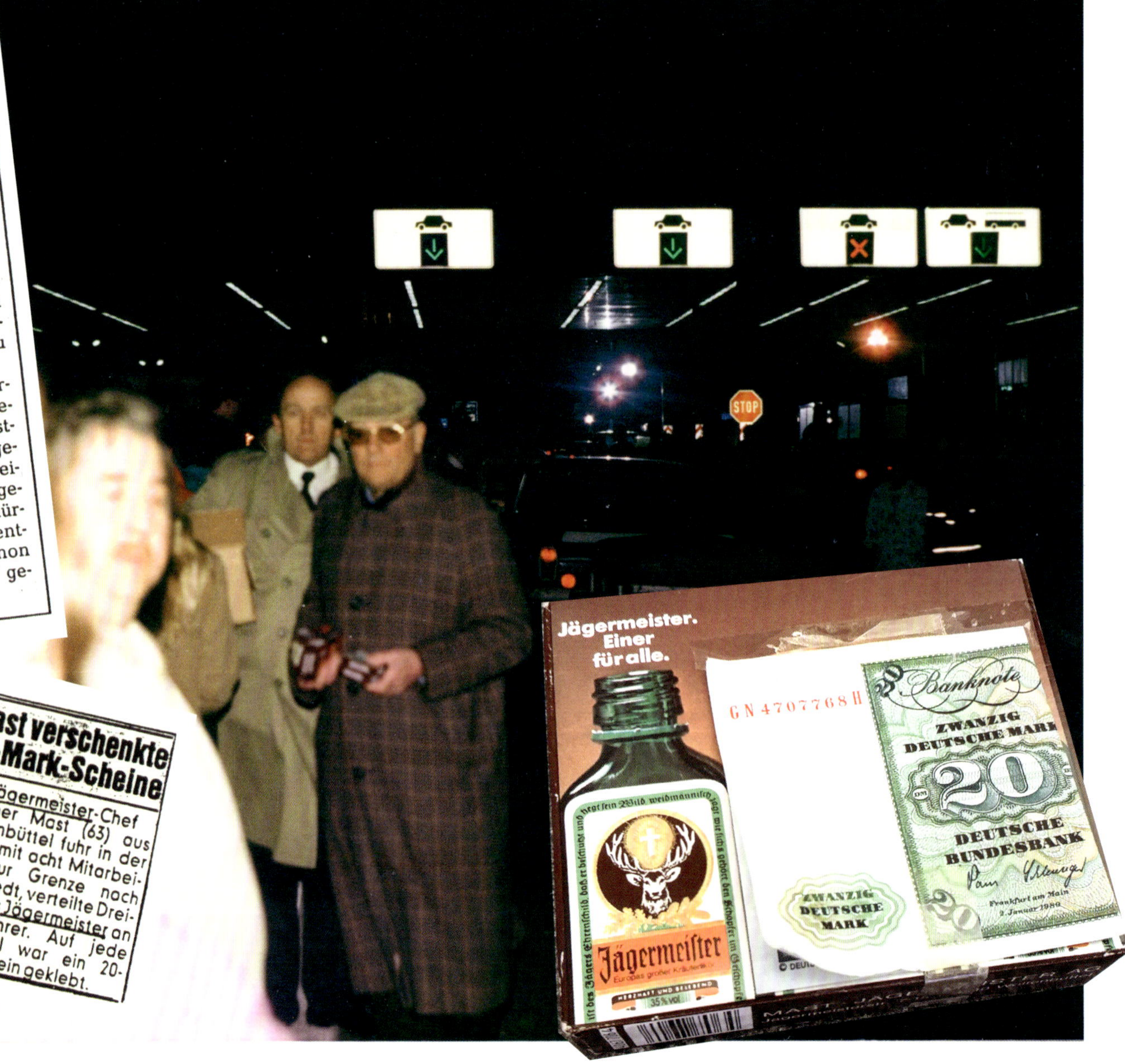

Günter Mast verteilte in Helmstedt Päckchen

WOLFENBÜTTEL (-pp-) Für eine weitere Überraschung in Helmstedt sorgte im Verlauf des Freitags der Wolfenbütteler Unternehmer Günter Mast. Er entbot den vielen Bürgern aus der DDR, die mit ihren Fahrzeugen gekommen waren, Grüße in Form von vielen Päckchen, die er freizügig verteilte.

Eine tolle Überraschung erlebte auch ein Wolfenbütteler Taxifahrer, und das schon in den frühen Morgenstunden. Ein Kollege von ihm war schon am frühen Morgen losgefahren, um in die Lessingstadt zu kommen.

Unser Fotograf Detlev Splitt erhielt am späten Freitagabend ein Telegramm von Verwandten aus Ost-Berlin, in dem ihr Kommen angekündigt wurde. Wie lange sie bleiben wollten, war noch nicht mitgeteilt worden. Die Freude war natürlich bei unserem Fotografen entsprechend groß, nachdem er schon den ganzen Tag über „am Ball" geblieben war.

Mast verschenkte 20-Mark-Scheine

Jägermeister-Chef Günther Mast (63) aus Wolfenbüttel fuhr in der Nacht mit acht Mitarbeitern zur Grenze nach Helmstedt, verteilte Dreier-Packs Jägermeister an Trabi-Fahrer. Auf jede Schachtel war ein 20-Mark-Schein geklebt.

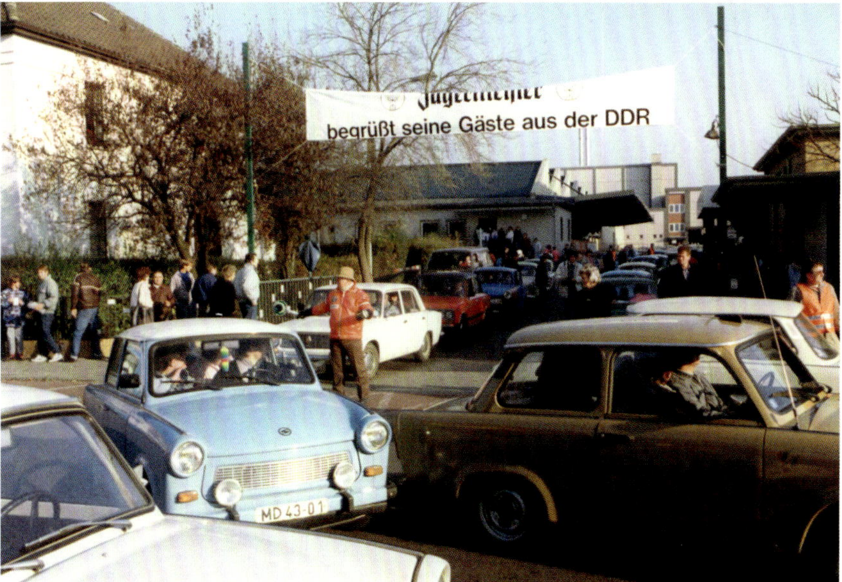

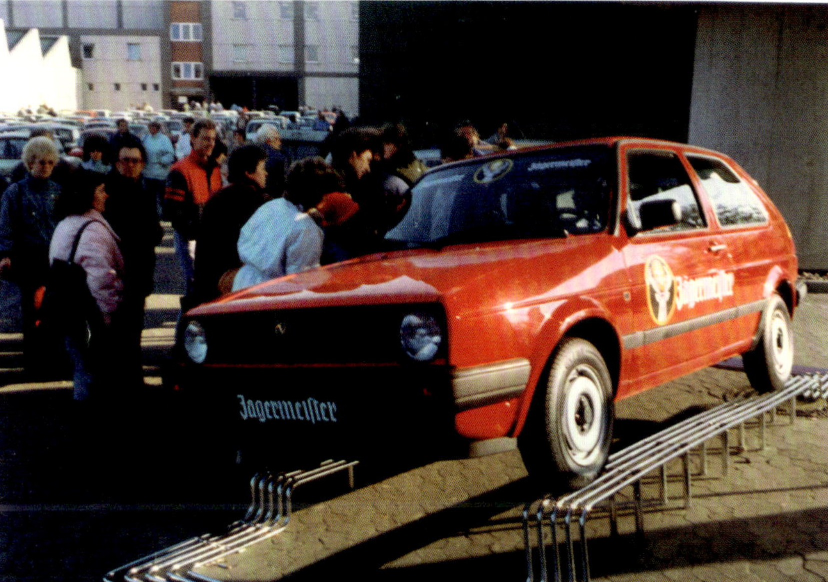

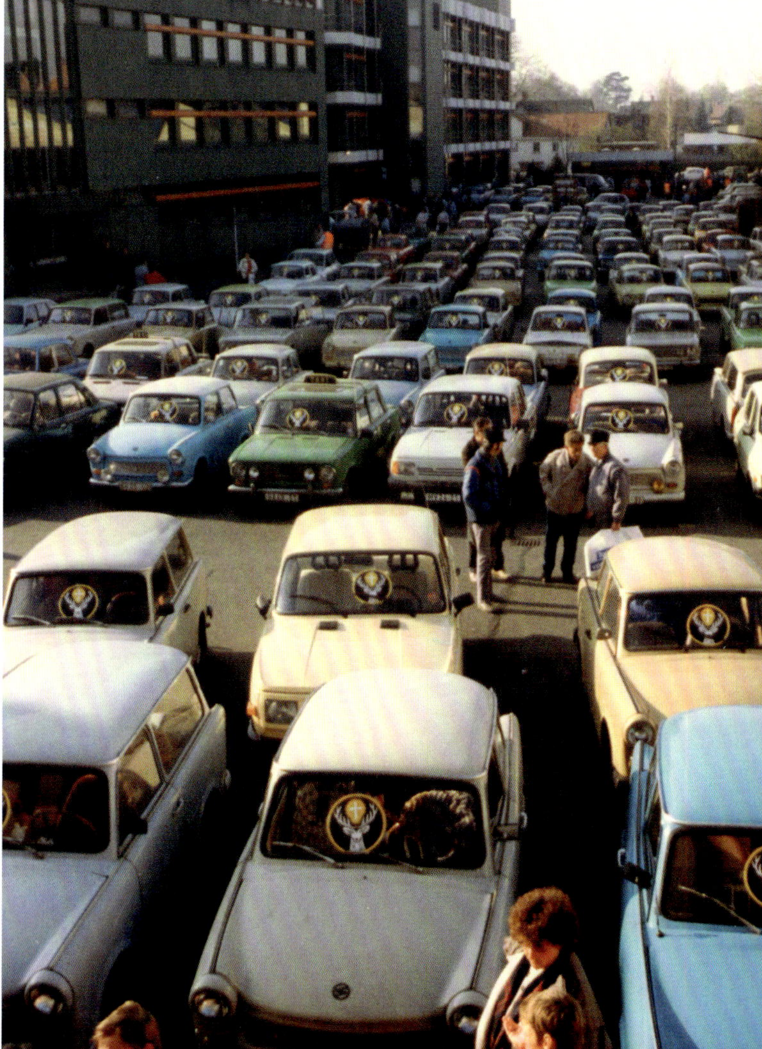

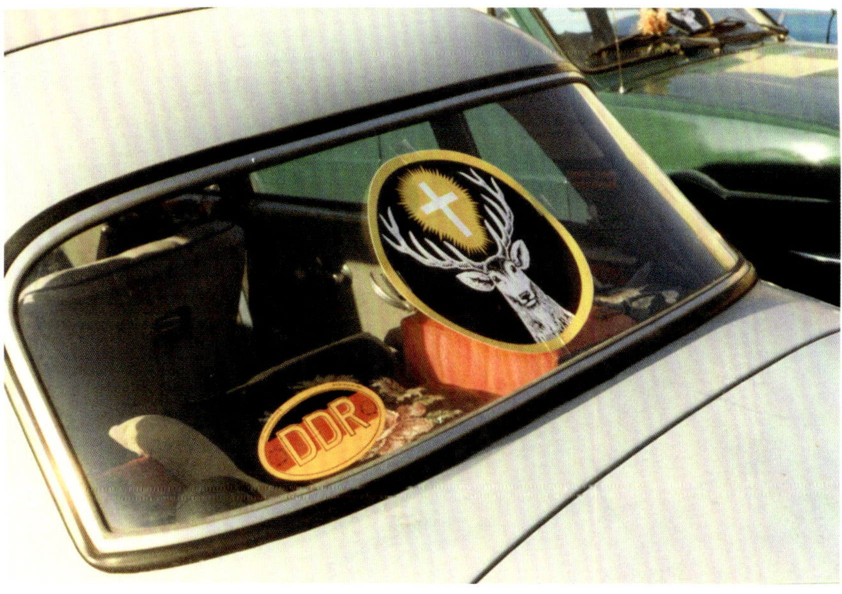

17 November

After the opening of the border in the West, Jägermeister
welcomed the GDR citizens with emotional messages.

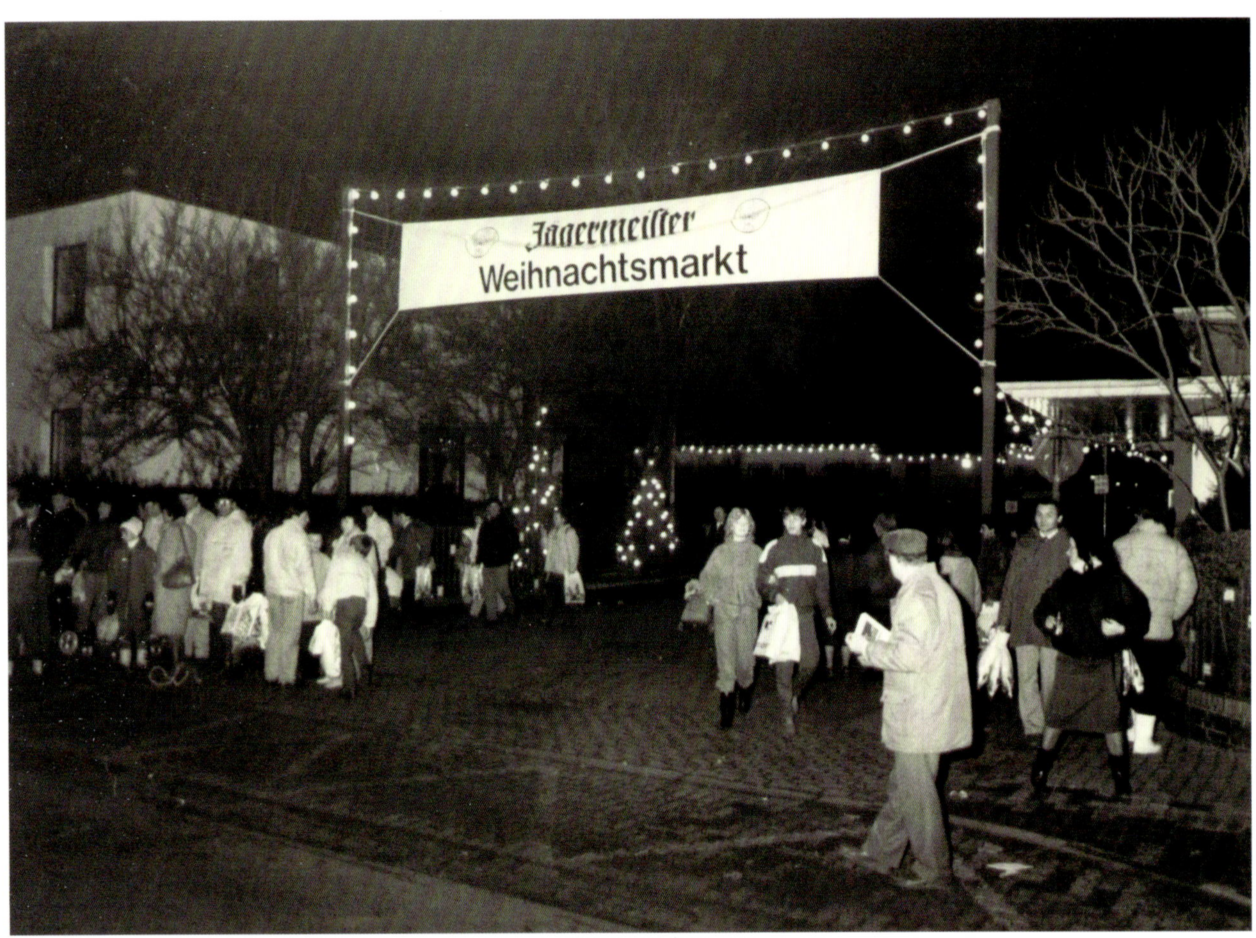

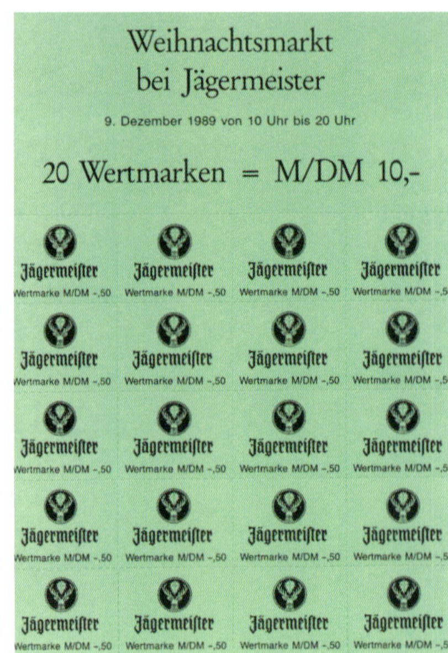

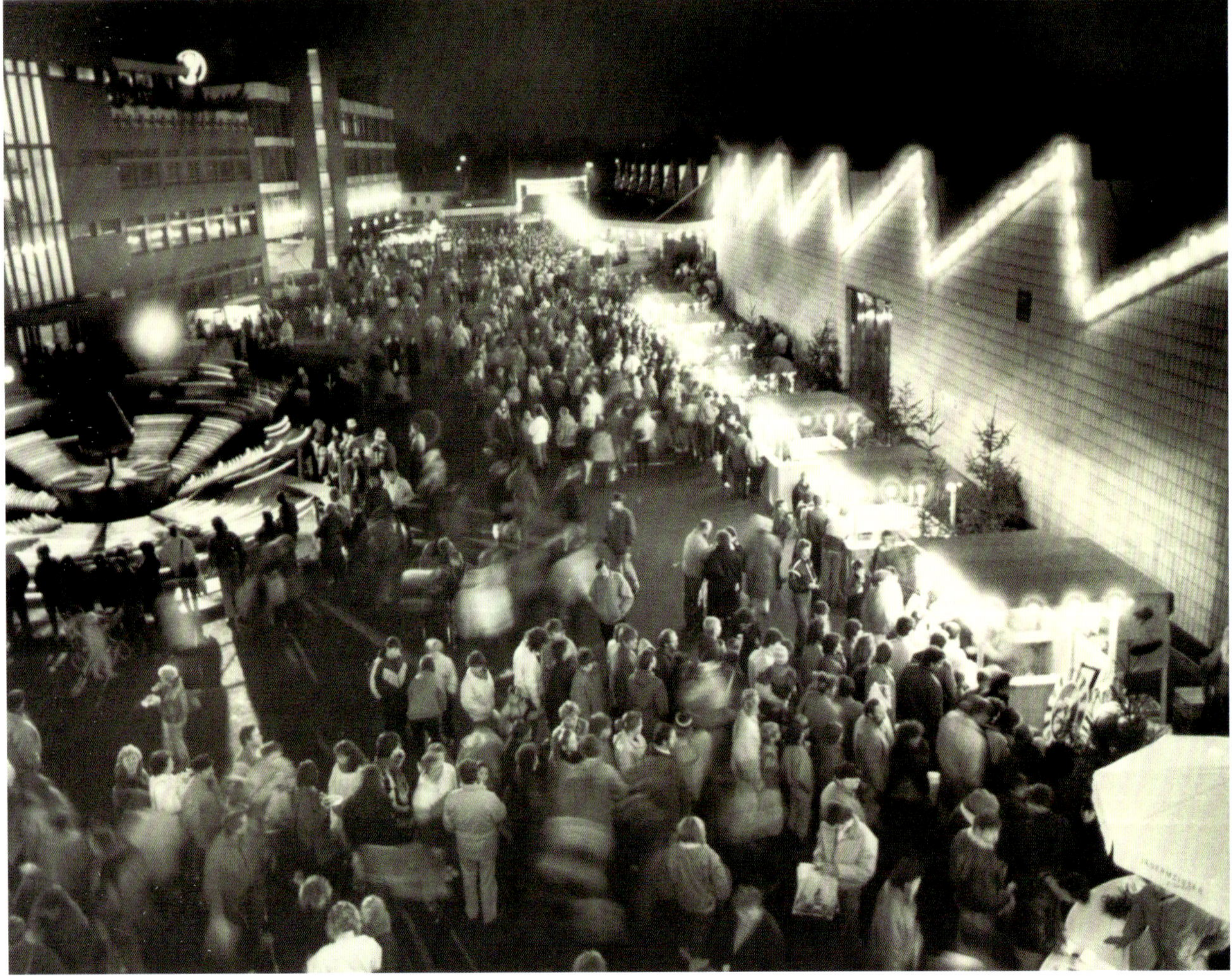

An incredible 25,000 visitors flocked to the big Jägermeister Christmas market on 9 December 1989 – a record for the city of Wolfenbüttel which has just 50,000 inhabitants.

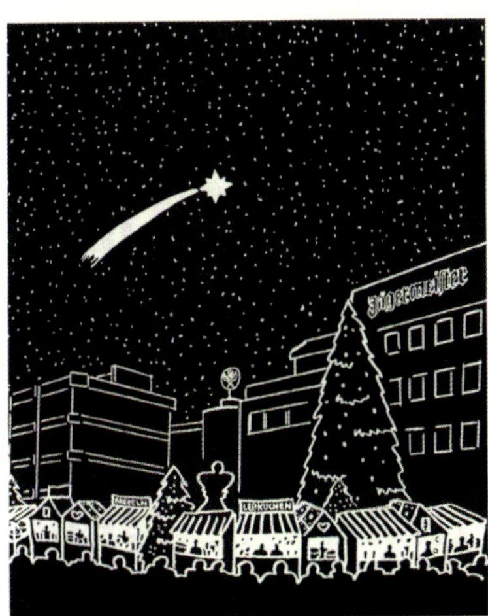

Weihnachtsmarkt
bei Jägermeister

Am 9. Dezember 1989 von 10 Uhr bis 20 Uhr

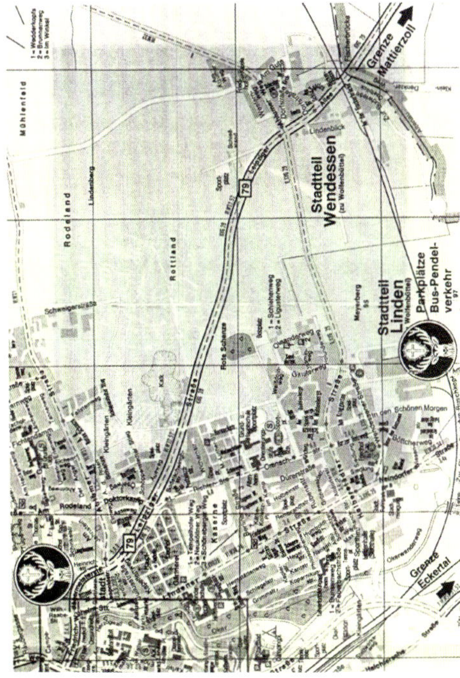

Wolfenbüttel, im Dezember 1989

Liebe Besucher aus der DDR!

Willkommen zum "JÄGERMEISTER-Weihnachtsmarkt"!

Am 9. Dezember 1989 findet für Sie von 10 Uhr bis 20 Uhr auf unserem Betriebsgelände in Wolfenbüttel ein Weihnachtsmarkt statt, zu dem wir Sie hiermit herzlich einladen.

Lichterglanz, Tannenduft und vieles mehr sorgen für vorweihnachtliche Stimmung:

— Gratis-Imbiß mit warmen und kalten Getränken

— Günstige Einkaufsmöglichkeiten für Südfrüchte, Süßigkeiten und elektronische Geräte durch Ausgabe von JÄGERMEISTER-Wertmarken

— Überraschungen für Ihre "Kleinen"

— Stündlich musikalische Darbietungen

Ausreichende Parkmöglichkeiten haben wir in unserem Werk **Linden** geschaffen (siehe Lageplan). Dort finden Sie auch einen kostenlosen Pannendienst. Folgen Sie bitte der Ausschilderung. Zwischen den Parkplätzen und dem Weihnachtsmarkt ist ein unentgeltlicher Bus-Pendelverkehr eingerichtet.

Wir freuen uns auf Ihren Besuch und wünschen Ihnen viel Vergnügen auf dem "JÄGERMEISTER-Weihnachtsmarkt".

Ihre
MAST-JÄGERMEISTER AG

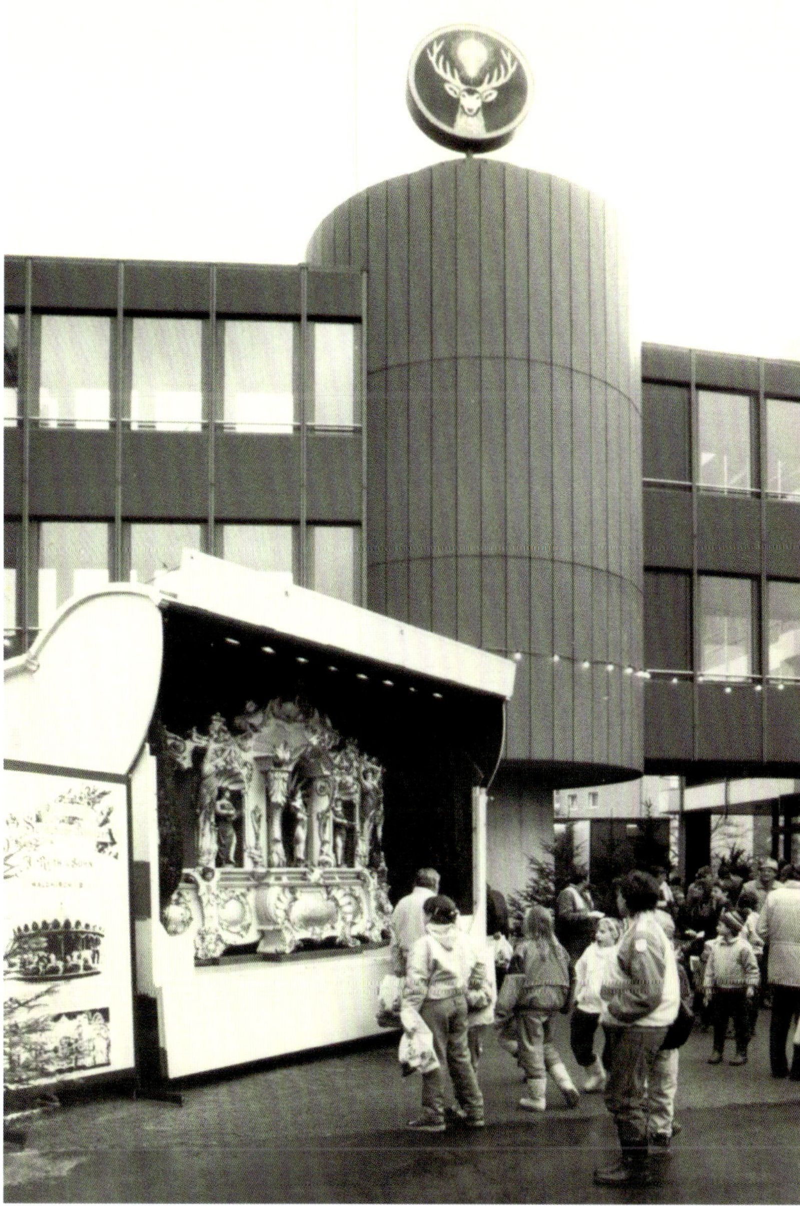

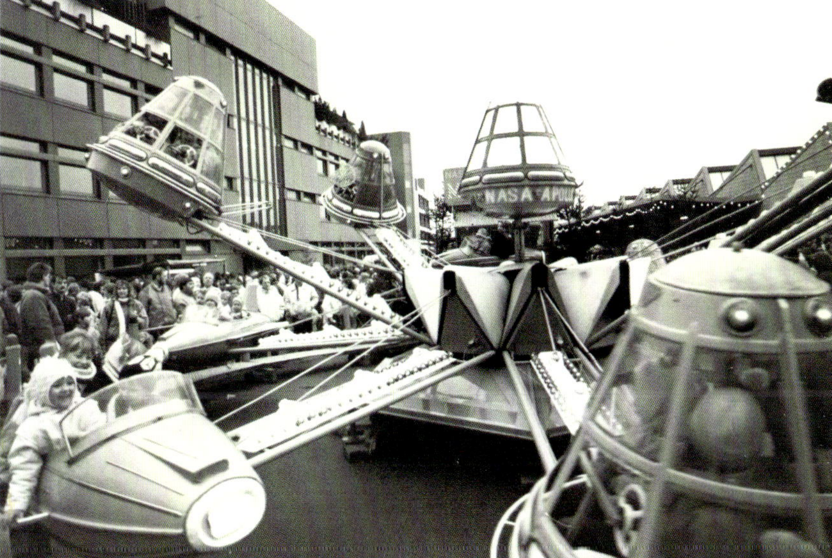

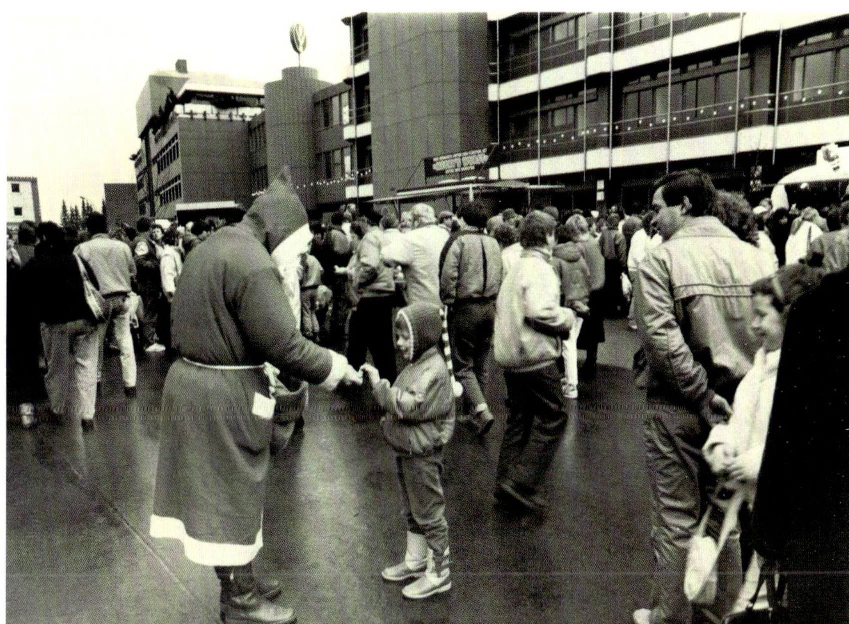

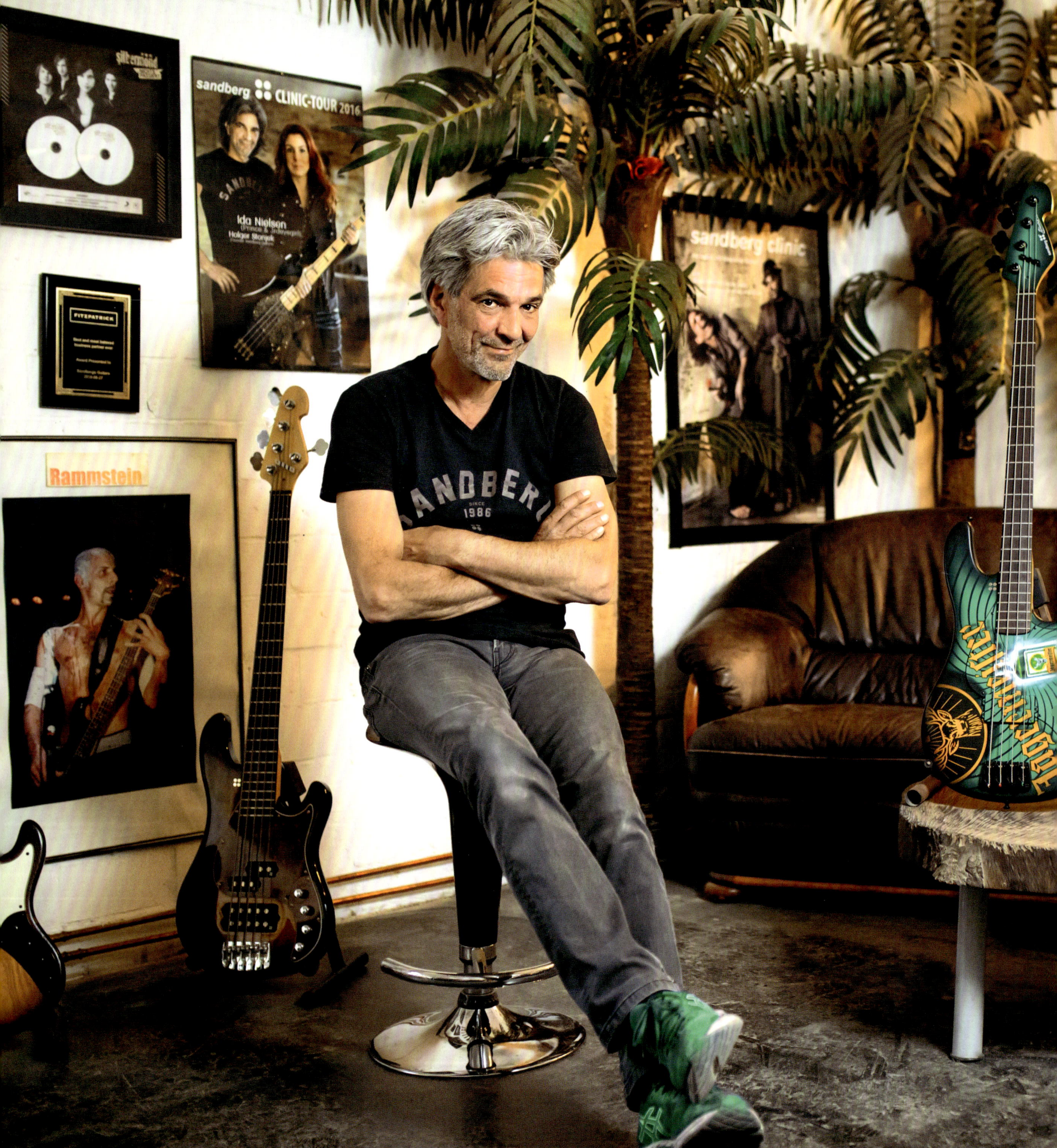

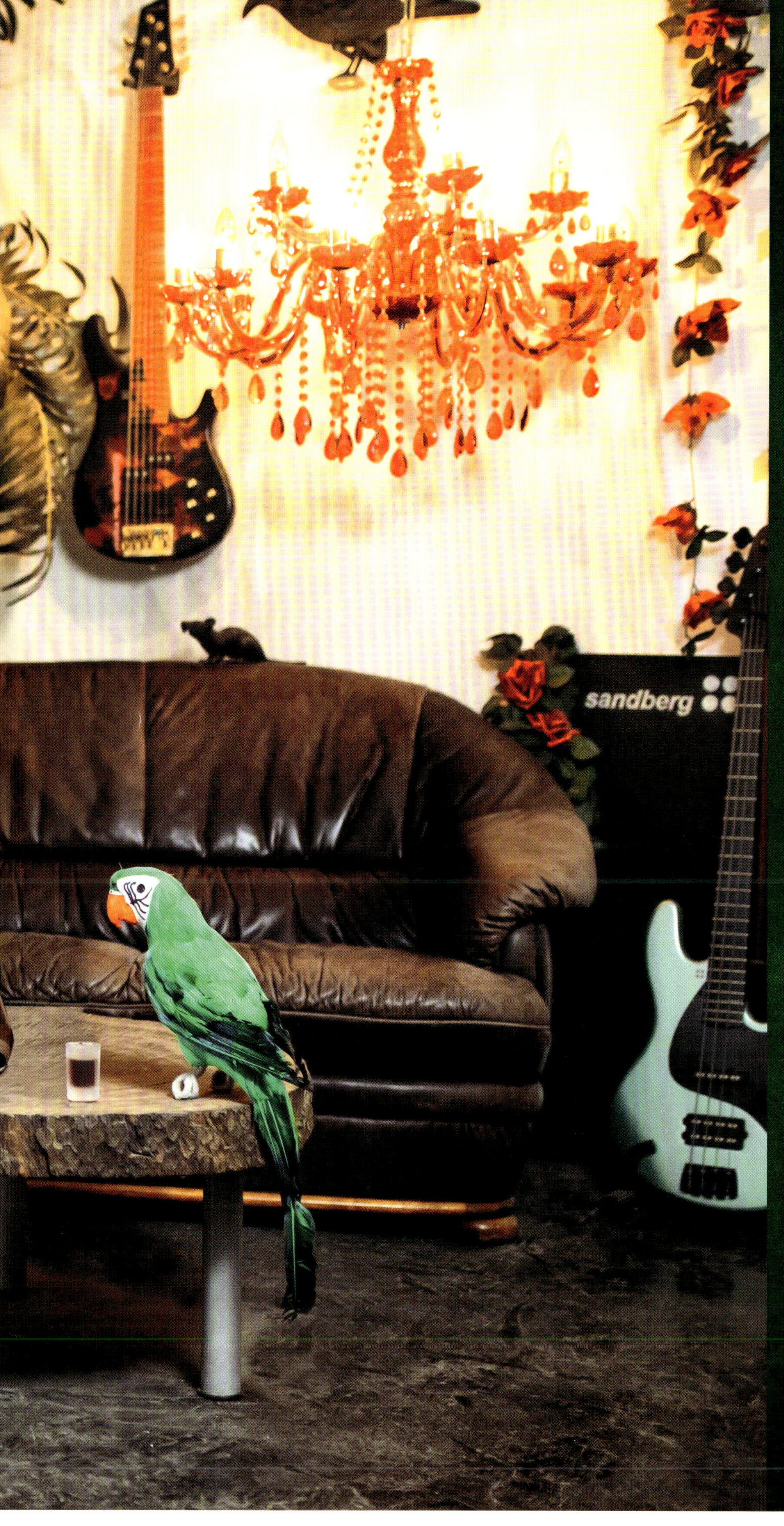

HOLGER STONJEK

'We always get a lot of international visits from musicians and business partners from all over the world,' he says, adding 'and there are always Jägermeister shots on the table when it starts getting dark outside and merry inside.'

THE BOSS OF THE BASS

J ägermeister contains 56 different herbs – almost exactly the same number of elements as a bass guitar's various hardware parts, pickups, electronics, glues, oils and of course a variety of woods. 'Depending on the wood used, there is apparently even some overlap with the ingredients in Jägermeister,' laughs Holger Stonjek. At least that's what someone told him many years ago. 'Wood that not only sounds good, but also tastes good!' Is it true? In any case, it's a nice idea that fits wonderfully with his equally fine idea of the Jägermeister Bass.

Holger Stonjek is the founder of Sandberg Guitars, a renowned Braunschweig-based company that makes electric guitars. Prominent musicians known to use Sandberg basses include Oliver Riedel from Rammstein, Ken Taylor (studio and live bassist for Peter Maffay and Bruce Springsteen), Steffi Stephan (bassist in Udo Lindenberg's Panic Orchestra), Flux from Oomph!, Ida Kristine Nielsen (Prince's former bassist) . . . The list goes on and on, and boasts bands such as Silbermond, Bullet for My Valentine, Helloween, Dimmu Borgir, Truckstop, Tokio Hotel and the Beatsteaks. Holger Stonjek knows all of them. 'We always get a lot of international visits from musicians and business partners from all over the world,' he says, adding 'and there are always Jägermeister shots on the table when it starts getting dark outside and merry inside.'

Whether it be the promise, the taste, the consistency, the image, the advertising or the geographic proximity to Jägermeister's hometown of Wolfenbüttel, the affable Braunschweig native has many reasons to like Jägermeister. 'I have great respect for the Jägermeister company philosophy; they have achieved massive worldwide success with just a single product, and I don't know of any other company that has managed to do anything like that,' he says. 'They have been able to build an incredibly strong image for their product.

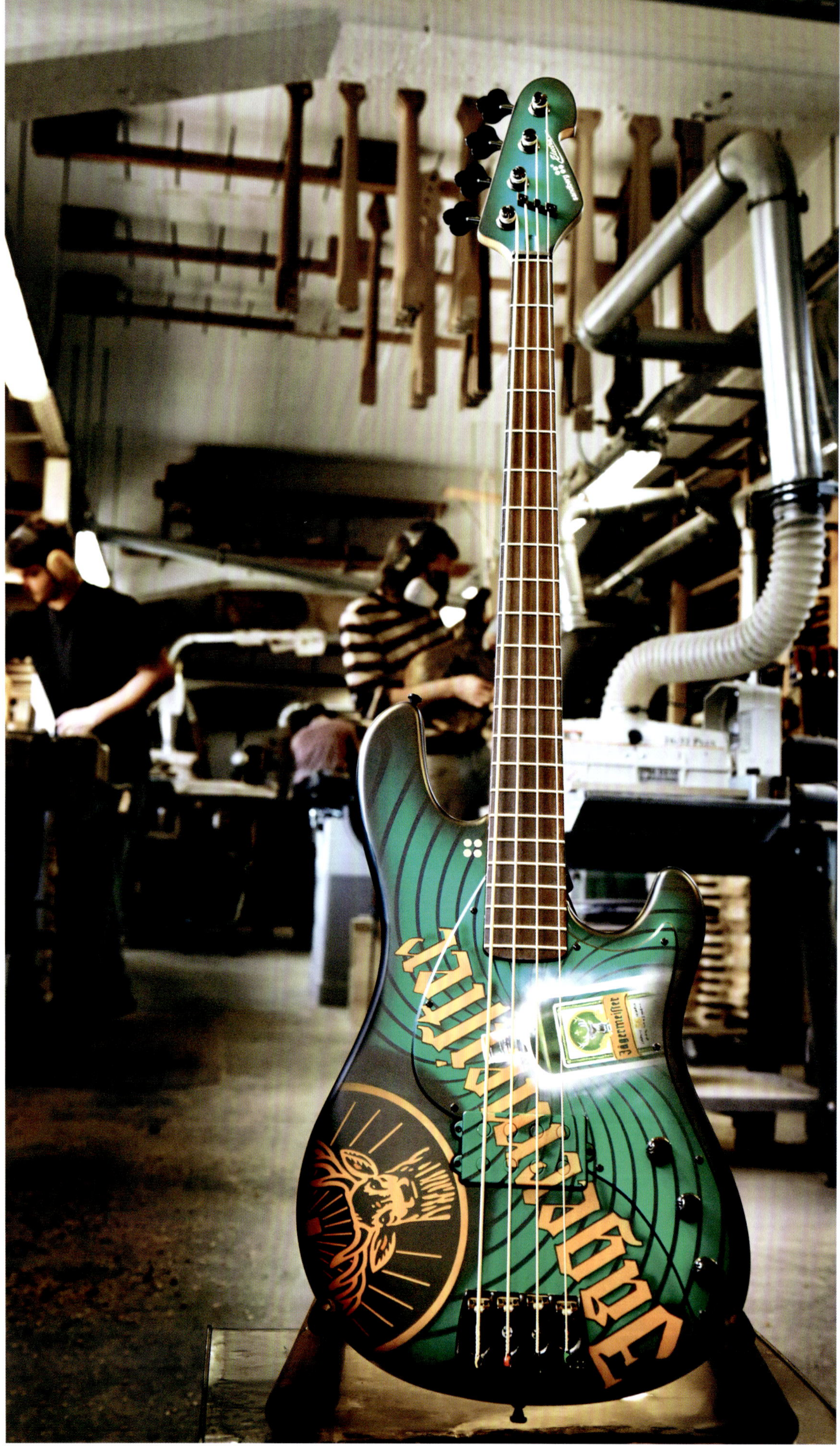

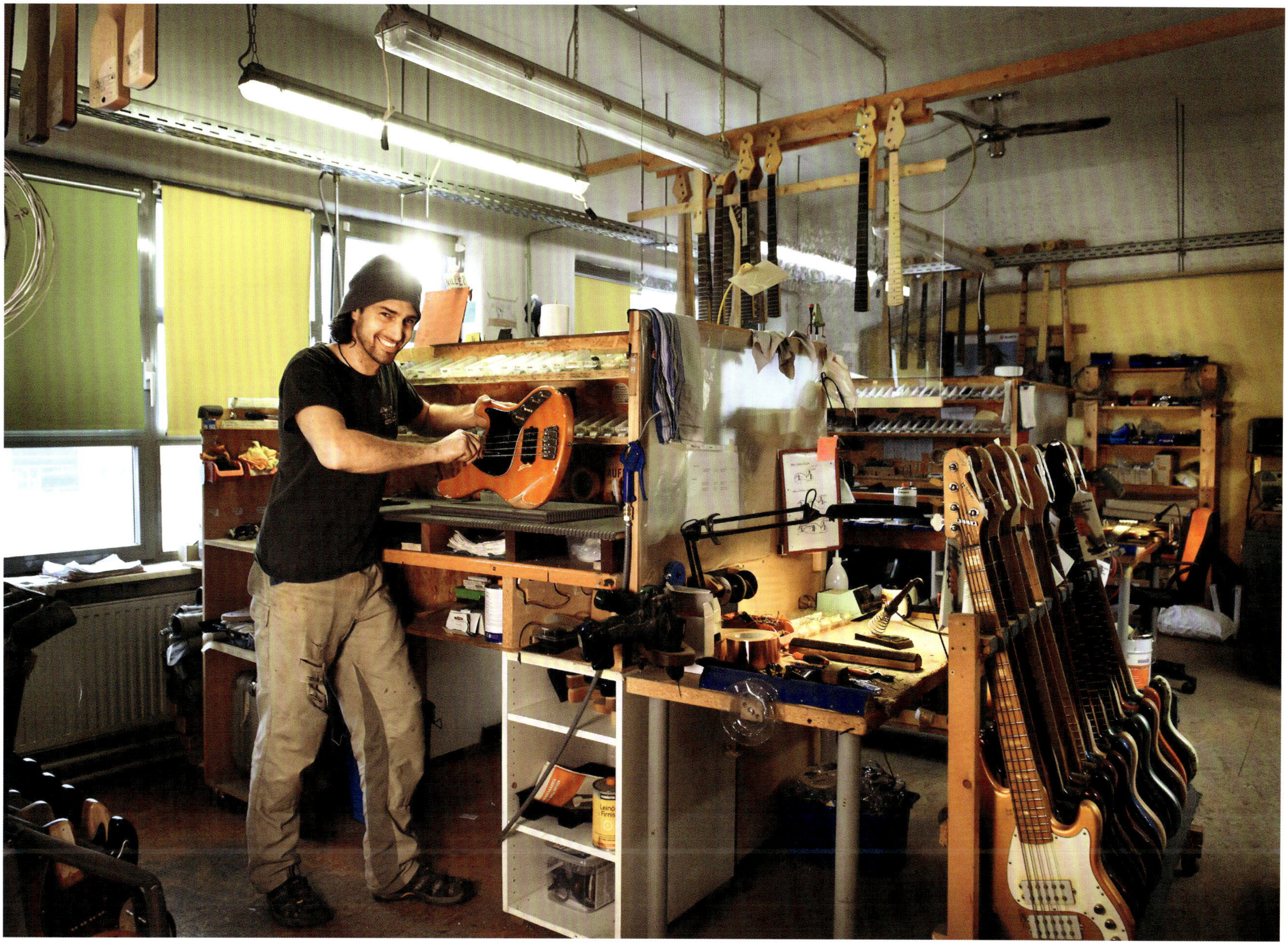

Jägermeister is not just a herbal liqueur; it's a cult.' As a widely travelled instrument-maker, he knows what he's talking about. 'I frequently go to music fairs and events and music stores around the world. Whenever I mention that Jägermeister are basically our neighbours, I am met with surprised, happy faces. Whether it be in Los Angeles, Singapore, Sydney, Montreal, Chicago, Tokyo or wherever, Jägermeister has an amazing reputation all over the world.'

As a tribute to, and gift for, the company, Sandberg Guitars created the Jägermeister Bass, elaborately designed by airbrush artist Frank Walter. There was a sense of bewilderment at Jägermeister when Holger Stonjek presented them with the exquisite, hand-crafted present. 'I brought the bass over myself,' he says. 'It was a lovely meeting, but I think they really were very surprised.' So surprised that he was asked how much he wanted for it. 'I had to laugh and say,

"You don't charge anything for a gift – otherwise it wouldn't be a gift!"' The many wonderful Jägermeister nights are more than enough repayment for this creator of bass guitars. He also believes that Jägermeister has managed to achieve something no other alcoholic beverage has: 'When you drink it, you just know things are going to somehow get cool, fun and merry.' ❦

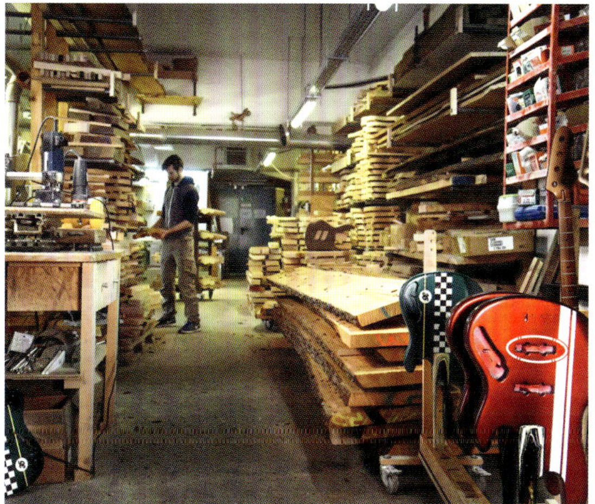

Sandberg Guitars was founded in 1986 by Holger Stonjek and Gerd Gorzelke. Today the Braunschweig company is the second largest manufacturer of electric basses and guitars in Germany and produces around 1,500 instruments a year – in its own custom shop, and also on individual customer request. Incidentally, the Jägermeister bass was also created here.

RAZ DEGAN

'THE FAIRYTALE OF MY LIFE'

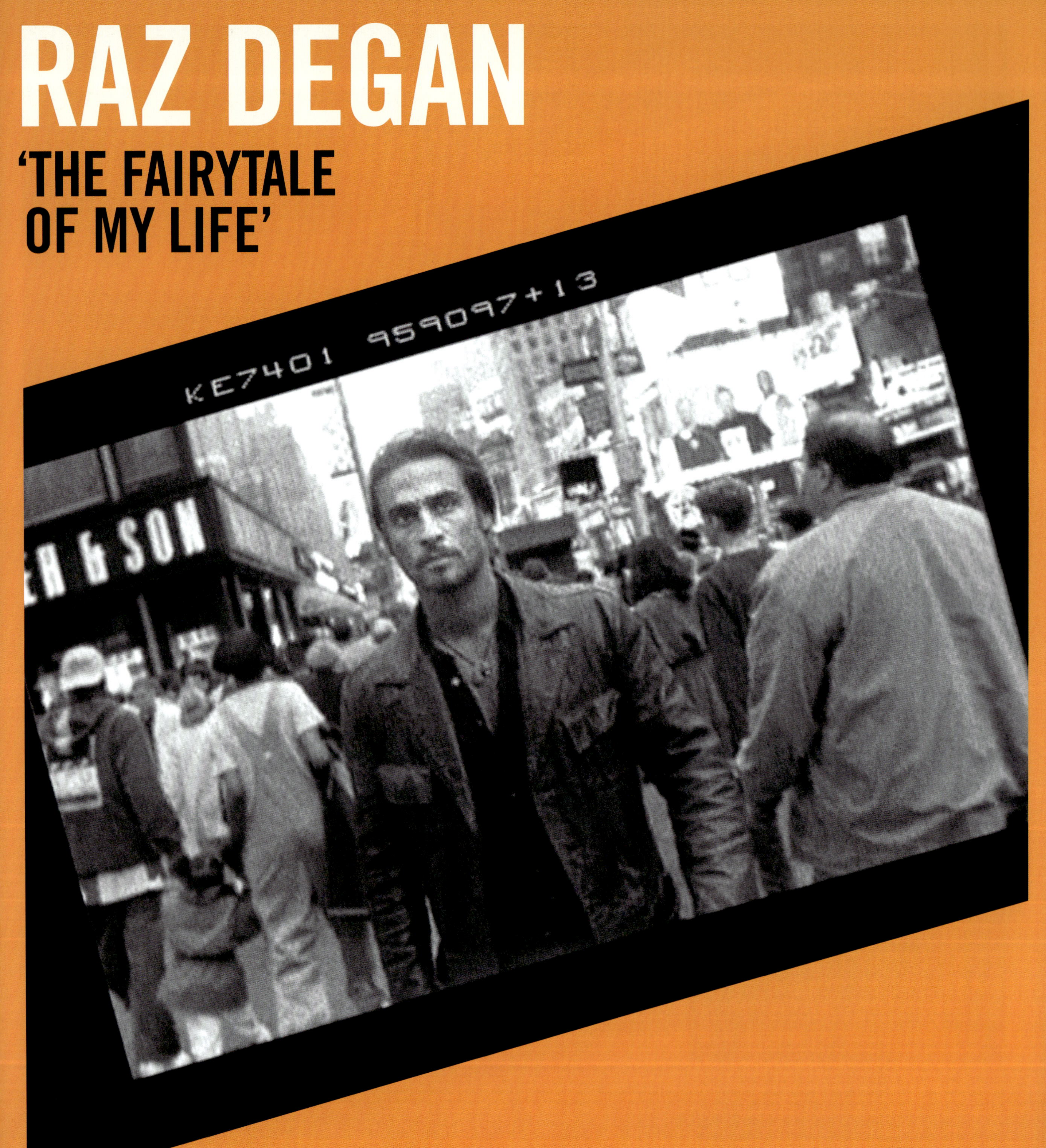

I

t was the mid-1990s in New York, and a young, good-looking actor from Israel, self-described as 'wild and hungry for life', was looking for his big break in the city that never sleeps. He knocked on countless doors, made contacts in the film industry and at modelling agencies, tried to get by somehow, took on small movie roles and appeared in TV commercials. One of them ended up taking him to Milan. 'I had no idea that this commercial would be my trampoline,' he recalls. 'To me it was just another day at the office back then.' Only many months later would he realize that his day in Italy was the magical start of a fairy-tale success story.

The clip, which took 45 minutes to film, shows a fiery-eyed, long-haired Raz Degan pacing back and forth in front of the camera. *'Non bevevo Jägermeister!'* ('I have not drunk Jägermeister!') he says. *'Perché, perché, non so perché!'* ('Why, why? No idea!') He gesticulates energetically; on the floor are a glass and a bottle of Jägermeister. Then, grinning, he says, *'Sono fatti miei'* ('That's my business'). Although the 26-year-old didn't speak Italian at the time, his performance was very convincing. 'They wrote the words in English on a board,' he says, laughing. 'But it was more about the energy than about the words itself; the energy that was to emanate from the brand.' Even during filming, it was clear that this commercial was different and somehow cooler than all the other TV commercials around at the time. 'Its structure and the way the brand

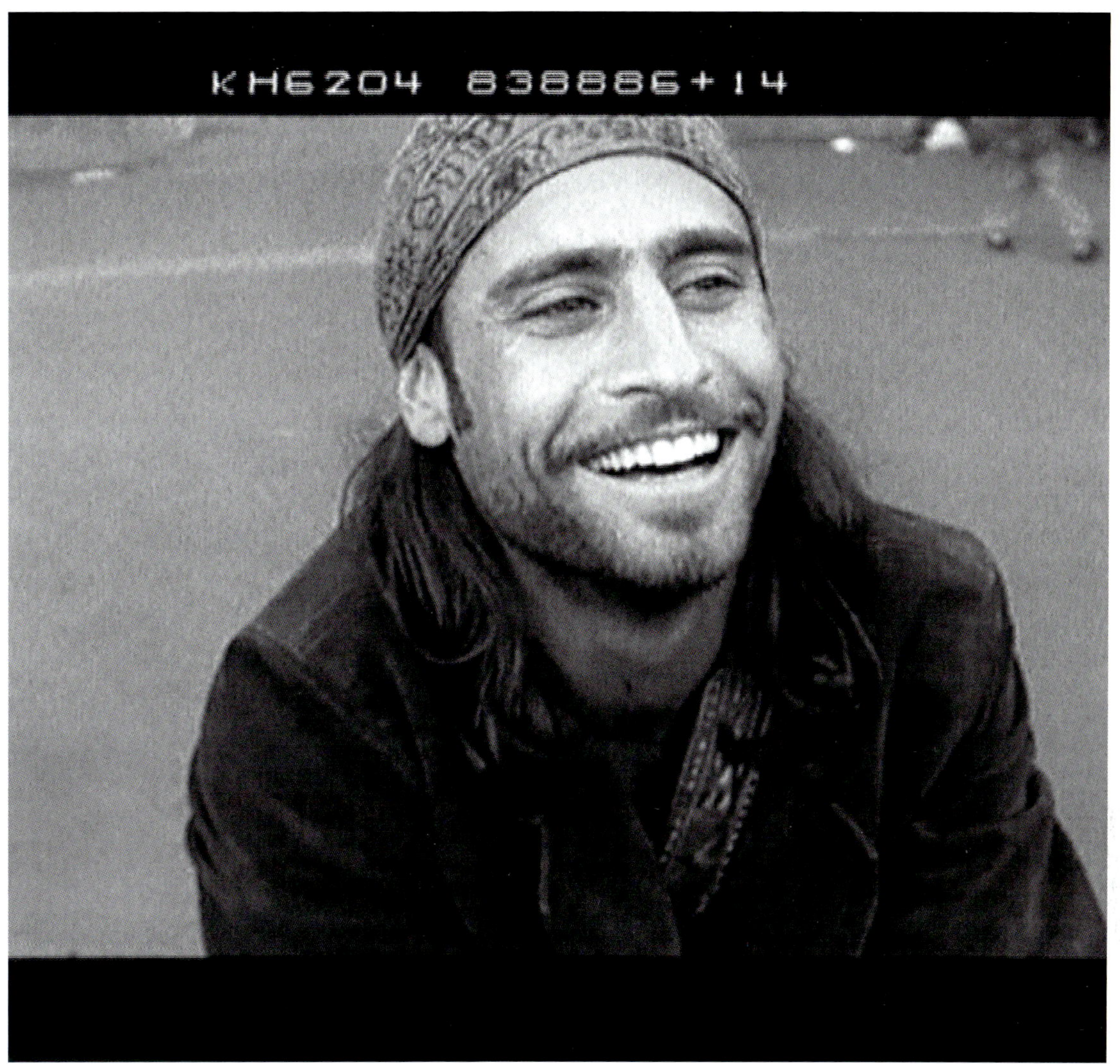

'I had no idea that this commercial would be my trampoline.'

'I remember exactly – it was on 15 November 1995. From that day on, after being on the talk show, people not only knew my face, but also my name.'

was presented was very innovative, experimental, raw and authentic for that time,' says Degan with a smile.

Satisfied with his work, he returned to New York the next day, then travelled onwards to Sydney, where he was living at the time – far away from the hype his herbal-liqueur commercial would soon generate in Italy. For months, Degan had absolutely no idea that it had made him famous. 'When you filmed a piece in one country, it didn't automatically become a part of your life, the way it would now in these times of social media,' the actor explains. So it was difficult to track down the nameless Jägermeister guy with whom Italian audiences had become so infatuated but, after an Italian television crew spent over a year looking for him, Degan finally took a fateful call in New York. The caller told him he was a star in Italy, and had been invited to appear on a talk show in Rome.

It was not until he got there that he saw just how popular he really was. Even the taxi driver at the airport recognized him as the Jägermeister man. 'I remember exactly – it was on 15 November 1995. From that day on, after being on the talk show, people not only knew my face, but also my name,' he says. That was the day his world was turned upside down. 'Neither myself, nor my agent, nor people close to me were prepared for any of it. I didn't speak the language, and I didn't know what people expected from me.' But someone as 'wild and hungry for life' as him was undeterred by this – indeed, it motivated him. 'You never know where life's paths lead you, but that is why it's so exciting,' says Degan today. 'If you stay true to your vision, the path usually leads you to your goal.'

And that also involves getting, recognizing and seizing opportunities. 'I moved into a hotel room in Milan and stayed there for the next three years,' says the actor. He was fêted by fans and media, and soon became one of the greats of Italian show business, both in front of and behind the lens, with his own television shows and documentary series. He worked with renowned directors, including Italian maestro Ermanno Olmio and cult director Oliver Stone, in whose blockbuster *Alexander* (2004) he appeared alongside Colin Farrell and Angelina Jolie. He was also responsible for the screenplay and direction of the Net-

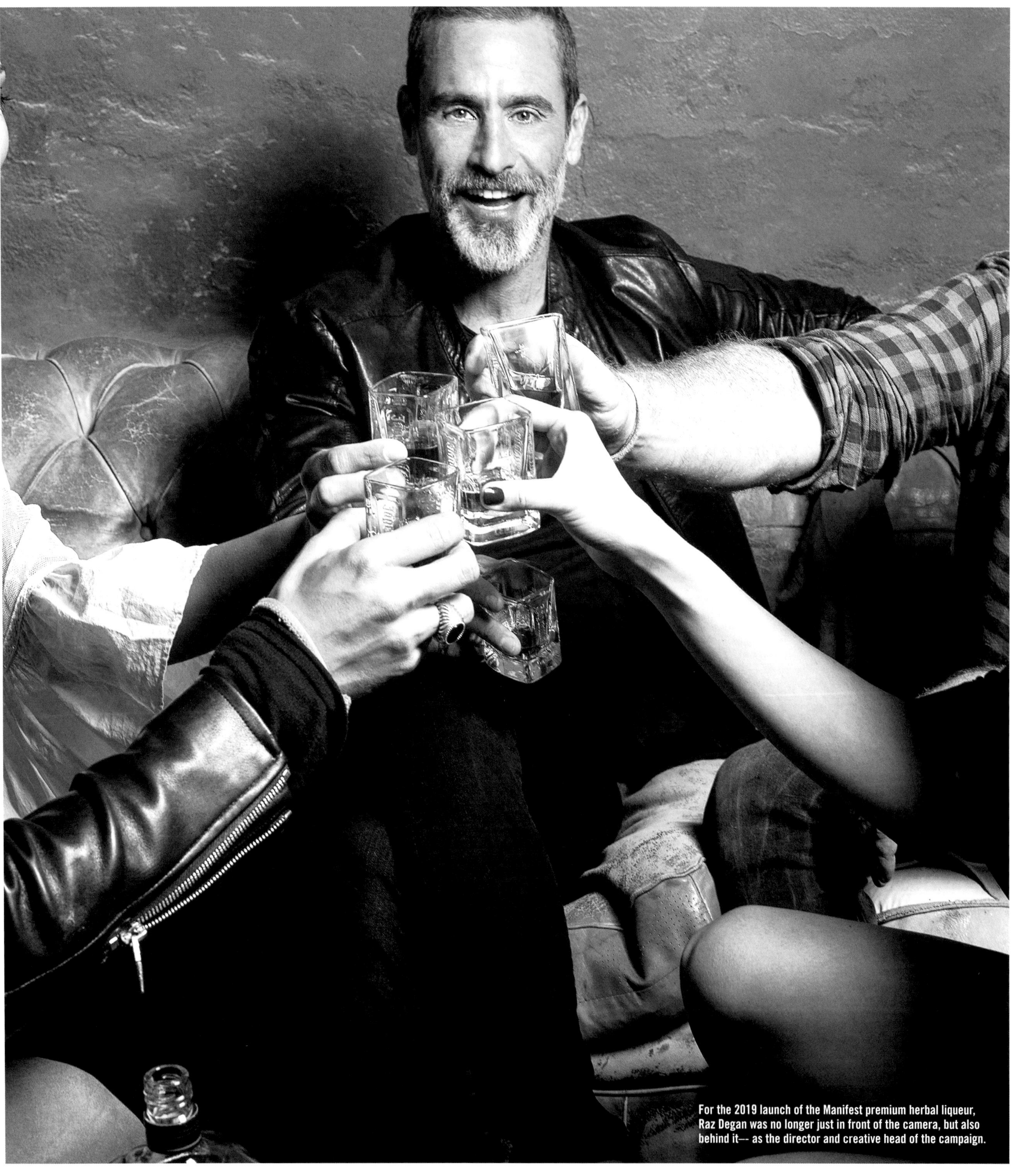

For the 2019 launch of the Manifest premium herbal liqueur, Raz Degan was no longer just in front of the camera, but also behind it— as the director and creative head of the campaign.

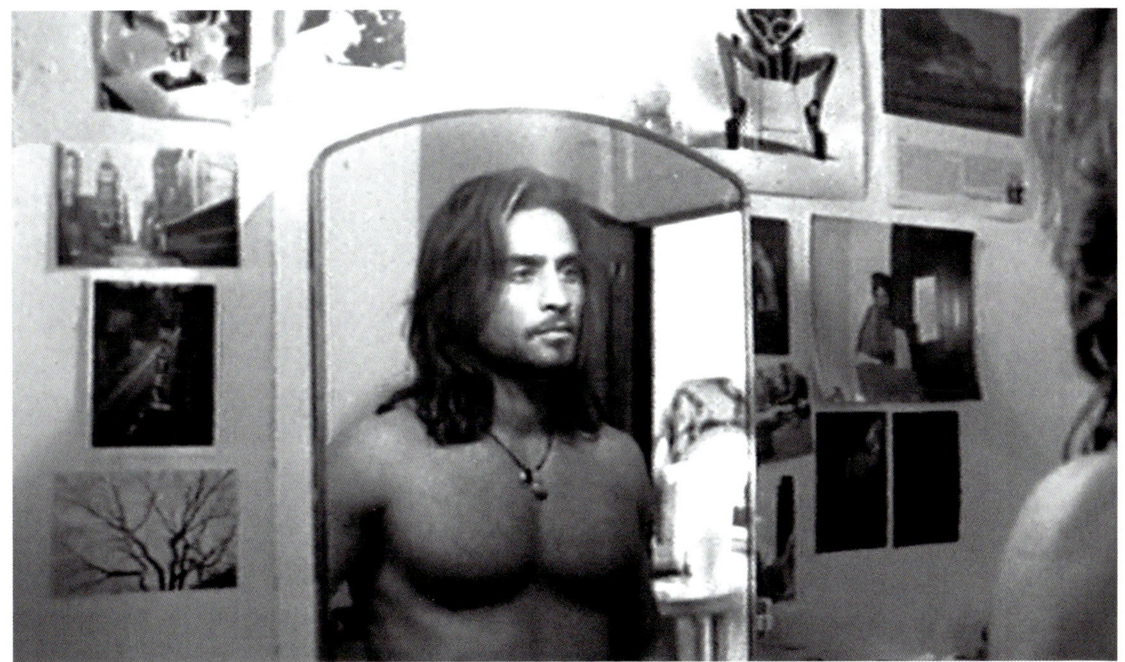

Sono fatti miei, his legendary phrase
from the commercial, has long become
a figure of speech in Italy, and continues
to be associated with him and Jägermeister.

able to tell it himself – not as an unknown face, but as a charismatic brand ambassador, director and creative mind behind the fascinating campaign video that premièred in October 2019.

'You cannot rationally explain such things, you cannot plan them, they just happen – particularly when passion and creativity are the fuel,' says Degan. And once again there is the question of whether it is chance or fate, especially as old archive material that Degan was able to use for the film unexpectedly became available. 'It was fascinating to see how the puzzle pieces all of a sudden fitted perfectly,' he enthuses. 'That is what made it so authentic. The material represented an amazing time in my life, when I lived my dreams. And every time I see the bottle with the stag, I think about that young guy who was wild and hungry for life.'

Exactly 25 years after his first shoot with Jägermeister, the new film has become one of the highlights of Degan's career. 'I am incredibly thankful that our paths crossed back then, and again now. My relationship with Jägermeister is not merely the story of an evening or a night, but of my life.' That is why, for Raz Degan, Jägermeister is far more than just a drink. 'It represents opportunity, joy, thankfulness, willpower, affirmation.' And magic? 'Yes, you have to believe in magic, in something bigger than yourself. Then you might be able to inspire people to manifest their own dreams.' And nobody knows that better than Raz Degan – which is why the fairy tale may continue even further. 'Who knows,' he says laughingly, 'maybe 25 years from now, I will shoot the next Jägermeister film!'

flix documentary *The Last Shaman* (2017), for which Leonardo DiCaprio was executive producer.

'*Sono fatti miei*', his legendary phrase from the commercial, has long become a figure of speech in Italy, and continues to be associated with him and Jägermeister. Degan is now also more than at home in the country that so defined his destiny; he owns a property full of olive trees in Apulia. 'I never dreamt that would happen,' he says today. On the other hand, an adventurer and cosmopolitan such as him will never truly put down roots anywhere. 'I find it hard to sit still,' he says. 'I need to be moving in order to be inspired.' This is the reason he travels the world as an actor, filmmaker, producer or photographer. And he is accompanied by the brand that sparked his career. 'No matter where I go or what I do, the commercial is always present, has become almost iconic.' Laughing, he adds, 'Even today when I walk into a bar, people offer me a Jägermeister.'

But that isn't the end of the fairy tale involving the young man and the stag. Chance (or fate?) was once again at play when representatives of Jägermeister's partner, Campari, bumped into Degan at the 2018 Biennale in Venice. A conversation about old times and new projects laid the foundations for a wonderful idea that would potentially reunite two parties that had long belonged together. And, once again, the timing was just right: Jägermeister was in the process of planning the international launch of its super-premium herbal liqueur, Manifest, and Degan was the perfect embodiment of the new product. His story was the Jägermeister Manifest story, and, this time, he was

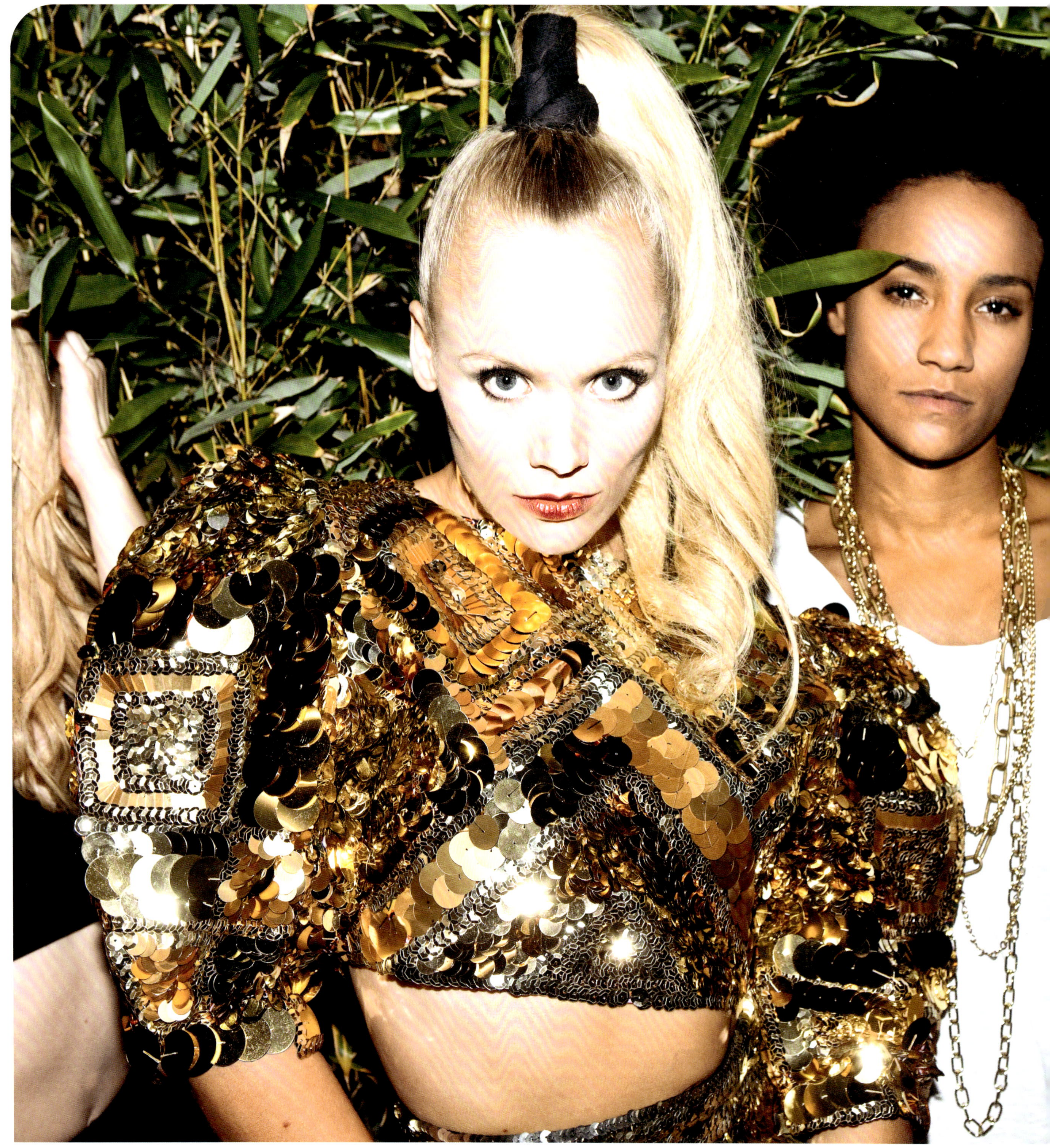

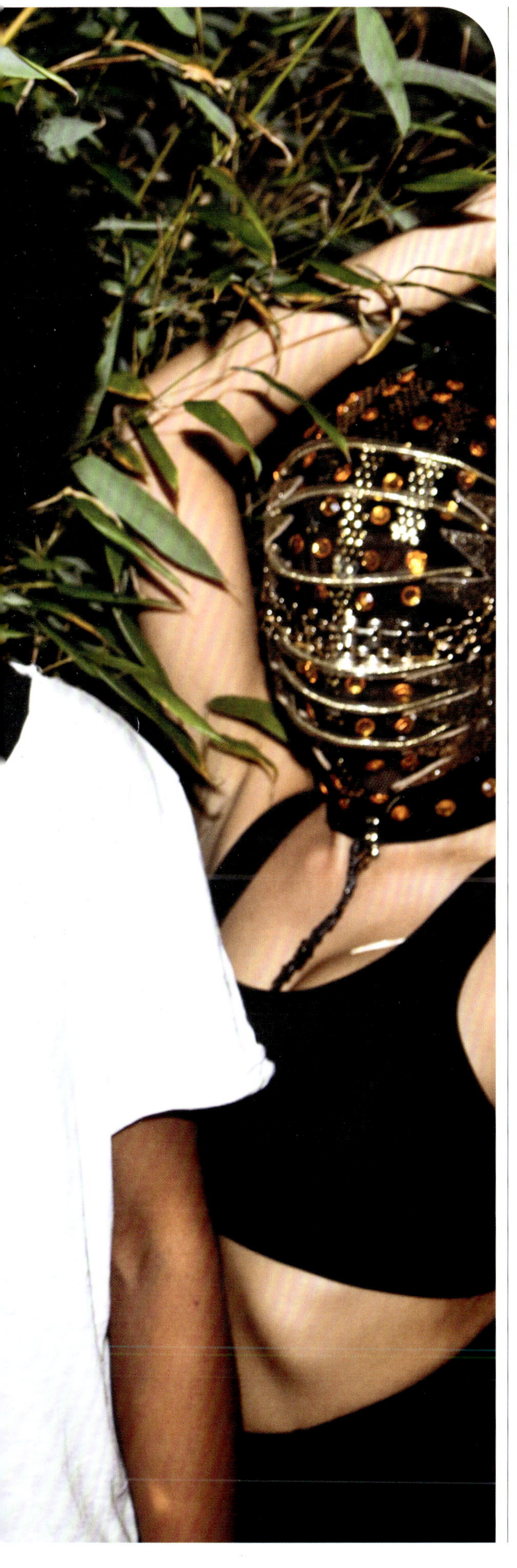

NO THROWING IN THE TOWEL WITH CHEFBOSS

When singer Alice Martin says 'Jägi' instead of Jägermeister, one thing immediately becomes clear: it's not just about business, but rather also about friendship, close bonds and affection. Dancer and choreographer Maike Mohr confirms this. 'This co-operation is different, more intense.' With their mix of dancehall and rap, energy-charged ticketed shows and exceptional flair for producing hits, Alice and Maike, alias Chefboss, have been causing a stir since 2014. And 'Jägi' is always on hand.

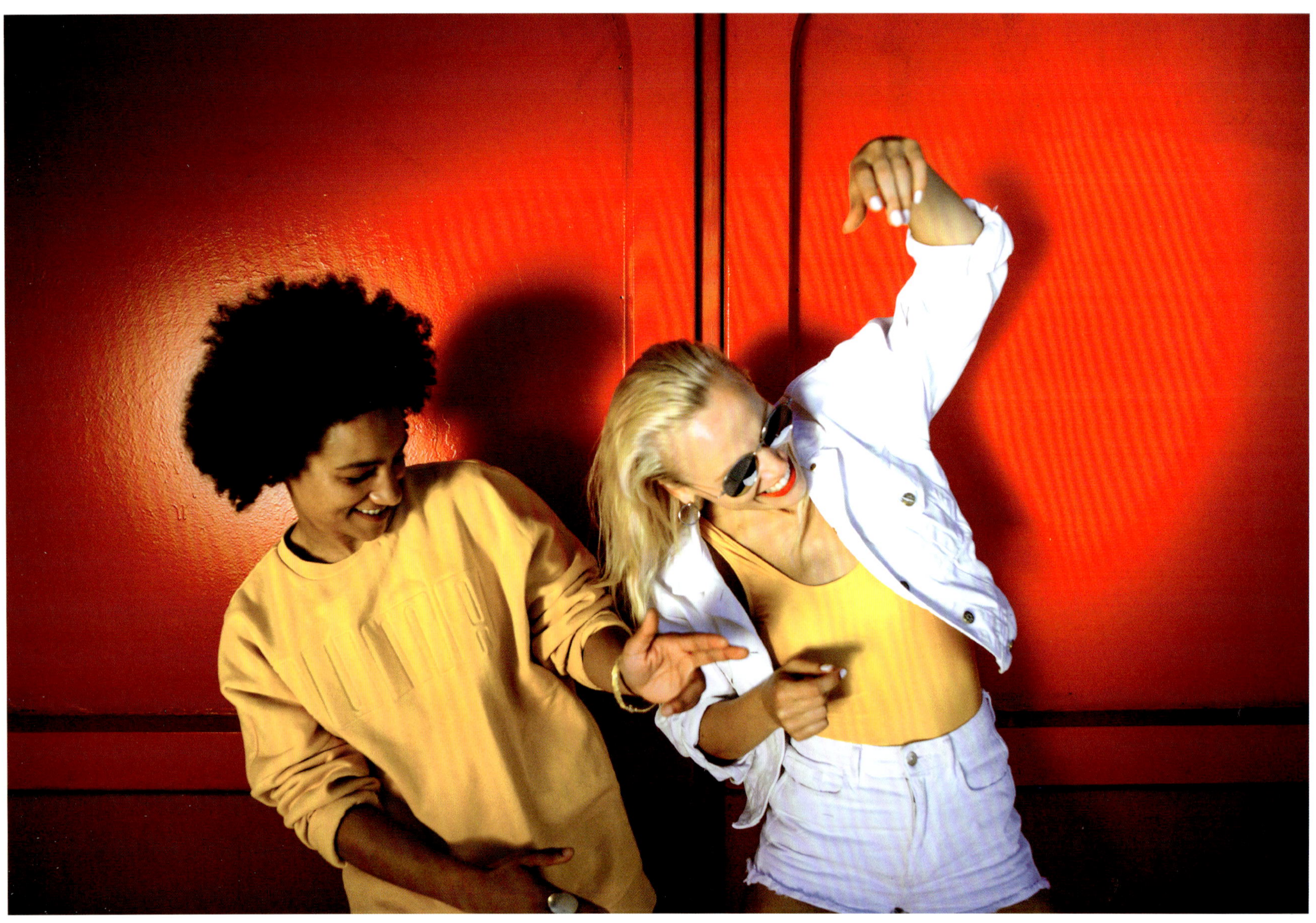

Maike still vividly remembers that first meeting. 'Before we brought out our very first song, we played a few secret gigs. One of them was at a Jägermeister event in Berlin,' says the artist. 'We got talking and quickly realized we shared similar visions – and that's how our friendship began.' One of the

visions, barely more than a dream for a newbie band, was to go on tour in a sleeper bus at some point. When Alice and Maike told the Jägermeister team about it, they never imagined their great dream would soon become a reality. 'Oh, it was insane,' says frontwoman Alice, who still lights up at the thought of that special moment. 'It was just before the start of our first tour, and we were sitting outside on the terrace giving an interview when this monster shadow was suddenly cast over us. A giant sleeper bus bearing the word "Chefboss" was coming around the corner. We completely freaked out.'

That moment also brought fond memories for Maike. 'We had no idea Jägermeister was going to surprise us with that. All our friends, the crew, our dancers were inside the bus – just awesome.' And, of course, the totally unexpected start of the tour on the sleeper bus called for a few shots. 'Before that tour, I was a regular Jägermeister drinker,' grins Alice, 'but afterwards, I couldn't handle it for a while.' These days, however, she once again enjoys toasting her loyal fans with a Jägermeister. 'You can honestly say this drink transcends genres and is rarely snubbed by music fans,' she adds with a happy smile.

But that first tour was just the kick-off for many unforgettable 'Jägi moments' the Hamburg duo have experienced throughout their career to date. Shortly before Chefboss released their debut album *Blitze aus Gold* (Flashes of Gold) in 2016, they were already playing gigs at the legendary Jägermeister Platzhirsch stage, which was making its festival première at the time. 'It was massively fun,' recalls Alice. 'Something happens to you as soon as you get in there. It's so intimate, you're closer to the audience.' She laughs and adds, 'We definitely got the place rocking. The people were amazing, waving their towels.'

Maximum physical exertion is the name of the game for Chefboss, and waving towels (an action known as the 'helicopter') is a permanent fixture at all their concerts – at least by the time the song *Rauer Wind* (Rough Wind) comes on. This was the case three years later when the band opened the Deichbrand festival on the giant Water Stage. But for this particular gig, Jägermeister and Chefboss had thought up an especially unusual promotional campaign. 'Our vision was to take everything to the extreme and set some kind of world record for towel-waving,' says Maike. Before the show began, 10,000 exclusively branded towels bearing the words Chefboss and Jägermeister were handed out among the audience – and, on cue with the song *Rauer Wind*, the biggest towel helicopter the world had ever seen spun into action in front of the stage. It was a logistical masterstroke and one of the band's absolute career highlights. 'Though we knew even before the show that everything had gone smoothly, that the pallets containing the towels had arrived at the grounds and everything had been distributed,' says Maike, 'when I think back to the moment we went on stage and saw that crowd of people waving our towels . . . I still get goosebumps.'

The towels are now cult objects and highly coveted collectors' items, regularly appearing at the band's concerts. 'It's a ritual for us to hang out and chat with people for a bit after the shows,' says Maike. 'And they often come with towels and a bottle of Jägermeister,' adds Alice. 'They want to take photos or drink a shot with us.' And not just at their German gigs. 'We once played in Namibia, where the people are crazy about Jägermeister,' recounts Alice, laughing. The Namibia trip had been organised by the Goethe Institute, but of course Jägermeister was also indirectly part of it, accompanying Chefboss on their African adventure. 'When we arrived, we were met by the local Jägermeister team at the airport,' recalls Maike. 'They had signs saying "Welcome, Chefboss!" – it was such a lovely moment.'

The gratitude Chefboss have for their partnership with 'Jägi' is unaffected, unmistakable and unashamed. For their five-year band anniversary, they released the *Kein Geld der Welt* (No Money in the World) EP together with a book of the same name, in which they tell their story. 'Without Jägermeister's support, we would never have been able to get to enjoy things like the tour bus, the club tour and the extatic concerts at the Platzhirsch,' Alice writes in the book. 'Seriously, it doesn't feel like a business co-operation; it's a friendship-based collaboration.' The pair continue to be visibly mind-blown when they think back to that intense period. 'Many highlights of our career to date were only possible because of Jägermeister – and they really were great moments,' says the singer. 'You have to stop and process it all. It's so awesome that it all actually happened and went off successfully.' Can it be topped? 'It would be hard to get any better, wouldn't it?!' Not necessarily. Because one thing is as certain as the clinking of glasses in the tour bus: they have always been able to count on 'Jägi'. ⚹

'Seriously, it doesn't feel like a business co-operation; it's a friendship-based collaboration.' The pair continue to be visibly mind-blown when they think back to that intense period. 'Many highlights of our career to date were only possible because of Jägermeister — and they really were great moments.'

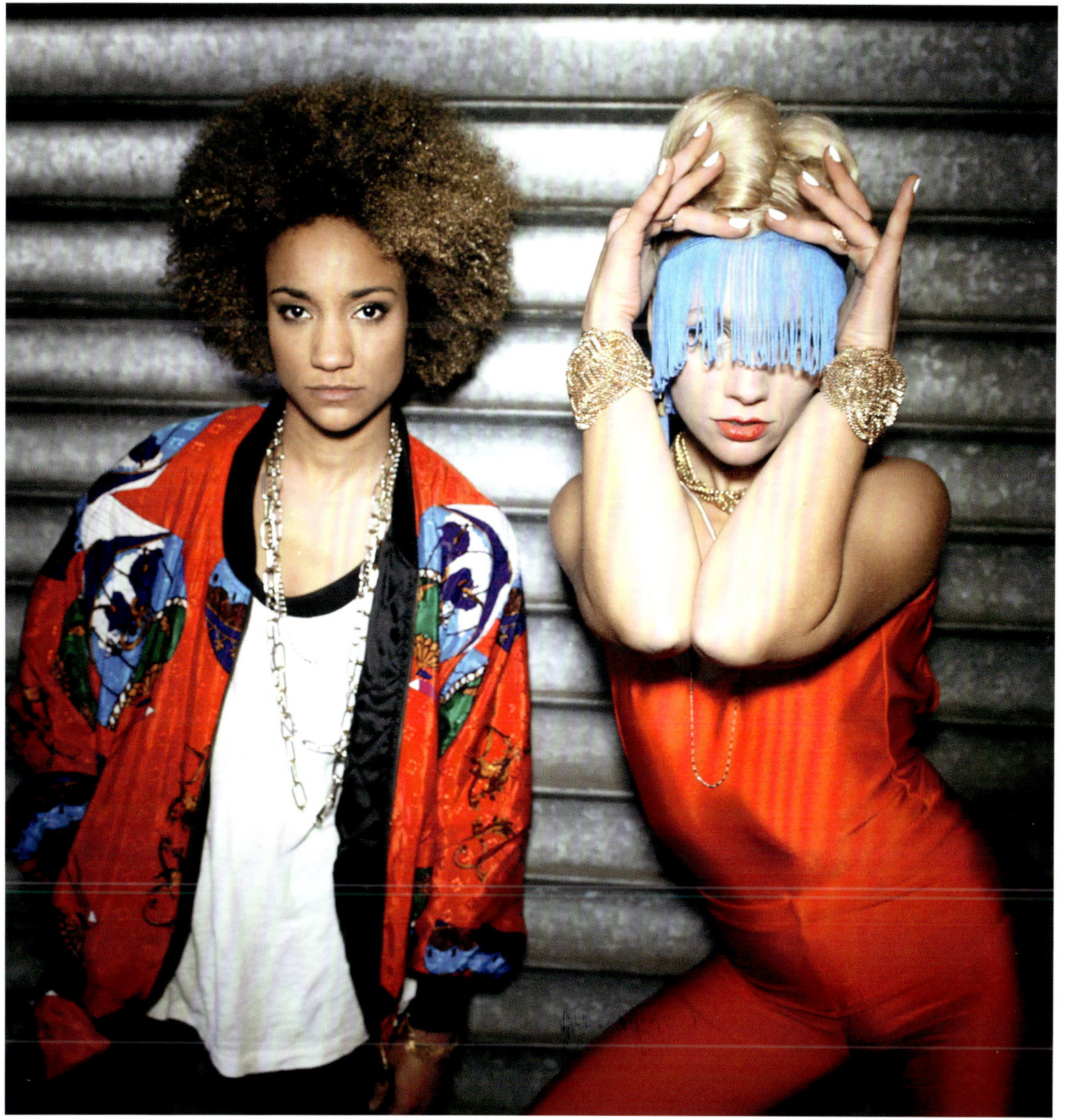

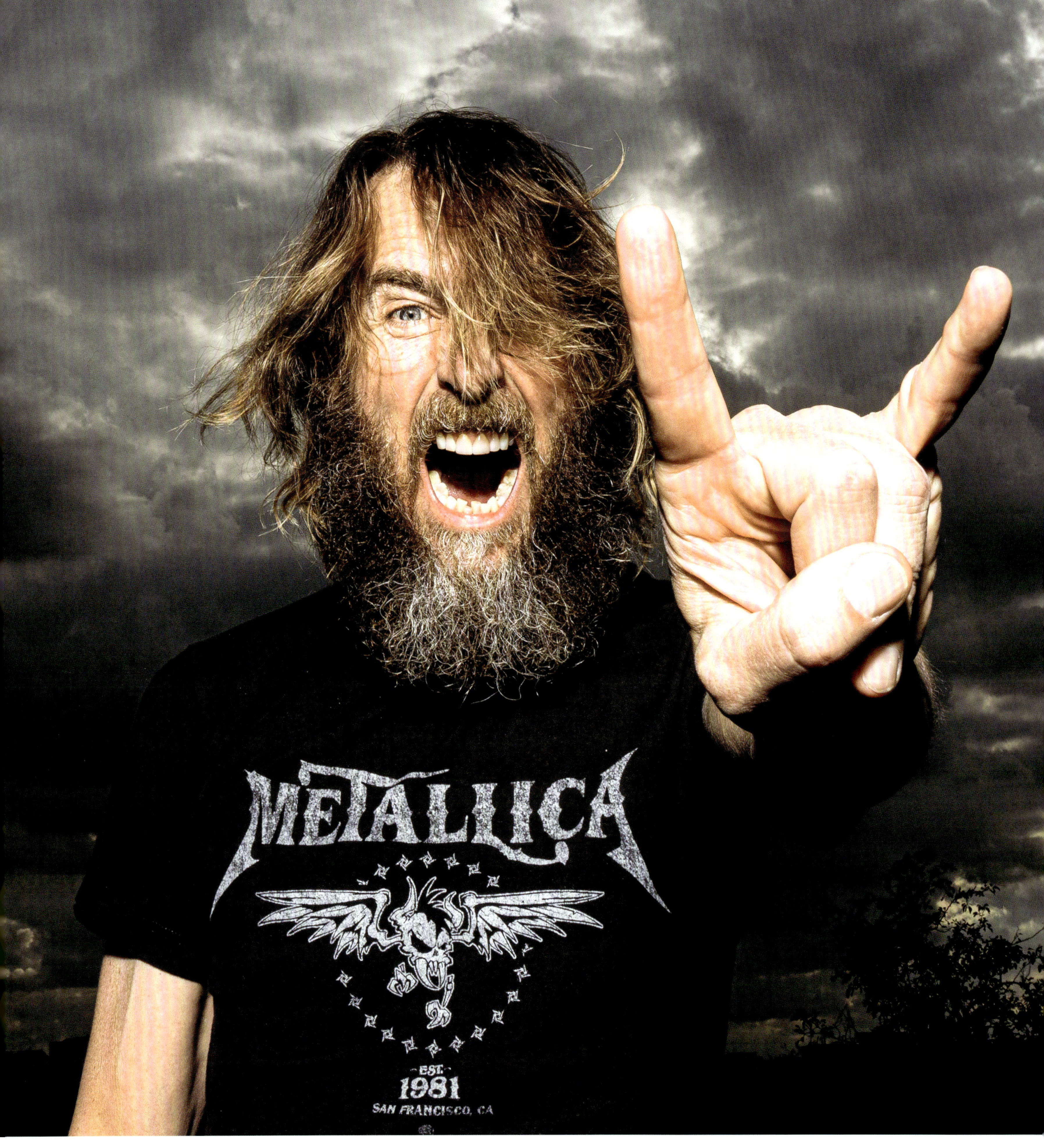METALLICA
EST.
1981
SAN FRANCISCO, CA

GREG KOCH
SOMETHING'S BREWING...

Opposites attract. Sometimes they even complement each other, like beer and a short drink. This is known as a 'boilermaker' in the land of beer-drinkers. 'It's a term I should actually know!' laughs Greg Koch. With his shoulder-length hair, full beard and relaxed manner, he could be an artist, or an environmental activist, or the guitarist of a rock band. He's probably got a bit of all of them in him – and that is probably why 'The Beer Jesus from America', as he is called in the 2019 documentary of the same name, is one of the most successful brewery owners in the United States.

Stone Brewing, one of the country's largest and most reputable craft-beer breweries, was founded in 1996 by Greg Koch and his business partner, Steve Wagner, in San Diego County, California. Online platform BeerAdvocate has already twice named Stone Brewing the world's greatest brewery of all time – which Koch considers a highlight of his career to date. 'The list consists of ratings by ordinary beer-drinkers; people who have bought the beer at a store or drunk

'If someone hands you a glass of Jägermeister, you'll be able to identify it even if blindfolded in a darkened room. All you need is your nose.'

it at a pub or restaurant.' He can be proud of this, because all his beers are designed for refined gourmet tastes. Koch occasionally even invents an entirely new style; his fruity-hoppy Stone IPA, the first iconic West Coast IPA, is Stone Brewing's most frequently sold beer variety and was part of Jägermeister's Deer & Beer campaign.

Deer & Beer first started in the United States in the summer of 2019 as a collaboration between Jägermeister and five independent craft-beer breweries. Because, as Jägermeister proclaims, 'it's not just Jägermeister fans who enjoy being out with friends. The liqueur itself is often found in the company of its best friend, beer.' The idea of the campaign was to select and recommend beer varieties that perfectly complemented the different flavour nuances of the Wolfenbüttel-made herbal liqueur. Conversely, Jägermeister also accentuated certain notes of the respective beers to create a perfectly balanced boilermaker. There is nothing, as even the Stone Brewing founder stresses, that involves watering down the flavour of the shot with beer.

It took much experimentation and sampling to

find the right pairings. 'If you focus on it, you'll discover all these characteristics, all these new alliances and all these flavours,' says Greg Koch, adding, 'I like having a sip; I've always had a weakness for anything with intense flavour.' This is also why he is so fond of the herbal liqueur.' If someone hands you a glass of Jägermeister, you'll be able to identify it even if blindfolded in a darkened room. All you need is your nose.' He loves such distinctiveness. 'Whether it be beer, wine or other spirits, if they're made right, they're like art,' he says, philosophically. 'Art is best appreciated from a clear standpoint. And Jägermeister definitely has a strong, distinct standpoint – just like our beers.'

Greg Koch loves drawing analogies between art, music and beer-brewing. 'Developing a recipe is very similar to writing a song, I'm convinced of it,' he says. Before founding Stone Brewing in the mid-90s, the 'wannabe musician', as he once called himself, ran a rehearsal studio in Los Angeles which would host bands such as Fear Factory, Fishbone and the Melvins. 'The best music, the best art, the best beer – the best of anything is always produced with creativity and feeling, with heart and soul,' he says. 'The worst is music that follows a set formula, like pop or elevator music.' Greg Koch has never really 'got' the mainstream – not as a fan, nor as a creator. 'It's great to produce something that so obviously does not fit mainstream tastes, and to see people still accepting it,' he says. 'Jägermeister is not mainstream. Nor are we.'

And nor is one of Greg Koch's favourite bands, Metallica. Both came 'from outside the mainstream', says Metallica drummer Lars Ulrich. 'But then the mainstream came to us.' So it was no surprise, but a coup nonetheless, when, in spring 2019, Metallica and Arrogant Consortia, a subsidiary of Stone Brewing, launched Enter Night Pilsner – a cold-hopped Pilsner with modern, aggressive riffs, ahem, notes. But the Metallica Pilsner was also part of Jägermeister's Deer & Beer campaign, and was even brewed in Berlin, where Stone Brewing ran a production plant for four years.

In October 2019, Greg Koch finally presented his masterpiece to the world: a beer brewed with Jägermeister ingredients. He spent nearly three whole years working on this very first 'liquid collaboration' with Jägermeister, travelling to Wolfenbüttel and being heavily involved in the creative process. Did Jägermeister disclose its carefully guarded trade secret, the ingredient list, for this very first liquid collaboration?

'I was given a lot of insights,' says Koch, remaining vague in his response. He does not reveal how many or indeed which of the 56 herbs and ingredients he used for the Jägermeister beer. What is for certain, however, is that it wasn't just any old beer that was used as the base for this cult drink! 'You're not worthy' is the slogan that Arrogant Bastard Ale has been using provocatively for over 20 years. It is part of the brewery's standard range, thrills craft-beer fans all over the world and boasts a long-time cult status. 'Combining it with Jägermeister was probably the obvious choice,' says the beer visionary, 'but it was absolutely the best choice for creating something extraordinary, something polarising.' And what does the fusion taste like? 'Impossible to describe,' says

Deer & Beer first started in the United States in the summer of 2019 as a collaboration between Jägermeister and five independent craft-beer breweries.

Koch, adding that the combination of flavours of the Jägermeister Arrogant Bastard Ale is unique. Just like Greg Koch, just like Jägermeister, and just like their shared story. Opposites attract. Sometimes they even complement each other. And sometimes something special even starts brewing. ⚜

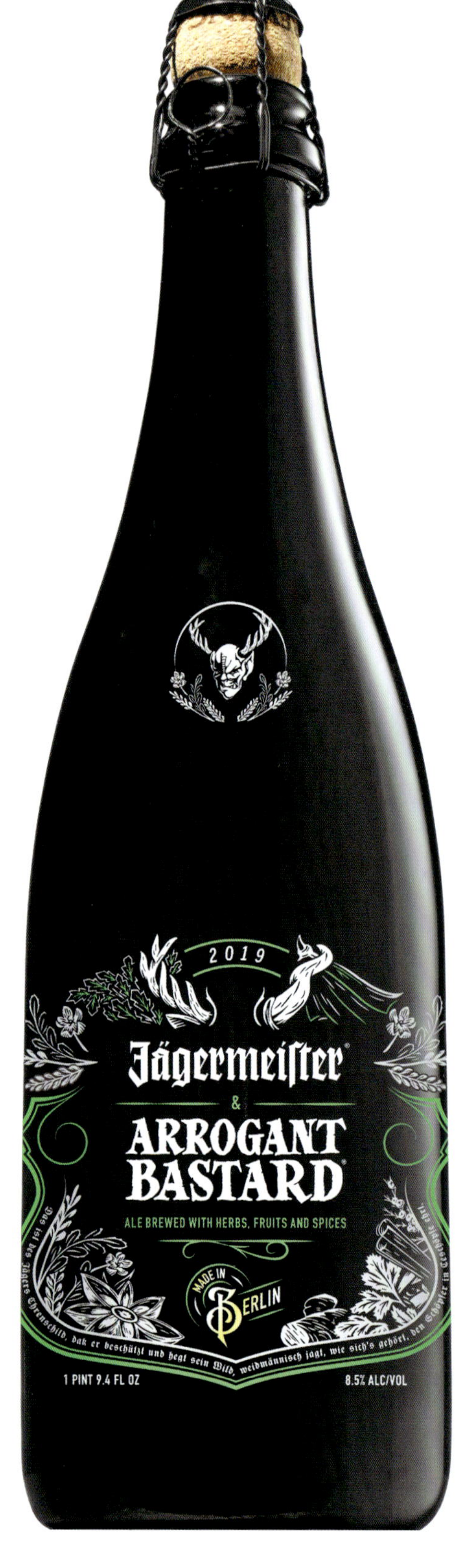

And what does the fusion taste like? 'Impossible to describe,' says Koch, adding that the combination of flavours of the Jägermeister Arrogant Bastard Ale is unique.

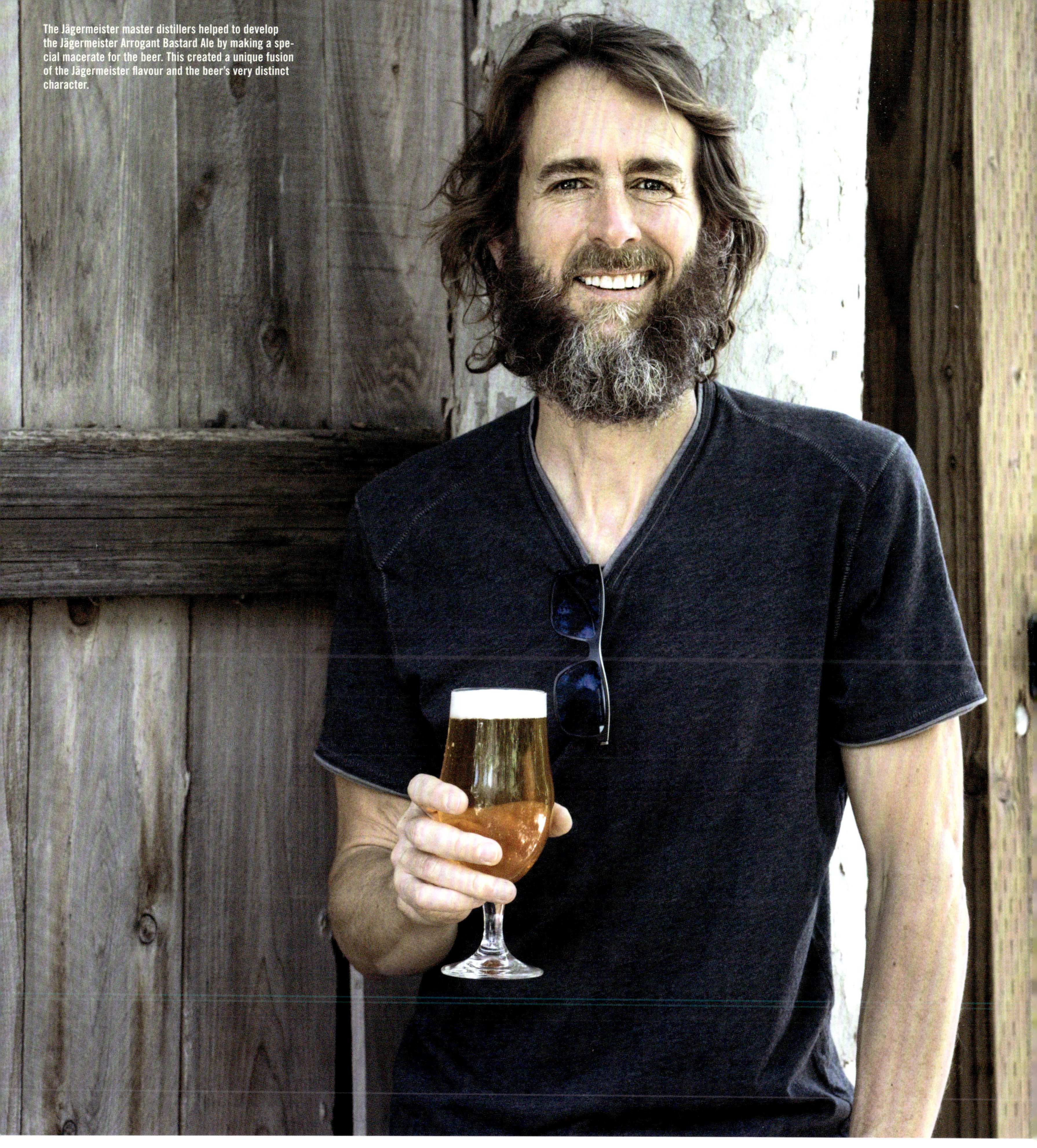
The Jägermeister master distillers helped to develop the Jägermeister Arrogant Bastard Ale by making a special macerate for the beer. This created a unique fusion of the Jägermeister flavour and the beer's very distinct character.

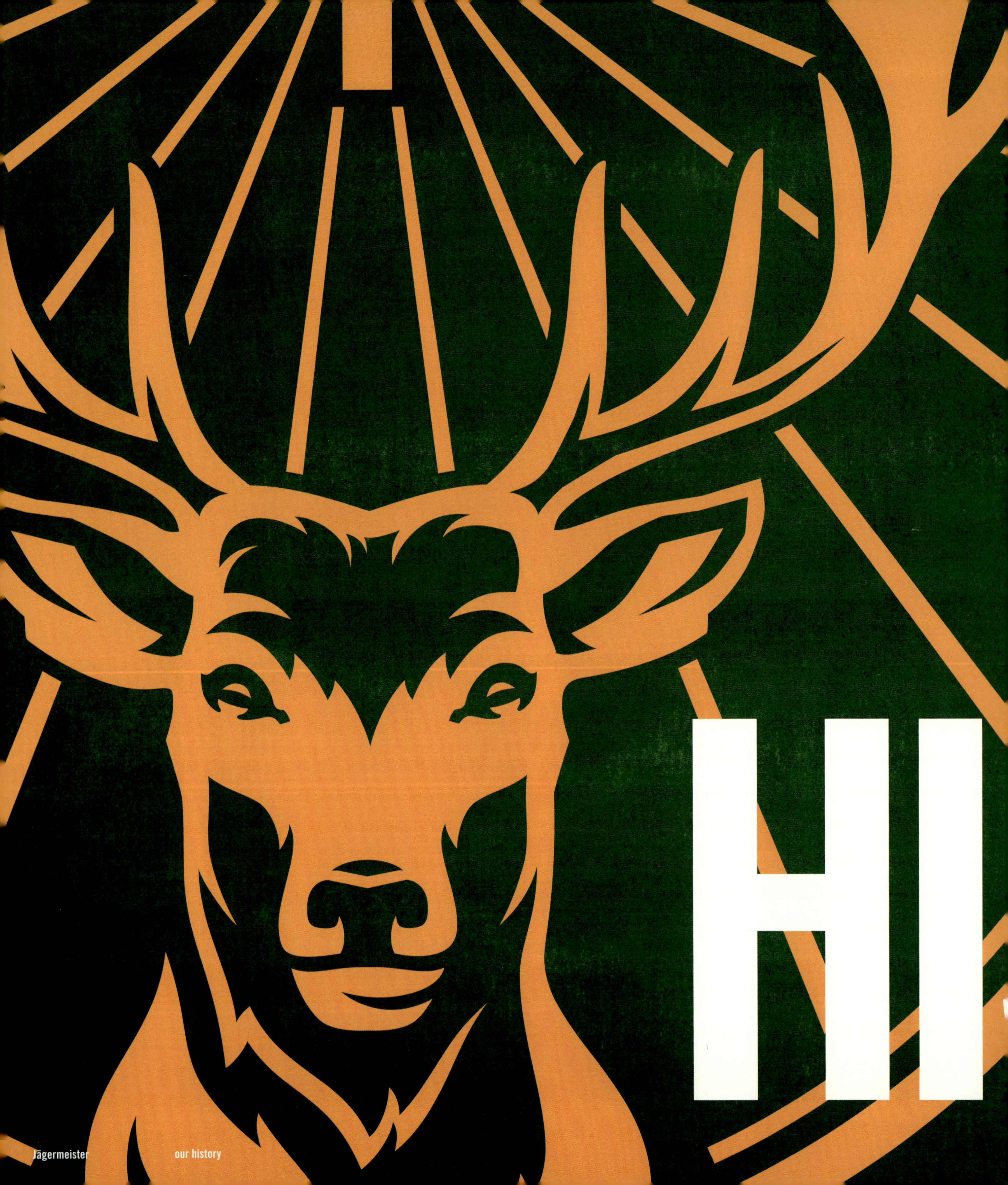
HI.

OUR STORY

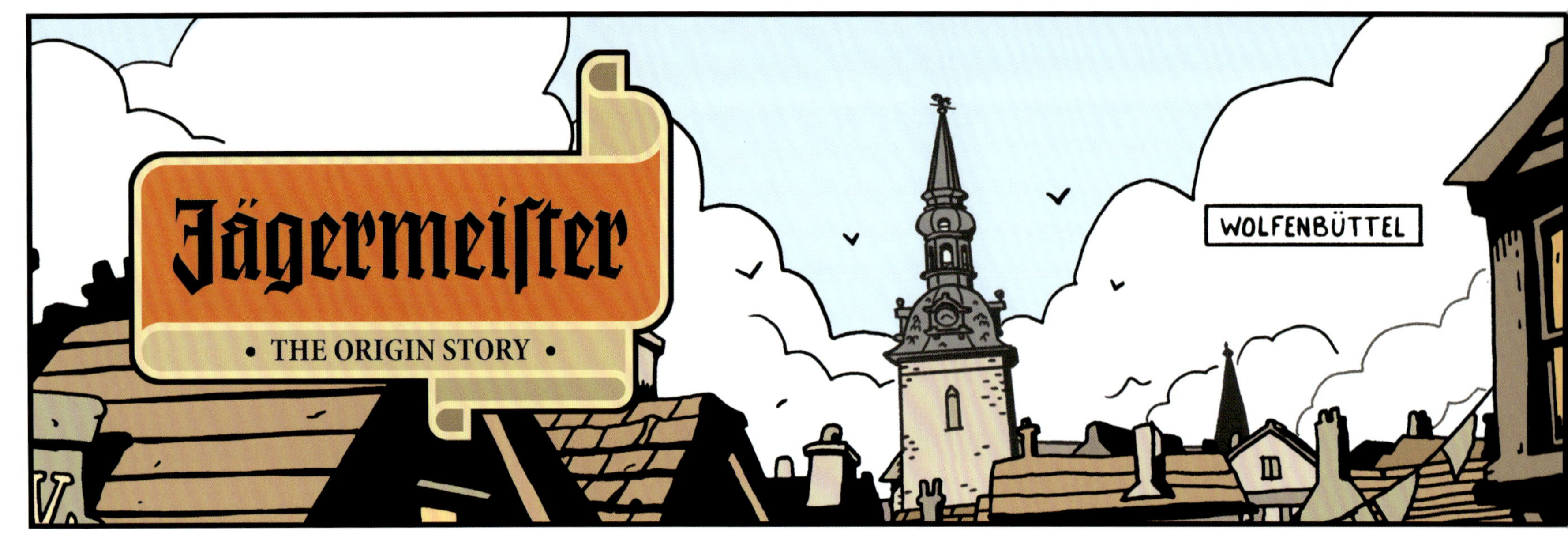

Jägermeister
• THE ORIGIN STORY •
WOLFENBÜTTEL

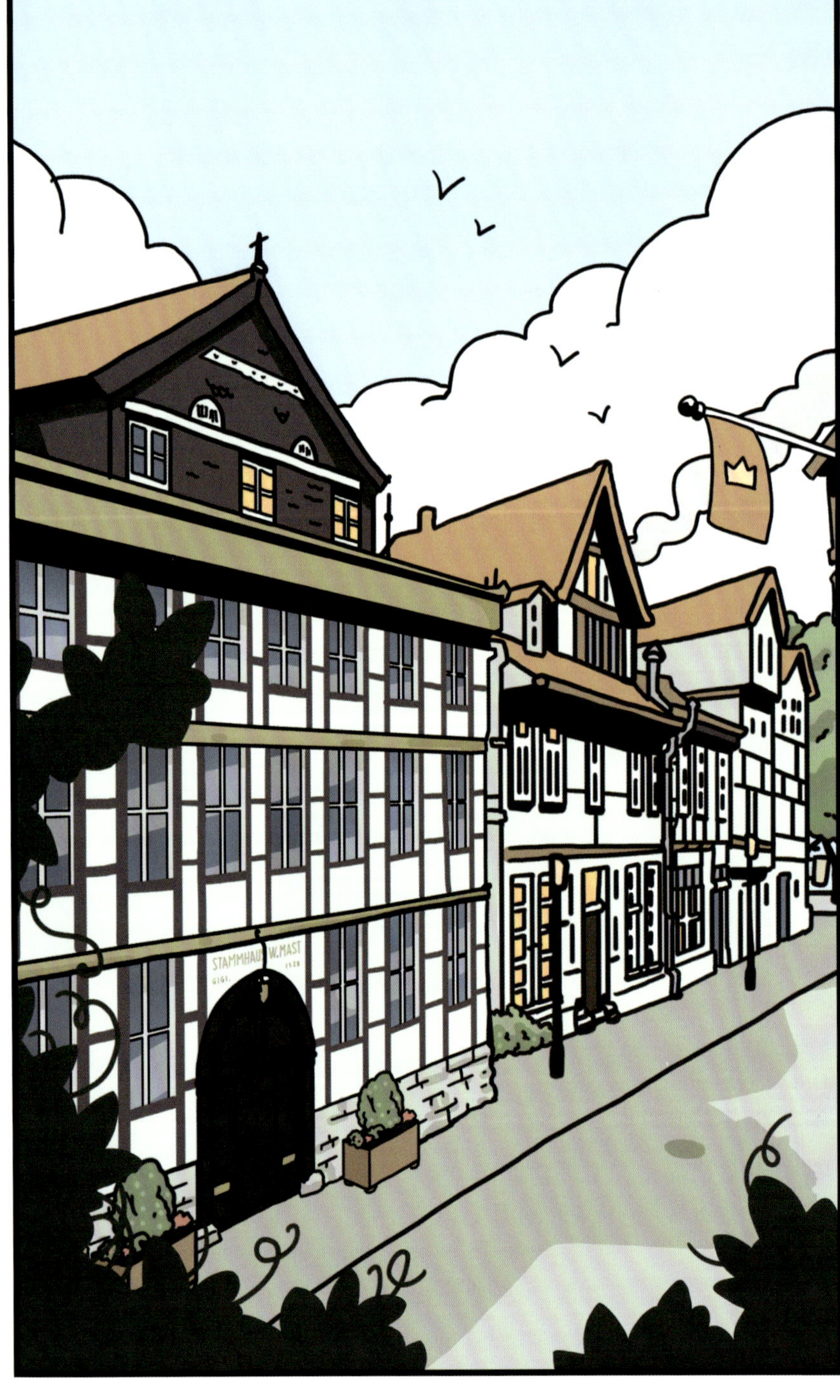

STAMMHAUS W. MAST

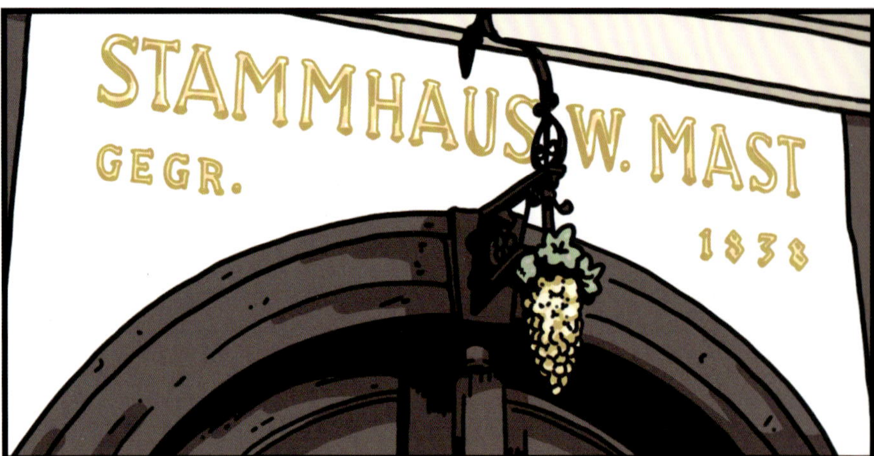

STAMMHAUS W. MAST
GEGR.
1838

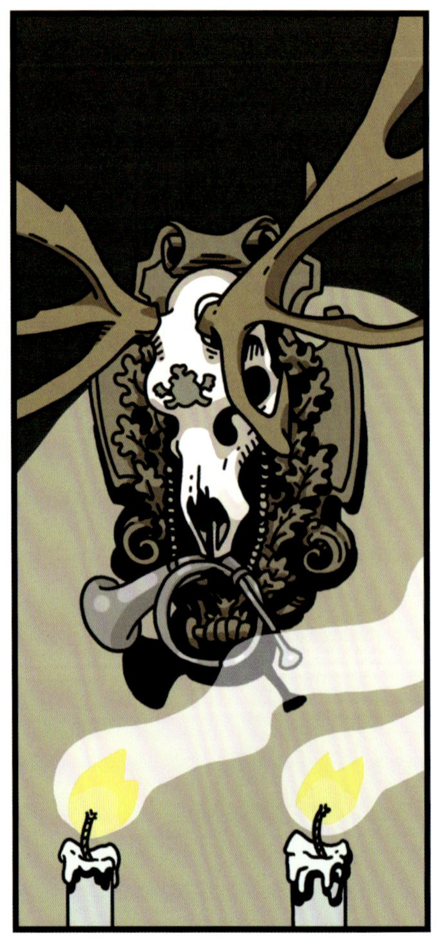

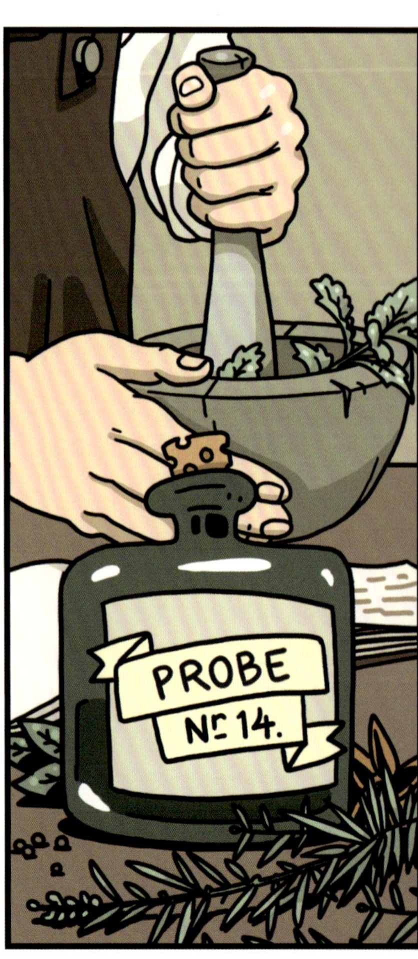

PROBE
Nº 14.

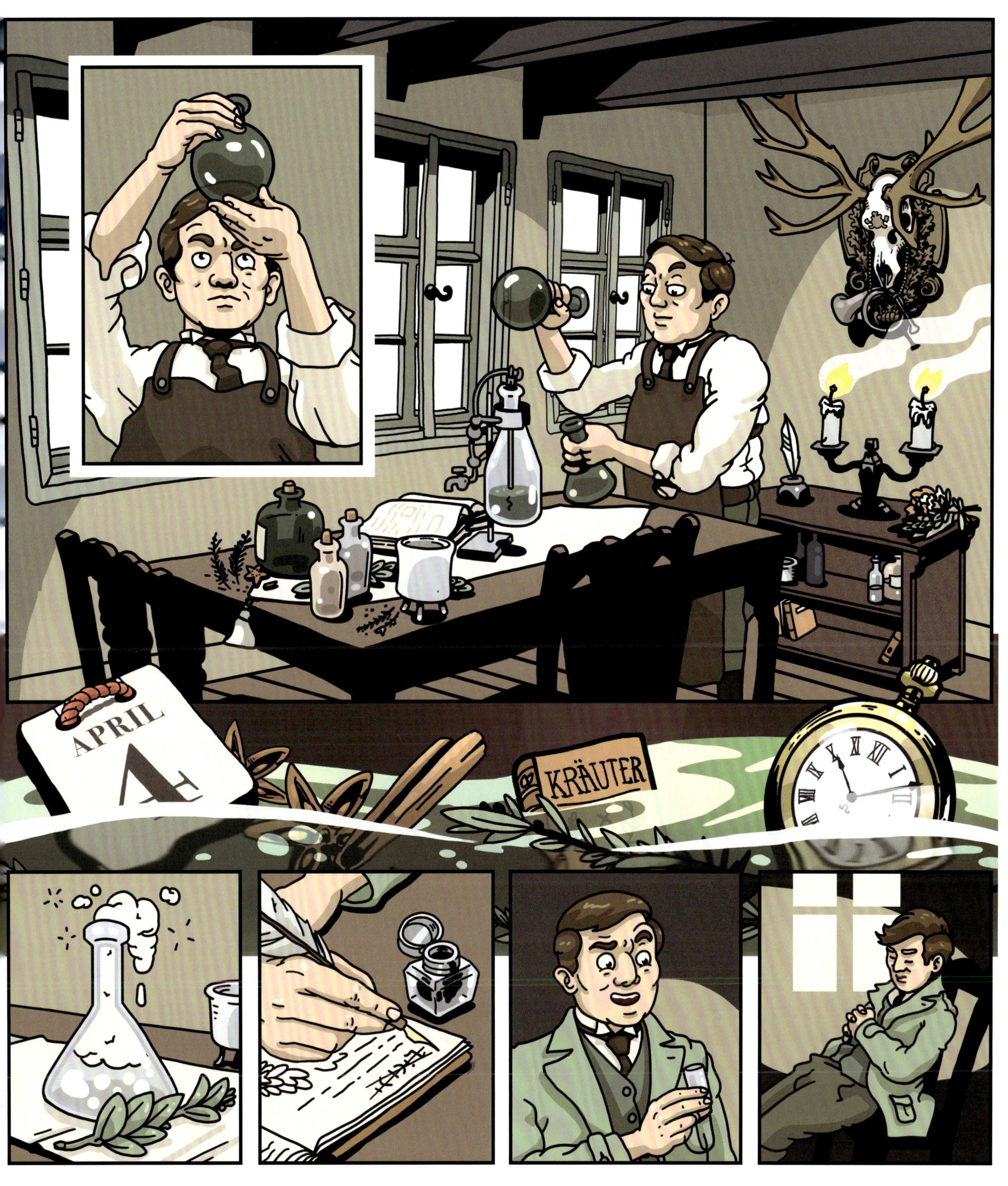

APRIL
4
KRÄUTER

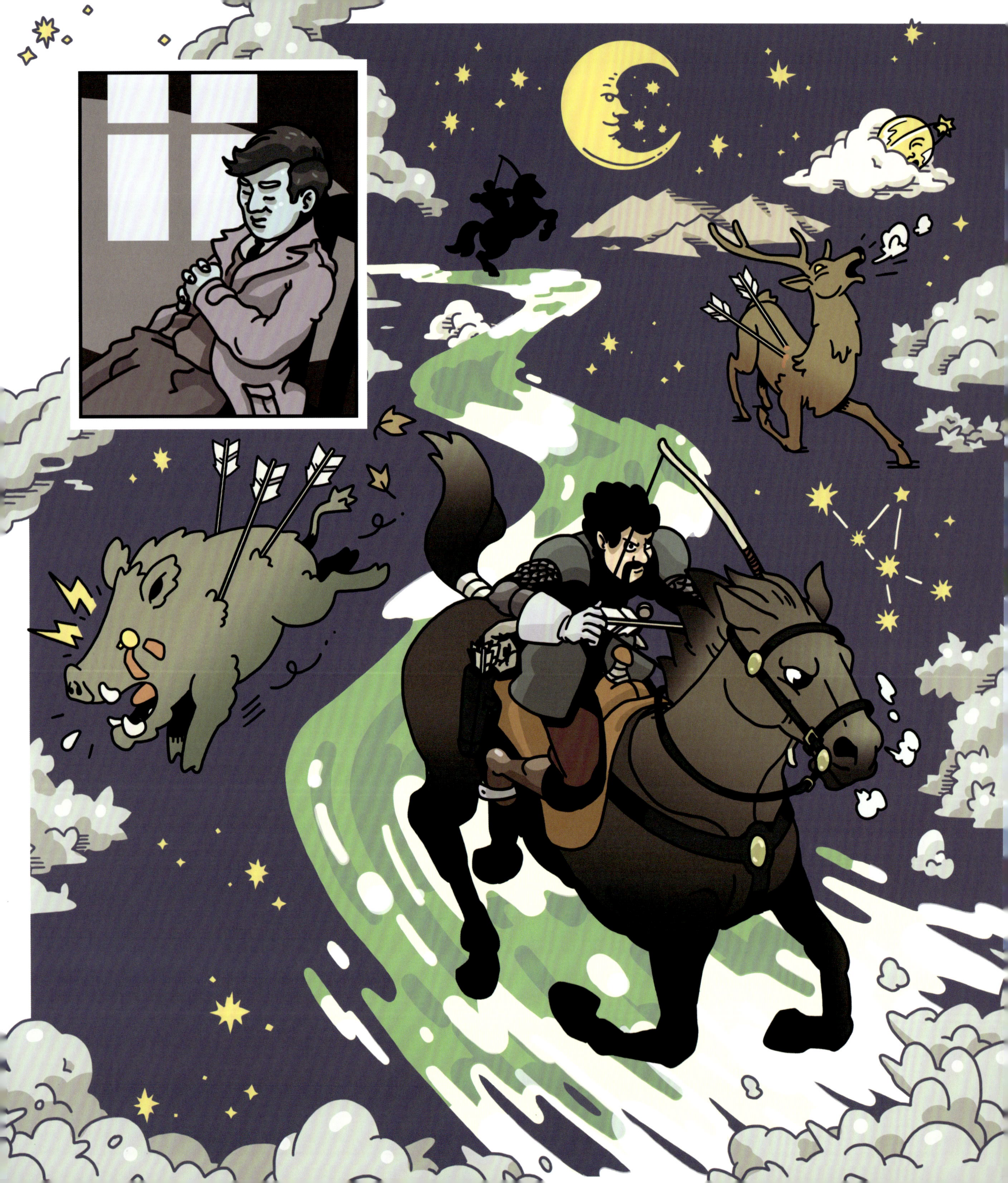

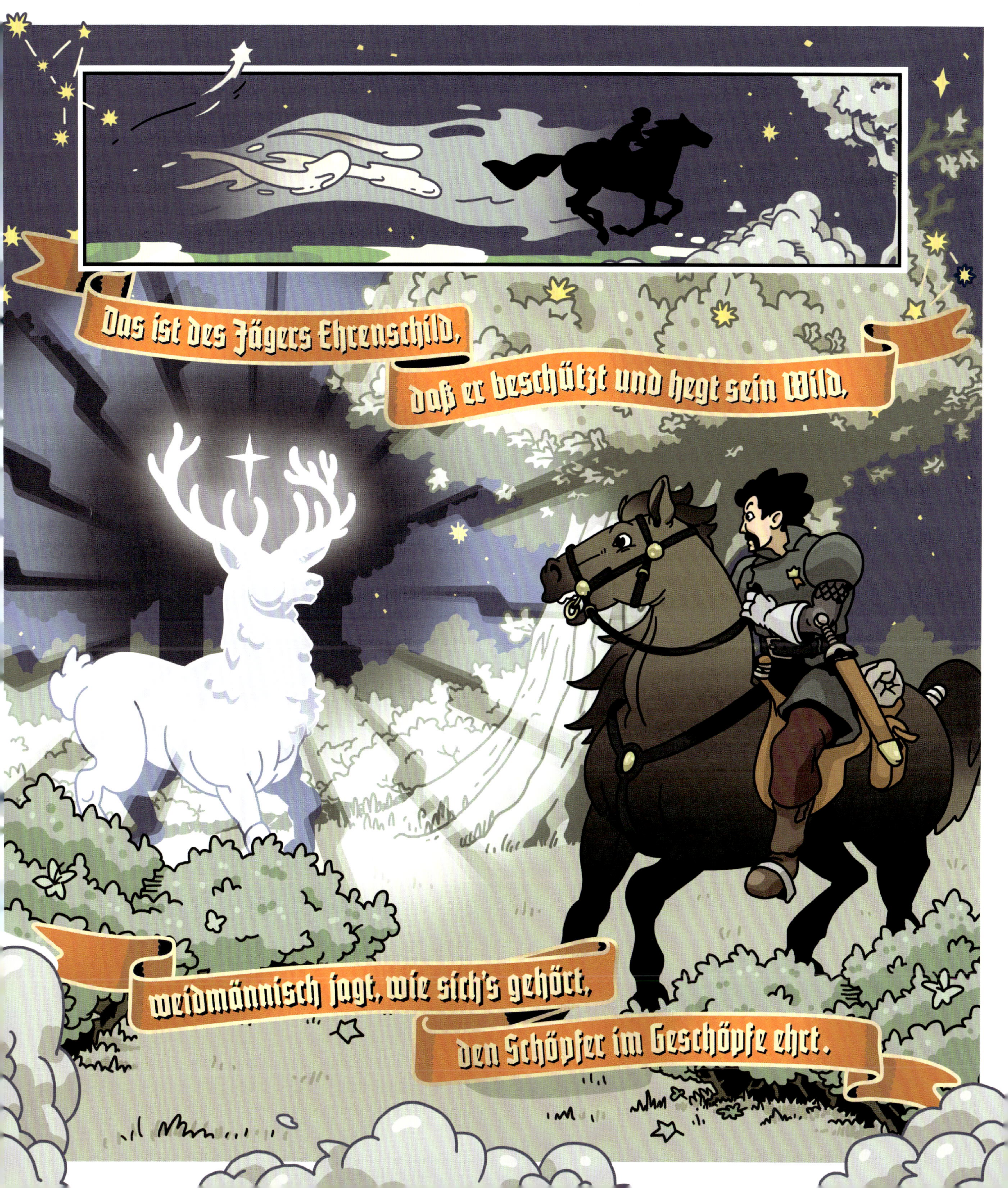

Das ist des Jägers Ehrenschild,
daß er beschützt und hegt sein Wild,
weidmännisch jagt, wie sich's gehört,
den Schöpfer im Geschöpfe ehrt.

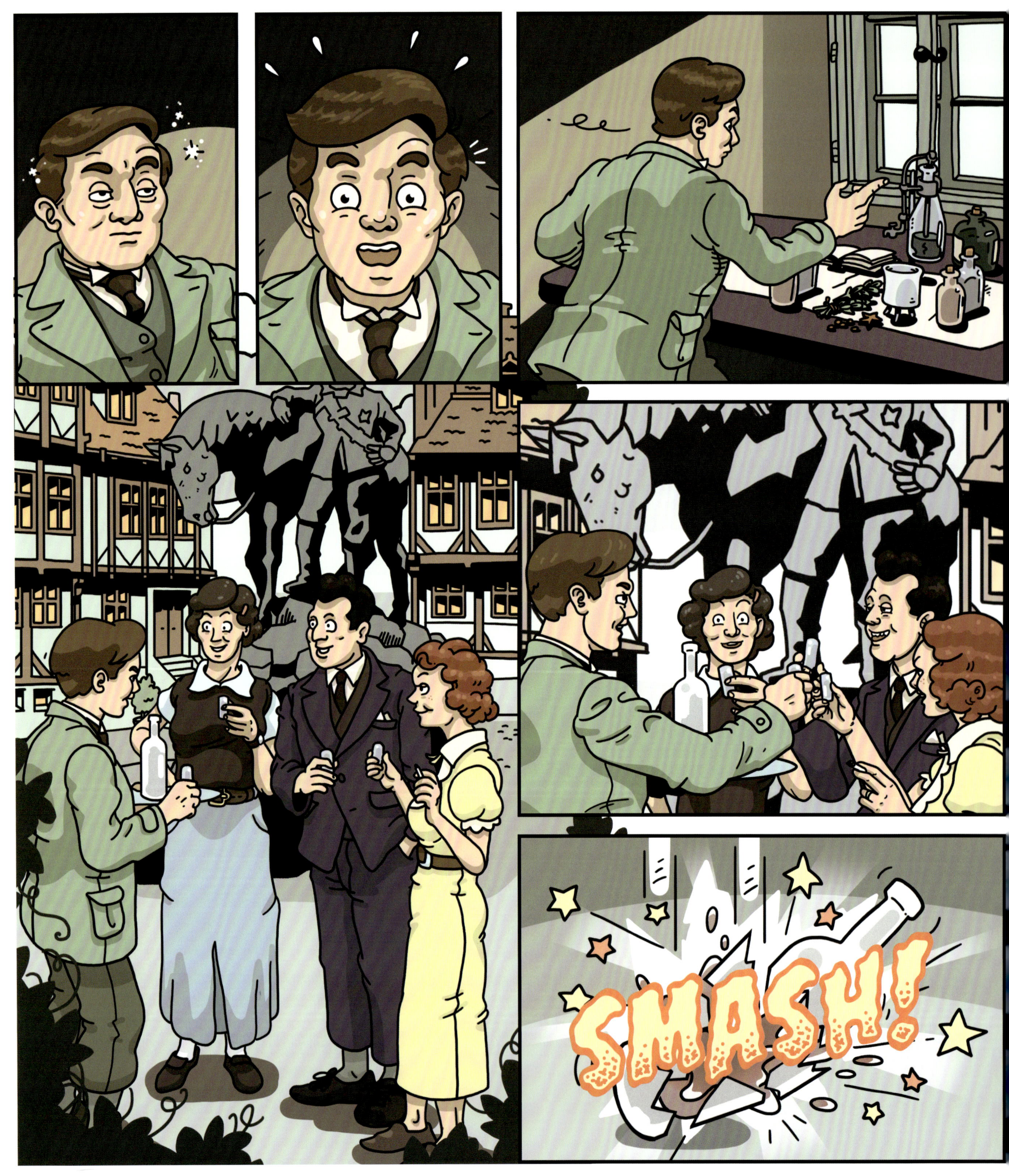
SMASH!

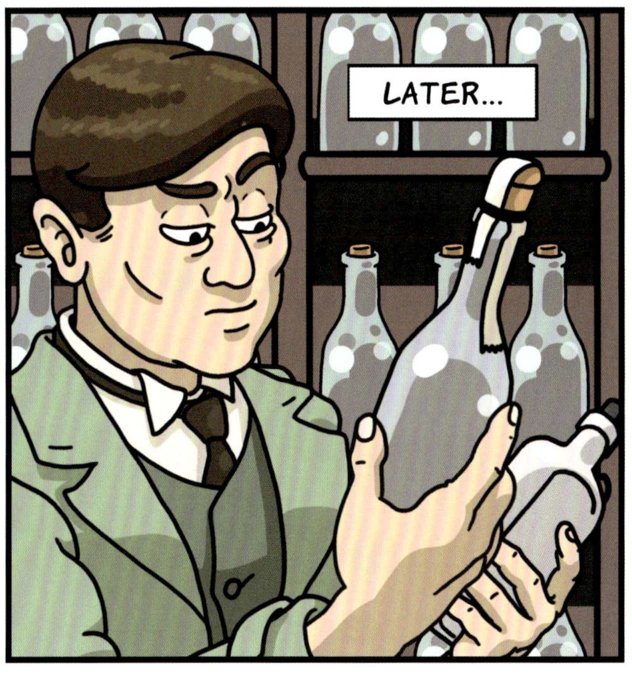

LATER...

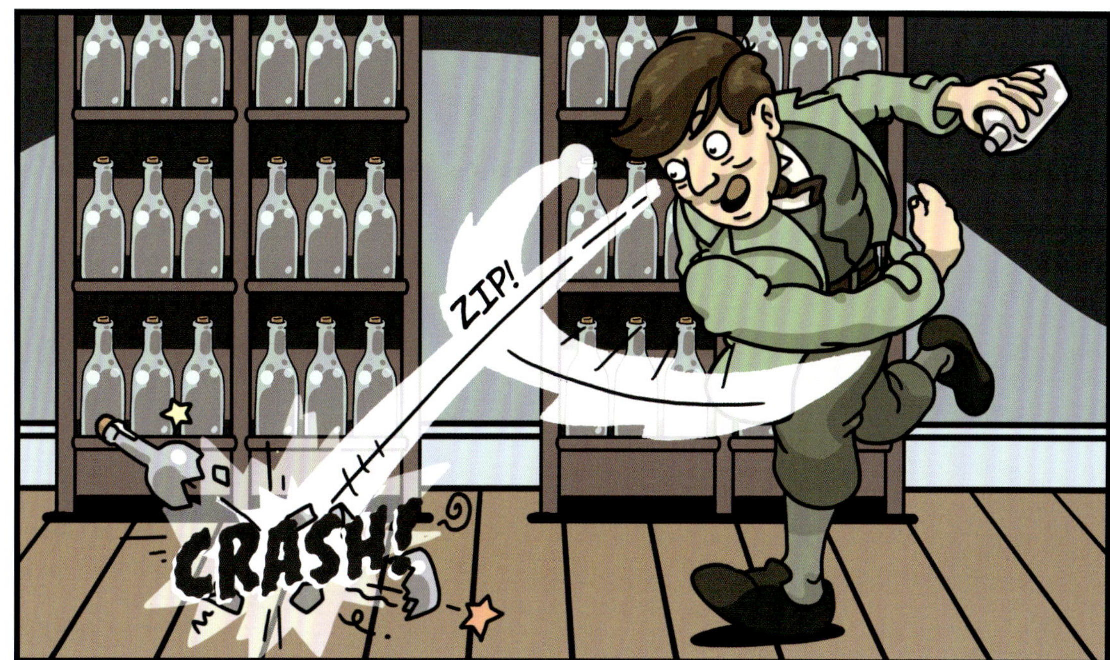

ZIP!
CRASH!

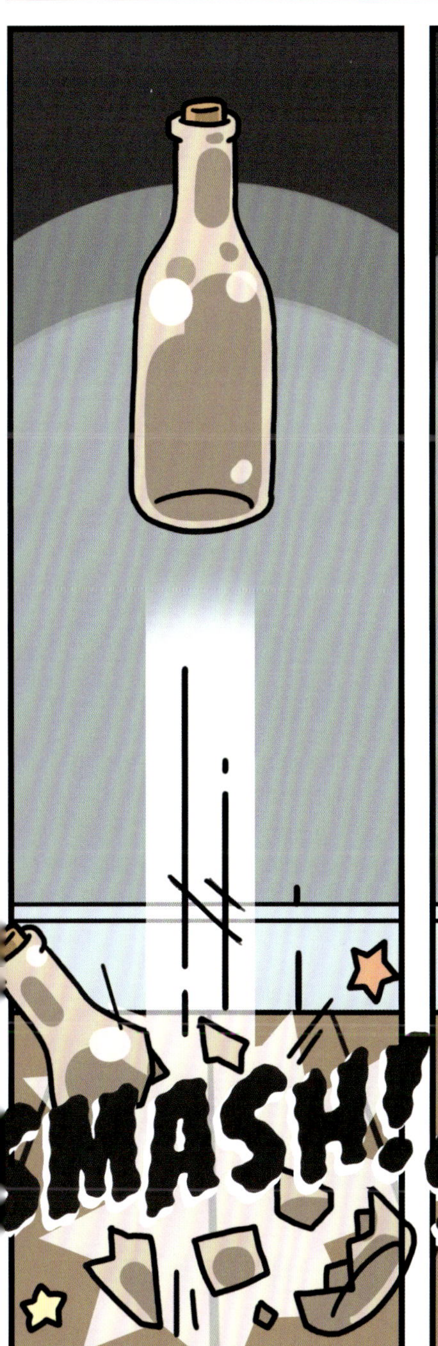

SMASH!

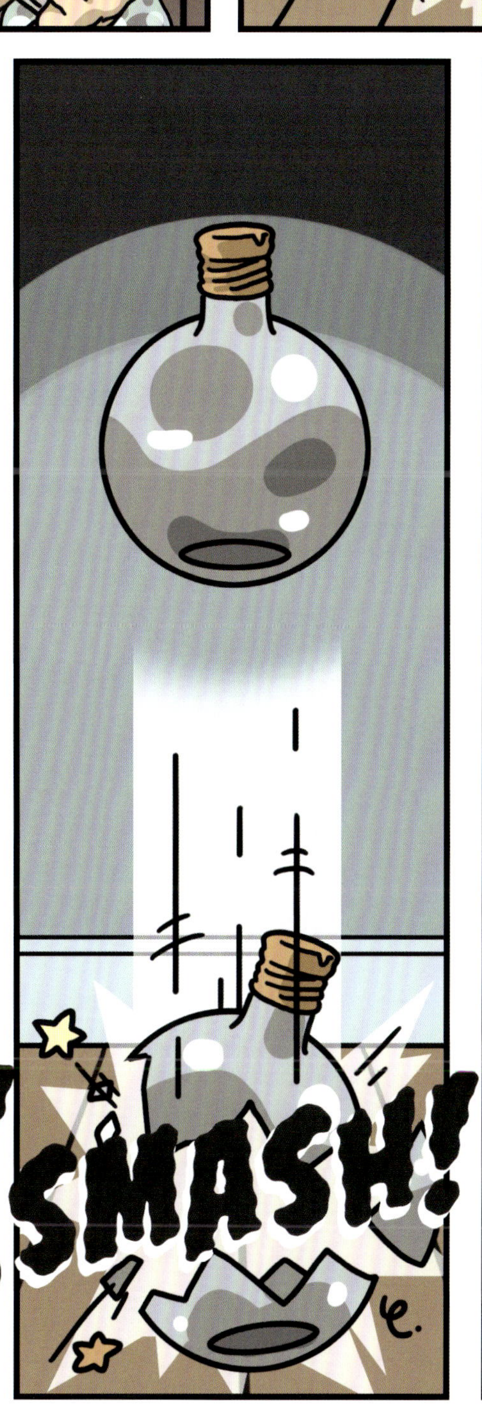

SMASH!

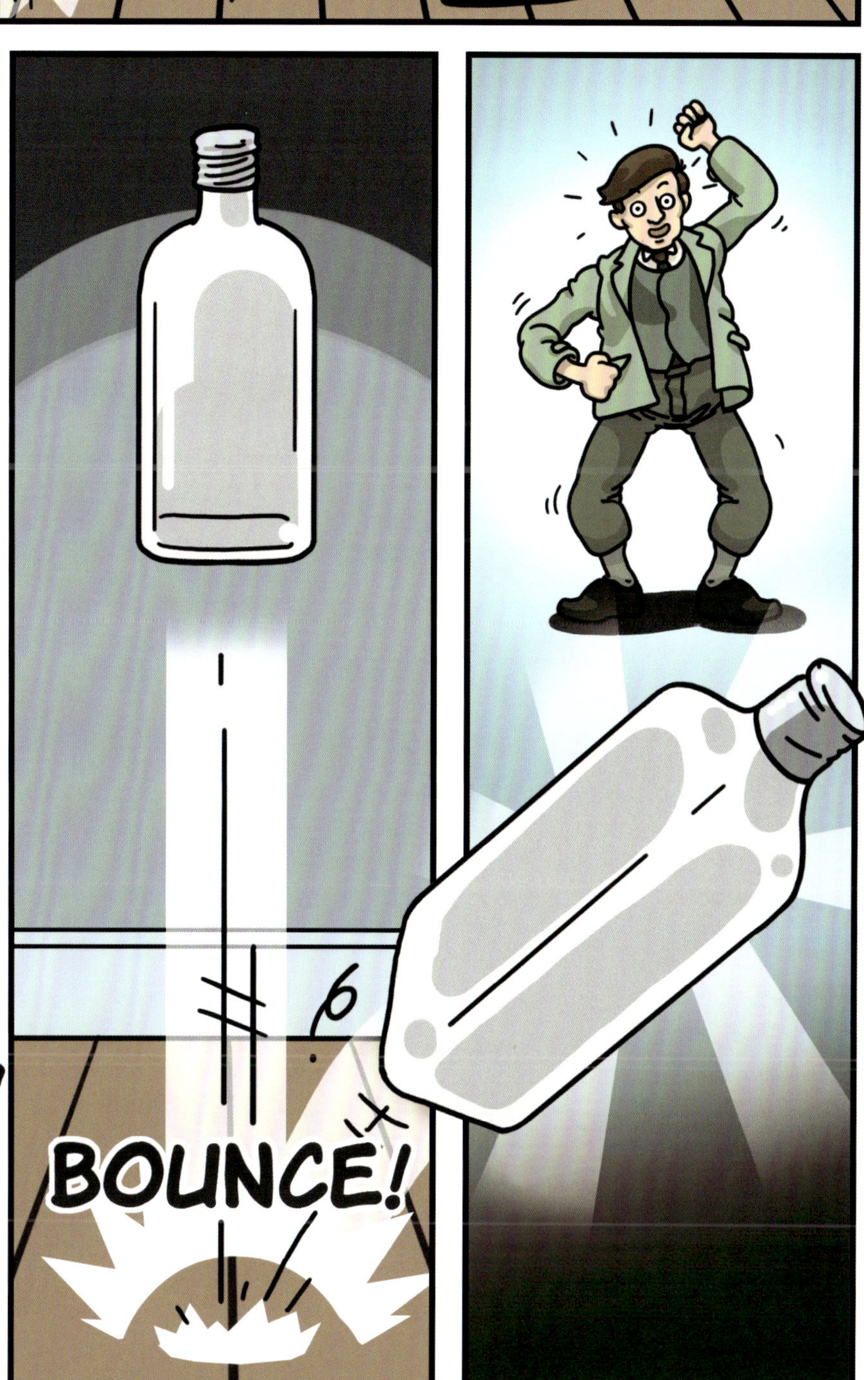

BOUNCE!

BOTTLE ANATOMY

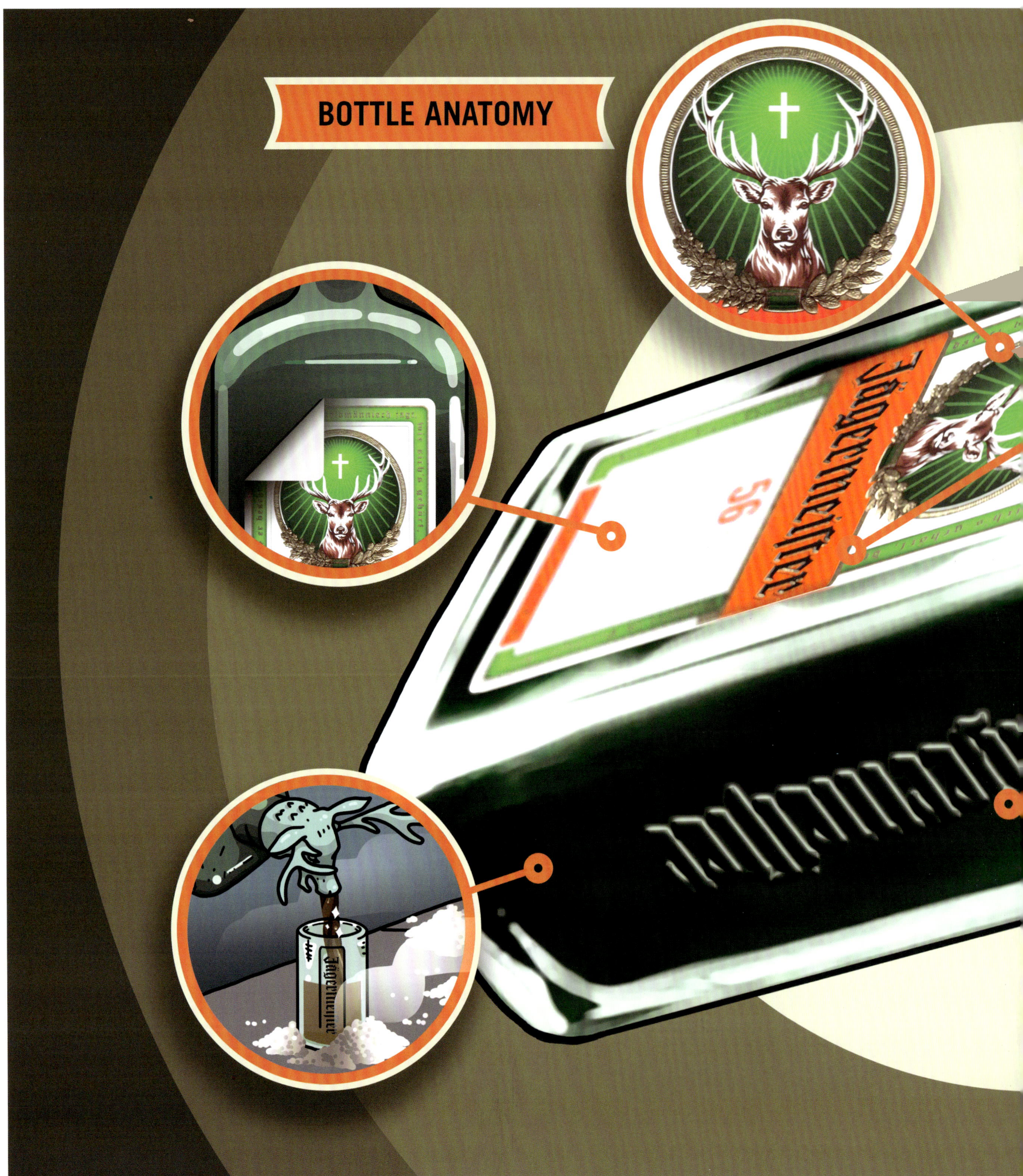

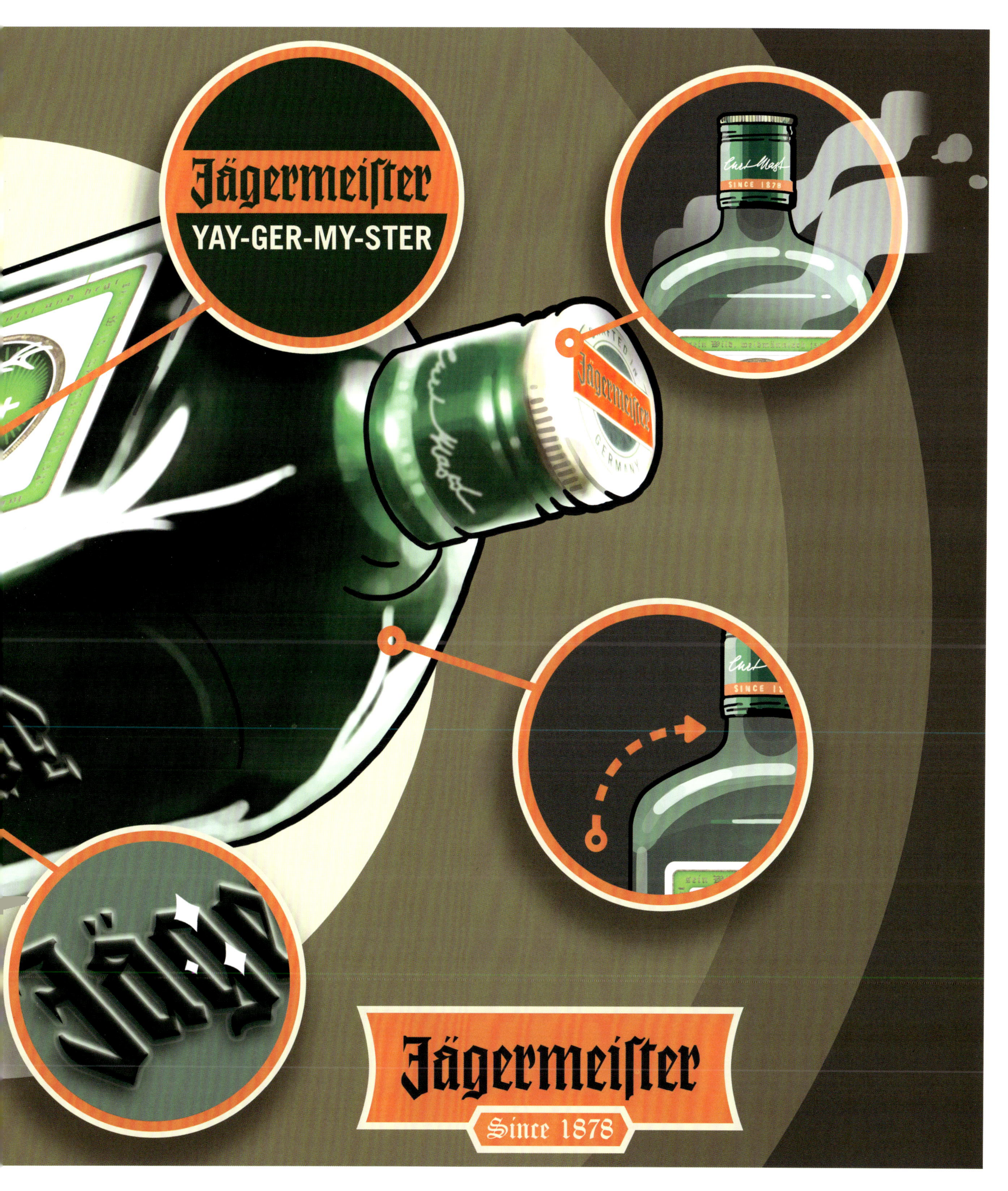

Jägermeiſter
YAY-GER-MY-STER
Jägermeiſter
Since 1878

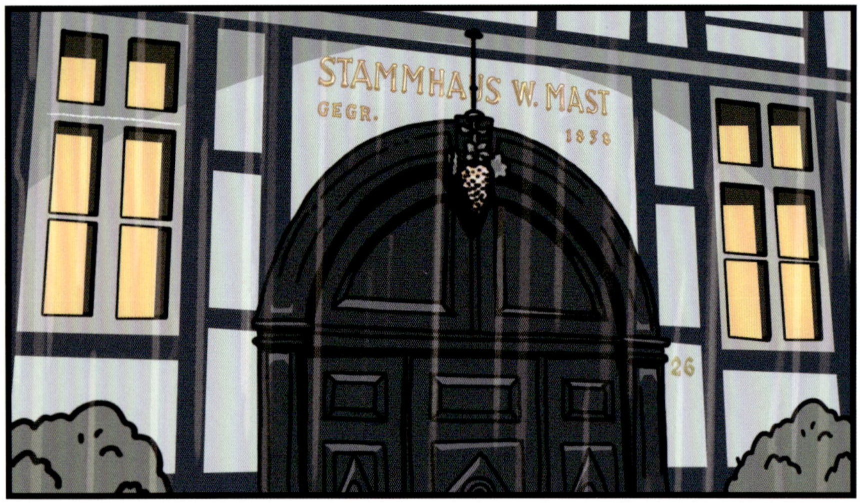
STAMMHAUS W. MAST
GEGR.
1878

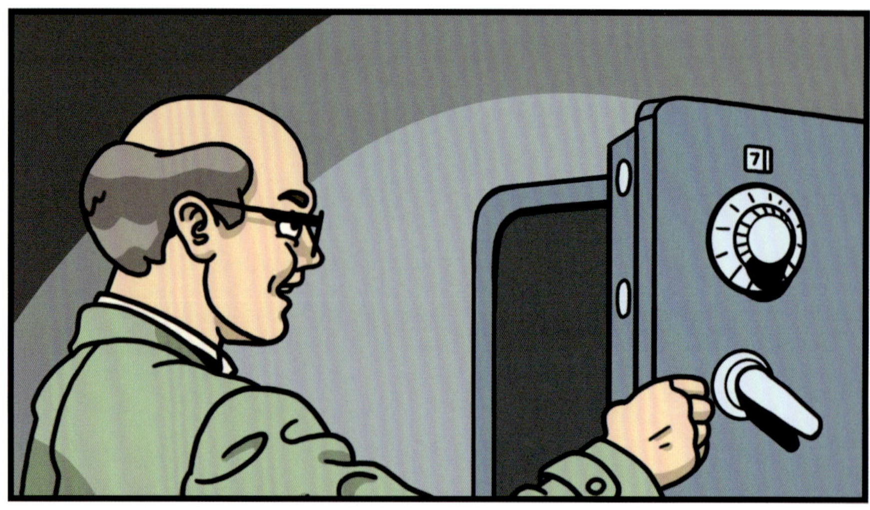

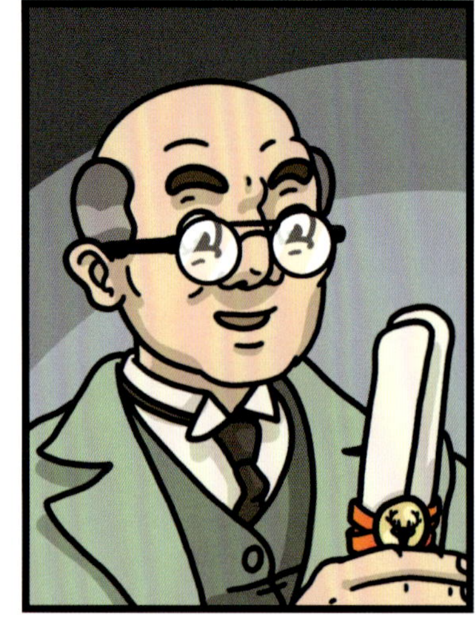

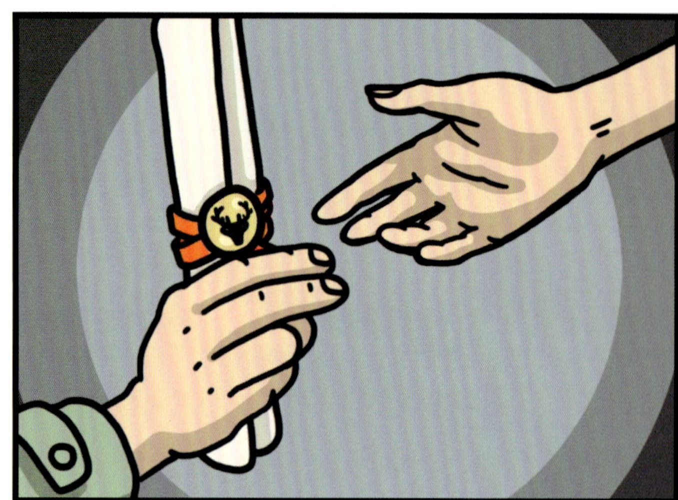

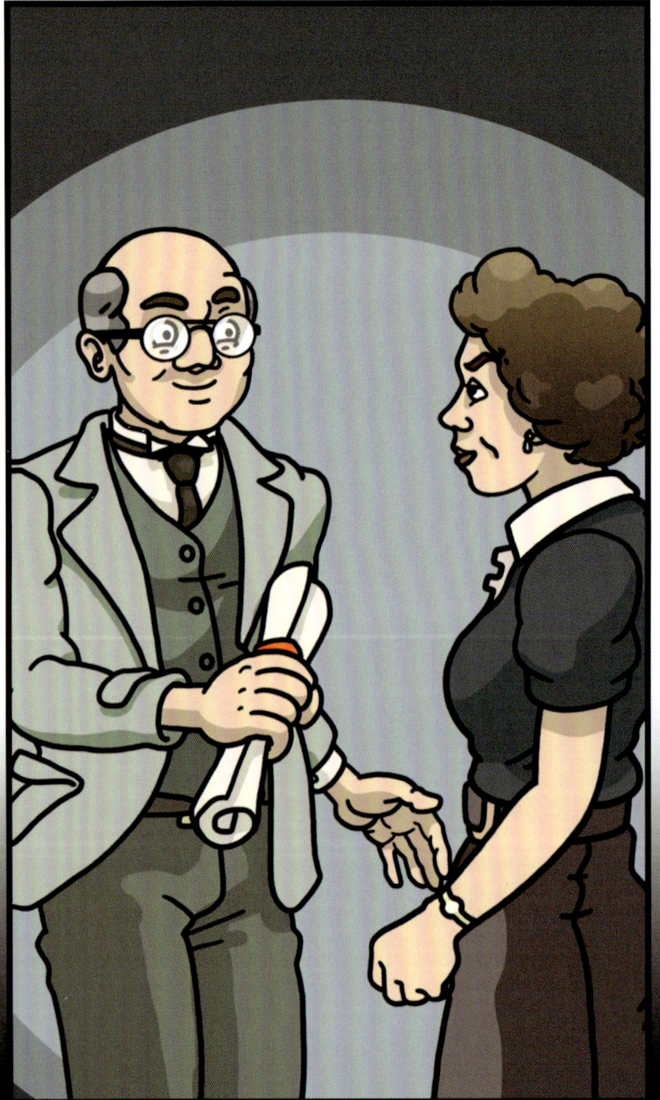

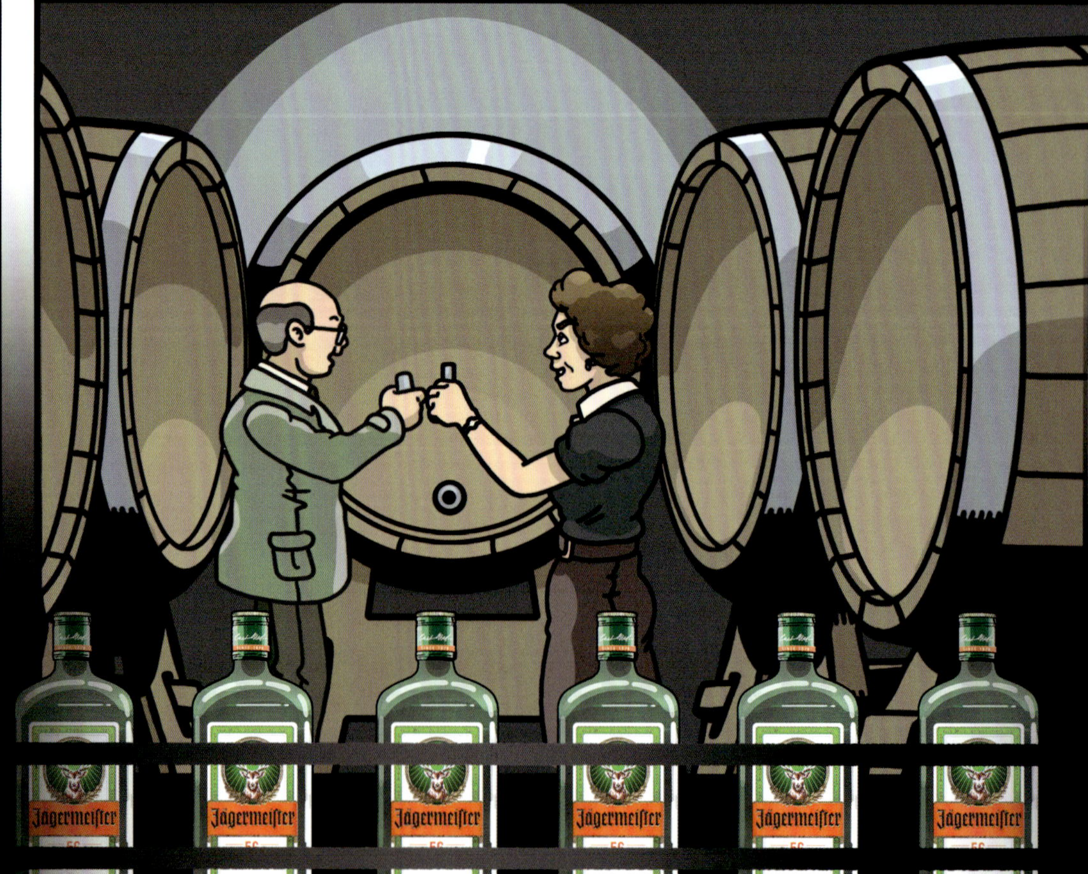
Jägermeifter
Jägermeifter
Jägermeifter
Jägermeifter
Jägermeifter
Jägermeifter

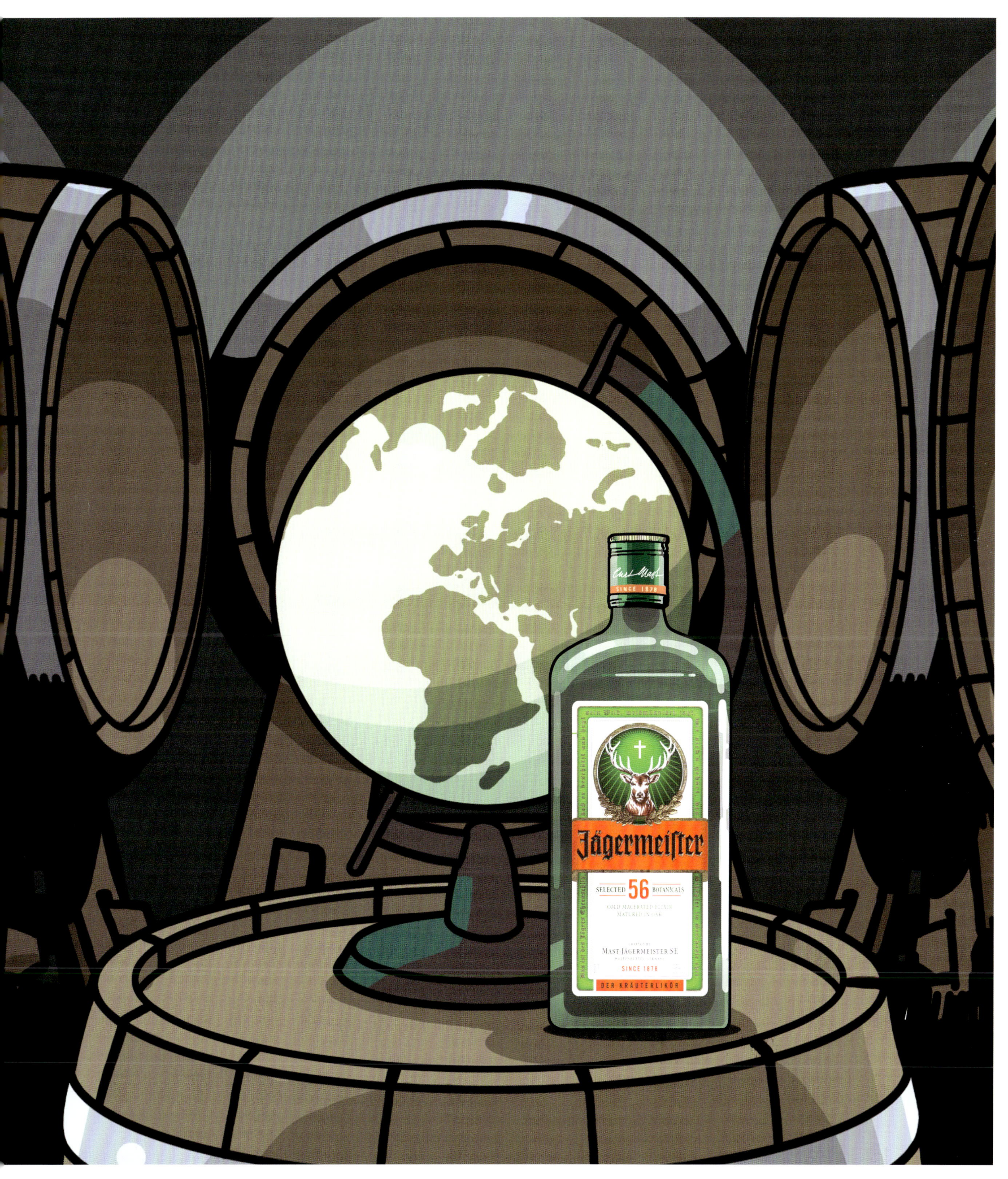

SINCE 1878
Jägermeiſter
SELECTED 56 BOTANICALS
COLD MACERATED ELIXIR
MAST-JÄGERMEISTER SE
SINCE 1878
DER KRÄUTERLIKÖR

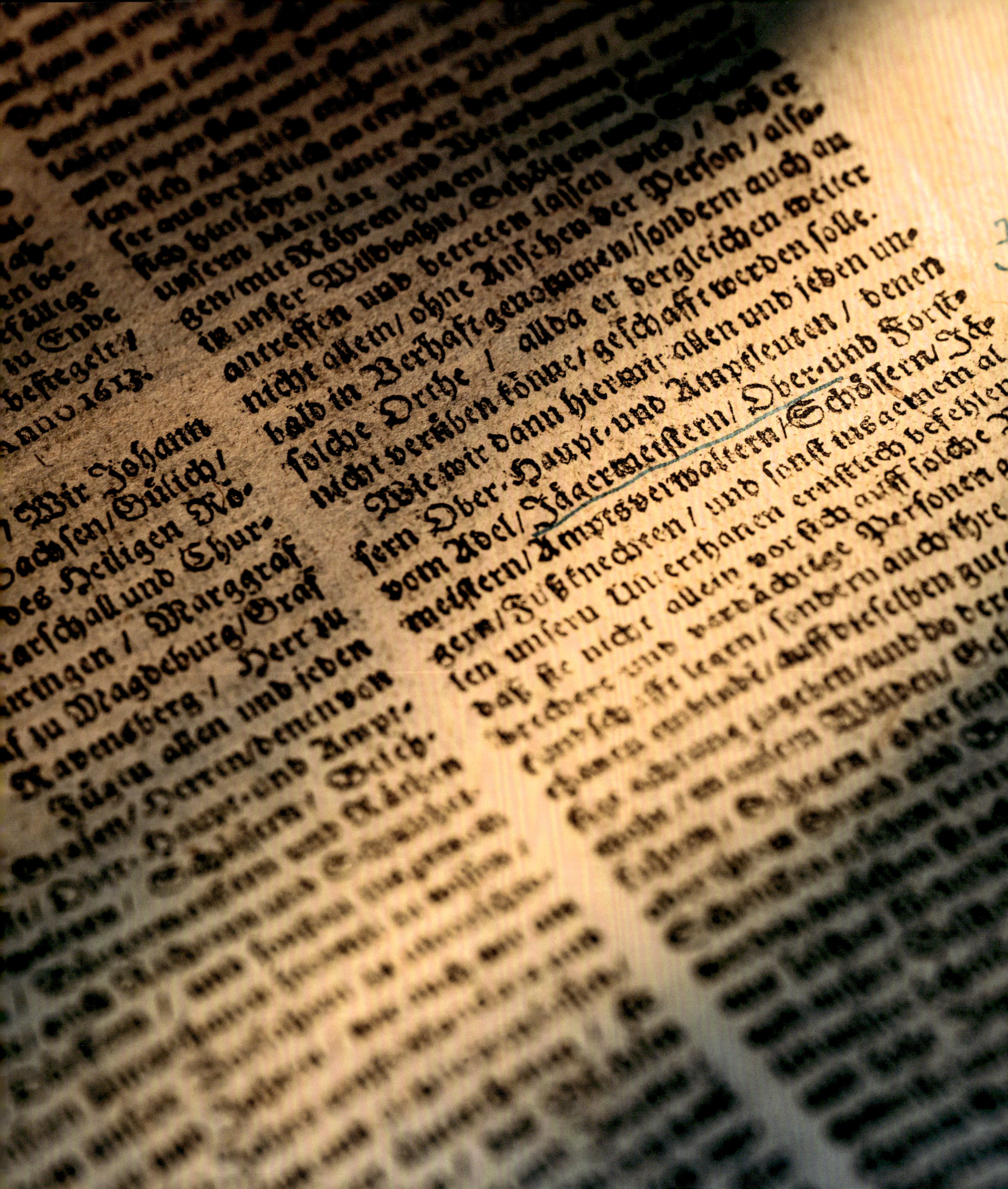

Anno 1613.

Wir Johann

Sachsen/ Gülich/

des Heiligen Rö-

...rschall und Chur-

...ringen / Marggraf

...zu Magdeburg/ Graf Herr zu

Ravensberg/

...en allen und jeden

...fen/ Herren/ denen von

...ur- Haupt- und Ampt-

Gülich.

...Gen/ mit Röhren/Schießen und / daß er

in unser Wildbahn und betreten laſſen wird / ...

antreffen und betreten laſſen wird /

nicht allein/ ohne Anſehen der Perſon / alſo

bald in Verhaft genommen/ ſondern auch an

ſolche Orthe / allda er dergleichen weiter

nicht verüben könne/ geſchafft werden ſolle.

Wie wir dann hiermit allen und jeden un-

ſern Ober-Haupt- und Amptleuten / denen

vom Adel/ Amptsverwaltern/ Ober- und Forſt-

meiſtern/ Jägermeiſtern/Schöſſern/ Jä-

gern/ Fußknechten/ und ſonſt insgemein befehlen

...en unſern Unterthanen ernſtlich befehlen

daß ſie nicht allein vor ſich auff ſolche Perſonen...

...brecher und verdächtige Perſonen auch ihre...

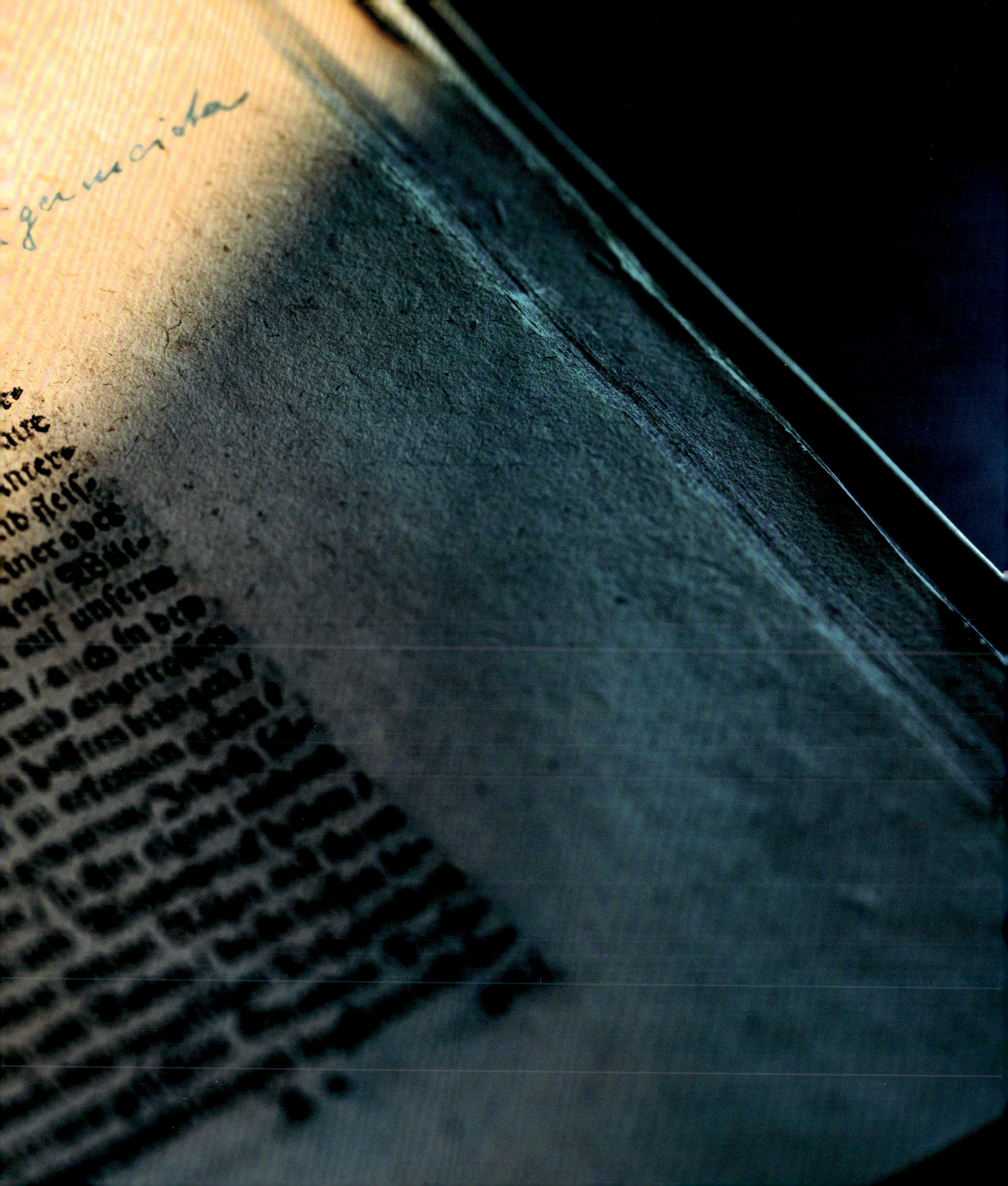

A FAMILY SINCE 1878

The company's beginnings: a Wolfenbüttel-based family business writes an international success story

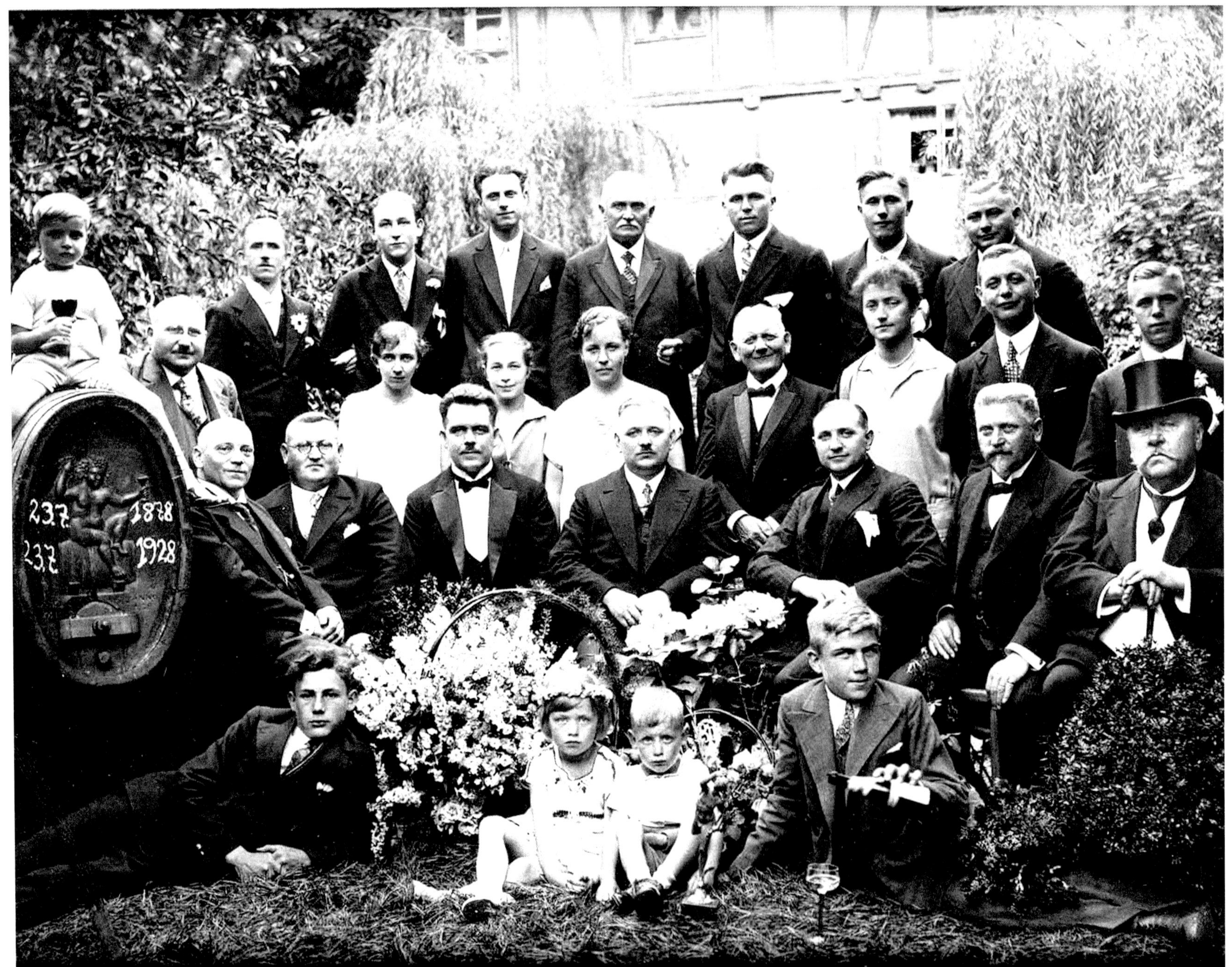

The company's 50th anniversary on 23 July 1928; in the middle is Jägermeister inventor Curt Mast, with his children Annemarie and Wilhelm in front.

Precision craftsmanship and natural ingredients, born in Wolfenbüttel and at home in the world, community-focused yet bold enough to go its own way – this is all in Jägermeister's DNA, and has been for decades. One family has been the driving force here since the very beginning, passing the secret Jägermeister recipe down from generation to generation.

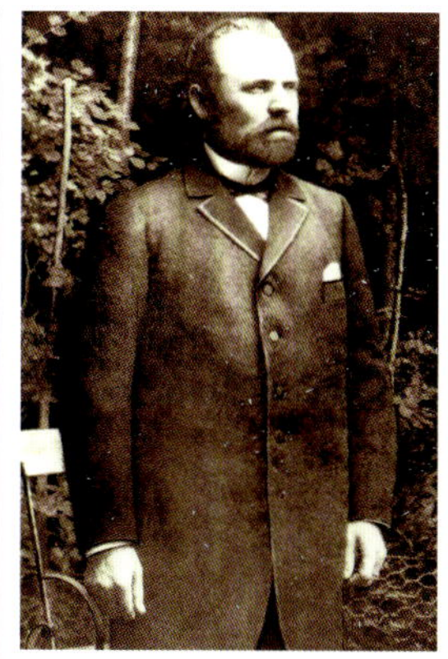

Company founder Wilhelm Mast in the 1890s.

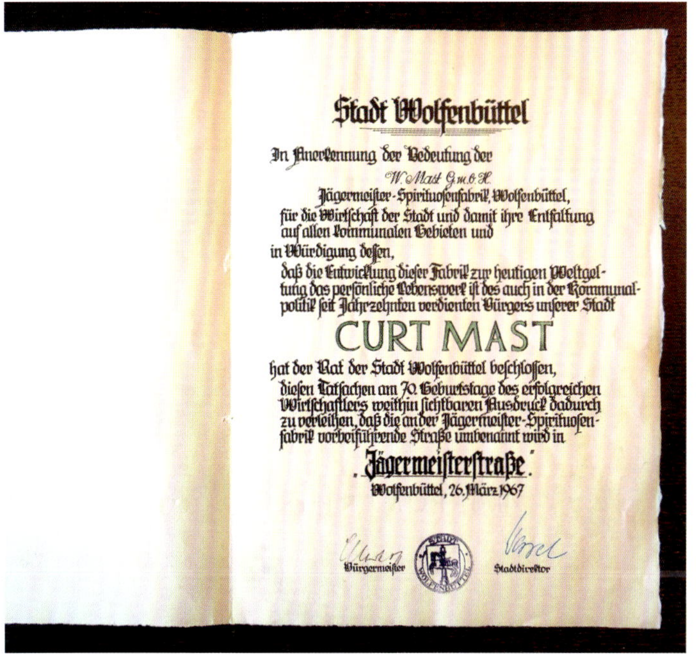

The date 26 March 1967 was a special day for Curt Mast. For his 70th birthday, his hometown of Wolfenbüttel surprised him with a very unusual present, by changing the name of the street on which his company was – previously known as Hermann-Korb-Straße – to Jägermeisterstraße. The Jägermeister that Mast had developed over three decades ago had now become one of the most popular herbal liqueurs in Germany, and for Mast, the Jägermeister story had reached a climax. What he couldn't have imagined that day was that the brand was in reality only at the start of its international success story, despite the fact that the company founded by his father was already nearly 90 years old at the time.

BOLD DECISIONS

On 23 July 1878, young businessman Wilhelm Mast, Curt Mast's father, signed the purchase agreement for a half-timbered building at Großer Zimmerhof 26 in Wolfenbüttel. As the 'Stammhaus', this place would still amaze visitors to the city and also fans of the brand even decades later. The Mast family originally hailed from the south-

ern part of the mid-mountain Harz region. Wilhelm had come to Wolfenbüttel as early as 1865 and instantly liked the city, with its traditional streets, historic half-timbered structures and friendly people. When he first commenced business operations upon purchasing the 'Stammhaus', he initially produced vinegar, because this could be used both to preserve the fruit and vegetables grown in Wolfenbüttel's many market gardens, and for mining in the nearby Harz mountains.

His son Curt had been interested in his father's business even as a child, and was fascinated by the precision craftsmanship and the work involving the various natural ingredients which he was able to observe at his father's company. Wilhelm Mast loved to tinker around, and he had a particular penchant for 'producing all kinds of spirits'. Curt soon began sharing this passion for experimentation and development with his father.

Curt had to make his first brave decision at a very young age: his father Wilhelm fell seriously ill in 1913, prompting Curt, aged just 16, to start working in the business, eventually taking over full management a few years later in 1917. The years that followed were turbulent ones for the company. In 1922, Curt ceased the production of vinegar, having already shifted the focus of the business to the wine and spirits trade. Soon after, he would become utterly captivated by another, much greater idea that would combine all his passions in one.

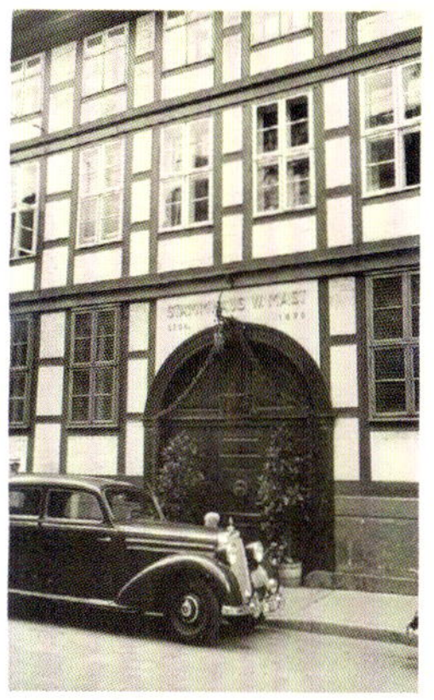

A historic image from the 1950s: the Mast-Jägermeister 'Stammhaus'.

INVENTING JÄGERMEISTER

In his free time, Curt Mast enjoyed being out in the middle of nature. He was an avid and responsible hunter who loved going deerstalking in the forest, so he was certainly well acquainted with the convivial moments pre- and post-hunt, when people would stand around chatting and laughing together. It is possible that it was precisely these moments that gave him the idea of developing a herbal liqueur to make these gatherings even more social.

He tinkered with his new idea for months, testing out various new creations. Sometimes he noted that the mixture was 'too sweet', other times the fruity 'citrus note too weak'. But 1934 was the year.

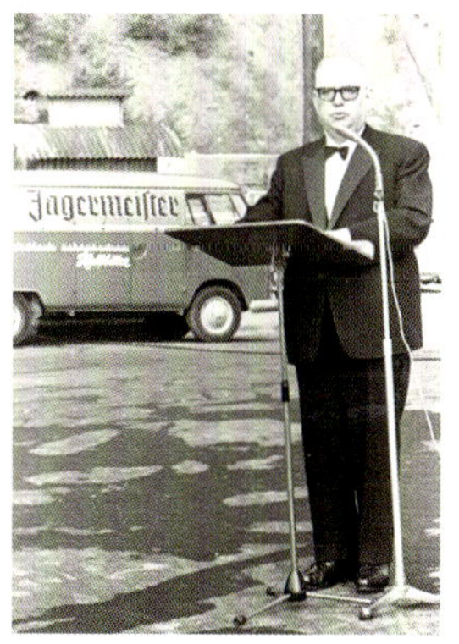

The inventor of Jägermeister, Curt Mast, in 1968.

Left: Hermann-Korb-Straße in Wolfenbüttel was officially renamed Jägermeisterstraße in 1967 in honour of Curt Mast's 70th birthday. The company's headquarters are still located on that street.

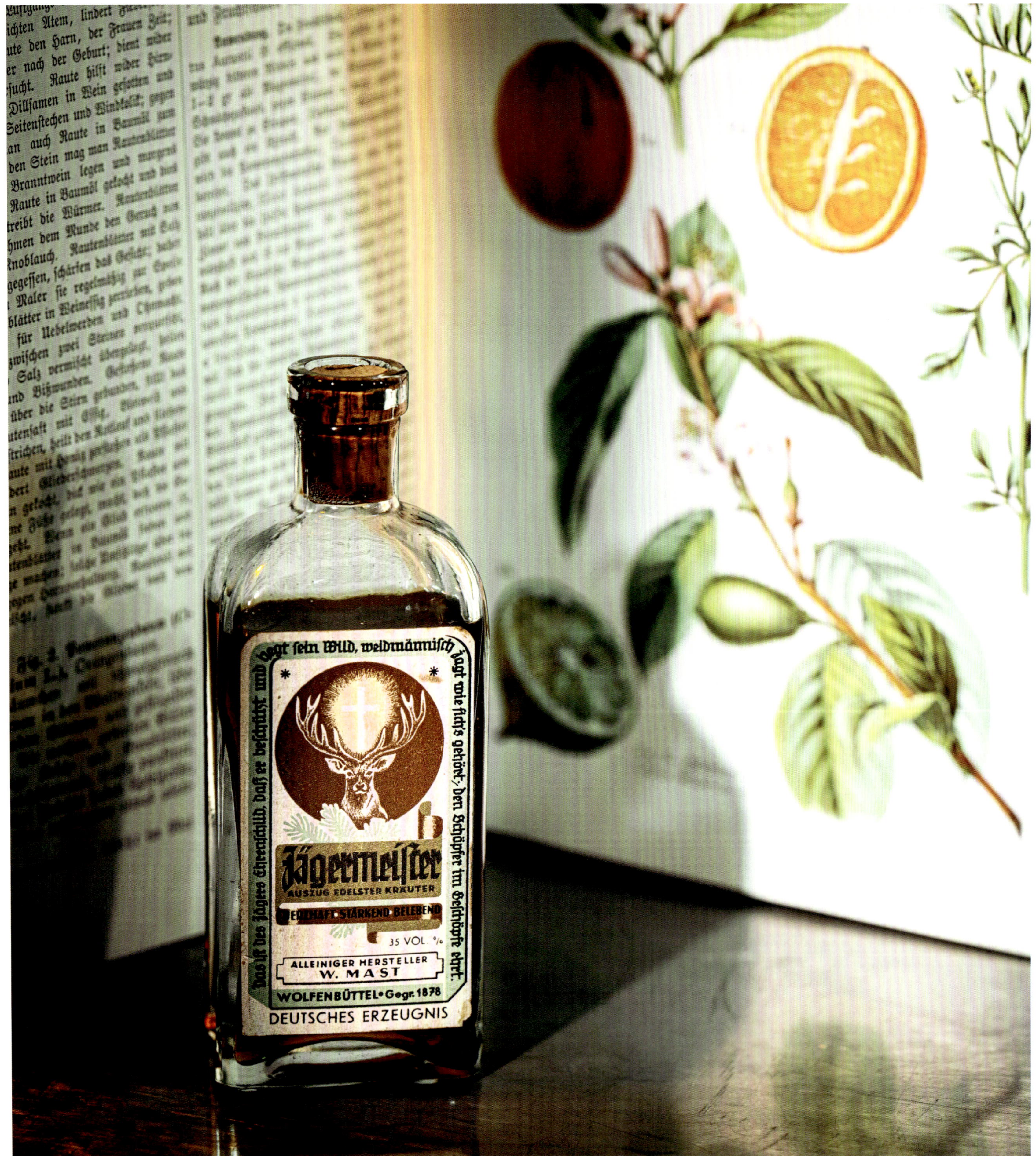

One of the oldest preserved Jägermeister bottles, at that time still corked and made of clear glass.

From a combination of 56 herbs, blossoms, roots and fruits, coupled with 35% alcohol by volume, he had found the perfect recipe for his herbal liqueur, which he called Jägermeister ('master hunter').

Even before launching it on the market, Mast was already devising a unique brand personality that would live on for decades. Both the bottle and the label had to be distinct. Together with graphic designer Günther Clausen, he developed the world-famous label featuring the legendary Hubertus stag – a symbol of St Hubertus, the patron saint of all hunters. According to legend, Hubertus was a reckless, irresponsible hunter – a poacher – who went hunting one Good Friday and came across a mighty stag. Just before he was able to fire the arrow from his crossbow, the stag turned around and a dazzling cross appeared between its antlers. Hubertus took this as a divine sign, threw his weapon away and knelt down in reverence, promising to be better. From that day on, he radically changed his life, championing the respectful treatment of nature and becoming the patron saint of hunters.

Curt Mast was familiar with the Hubertus legend from his own passion for hunting, and he saw to it that its moral of handling nature and its resources responsibly and sustainably was celebrated on the

> Das ist des Jägers Ehrenschild,
> daß er beschützt und hegt
> sein Wild, weidmännisch jagt,
> wie sich's gehört, den Schöpfer
> im Geschöpfe ehrt.

Jägermeister label:
This roughly translates as: This is the hunter's badge of honour, that he protect and nourish his game, hunt sportingly, as is proper, and honour the Creator in creation.

These lines have featured on every bottle since Jägermeister was first produced, and come from the first verse of a poem by hunter and writer Oskar von Riesenthal. The bottle itself was designed to protect the precious contents, for Mast was very aware of the quality of his herbal-liqueur creation, even considering his Jägermeister 'too good to be sold'.

His chosen style of bottle for Jägermeister, which would later gain fame with its striking rectangular shape, was the only one to pass his hard break test without any damage. In this test, Mast dropped many different types of bottles, one after another, onto his wooden kitchen floor from a set height. He finally launched Jägermeister on the market in 1935, a perfect herbal liqueur with a distinct design and unique, highly robust bottle.

Driven by curiosity and a love of trying out new things, Curt Mast soon turned his hand to developing other spirits. Just a few years after the birth of his Jägermeister, he launched new specialities on the market, from Aprikosen Sonnengeist (Apricot Sun Spirit) to Schlehen mit Rum Jägergeist (Slose with Rum – Jäger Spirit).

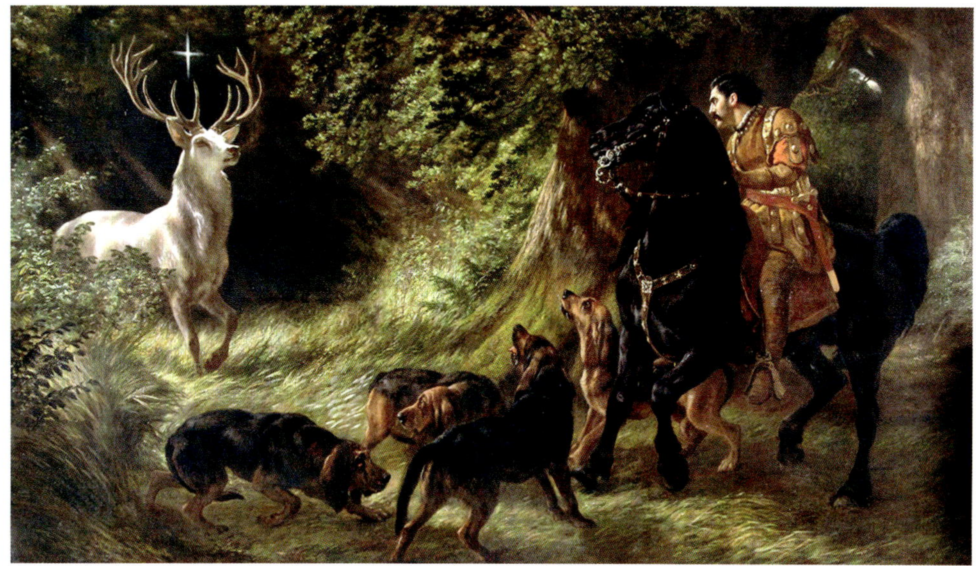

The iconography of the Jägermeister brand is based on the Hubertus legend, which tells of an encounter between Hubertus, Hubertus, who later became the patron saint of hunters, and a mighty stag, between whose antlers a dazzling cross lit up the night.

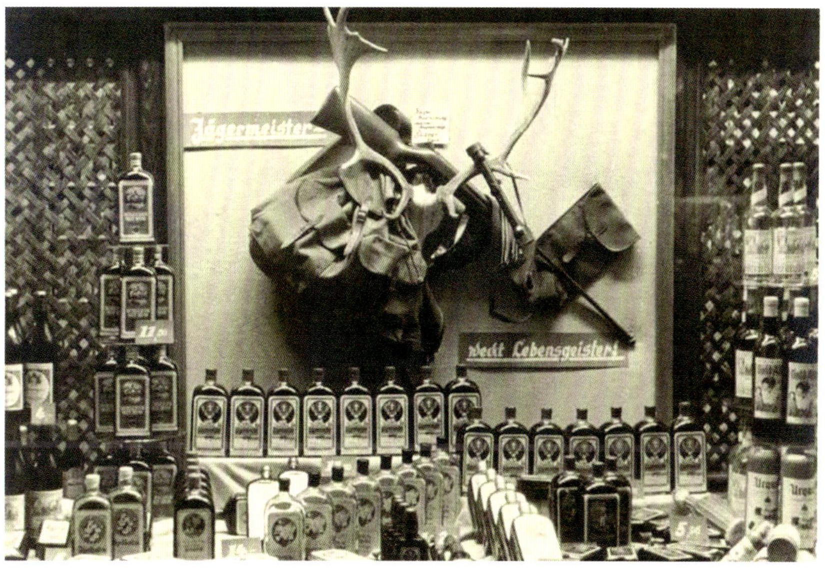

A historic trade-fair stand from the 1950s.

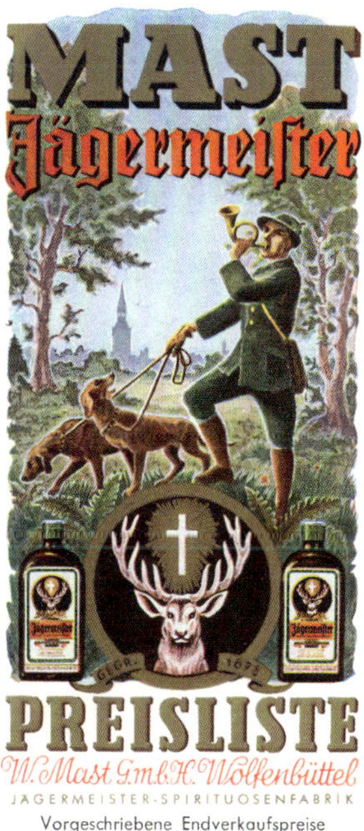

A Jägermeister price list from 1954.

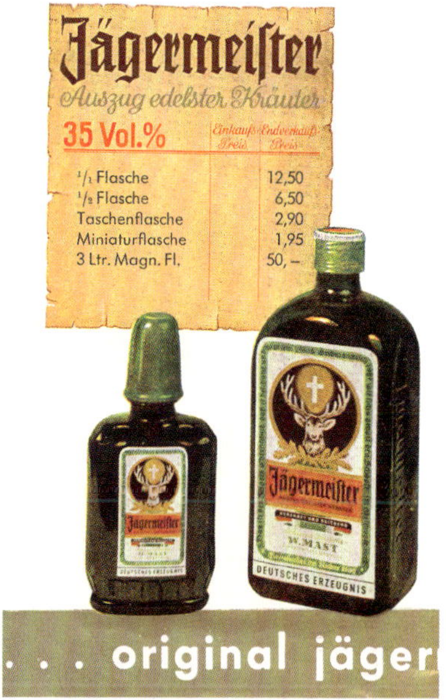

GÜNTER MAST STEPS UP

In 1952, Curt Mast gained some active support at the company. His nephew, Günter Mast, only in his mid-20s at the time, was a very confident individual who had an extraordinary flair for innovative and spectacular marketing concepts. He recognized the brand's tremendous potential and helped Jägermeister to gain huge popularity. But there was one thing he asked of his uncle Curt: to unconditionally focus on Jägermeister in order to keep increasing the core brand's profile with unique advertising ideas. So it was back to basics: concentrating the product range purely on Jägermeister – quality over quantity!

It was thanks to Günter Mast's ideas that Jägermeister was virtually omnipresent in Germany in the 1960s and 70s. Buses, trams and suburban trains bearing orange Jägermeister advertising had been in operation throughout nearly all of Germany's major cities since the early 60s. And Günter Mast's other concepts would go down in German advertising history. The first car of the Jägermeister Racing Team took to the racetracks in 1972, with none other than two-time Formula 1 World Champion Graham Hill at the wheel. The following year not only marked the start of the legendary 'I'm drinking Jägermeister because . . .' campaign, but it also hailed the first jersey sponsorship in German Bundesliga history at Eintracht Braunschweig.

It is fair to say that Günter Mast caused the odd scandal with all these ideas – but he was fine with that, because it meant everyone was talking about Jägermeister. Günter Mast loved provoking and polarizing, and it was this attitude that enabled him to create the brand's unique, bold and unconventional image. Meanwhile, Curt Mast's perfection in relation to craftsmanship and product quality was equal in intensity to his nephew's flair for building the right profile.

SCIENTIFIC PRECISION

That same year of 1952, when Günter Mast signed on to his uncle's company, also saw another young man in Wolfenbüttel open his first practice: Günther Findel. The doctor from Berlin had married Curt Mast's daughter Annemarie in 1947, and the young couple shared an interest in sciences and the arts, as well as a love of nature. As such they became involved in Wolfenbüttel's Herzog August Library, as well as in forestry and agriculture, and started their own herb garden. In addition to his work at his own practice, Günther Findel had also been the company doctor at the Jägermeister spirits factory run by his father-in-law since the early 1950s, and fitted in at Jägermeister like the proverbial hand in glove.

Günther Findel was a level-headed scientist, and a duly meticulous worker. And, just like his father-in-law, he also got a thrill out of exquisite product and production quality and working with natural raw materials. So it was that, during the 1960s, he devoted himself entirely to the company, taking over technical direction, and thus responsibility for production and technical operations, in 1970.

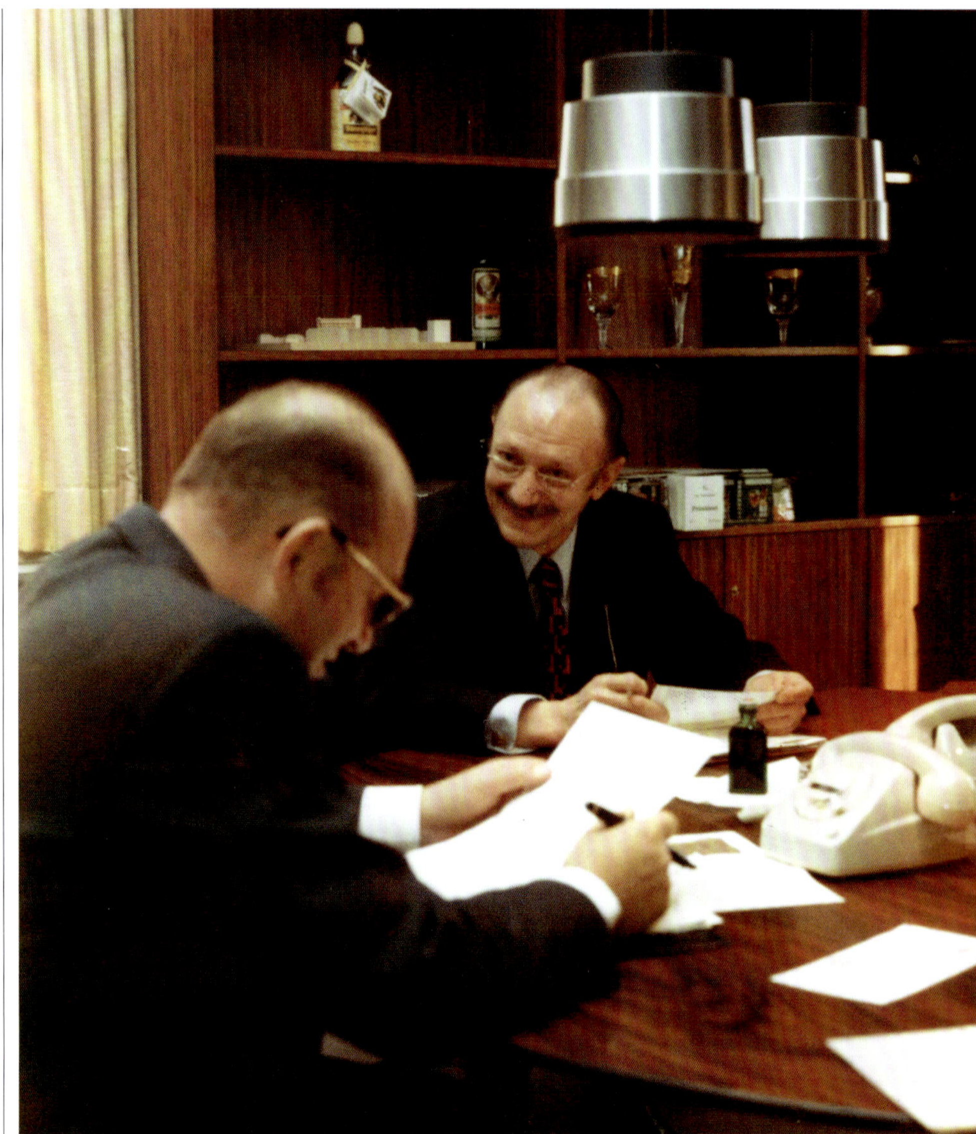

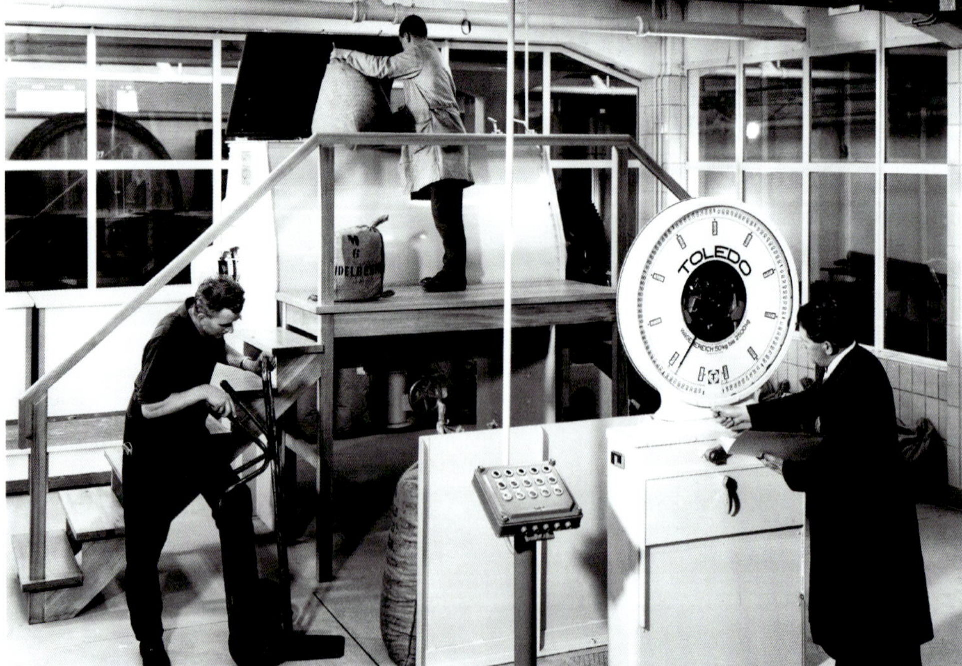

Top: Günter Mast and Günther Findel during the 'mail meeting' in 1973. At this regular meeting, the managing directors would inspect all the business mail coming and going from Jägermeister. Below: The herbs are weighed and ground with utmost care and precision during the production process (historic photos from the 1960s).

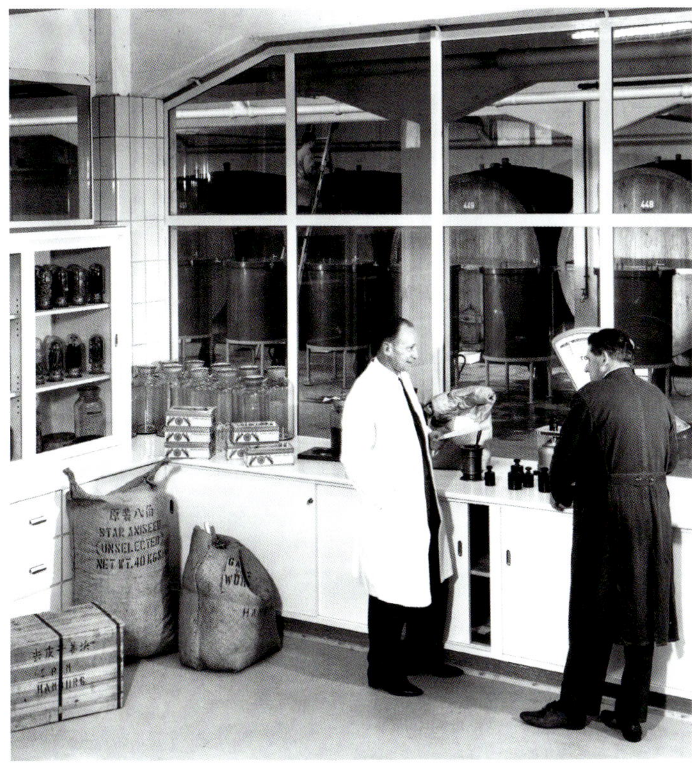

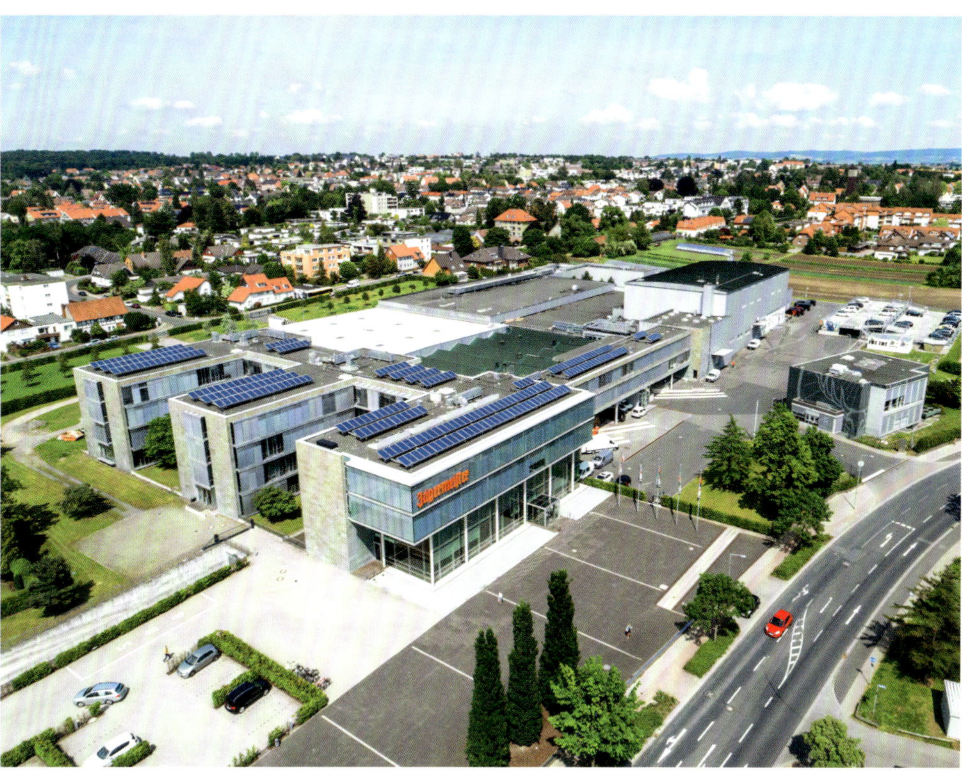

While Curt Mast had achieved Jägermeister's quality and manufacturing methods through persistent testing and experimentation, Günther Findel was now professionalizing it using a scientific approach, without changing the Jägermeister recipe.

Günter Mast and Günther Findel were like a congenial pair since the 1960s. What Mast was to Jägermeister marketing, Findel was to product and manufacturing quality. It was a time of their lives that saw Annemarie Findel-Mast and Günther Findel forge strong ties with Wolfenbüttel.

FROM WOLFENBÜTTEL TO THE WORLD

Its quality, unique taste and spectacular promotional and marketing campaigns saw Jägermeister become one of the best known and most successful herbal liqueurs in Germany in the 1960s. It was at this time that, encouraged by these successes, the brand made its first forays abroad. In 1960, Jägermeister could initially only be bought in Germany, Austria, Belgium and Luxembourg; ten years later, it was already available in France, Italy, the Netherlands, Denmark and Switzerland. And from the mid-1970s onwards, things would also pick up on the other side of the Atlantic, with US importer Sidney Frank signing the first Jägermeister distribution agreement in 1975. His stunning ideas – tap machines, Jägerettes and extensive involvement in the music scene – meant Jägermeister had become a genuine cult brand in the USA by the 1990s, and the brand reaped the advantages of this all over the world.

The vision of successfully selling Jägermeister internationally can largely be attributed to Walter Sandvoss. He had begun his traineeship at Jägermeister on 1 April 1958, eventually becoming sales manager. He firmly believed Jägermeister could inspire fans not only in Germany, but right across the globe – and he was right. Jägermeister is now sold in over 150 different countries, with the United States, Great Britain, China, Russia, the Czech Republic and of course Germany all among its most important markets. The Jägermeister base, however, continues to be produced exclusively in Wolfenbüttel. The name of its hometown is on the label of every bottle. Born in Wolfenbüttel, it is at home all around the world.

JÄGERMEISTER AND ITS ROOTS

The roots of Jägermeister's tremendous success lie in the tradition and history of the brand itself. Visions and bold decisions are what enabled the company, founded in 1878, to keep operating; they led to the invention and launch of the Jägermeister brand, and to the distinct and unconventional image it is known for today. Whether it be in the underground or on a camp site, at the pub on the corner or a hip club, in a German village or New Orleans – Jägermeister is at home everywhere.

To this day, it is based on craftsmanship and quality. And, though the brand has repeatedly been able to reinvent itself and set trends beyond its own industry throughout the course of its history, its essence has remained untouched. The Jägermeister recipe has never been changed – which is one of the reasons it is among the world's best-kept company secrets. ⍊

Above left: Günther Findel (on the left) was head of production and technical operations from the 1960s onwards. **Above right:** Jägermeister's current head office in Wolfenbüttel. **Below:** Curt Mast's daughter Annemarie Findel-Mast in 1962/63.

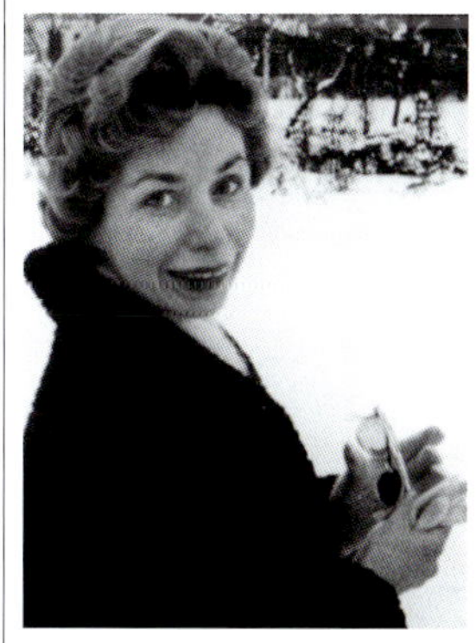

Below: Walter Sandvoss' vision was Jägermeister's international sales success.

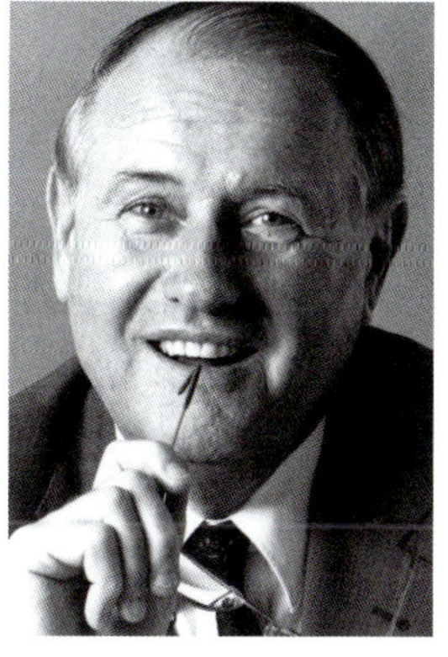

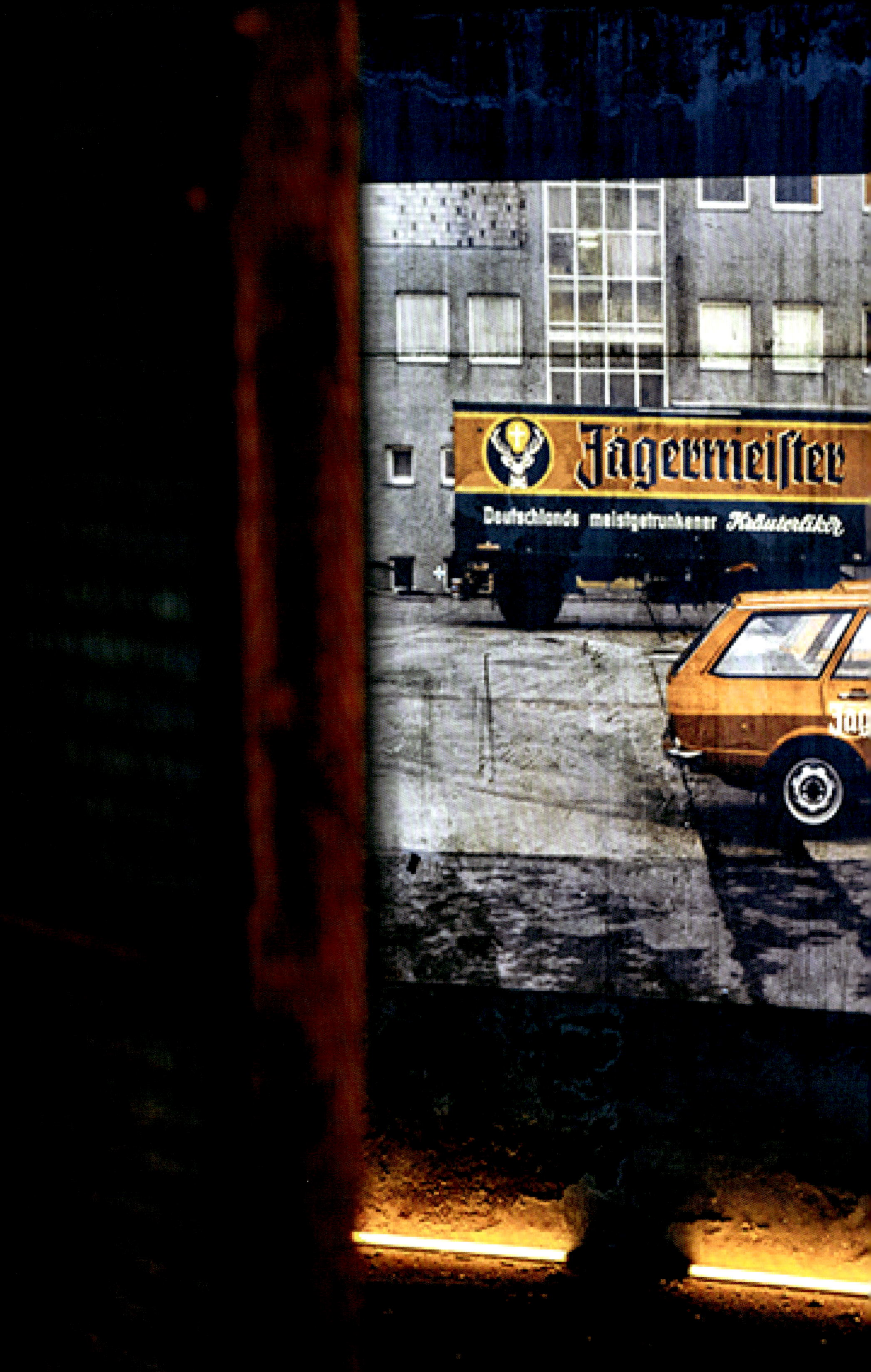
Jägermeister
Deutschlands meistgetrunkener Kräuterlikör

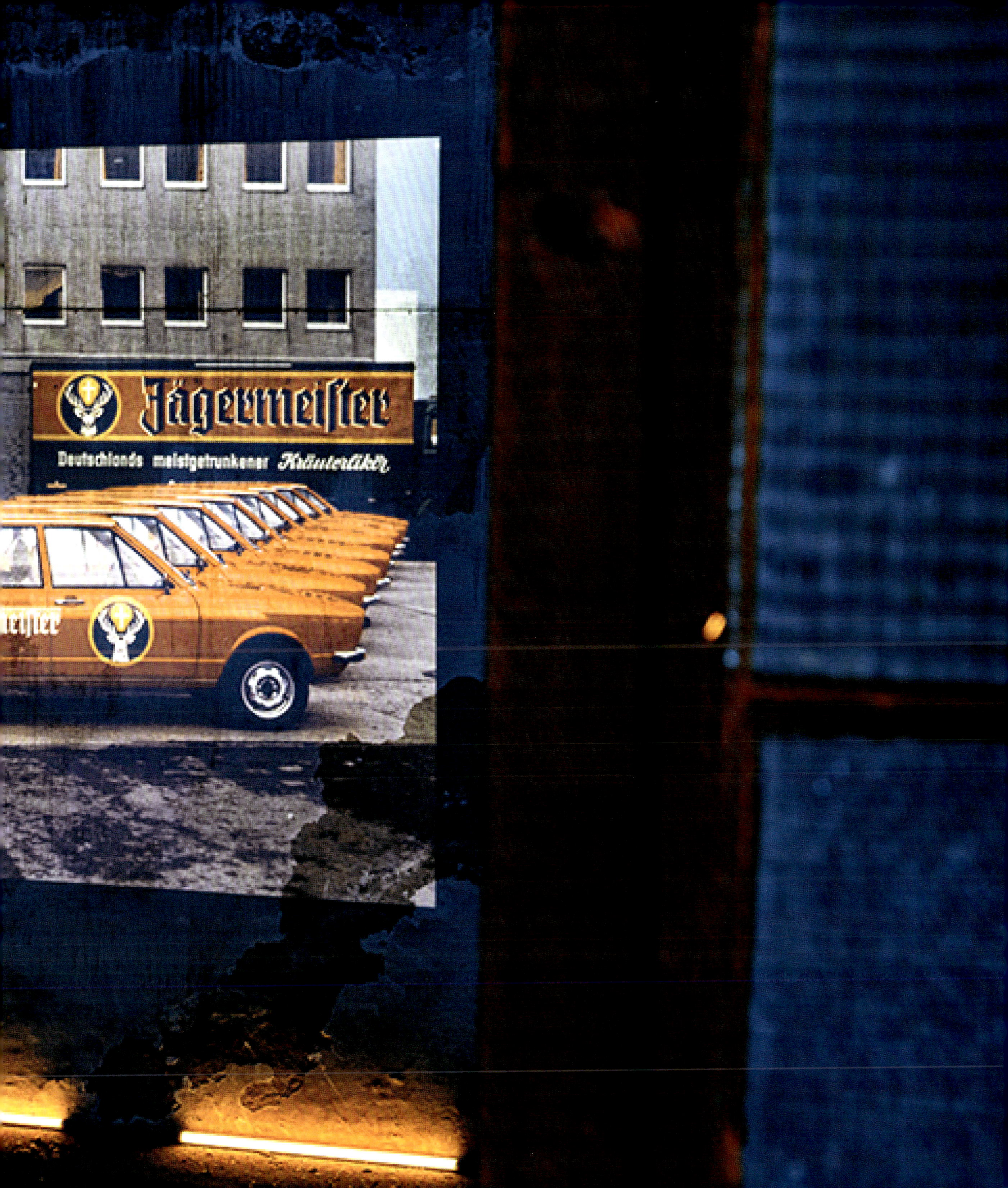

Jägermeister
Deutschlands meistgetrunkener Kräuterlikör

The period between 1972 and 2000 saw Jägermeister write motorsport history and stories that people still enjoy hearing to this day. The orange race cars bearing the Hubertus stag were a permanent fixture for three whole decades, in many race series and on countless racetracks.

IN THE FAST LANE

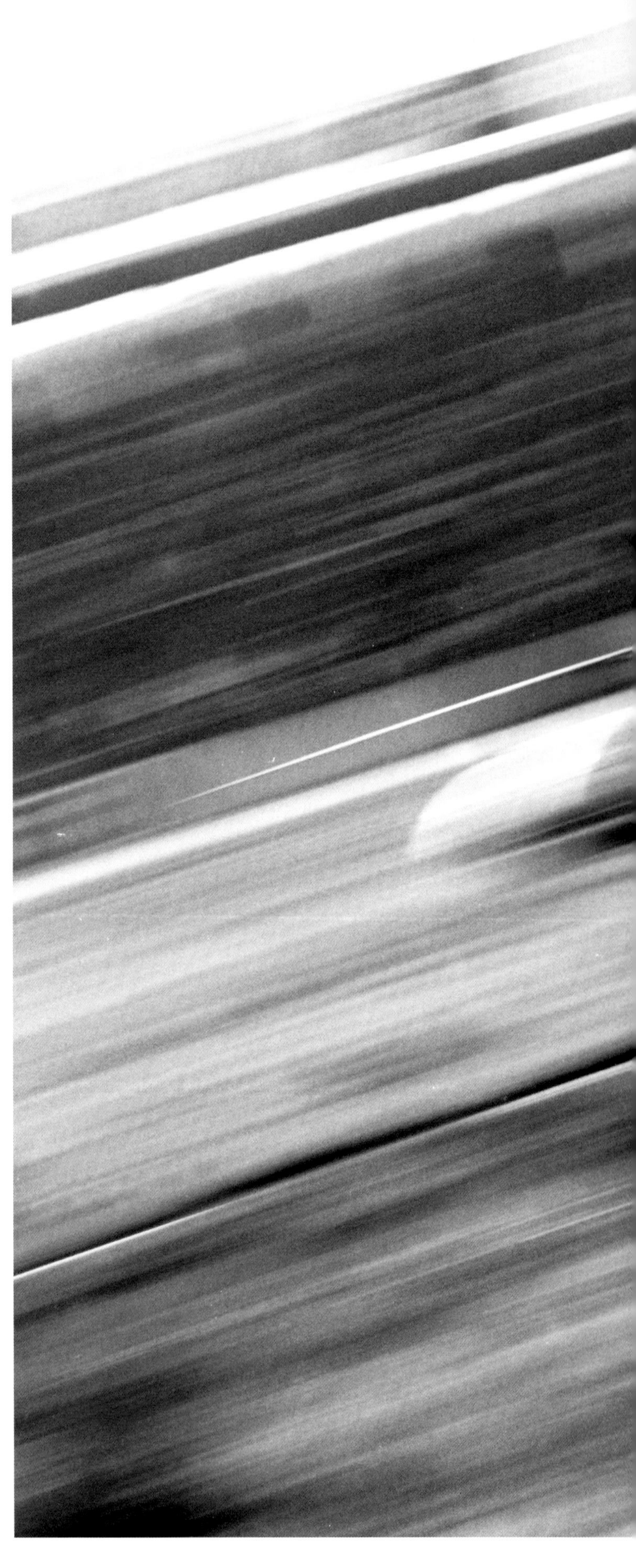

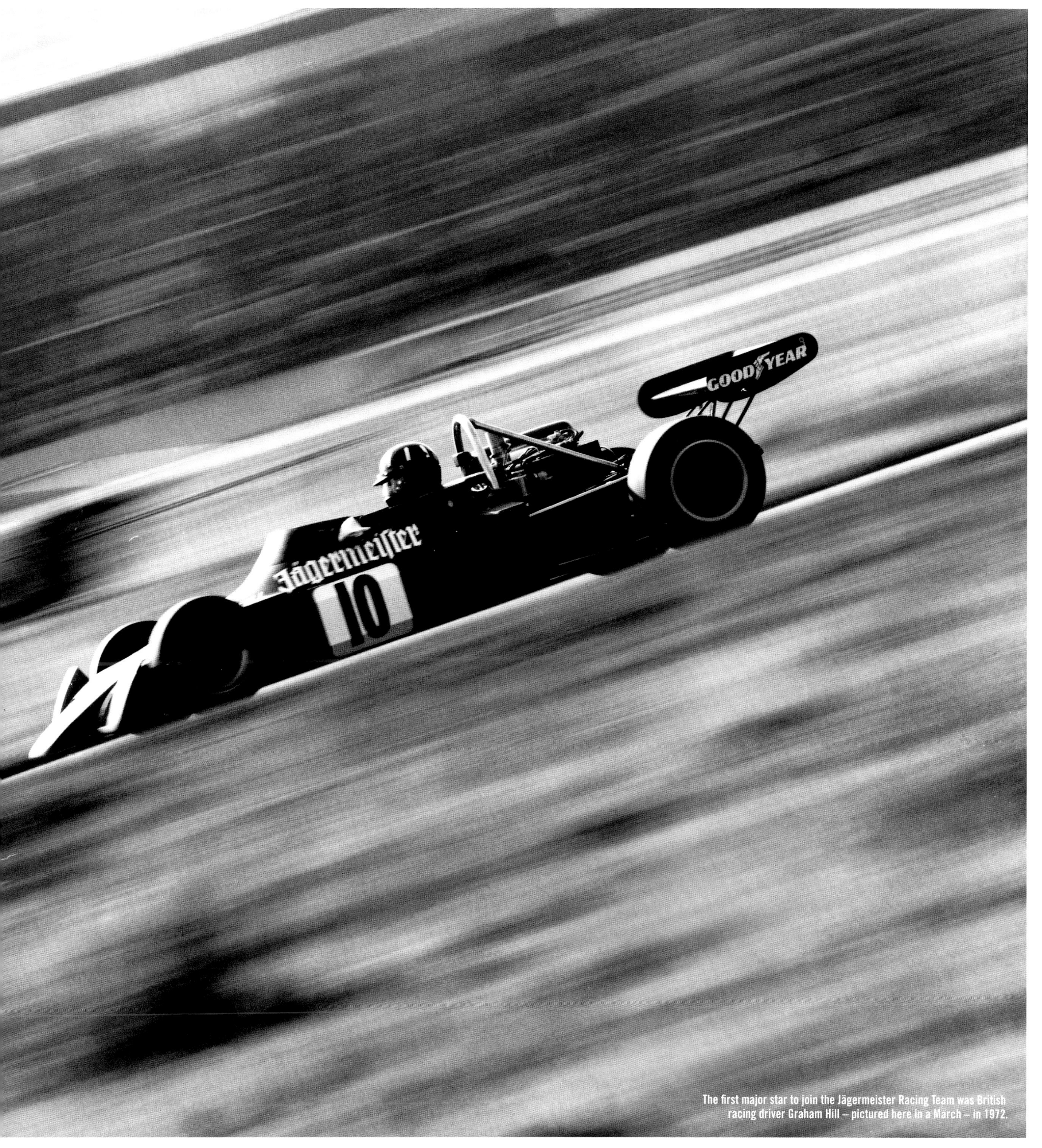

The first major star to join the Jägermeister Racing Team was British racing driver Graham Hill – pictured here in a March – in 1972.

'Buy yourself a decent race car at our expense, find a world-class athlete, and network with drivers. I want to have a whole fleet of race cars starting in Jägermeister colours.'

It was with these words, uttered to his cousin, journalist and racing driver Eckhard Schimpf, that Günter Mast launched Jägermeister Racing in 1972. Jägermeister shot out of the blocks as one of the first major motorsport sponsors; Eckhard Schimpf managed the team for nearly three decades, even enjoying getting behind the wheel himself – and doing so successfully.

The first car used was a Porsche 914/6 in Jägermeister green. But that wasn't sufficiently spectacular for Günter Mast: 'It's not striking enough. Don't we have a better colour?' After a brief period of deliberation, it was decided that the orange stripes of the Jägermeister bottle would be the new racing colour – and this would adorn the Jägermeister cars for nearly 30 years. More than 150 of the world's best racing drivers graced the cockpits of the Jägermeister Racing Team during this time, including greats such as Niki Lauda, Gerhard Berger, Jacky Ickx, Hans-Joachim Stuck, Graham Hill, Stefan Bellof, Klaus Ludwig and Jochen Mass. From 1972 to 2000, the orange racing cars with their famous trademark were at home on all the tracks of Europe, from the German Touring Car Championship to Group C, and even Formula 1.

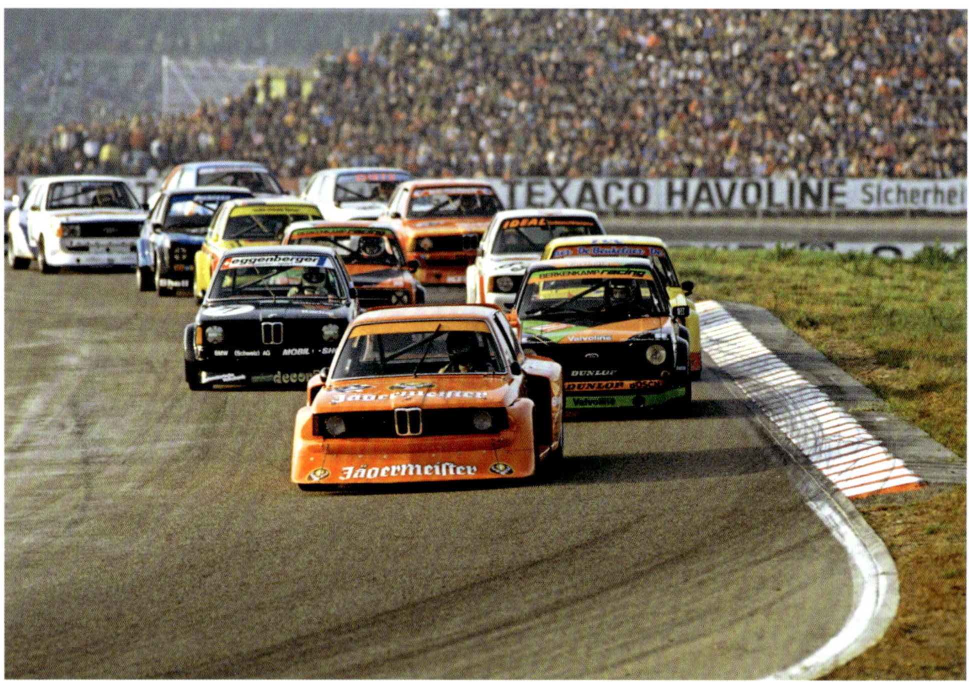

Above: Motorsport legend Hans-Joachim Stuck during a DRM race in a BMW 320.
Left: 'Strietzel' Stuck (in the background holding an umbrella) and crew shortly before the start of a Formula 2 race in a March BMW.

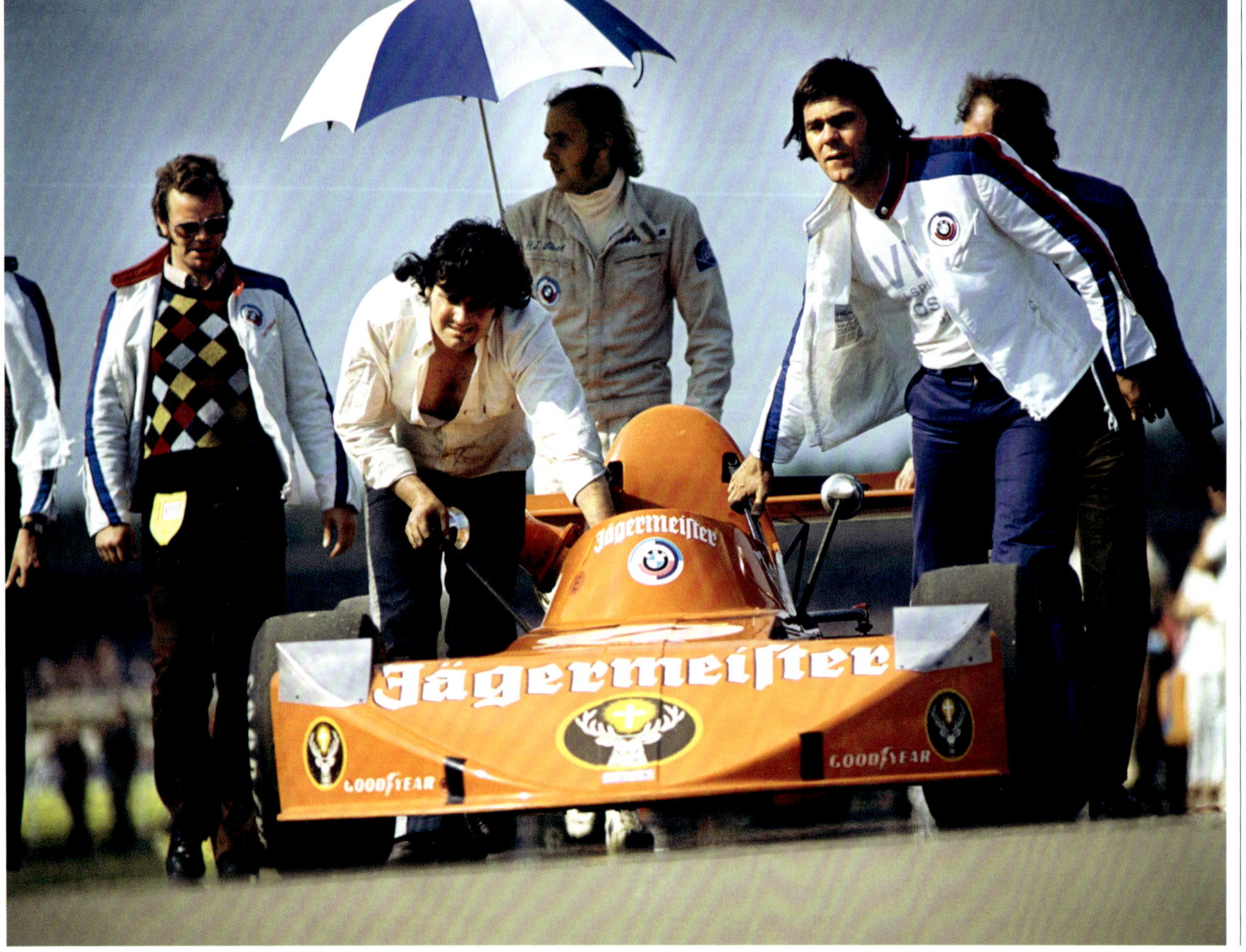

Right: Niki Lauda, who would later become a Formula 1 World Champion, raced for Jägermeister in an Alpina BMW 3.0 CSL in 1973.

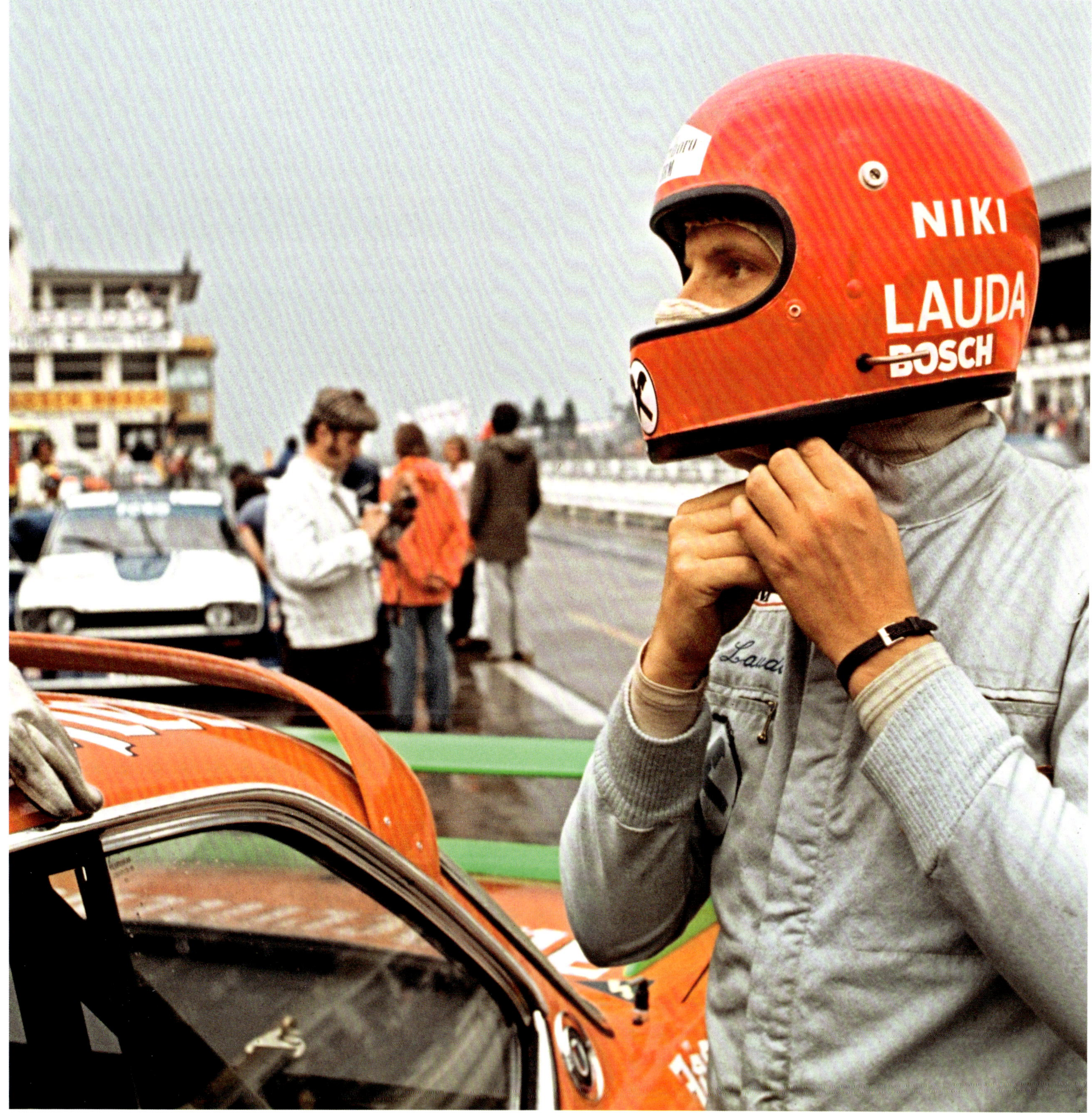

The first car used was a Porsche 914/6 in Jägermeister green. But that wasn't sufficiently spectacular for Günter Mast: 'It's not striking enough. Don't we have
a better colour?'

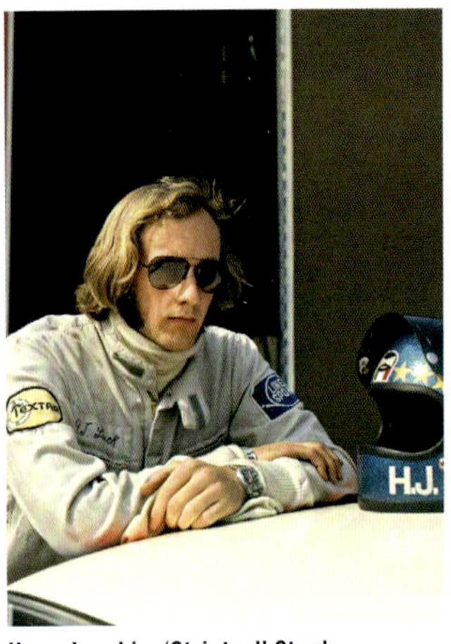

Hans-Joachim 'Strietzel' Stuck

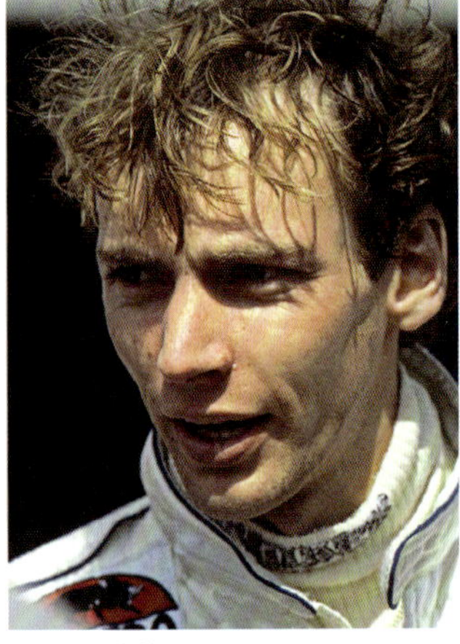

Stefan Bellof

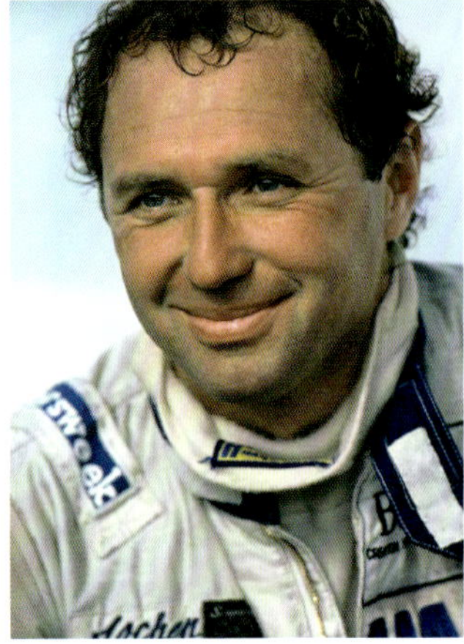

Jochen Mass

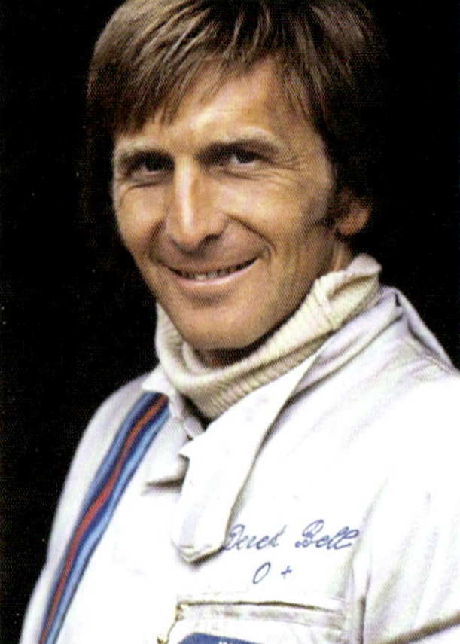

Derek Bell

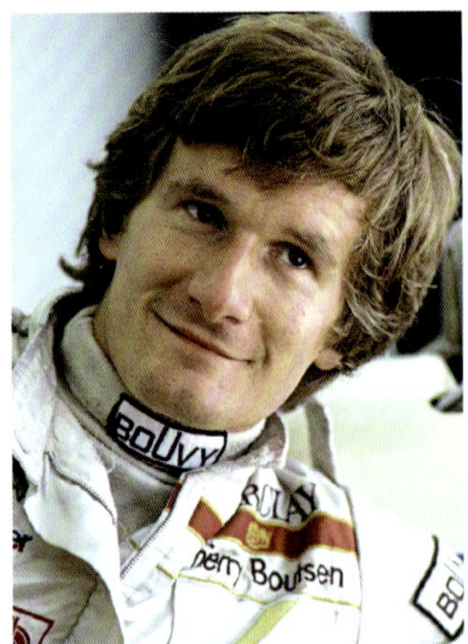

Thierry Boutsen

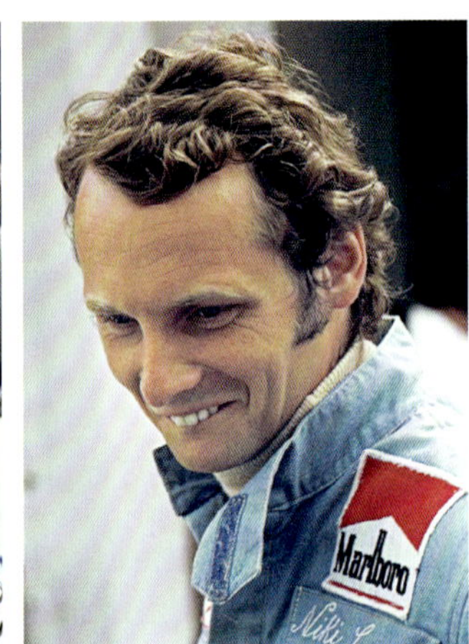

Niki Lauda

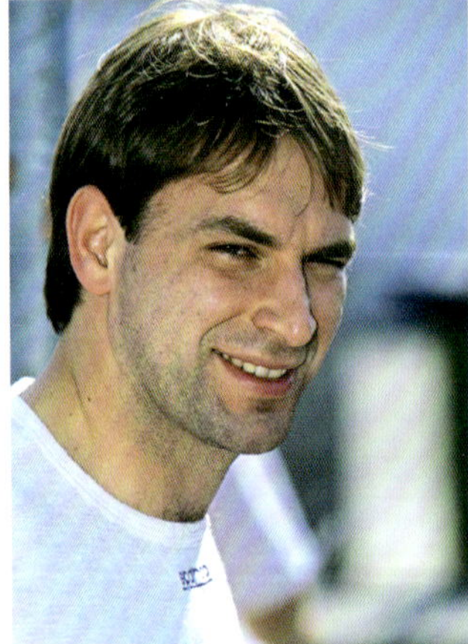

Michael Bartels

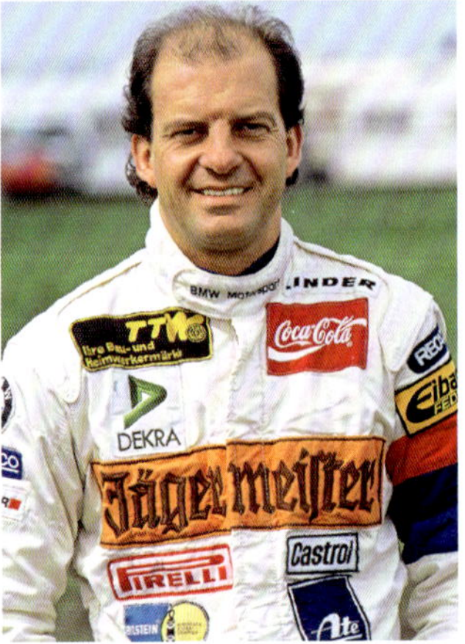

Armin Hahne

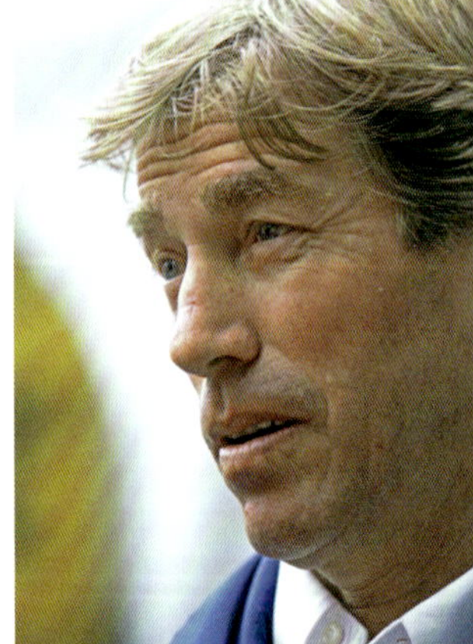

Prince Leopold of Bavaria

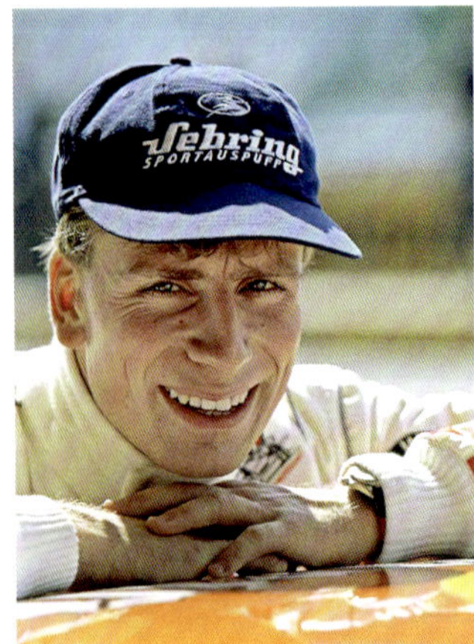

Christian Menzel

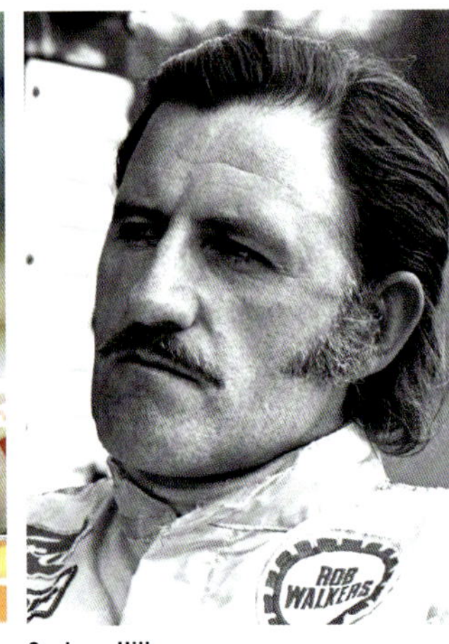

Graham Hill

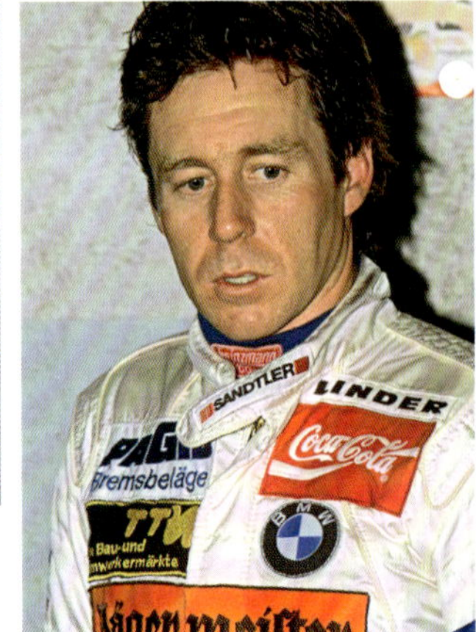

Wayne Gardner

In the 1970s, Europe's Touring Car Championship and the German Racing Championship (known as DRM for Deutsche Rennsport Meisterschaft) were veritable crowd-magnets. It was here that the Alpina BMW 3.0 CSL/3.6 CSL, featuring the Jägermeister stag, served as the springboard for Niki Lauda to join Ferrari. In 1973, he won the Nürburgring 24 Hours race in the Jägermeister BMW, and for the DRM race in Diepholz, he was even flown between the practice sessions and the Formula 1 race at Silverstone in a Cessna chartered by Jägermeister. All the bales of straw used for the chicanes on the start/finish straight were quickly cleared away for the plane to land – something that is difficult to comprehend from a modern-day perspective.

But the first star at the Jägermeister Racing Team was a British racing legend. In 1972, Graham Hill drove a bright orange Brabham-BT 38 in Formula 2, and although he was only able to win one race, the two-time Formula 1 World Champion was a crowd favourite. Between 1974 and 1977, however, Hans-Joachim 'Strietzel' Stuck generated waves of excitement with his Formula 2 appearances in the Jägermeister BMW and Jägermeister March – especially at the Hockenheimring. As soon as his car got going, firecrackers would be set off across the sold-out racing circuit. He won there eight times in three years, so it's no wonder the fans fêted him as the 'King of Hockenheim'.

The 80s were all about Group C. From 1984 to 1990, Jägermeister partnered Swiss team Brun and achieved great success, with the partnership securing the World Sportscar Championship in 1986. At the time, Stuck, Bellof, Mass and many other drivers were pitting their pancake-flat Porsche cars against the likes of Jaguar, Sauber-Mercedes and Lancia. Stuck had already competed for the Brun team in a Jägermeister BMW 635 CSI at the first edition of the German Touring Car Championship (DTM) two years before, and had been runner-up in the German Sportscar Championship in a Porsche 956 in 1985. In the Supercup, meanwhile, a varying array of drivers, including Derek Bell and Thierry Boutsen, took to the wheel of the Jägermeister Porsche.

The DTM reached its peak in the 1990s – and, consequently, so too did the orange Jägermeister cars, initially with the BMW-Linder team in 1992. Alongside Armin Hahne, world motorcycle champion Wayne Gardner also conquered the hearts of racing fans there with his gutsy driving style, and, in 1994, Michael Bartels similarly engaged in exciting duels with Mercedes and Opel in his Alfa Romeo. When the DTM was stopped in 1996, the motorsport side of things at Jägermeister also fell quiet. The year of 1997 was the first since 1972 in which the company had not been involved in any motorracing activities – until Christian Menzel and Prince Leopold 'Poldi' of Bavaria entered the Super Touring Championship with their Jägermeister BMW 320i in 1998. When a new DTM series was established in 2000, Jägermeister was initially part of it with Opel, before pulling out of German motorsport soon after.

Fans and enthusiasts can today still admire the famous race cars, with the Hubertus stag head, at historic racing events – no longer in the fast lane, perhaps, but certainly under the fabled and emotive banner of '72STAGPOWER – The Spirit of Jägermeister Racing'. Many of the Jägermeister race cars have found their way back home here. Fancy a sneak peak? www.72stagpower.com. ❦

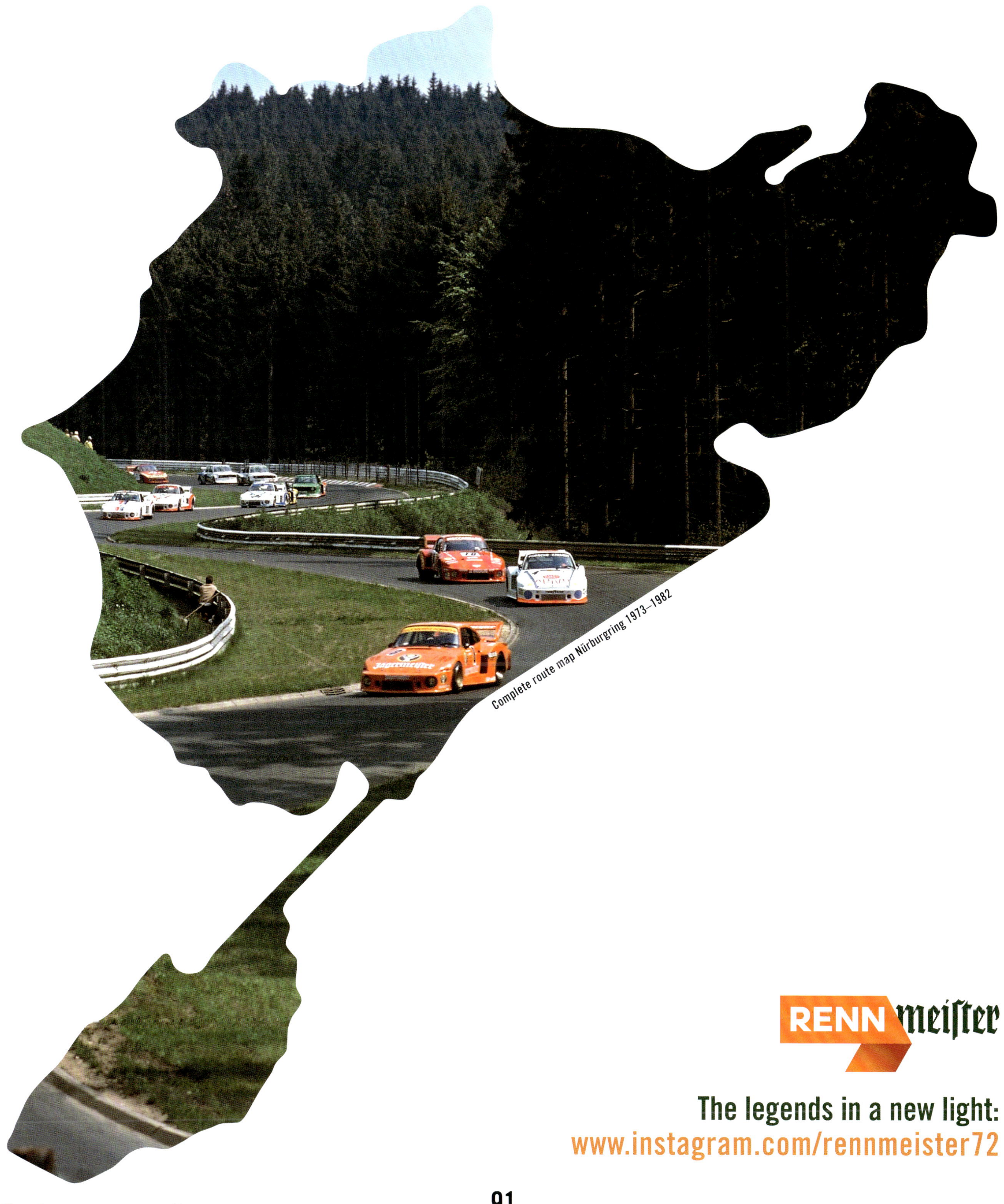

Complete route map Nürburgring 1973–1982
RENN meifter
The legends in a new light:
www.instagram.com/rennmeister72

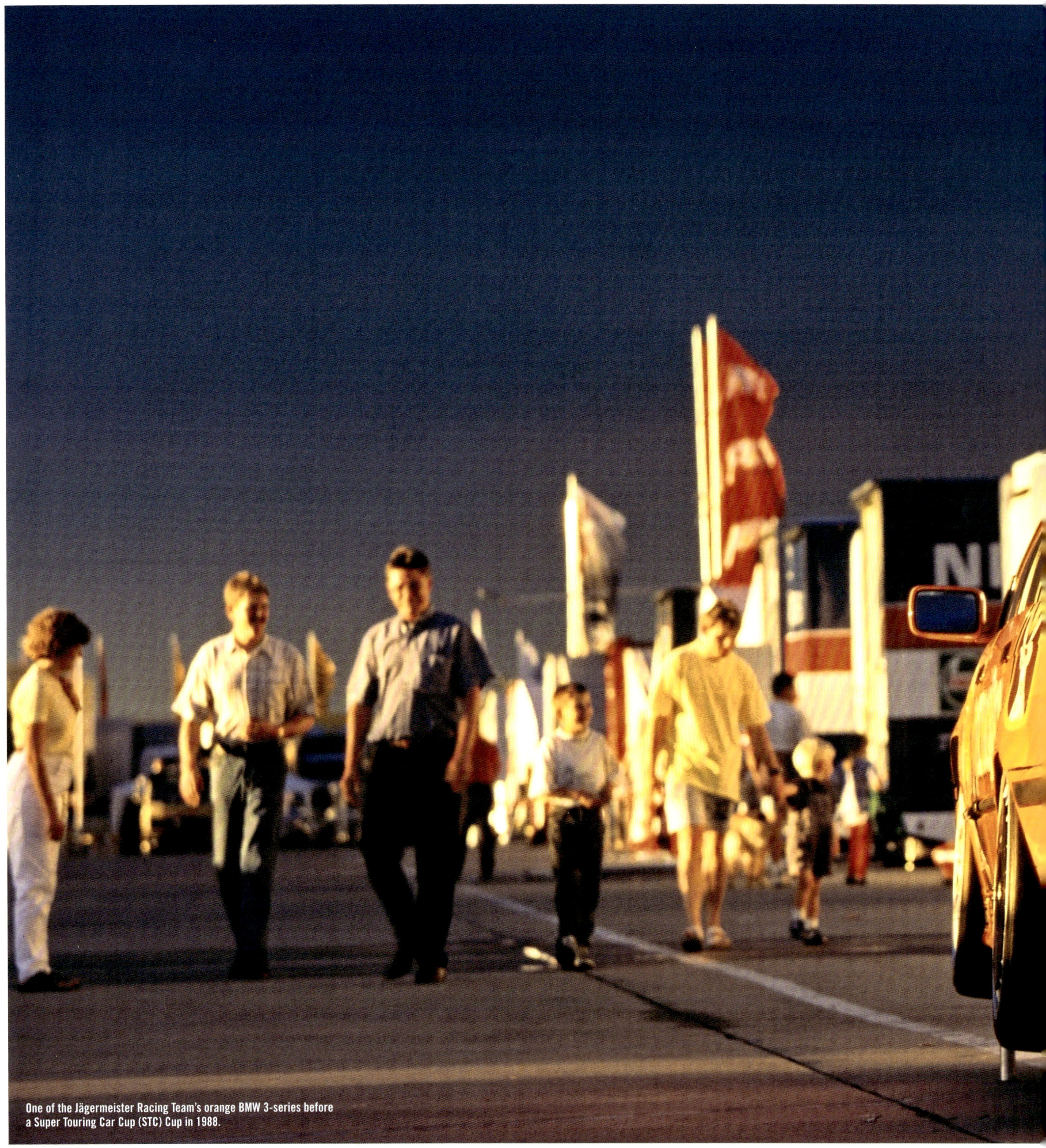

One of the Jägermeister Racing Team's orange BMW 3-series before
a Super Touring Car Cup (STC) Cup in 1988.

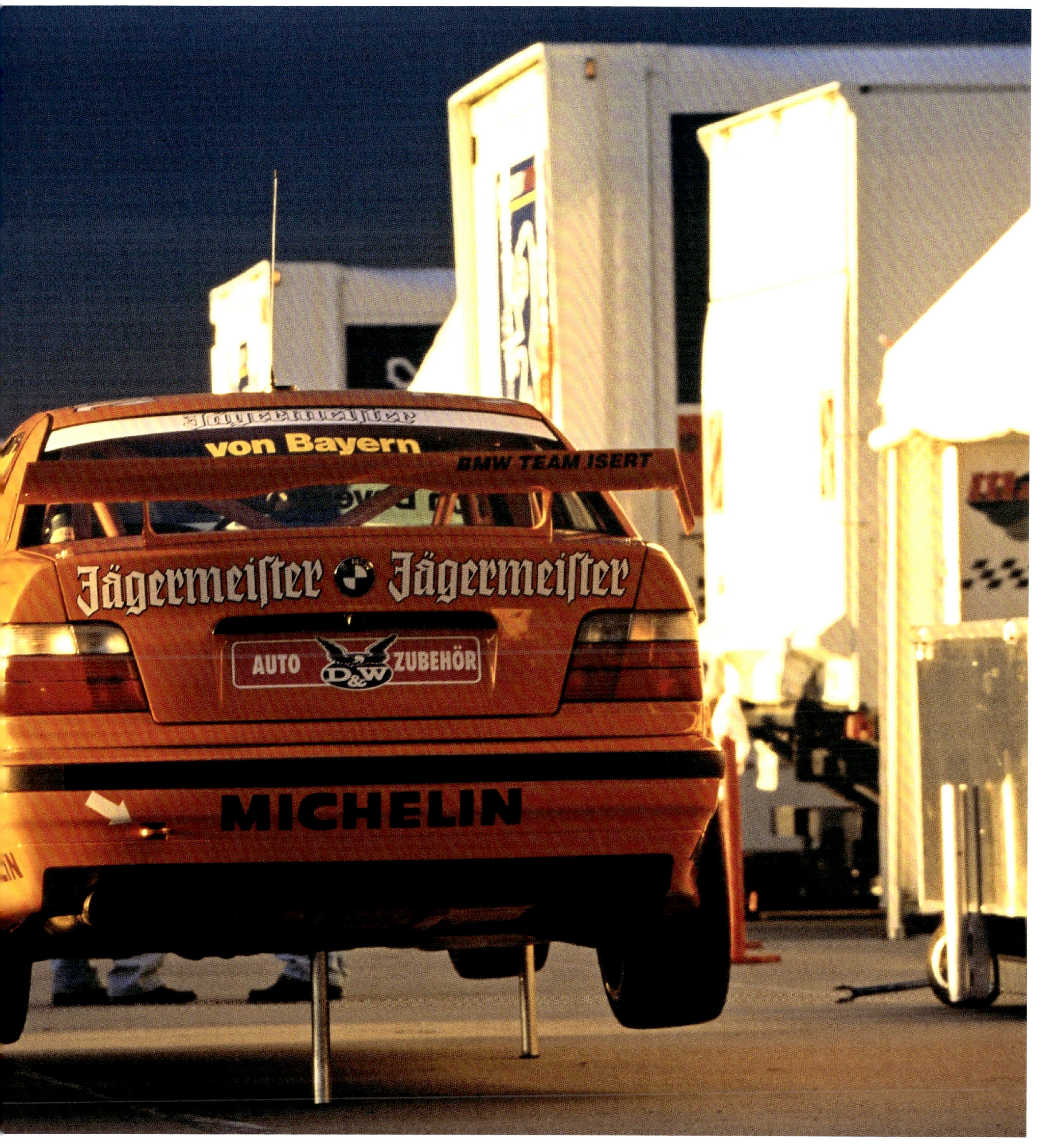
Jägermeister
von Bayern
BMW TEAM ISERT
Jägermeister Jägermeister
AUTO D&W ZUBEHÖR
MICHELIN

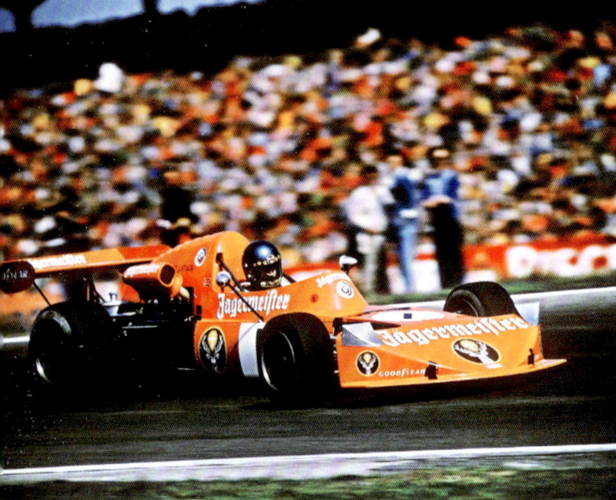

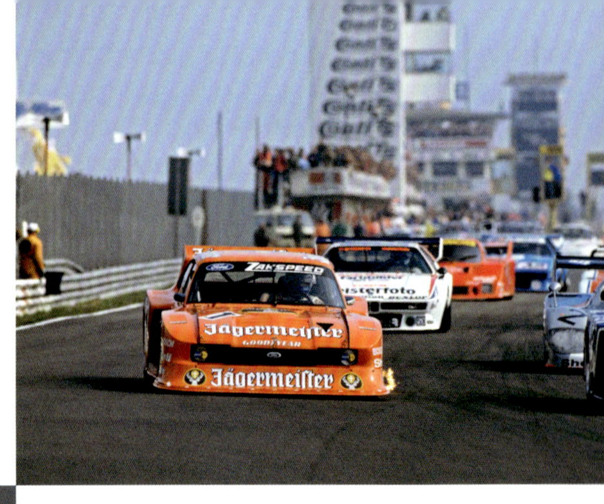

▼ Car: BMW M3
               Type Sport Evolution
Weight: 980 kg
Capacity: 2.5 litres
Power: 370 bhp
Top speed: 180 mph

▲ Car: March-BMW
Weight: 450 kg
Capacity: 2.0 litres
Power: 205 bhp

▲ Car: Ford Super Capri
Weight: 780 kg
Capacity: 1.7 litres
Power: 600 bhp
Top speed: 186 mph

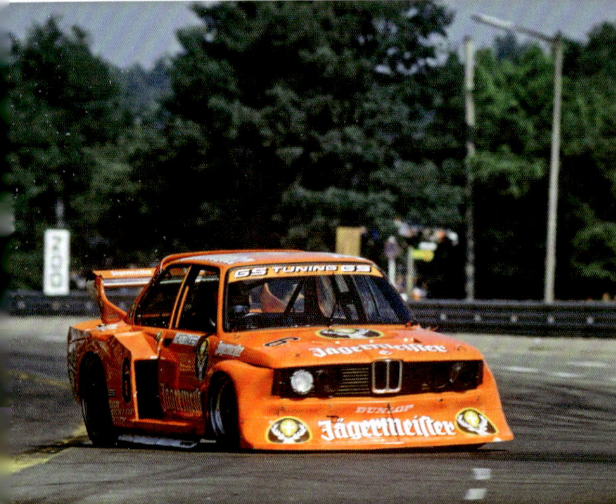

▼ Car: Porsche Carrera RSR
Weight: 800 kg
Capacity: 3.0 litres
Power: 350 bhp
Top speed: 180 mph

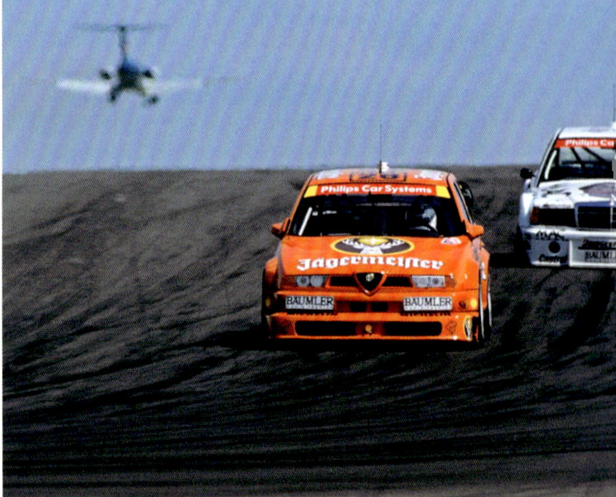

▲ Car: BMW 320 Group 5
Weight: 765 kg
Capacity: 2.0 litres
Power: 205 bhp (Sauger) – 341 bhp (Turbo)

▲ Car: Alfa Romeo 155 V6 TI
Capacity: 2.5 litres
Power: 500 bhp
Top speed: 186 mph

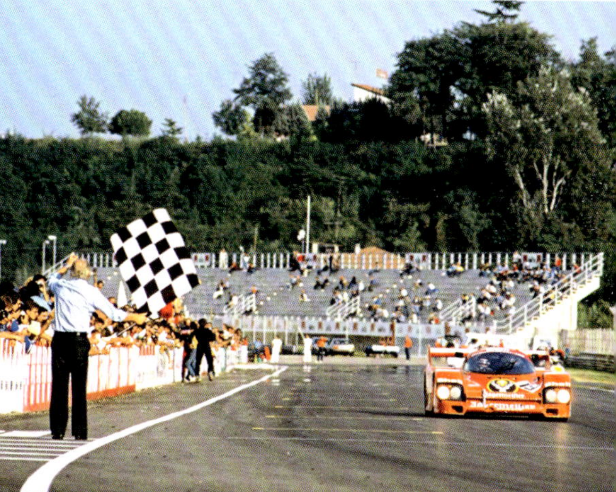

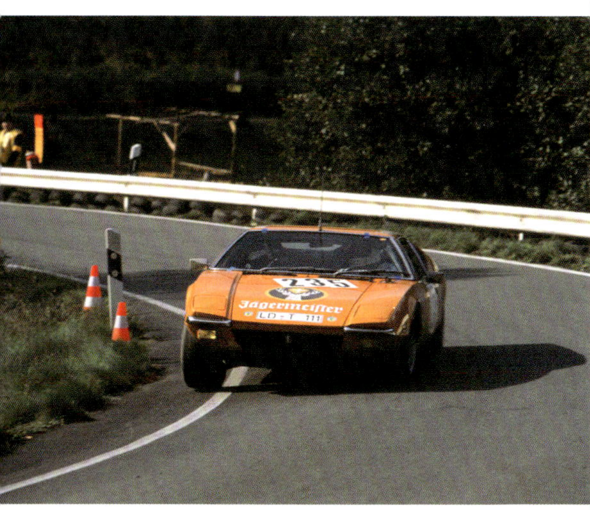

▲ Car: BMW 3.0 CSL
  Weight: 1080 kg
  Capacity: 3.3 litres
  Power: 360 bhp
  Top speed: 177 mph

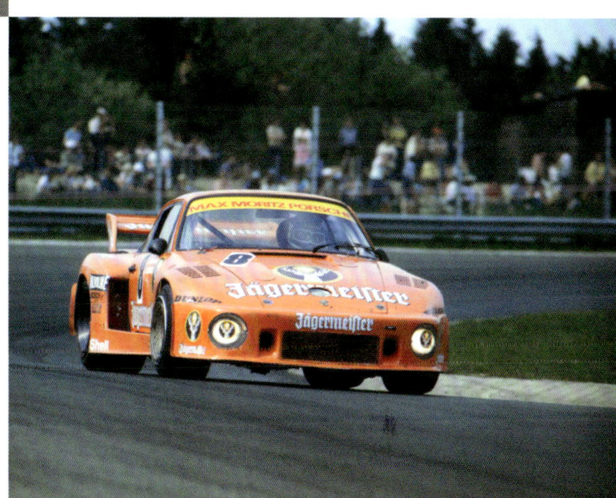

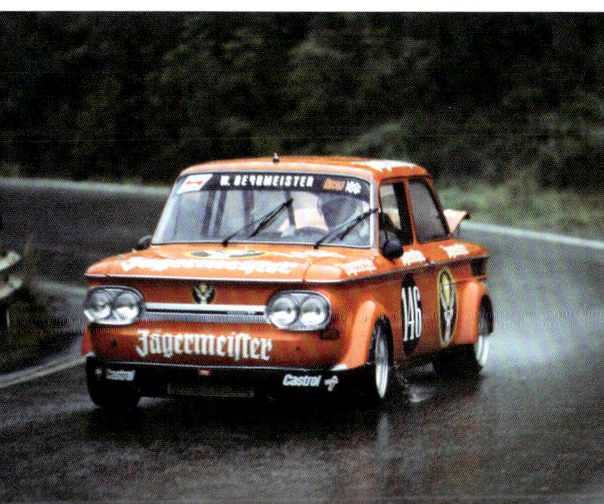

▲ Car: Porsche 935 Turbo
  Power: up to 800 bhp
  Top speed: up to 239 mph

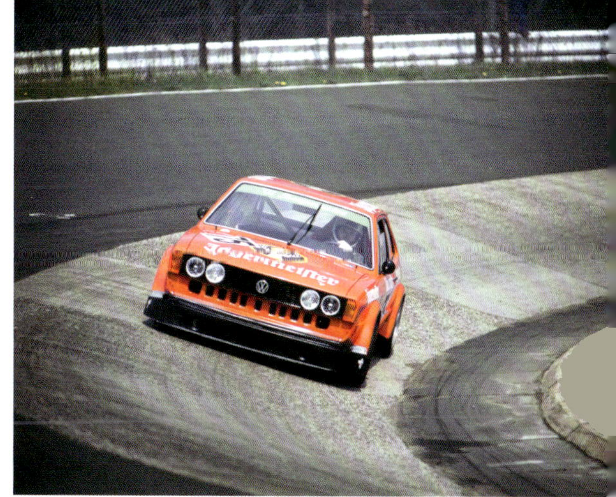

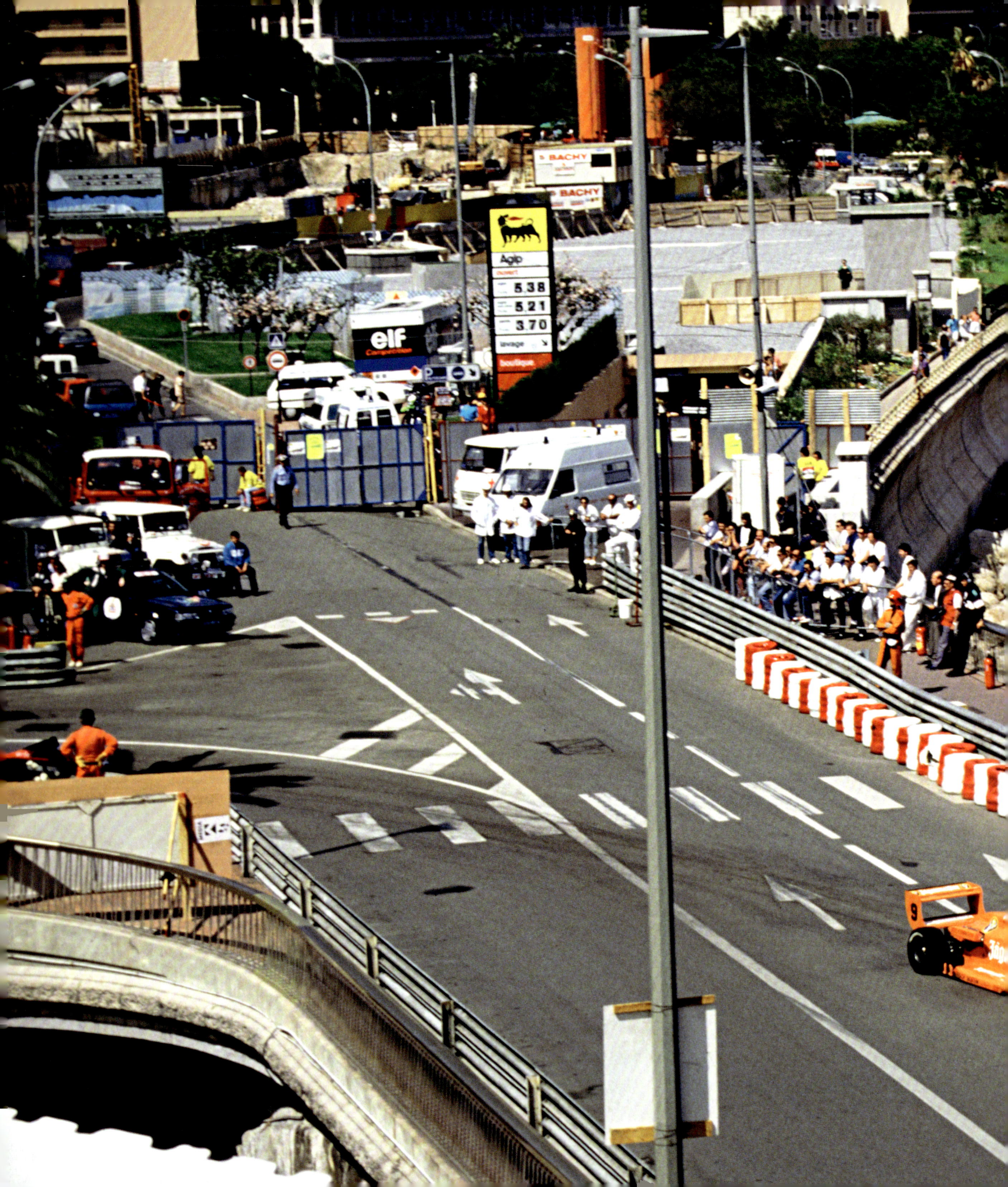

BACHY
BACHY
Agip
5.38
5.21
3.70
lavage
boutique
elf
9

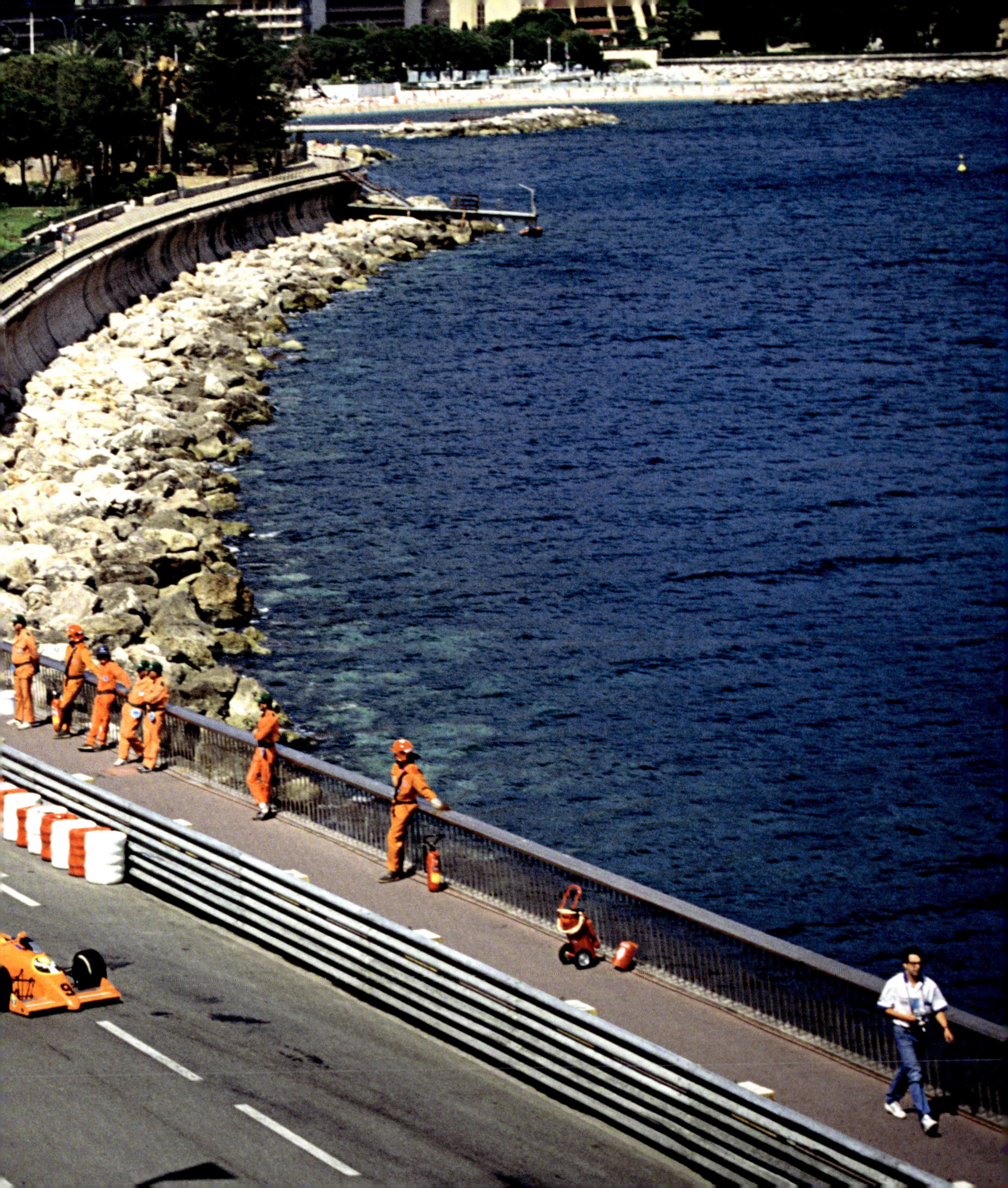

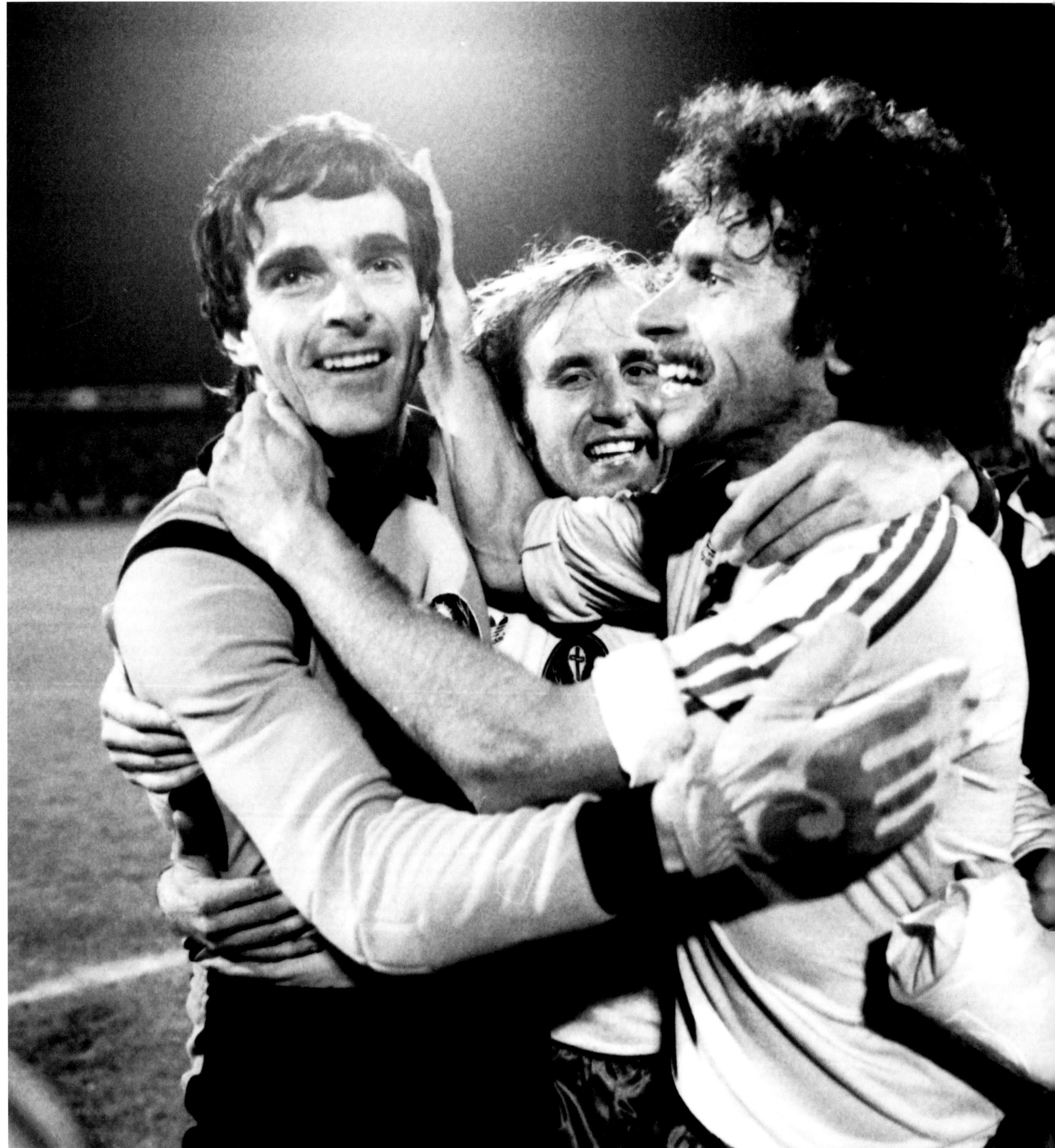

'WE NEED TO RAISE DUST WITH OUR ADVERTISING'

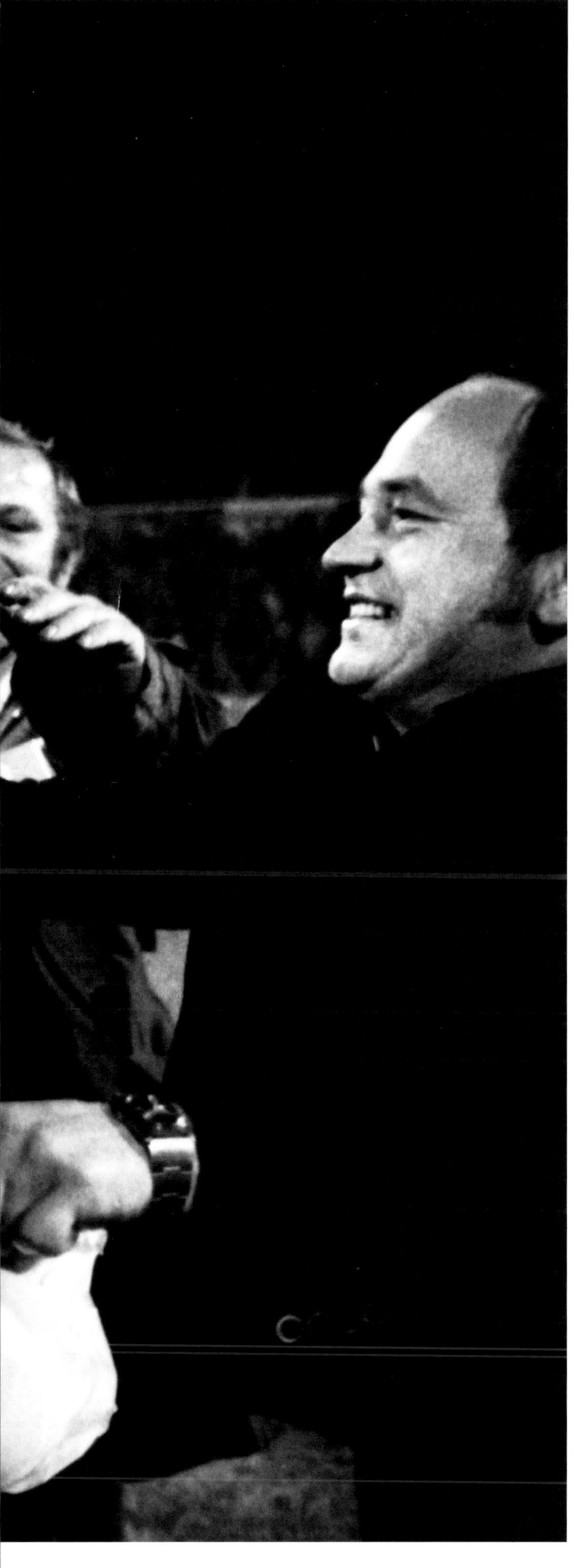

Radiant victors: Bernd Franke (left) was the goalkeeper at Eintracht Braunschweig for 14 years, while global star Paul Breitner (middle) was signed by Jägermeister boss Günter Mast for the 77/78 season.

Sport sponsorship is a million-pound business, and jersey advertising is one of the most important sources of income in modern football. In times past, however, players' chests were considered an advertising-free zone – until Günter Mast sparked a commercial revolution with Jägermeister and Eintracht Braunschweig in 1973.

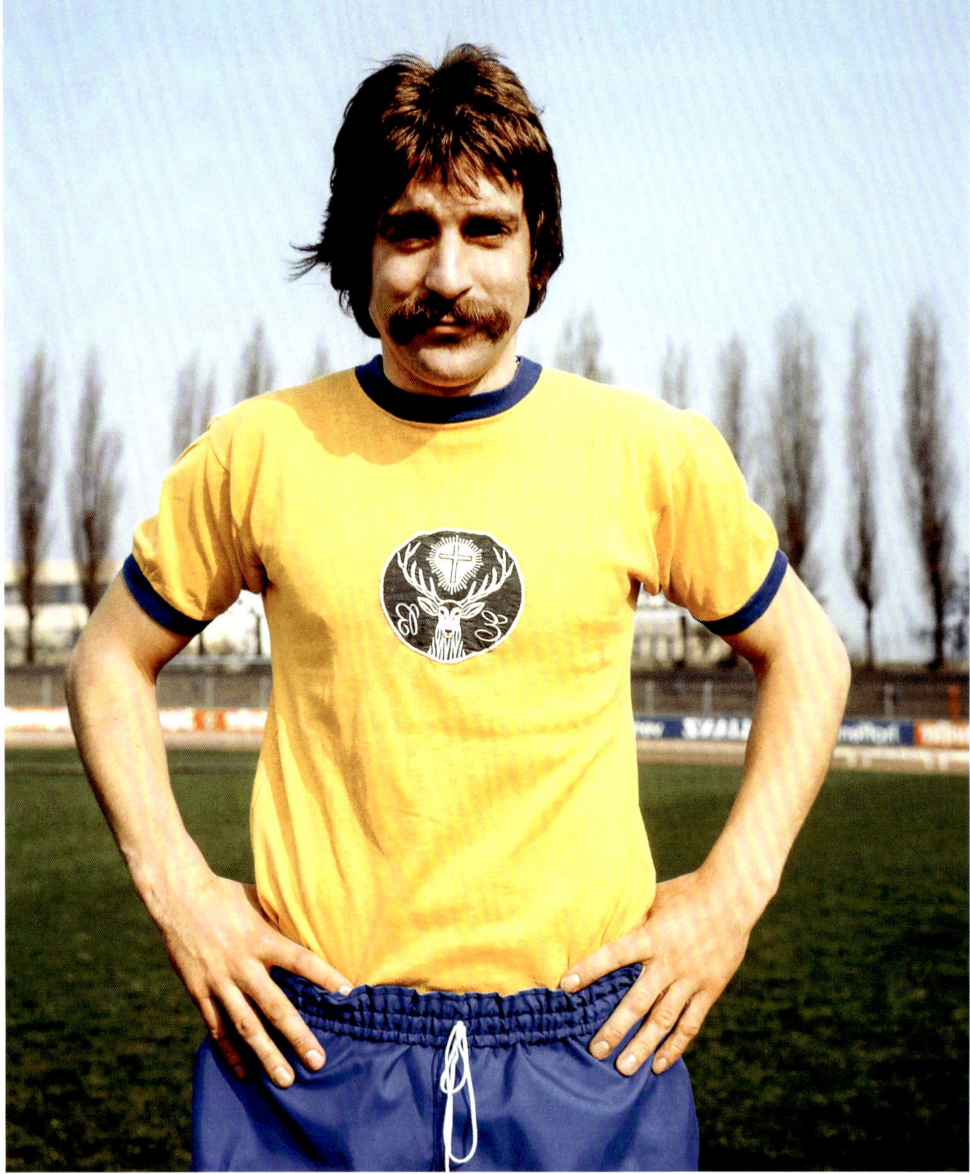

It all began on a summer's day in Braunschweig. Jägermeister's lawyer had called his most important clients to a barbecue – 'Seven, eight men from the business world,' Günter Mast later recalled in the *Süddeutsche Zeitung*. 'I noticed one left and didn't come back. Another also left and failed to return.' The Jägermeister boss, who had no interest in football, eventually found his guests in front of the television in the kitchen, eagerly watching Germany's game against England. 'That's when I realized that my view that football was only for the lower classes in Germany was incorrect. Football can be used to reach all levels of society,' says the entrepreneur. A few years before, he had already declared that 'we need to raise dust with our advertising'. And even back then, in 1967, he firmly believed that whether or not the masses reacted positively was not a critical factor. 'The impact of our advertising most certainly gets reinforced as long as the advertising is simply talked about more.' But even he, the marketing visionary, could never have imagined then that his observations during a barbecue would lay the foundations for commercializing Germany's football Bundesliga.

In 1972, a chance encounter with Ernst 'Balduin' Fricke, president of Bundesliga club Eintracht Braunschweig, ultimately saw the company patriarch's vision become a business model. An ominous hole in its funds had forced the club to tap into new business fields and consider co-operating with companies. That same year, Braunschweig-born Günter Mast finally agreed to inject 100,000 German marks' worth of funding every year for the next five years. But not without asking Fricke for something in return.

He wanted the Braunschweig lion that had adorned the Eintracht players' jerseys for 78 years to be replaced by an 18cm Hubertus stag. This request was as brazen as it was shrewd. While the German Football Association had expressly prohibited another club from featuring advertising on its jersey just five years earlier, the master had found a loophole: nowhere was there any mention of club coats of arms. To secure its future, the indebted club agreed to participate in the bold coup. And, on 9 January 1973, the members approved, by majority vote, the request to change the Eintracht coat of arms. Three weeks later, the new jersey would make its debut appearance at the club's home game against Kickers Offenbach. But Mast and Fricke, who had now become close friends, had proceeded without the all-clear from the association. Shortly before kick off, the German Football Association sent a fax banning the use of the Jägermeister logo – and the Eintracht players had to take to the field in their old jerseys.

This veto by the German Football Association triggered a fierce power struggle that had the entire country talking for weeks. The association stipulated that the club's initials had to be integrated into the Jägermeister logo, and that the maximum diameter of the

He wanted the Braunschweig lion that had adorned the Eintracht players' jerseys for 78 years to be replaced by an 18cm Hubertus stag.

Hubertus stag had to be reduced to 14cm – which, in Günter Mast's view, was 4cm too small. The dispute was perfect fodder for the media – and, as such, the

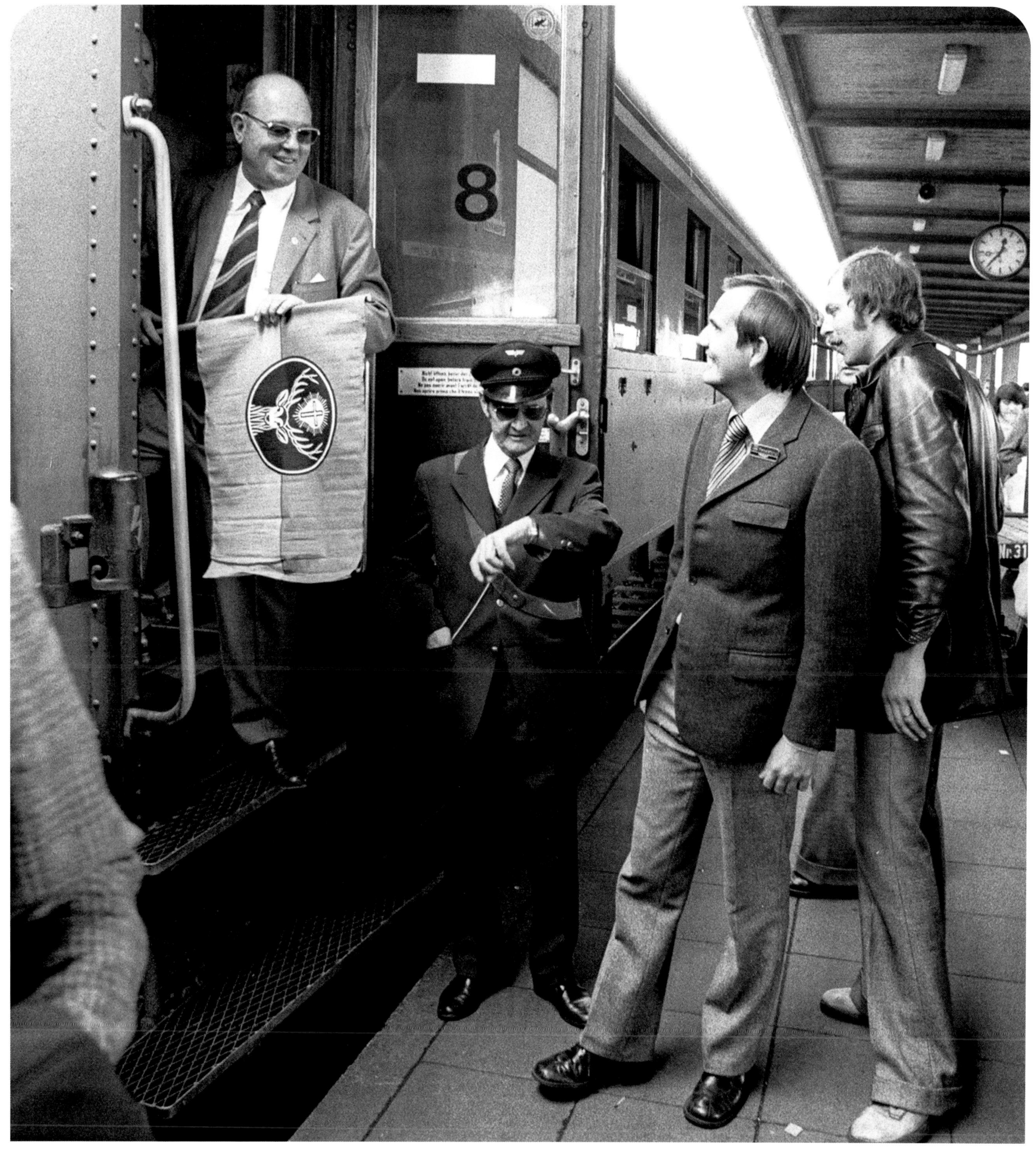

All aboard! Günter Mast paid for 4,000 Eintracht fans to make a special train trip to the team's away game against Nuremberg on 9 May 1974. <u>Left:</u> Casual but ambitious: Eintracht captain Bernd Gersdorff.

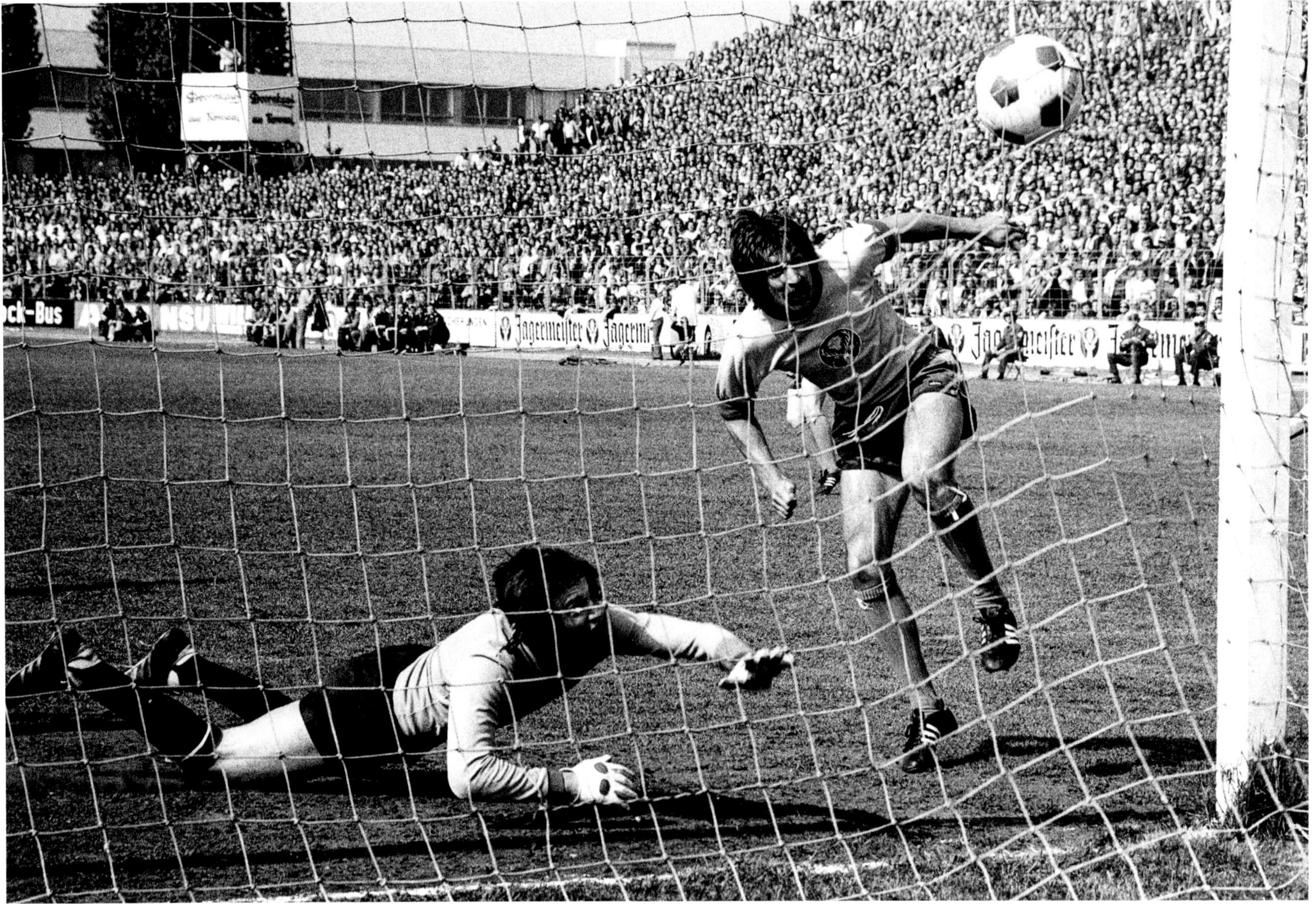

Legendary centre-forward Bernd Gersdorff kicked (and headed) Eintracht back into the Bundesliga with an incredible 35 goals in 19 games.
Right: Eintracht patron Günter Mast in his office.

entrepreneur had already won, creating quite the stir before a single Eintracht player had ever donned the stag-bearing jersey. After weeks of correspondence between Jägermeister, the club and the German Football Association, an agreement was finally reached. On 27 February 1973, the German Football Association grudgingly consented, and Eintracht Braun-

'[Mast] probably took the most ground-breaking step ever in the history of the Bundesliga,' said football legend Paul Breitner.

schweig created history as the first Bundesliga team to have a jersey sponsor. The Eintracht Braunschweig players first took to the field in their new jersey against Schalke 04 on 24 March 1973, with the club boss con-

fidently declaring that 'the Braunschweig example will catch on.'

He would prove to be right. Other clubs followed suit the very next season, and, five years later, every Bundesliga team had a jersey sponsor. 'In doing what he did, [Mast] probably took the most ground-breaking step ever in the history of the Bundesliga,' said football legend Paul Breitner, who was the source of Günter Mast's next coup. Mast managed to lure the superstar from Real Madrid to Braunschweig in 1977, for a transfer fee of 1.75 million German marks. For more than three decades, this would remain the highest transfer fee paid by Eintracht, and it is still probably the most spectacular. From then on, Mast increased his involvement in the club, including after the death of his friend and Eintracht president, Balduin Fricke, in 1978. The entrepreneur never made any secret of his motives. He was always about rais-

ing his brand's profile – and, ten years after his revolutionary jersey deal, he finally planned his next daring exploit: renaming Eintracht Braunschweig as SV Jägermeister Braunschweig.

Just a few weeks earlier, the German Football Association had decided that product names could not be incorporated into club names – but that didn't seem to bother the company boss. He was convinced he could get the matter settled in court. But the Regional and Higher Regional Courts backed the German Football Association's decision, so the businessman turned to the German Federal Supreme Court as a last resort. And this court surprisingly did end up ruling in Jägermeister's favour on 17 November 1986. Once again, Mast had managed to achieve something no one had thought possible. Once again, it attracted nationwide media coverage, and once again, the Jägermeister brand was on everyone's lips.

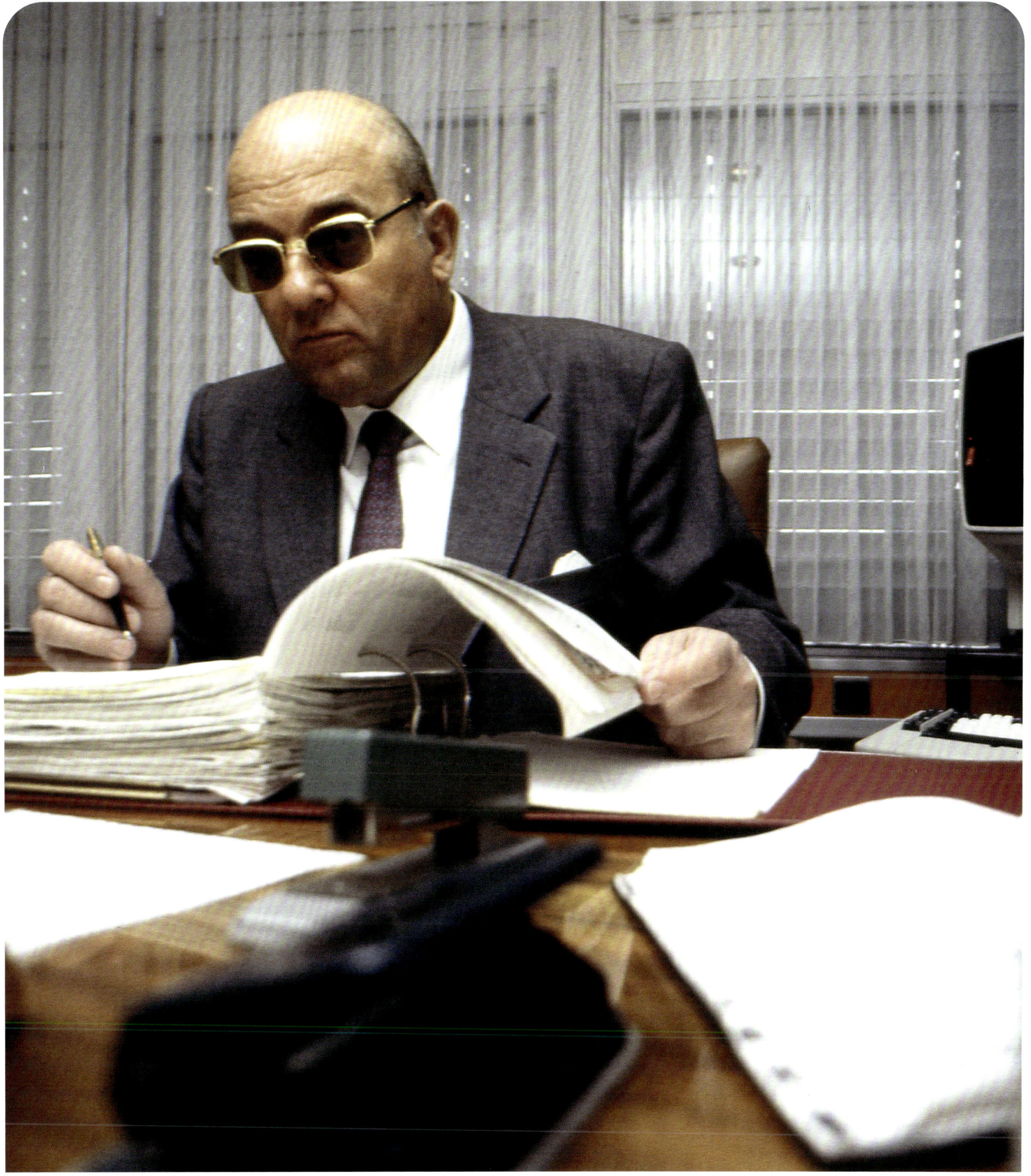

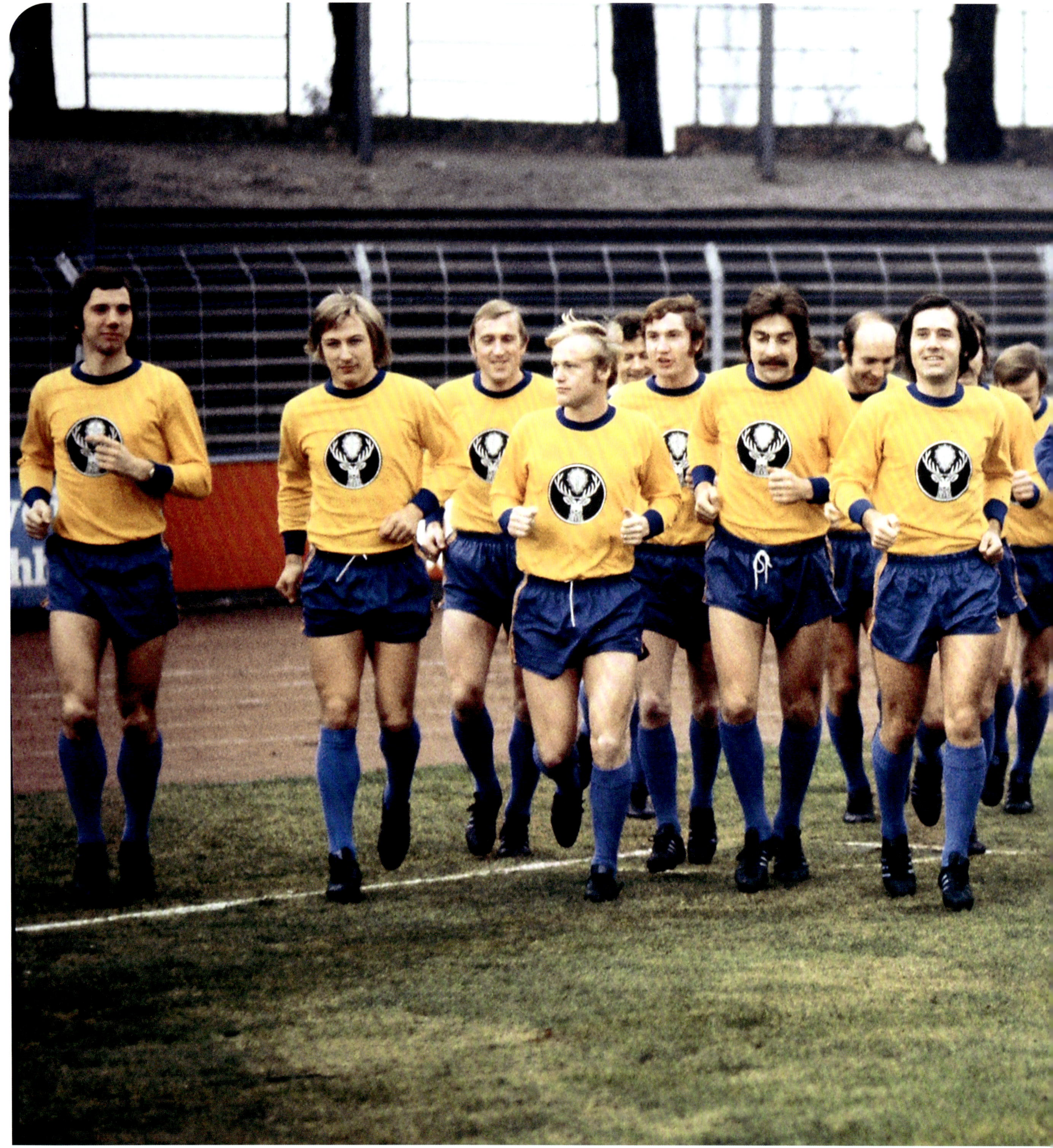

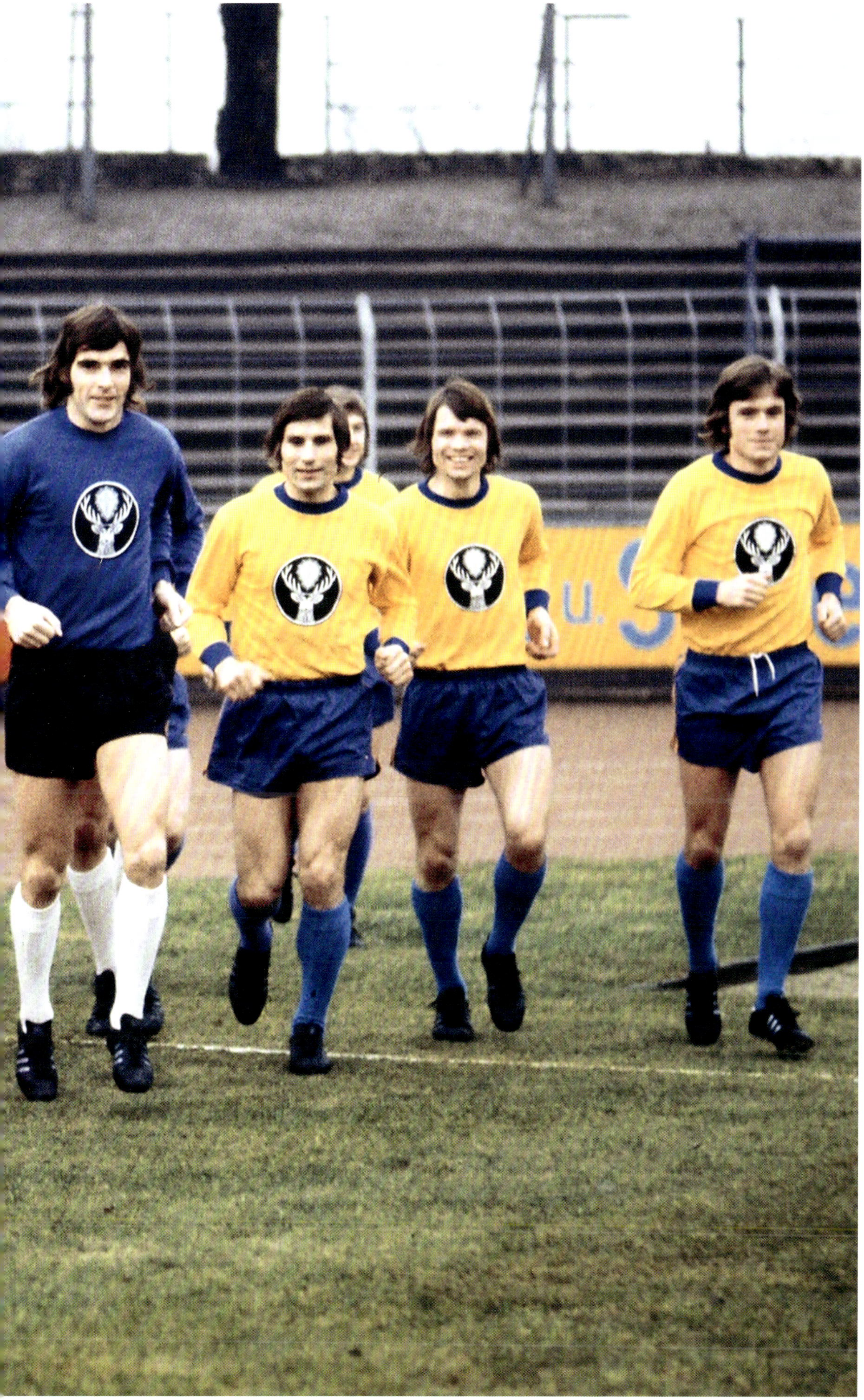

But the club name was never changed. The Lower Saxon Football Association had threatened to disqualify all junior teams of the future SV Jägermeister Braunschweig from playing. And Mast finally agreed to give up on the name-change idea. Vision, yes, but not at all costs. And although the journey shared by Jägermeister and Eintracht Braunschweig was not always easy, the pair were able to break completely new ground. No other team in history had committed itself so prominently to its sponsor, which remained with Braunschweig for a full 15 years. And no one else identified football's tremendous marketing potential as early and clearly as Günter Mast did. His advertising deal between Jägermeister and Eintracht Braunschweig not only saw the entrepreneur create a sensation, but it also laid the foundations for a successful formula that now earns Bundesliga clubs millions. A matter of course today, but revolutionary at the time. ⚜

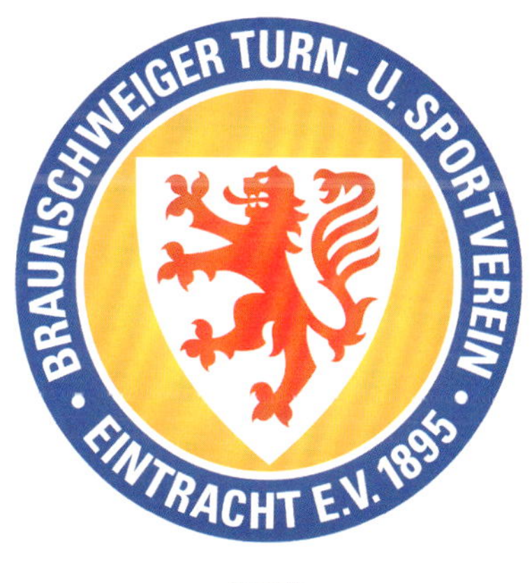

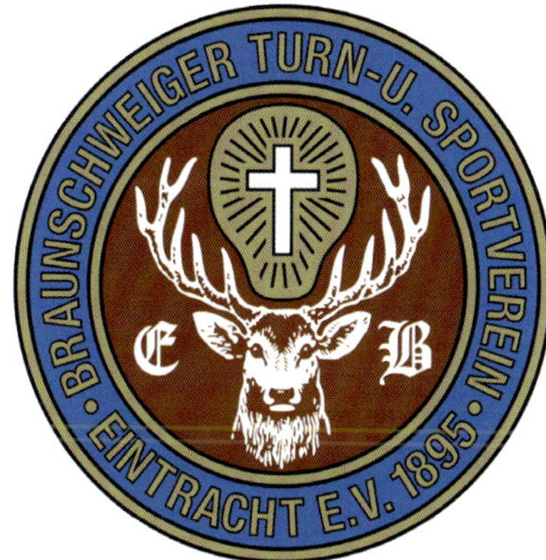

Sometimes a lion, sometimes a Hubertus stag: The Eintracht Braunschweig club logos. <u>Left:</u> The Eintracht team first wore their new Bundesliga jerseys in 1973.

Deutscher Fußball-Bund

6 Frankfurt/M. 90, Zeppelinallee 77
Fernruf: 770568
Drahtanschrift: Fußball
Postschließfach 900260
Bankverbindung: Dresdner Bank, Frankfurt/Main, Nr. 906992
Postscheckkonto Frankfurt/Main Nr. 87205-606
Fernschreiber 041-2500

Braunschweiger Turn- und
Sportverein
Eintracht v. 1895 e.V.

33 Braunschweig
Hamburger Str. 210

Ihre Nachricht vom	Ihr Zeichen	Unser Zeichen	Tag
		Str/bi	27. Februar 1973

Betr.: Vereinsabzeichen

Bezug: Ihre Vorlage mit Schreiben vom 21. 2. 1973

Lieber Sportkamerad Fricke,

unter Bezugnahme auf Ihr Schreiben vom 21. 2. 1973 sowie auf den
diesem Schreiben beigelegten Entwurf über die Ausgestaltung
Ihres Vereinsabzeichens - Wappenbild Hubertus-Hirsch - teilen
wir Ihnen im Auftrage des DFB-Präsidiums mit, daß hierfür die
Genehmigung erteilt wird.

Wir machen darauf aufmerksam, daß die Genehmigung davon abhängig
gemacht wird, daß die angegebenen Maße von 3 x 1 1/2 cm nicht
unterschritten werden. Sie werden gebeten, nach Anfertigung der
Originalvereinsabzeichen, so wie sie auf dem Trikot angebracht
werden, fünf Exemplare an die Geschäftsstelle des DFB einzu-
reichen.

Mit freundlichen Grüßen
DEUTSCHER FUSSBALL-BUND
- Der Generalsekretär-

Paßlack

D/DFB-Präsidium

An den
Braunschweiger Turn-
und Sportverein
EINTRACHT von 1895 e.V.

3300 Braunschweig
Hamburger Straße 210

Ma/Bs   Sekretariat 28. 3. 1973

Betr.: Hubertushirschkopf

Sehr geehrte Frau Martini!

Zur Weiterleitung an den DFB übersenden wir Ihnen in der Anlage
in fünffacher Ausfertigung eine exakte 1 : 1 - Wiedergabe des
Hubertushirschkopfes, wie er sich nunmehr auf den Trikots Ihrer
Bundesligamannschaft befindet.

Das zur Weiterleitung nach Frankfurt bestimmte Trikot ist ja
bereits bei Ihnen.

                        Mit freundlichen Grüßen
                        W. MAST   KOMMANDITGESELLSCHAFT
                        Jägermeister-Spirituosenfabrik

Anlagen

JÄGERMEISTER
OPEN AIR

With groovy rhythms, phat riffs, history-making shows, mud-wrestling, mosh-pits, thunder and lightning, the festival madness is off the charts. Every year, millions of party-going music fans around the world head to fields and outdoor venues to celebrate their heroes and the unique atmosphere, whether it be in front of a stage, at a camp site or over a Jägermeister – because the German herbal liqueur has been an institution at epic events

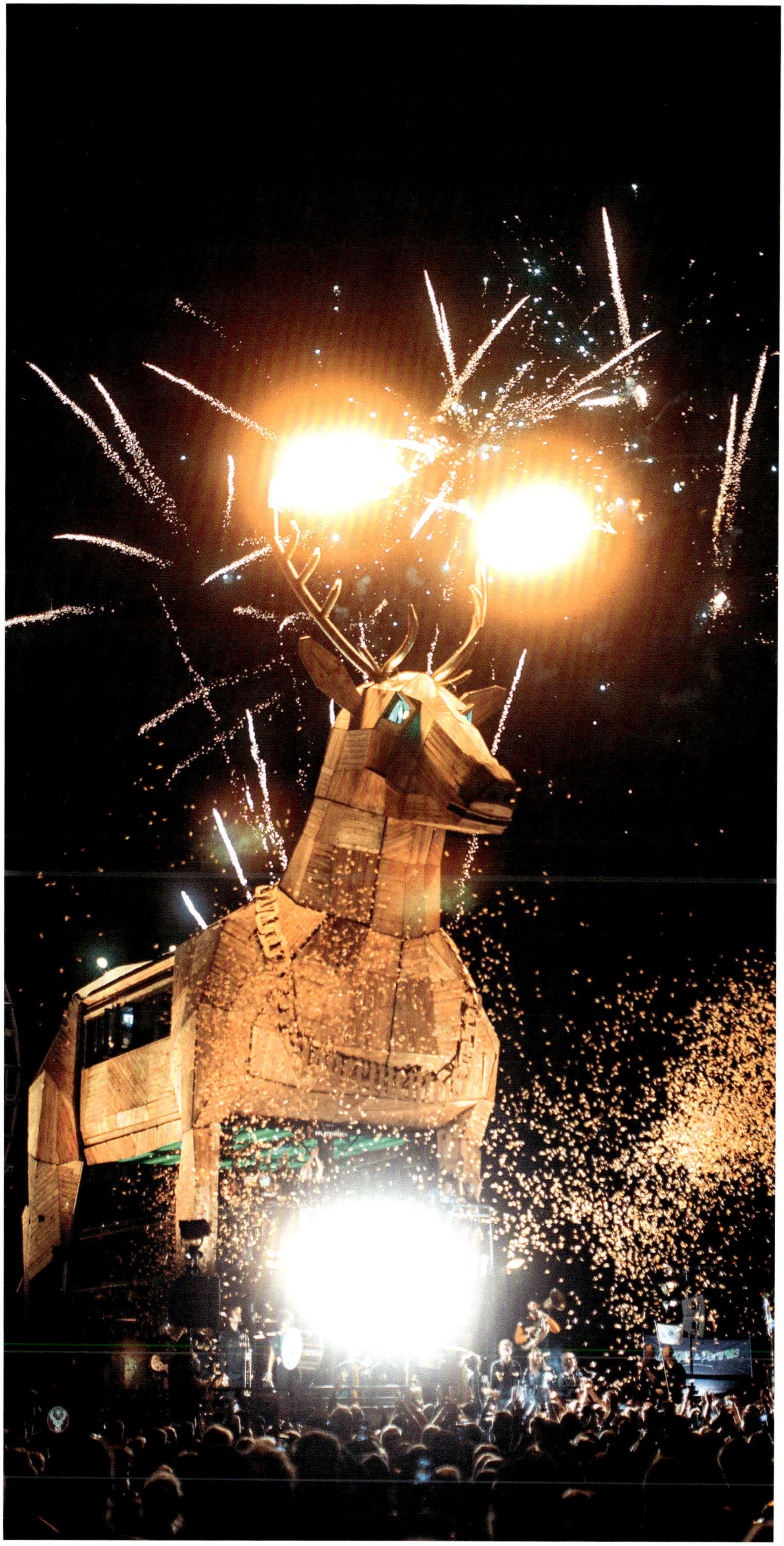

Festival veterans have long known that, while rocking out, grooving to the beat, or head-banging to Slayer is all well and good, the best stories — the ones you tell your children later on (once they are adults, obviously) — don't necessarily unfold in front of the stage. Anyone who attends a festival may indeed come home unwashed and tired, but they will also bring back with them a load of wild anecdotes and amazing memories.

Most certainly there will be a few special Jägermeister moments too, such as those involving the Jägermeister Platzhirsch, the beverage's emblematic animal, made of wood, with flaming antlers and smoking nostrils, towering 17 metres above the festival grounds at events like the Wacken Open Air — the 'world's hardest field'. 'Let's meet afterwards at the wooden deer' was a common phrase uttered by attendees there and elsewhere, before heading into the heaving crowd and temporarily losing sight of each other. The same can be said of the many other genius festival tools Jägermeister has used to cause a stir everywhere from Roskilde to Las Vegas, such as the Jäger Bodega, Haus 56 and the Jager Truck. All of these served as popular meeting points, but also as places where chance encounters resulted in genuine friendships.

In rain, mud or searing heat, to the tune of heavy metal or EDM, on the Wacken's field or at the Electric Daisy Carnival, festivals have to be enjoyed and celebrated regardless of what they throw up. And this is always best done with an ice-cold shot in hand.

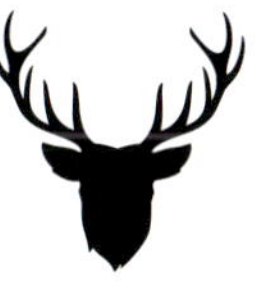

WACKEN OPEN AIR
BACK IN BLACK

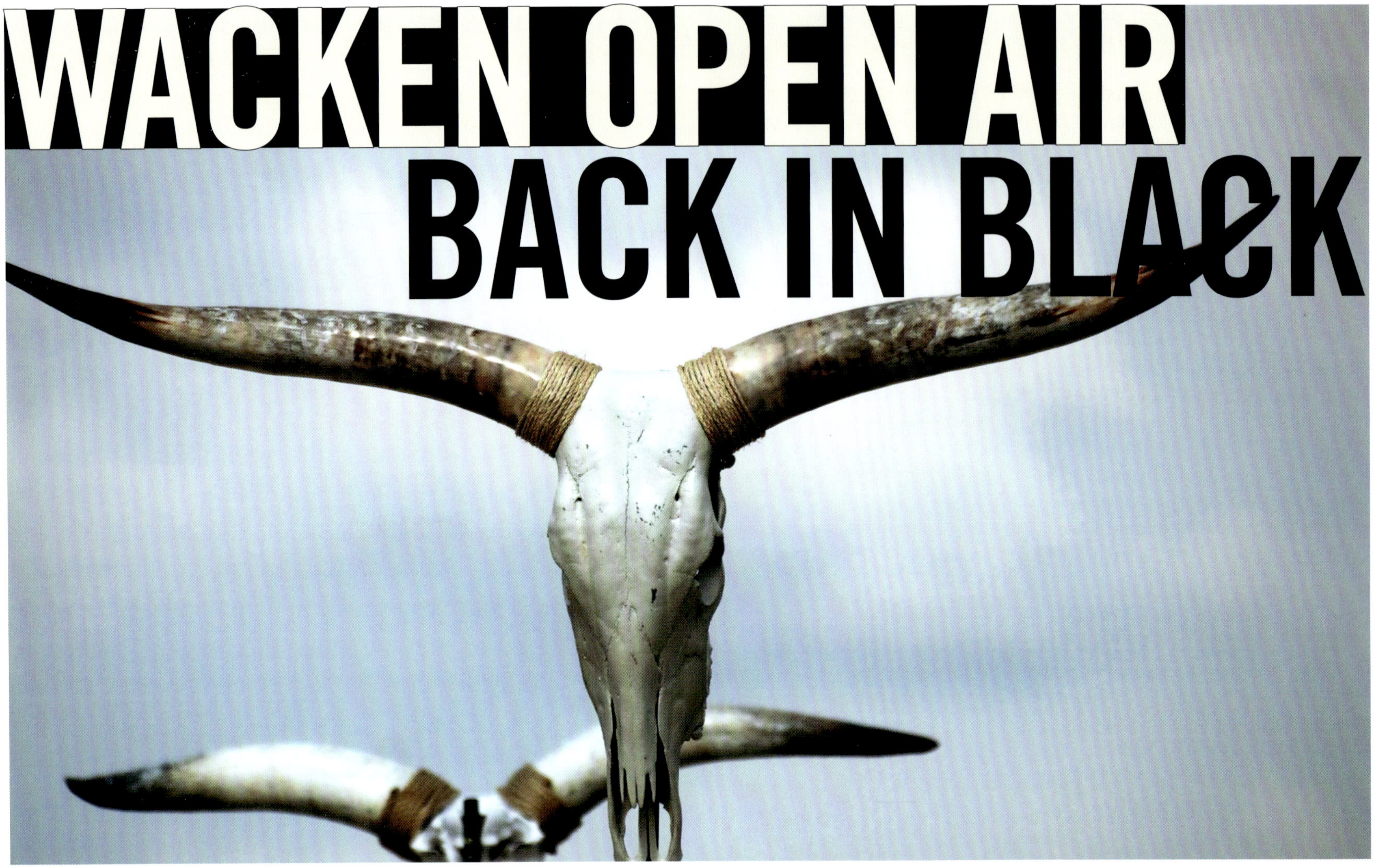

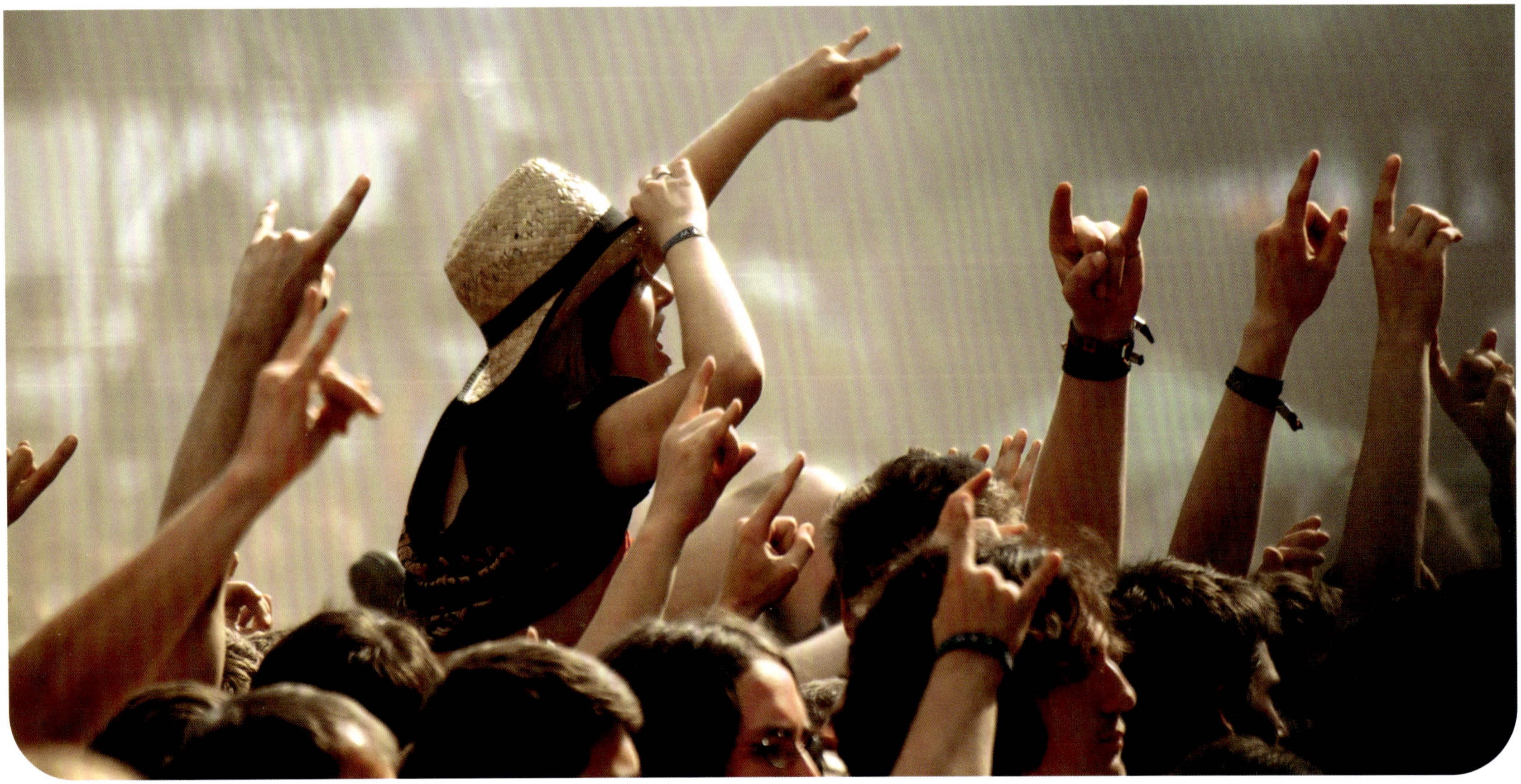

Orange is the new black? Metal-heads have a different view. 'You won't be coming to my festival with those orange things!' This is what organizer Holger Hübner is said to have exclaimed when Jägermeister first wanted to co-operate with Wacken Open Air in 2008. The co-operation failed, because the world's most famous herbal liqueur and the world's most famous metal festival simply did not have compatible colour schemes.

Black is the new black: when thousands of fans flock to Wacken for that one special weekend in August, the small municipality in Schleswig-Holstein is transformed into a global heavy-metal mecca. First organized in 1990 as a small rock festival with six bands and 800 attendees, today it proudly attracts 75,000 patrons from 80 different countries to the 'world's hardest field', which now spans a total area of 30 hectares. But anyone wanting to be part of it needs to be quick or lucky or have good connections, because once ticket pre-sale opens, they're soon all snapped up – often within a day. And is it any wonder, given what's on offer?

Today, more than 190 acts perform on the eight different stages under the banner of the Wacken bull, the motto being 'Faster, harder, louder'. Five venues, including the three main stages, are open to bands from all metal genres. Even the headliners play here, following in the footsteps of legendary performers such as Judas Priest, Slayer, Motörhead, Ozzy Osbourne and Iron Maiden. Those who prefer things a little cosier can head to one of the three themed stages; the Wackinger Stage is located in the medieval area of the festival grounds, and primarily features bands from the folk and medieval genres, while apocalyptic sounds ring out from the newest stage, the Wasteland Stage. The Beergarden Stage, the festival's on-site beer garden, is modelled on a typical folk-festival stage, and hosts popular regulars such as Mambo Kurt and the Wacken Firefighters. The marching band of the local fire brigade performs an opening concert there the night before the festival officially begins – and thousands of metal fans spend half the night dancing, roaring, partying and head-banging to booze-up hits and popular classics.

Another ritual takes place as soon as the gates open the next day, when head-bangers storm the festival grounds to the typical bellowing of a stag – giving Jägermeister an added acoustic presence. The signature sound of the Platzhirsch stag and pub sends a signal: 'Something's bound to happen when Jägermeister's involved, so get in there!' But as even the mightiest of bellows has no chance of being heard when AC/DC are busting out 'Back in Black' at 100 decibels directly opposite, the Jägermeister team promptly invented 'the best intermission of your life'. During the changeover between band sets, Jägermeister helps the fans kill time with free shots, surprise concerts, performances by the Jägermeister brass band and iconic giveaways.

It is thanks to those giveaways that Jägermeister and Wacken ended up co-operating. The organizer's original refusal forced the team to think up something new, and Jägermeister repositioned itself, including a new colour scheme for its branding. It marked the end of its orange events with flashy giveaways: hats and ponchos distributed in their thousands, turning the entire festival grounds into a sea of orange – and establishing Jägermeister as a festival heavyweight.

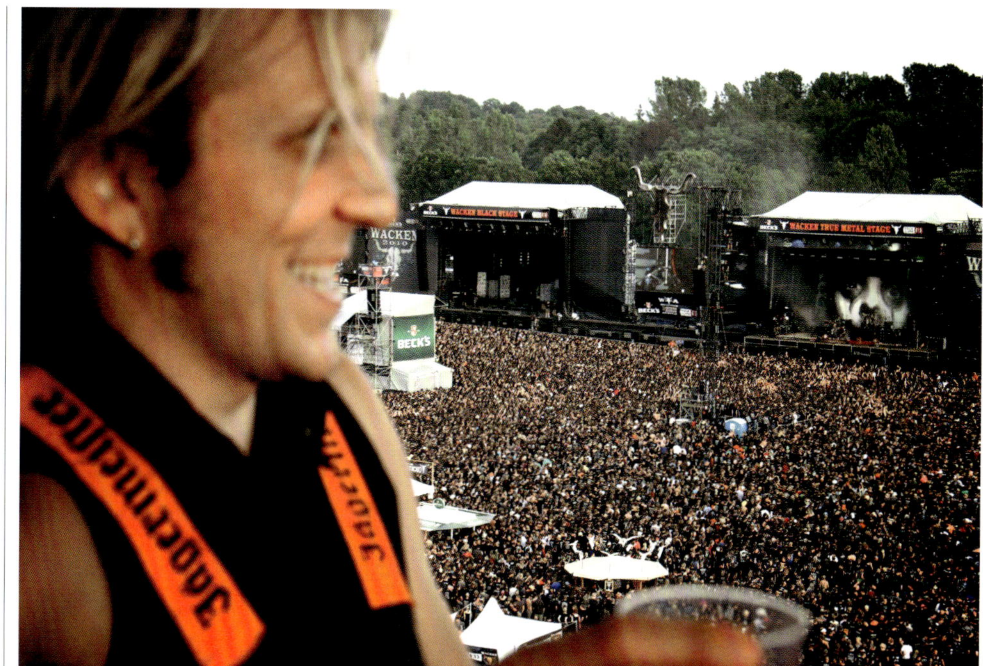

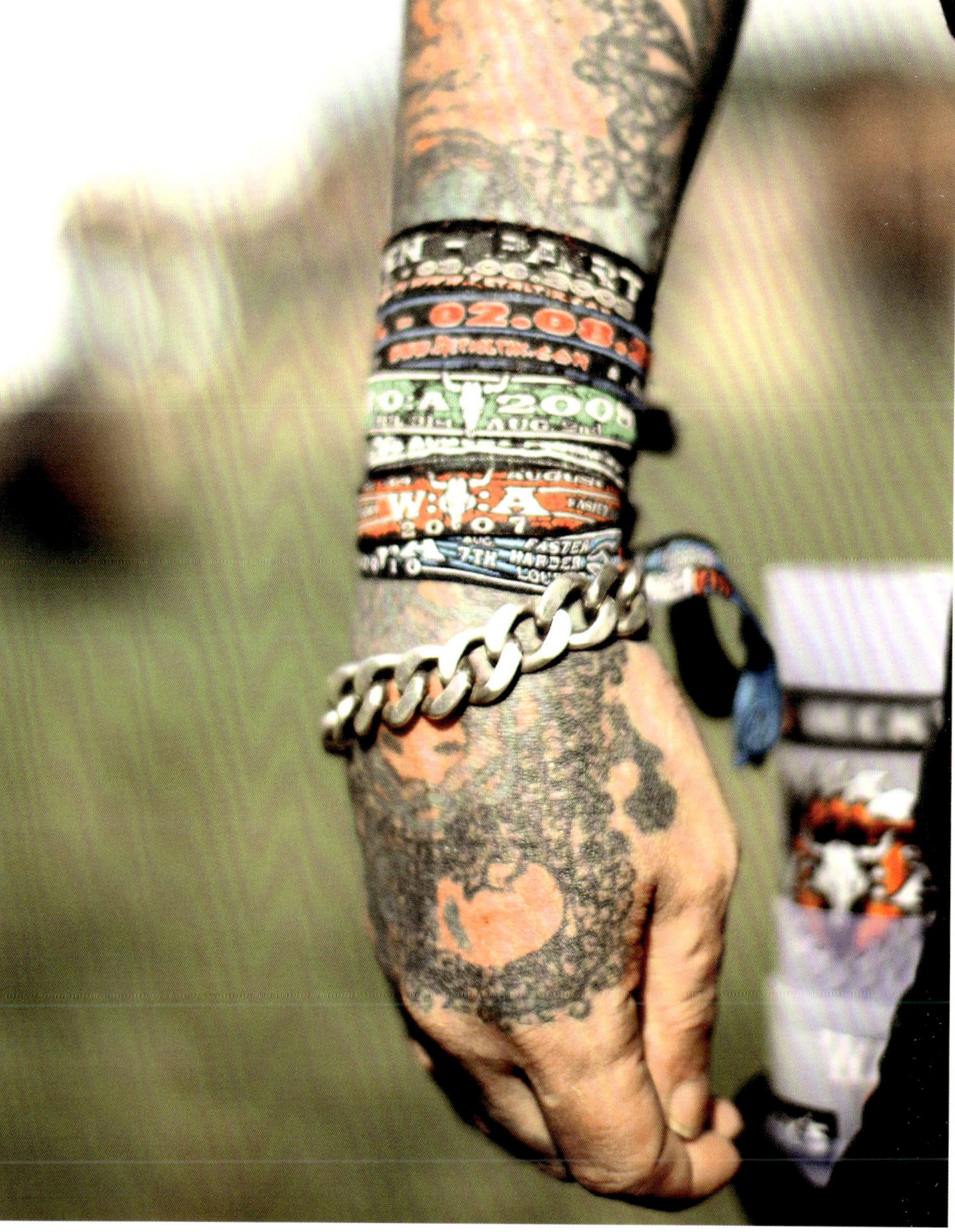

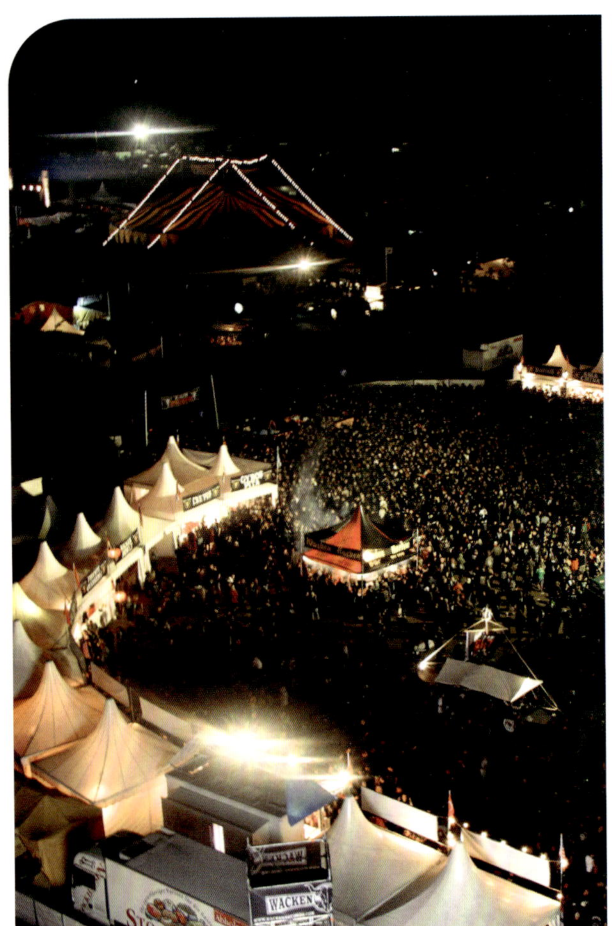

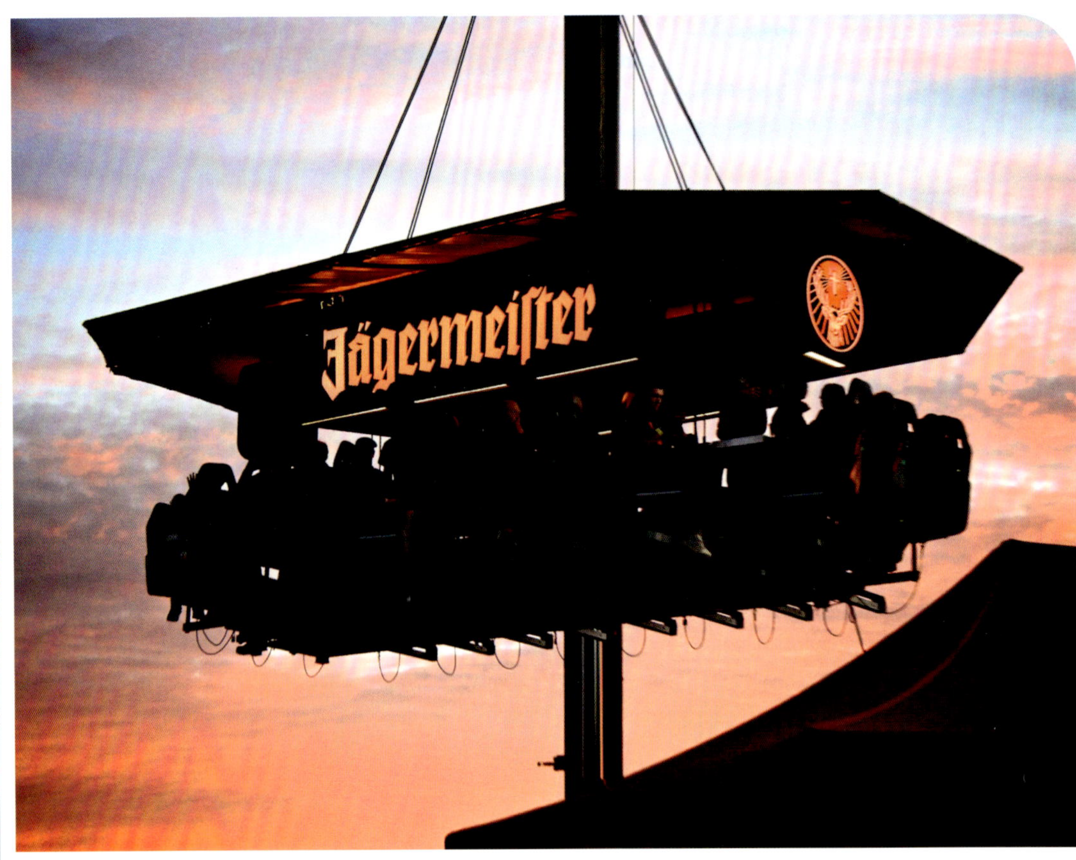

The Jägermeister perch was a popular attraction among visitors and bands alike — and not just at the Wacken Open Air. The 50-metre-high open-air bar provided a spectacular view of the festival grounds.

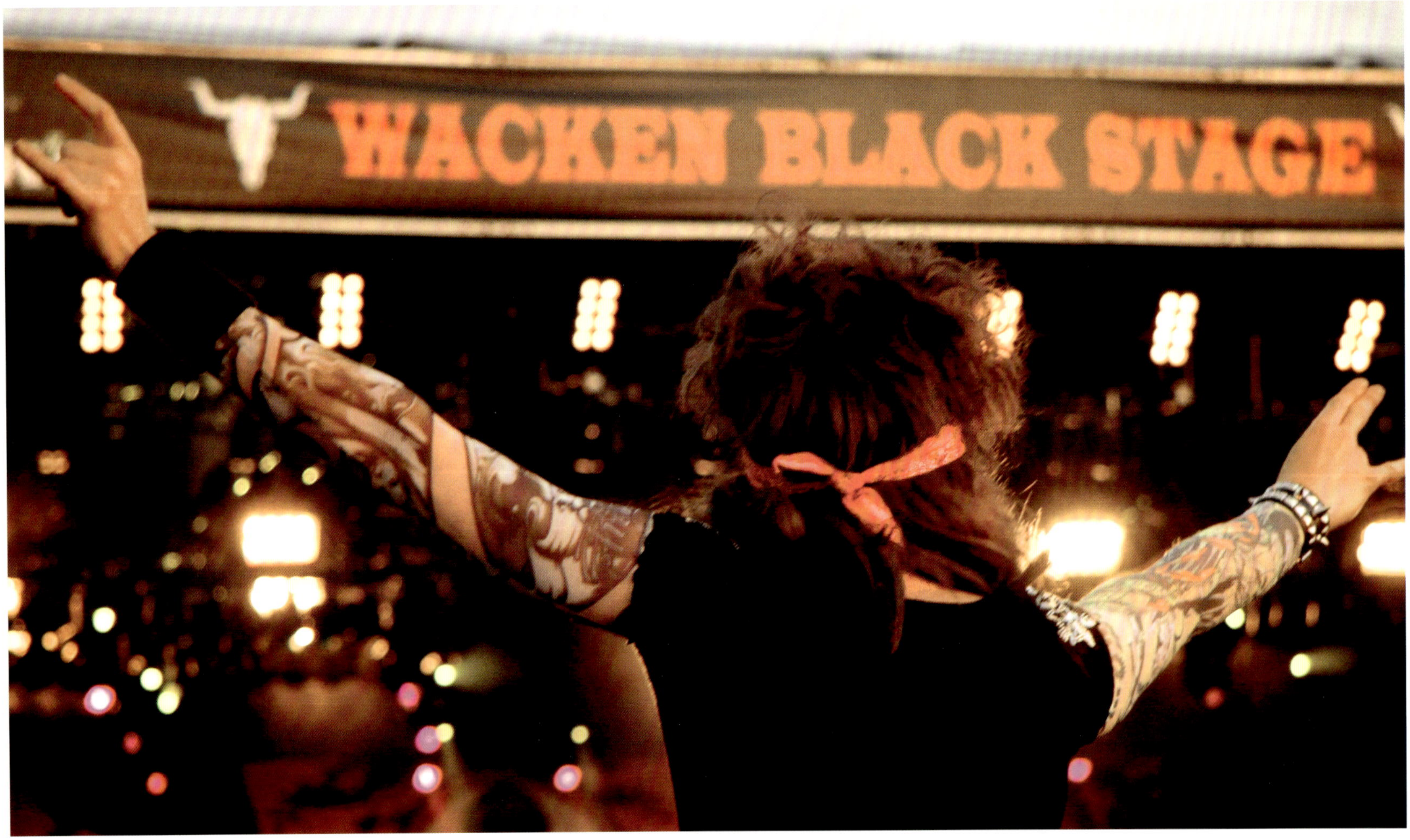

The end of the orange *Achtung Wild!* period and the start of the *Echt Jägermeister* (Genuinely Jägermeister) campaign meant even the giveaways were now predominantly black, and could be rolled out in Wacken for the first time. Particularly sought-after among metal-heads were the black caps with embroidered Jägermeister logo, which continue to appear at metal festivals and concerts to this day. Jägermeister later teamed up with the organizer to develop a strictly limited Wacken giveaway only available at the festival itself: a 0.33 litre bottle of Jägermeister, with a label that doubled as a peel-off iron-on patch, and still remains a coveted collector's item even today. Because, instead of the Hubertus stag, it depicted the Wacken bull, which appealed perfectly to the heavy metal fans' tastes. A giveaway as individual, lasting and emotive as this will always be kept, will continue to be worn and will serve as a lasting reminder of all the amazing (Jägermeister) moments.

Metal fans also celebrated many more of these exciting moments at the first major festival tool, which Jägermeister honed and promptly launched in Wacken in the form of the Jägermeister perch. Before fans could board and race skyward in the 50-metre-high open-air bar, their valuables and heavy items at risk of falling had to be handed in and stored in lockers at ground level. At other festivals, these objects were primarily keys, mobile phones, drink bottles and wallets, but the Wacken crowd posed a major challenge for the perch's ground staff, who had to store items such as studded belts, spikes, Viking pelts, chain-mail vests, drinking horns and even wooden swords.

And speaking of wood: The iconic Jägermeister Platzhirsch stag was a frequently photographed main attraction at the Wacken Open Air for four whole years. At its last appearance in 2019, marking the festival's 30th birthday, it was bid a fitting farewell in keeping with the metal vibe: under the motto of 'Your shirt for the stag', Jägermeister had called on fans and merchandise mail-order company EMP to send in their band shirts. These were then used to create a gigantic, black patchwork band shirt for the stag. After these amazing scenes in Germany, Wacken boss Holger Hübner would rather have just kept the Platzhirsch where it was and made it a permanent fixture at the grounds. His idea was for groups of visitors that came to the Wacken grounds the rest of the year to enjoy a fantastic view of the complex from the wooden colossus. But a retiree's life out in the wild? The Platzhirsch is still in too good a shape for that. Instead it roams the festivals in other countries across Europe.

Whether it be a stag made of metal shirts, or Roberto Blanco shouting 'We wanna have some fun' into the microphone to the screeches of an electric guitar; whether it be seniors' groups from the retirement

PACKIN' FOR WACKEN: MUST-HAVES FOR THE FESTIVAL

1 PONCHO, WELLIES, SUN HAT, SUNGLASSES AND SUN CREAM
Check the weather. Definitely make sure all other clothing is packed in waterproof plastic bags.

2 BUMBAG
A safe and practical way to store money, ID cards and disinfectant wipes (yes, these are unfortunately also a must).

3 FOLDABLE BOTTLES
You can take these into the grounds and fill them up at the drinking fountains (instead of buying mineral water at the stalls).

4 THICK SOCKS
To protect against cold nights and wet feet.

5 GAFFER TAPE
Patches up holes in boots, jackets or tents.

6 EXTRA TENT PEGS
These are often in short supply at the camp site – if the weather gets stormy, you can use them to further secure your tent.

P.S. To score a camp site not too far from the actual festival grounds, it is best to arrive early. Those who can afford to do so come two days before (Monday night or Tuesday morning) and make the Wacken weekend into a Wacken holiday. To ensure things don't get too hectic, don't plan to see too many bands, otherwise you just end up rushing from stage to stage. And be sure to head over to the festival grounds from the camping ground well in advance! Nothing is more frustrating than hanging around at the entry checks while Rammstein are playing.

home in Heide, who have been visiting for the last six years, or the joint show held by Rammstein and Heino – these are precisely the kinds of curiosities that repeatedly draw an amused, interested or irritated public's attention to the goings on in Wacken, along with the rain. But despite several years of the festival spectacularly ending up in the mud, there is yet to be a Wacken fan who has let bad weather spoil their fun. Instead, pop-up mud Olympics or mud slides have been set up on the camping grounds. It is no wonder that the *Hamburger Morgenpost* once described the festival as 'Germany's most popular mud party'. When the usual mud bath came under threat in recent summers due to the hot, dry weather, an artificial one was promptly created by attendees using a lawn sprinkler.

Granted, mud baths aren't everyone's thing, and they're not the reason for Wacken's legendary status anyway. Whoever you are, wherever you're from, whether attending alone or with friends, Wacken is a place where existing and would-be metal fans can spend a few days having the time of their lives. Because Wacken's real allure is more than metal, mud and madness: it's at its camping grounds, in its moshpits, over a Jägermeister, at beer pipelines or in the metal cinema that like-minded people meet, get to know each other better and forge friendships. And this is the reason previous attendees keep coming back – back in black. ⚹

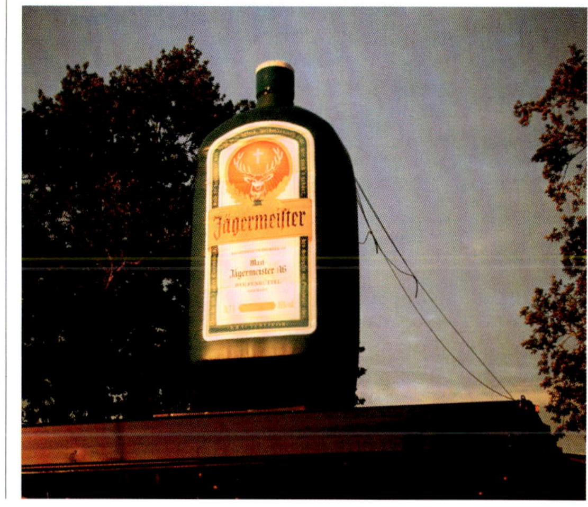

BLOODSTOCK OPEN AIR
RELEASE THE BEAST

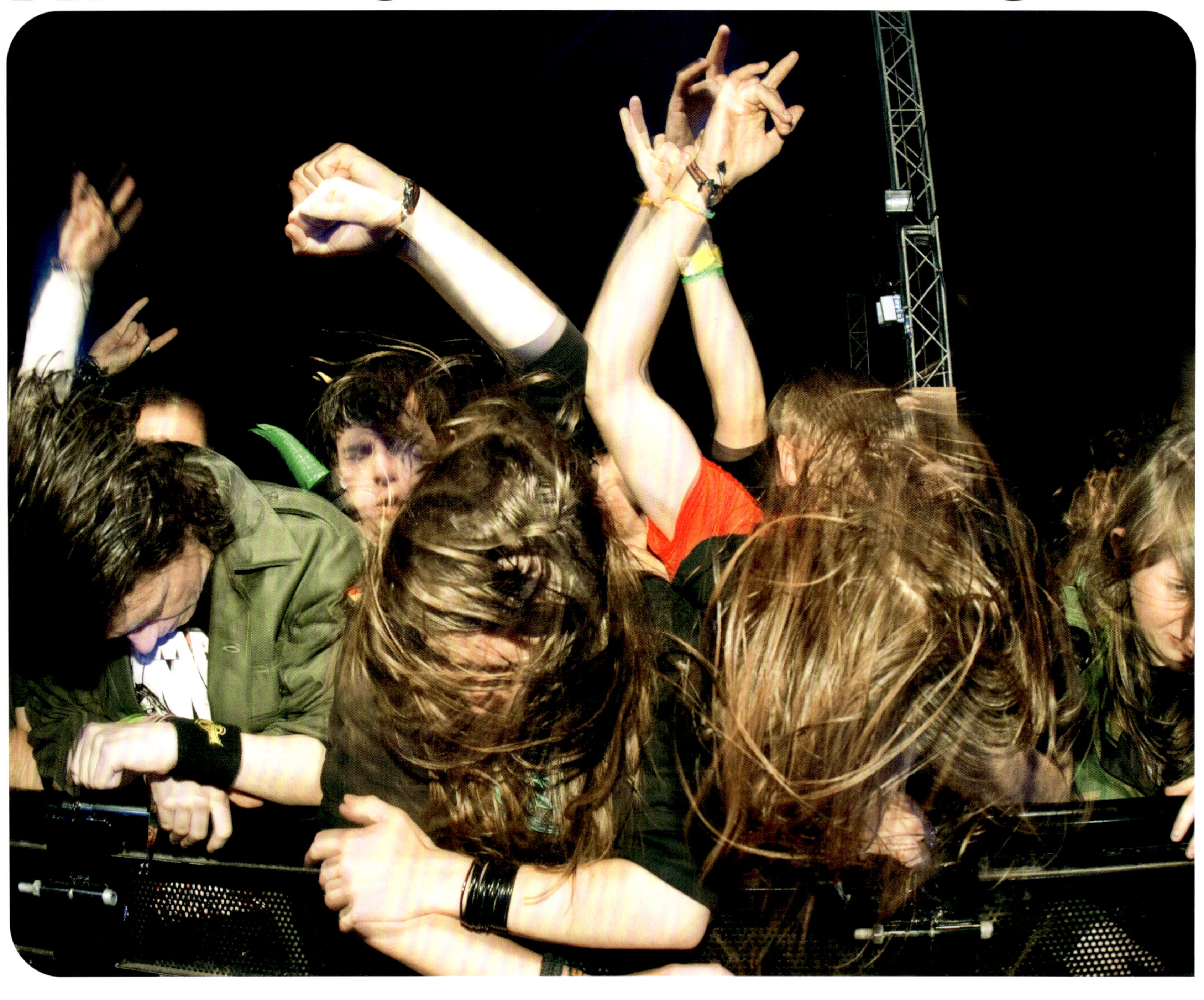

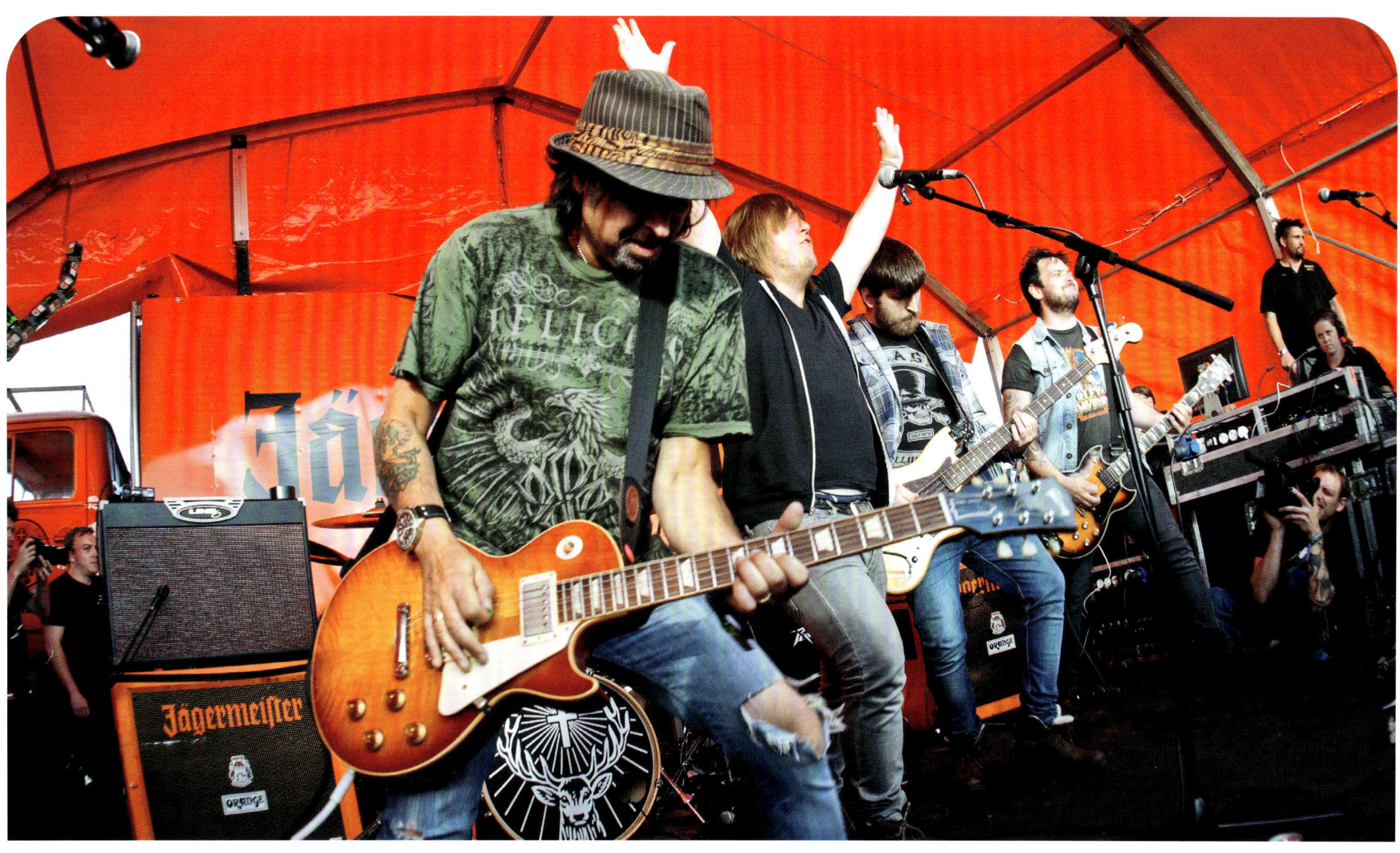

Hot nights, cool parties, hard-core music – when Jägermeister's involved, things can get wild, sometimes to the point of pain. It was a warm late afternoon at Catton Park, Derbyshire, in August 2009. At the Bloodstock Open Air festival, young people in tight jeans, cut-offs and wide studded belts head-banged to *Pleasure to Kill*, a classic by German thrash-metal band Kreator, who were performing it on the main stage. Meanwhile, at the orange 'Jager Truck' directly opposite, festival-goers partied over shots of Jägermeister – until the truck suddenly came under attack from a swarm of wasps. 'One even stung me on the neck,' recalls a member of the Jägermeister festival crew. It took the team great pain and effort to get rid of the wasps, as they had acquired quite a taste for the drink. 'We had to wait until they had knocked back enough of it,' he continues. 'Then we placed glasses over them and carried them away from the truck.' That day, the famous Jägermeister slogan of 'Release the beast' took on a whole new meaning for the Jägermeister crew when they discovered that wasps loved their herbal liqueur.'

Bloodstock was where it all started for Jägermeister, at least in England. Originally an indoor festival, which turned open air in 2005, the 2003 event marked Jägermeister's first presence at a festival in the UK. Eight years later, this figure had already reached 36 – a record that still stands today. And one constant throughout has always been the 'Jager Truck', which also serves as the backdrop for the Jägermeister Stage. Previously orange in colour, the former Soviet army vehicle, a Ural-4320 military truck, has today been painted matte black – for not only does this attract fewer wasps, it also fits perfectly with the martial melodies on the Bloodstock programme.

The BOA, as the 'UK's No. 1 Metal Festival' is also called for short, considers itself the successor event to the legendary Monsters of Rock, last held in 1996. It left a void that needed filling. The British metal scene was large, yet there were very few comparable heavy-metal festivals. And so it was that, within a few years, the BOA became a permanent major fixture, attracting some 15,000 attendees. The biggest names in prog, death, black, speed, thrash, gothic and heavy metal – Children of Bodom, Judas Priest, Ghost, Dimmu Borgir, Opeth, Motörhead, among others – vie for headliner spots on the main stage, the Ronnie James Dio Stage, while newcomers and smaller bands play on the Hobgoblin New Stage, Jägermeister Stage or Sophie Lancaster Stage. The last gets underway from the Thursday, and is named after a young British metal fan who was battered to death by a group of teenagers in summer 2007 because of her Goth-like appearance.

What many people fail to see, or interpret differently, from names such as Hatebreed, Slayer or Cradle of Filth is the fact that metal fans are a family, and that the metal scene is more cordial than its aggressive blast beats, brutal guitar riffs and dark lyrics suggest. The BOA is unanimously viewed as the country's friendliest festival. While karaoke and a mini fun fair are certainly a good place to start, Bloodstock founder Paul R. Gregory doesn't believe it's something you can force on people. 'It's something you have to earn over an extended period of time. Having fans who come back year after year has enabled us to build a relationship with them,' the music promoter, painter and metalhead told *The Independent*. 'We know what they want, because we are them and they are us' – people wanting to release the beast (or the wild stag, as it were) to guitar shreds, growls and gore. Just no wasps, please! ❦

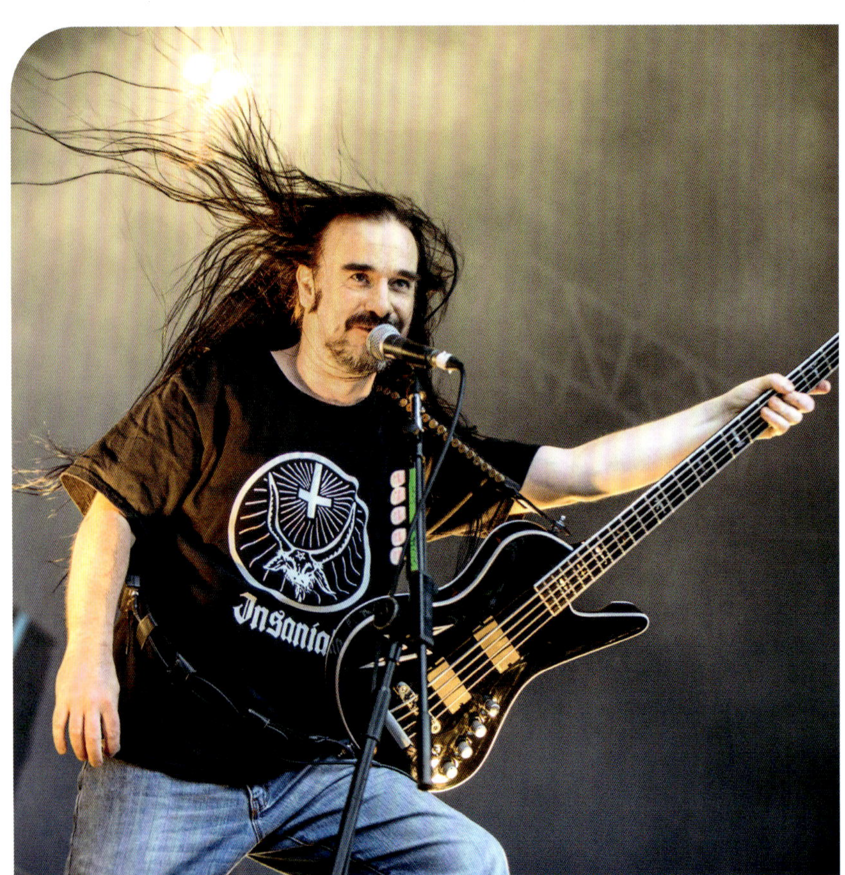
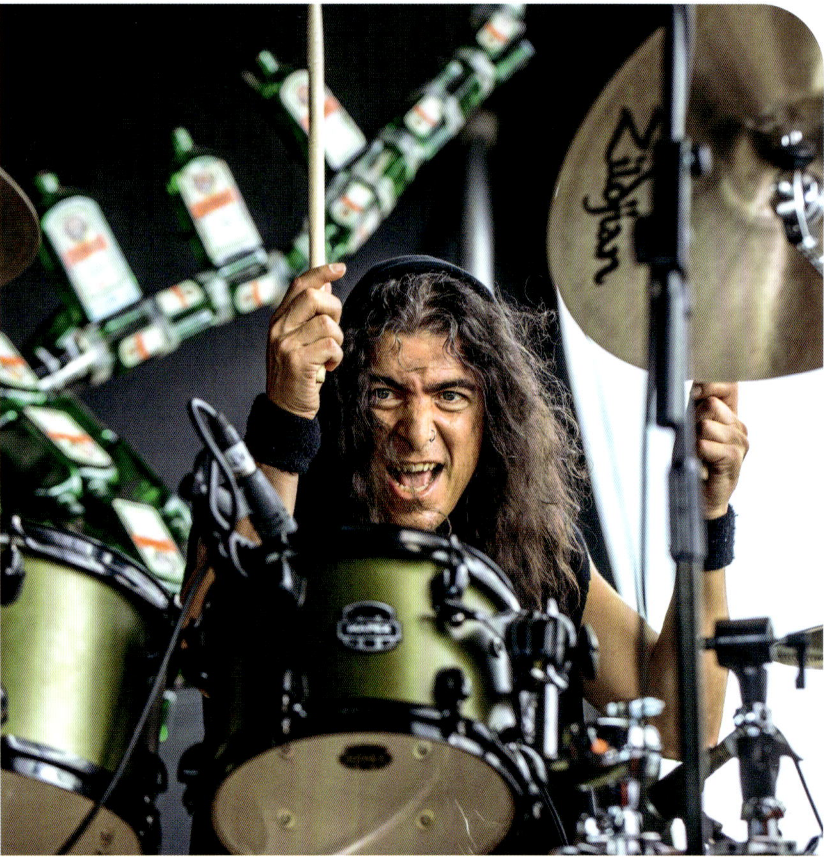

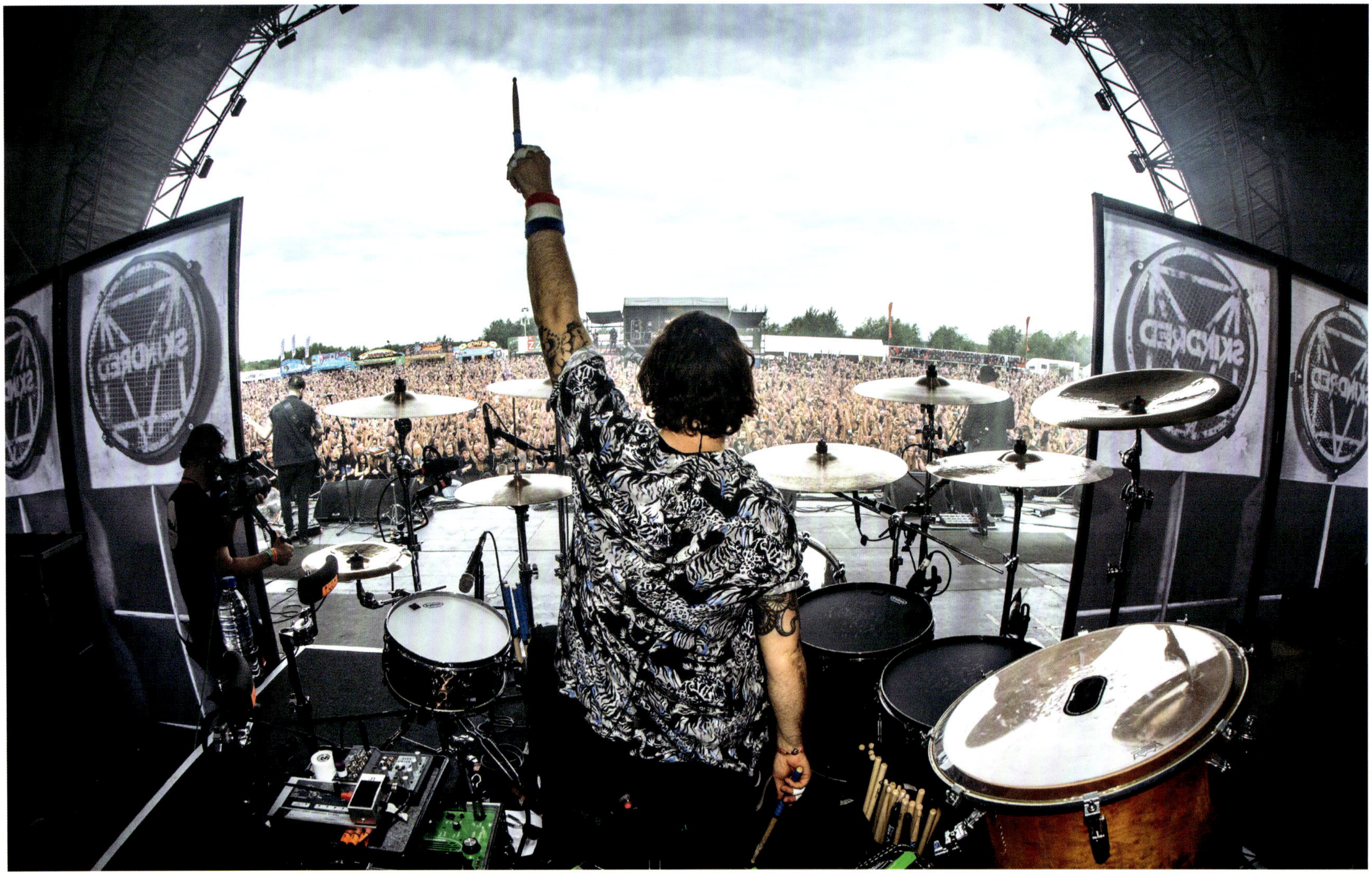

BOA! Five Bloodstock facts

🤘 While Bloodstock went open air in 2005, it was first held as a one-day **indoor festival in Derby** on a Monday in May **2001**, headlined by the godfathers of British heavy metal, Saxon.

🤘 The line-up of BOA bands to date span **55 different countries and protectorates** – from A for Argentina to Zimbabwe.

🤘 The BOA is best accessed **by train. Shuttle buses** taking attendees to the festival grounds operate between Lichfield City train station and the bus parking facility from Thursday to Monday.

🤘 The **names** of the Bloodstock **camping grounds** – Walhalla, Asgard, Ragnarök, Hel and Niflheim – are taken from **Germanic mythology**. The largest is Midgard, whose approximate literal translation is 'middle abode', meaning home in the centre of the world. So, in this case, the centre of the metal world.

🤘 Metal heart, metal art: one of Bloodstock's founding fathers, **Paul R. Gregory**, is an internationally acclaimed artist whose J.R.R. Tolkien-inspired fantasy paintings have graced the covers of books and metal albums. It is thanks to him that the **festival grounds** have included a real **art gallery**, the RAM Gallery, since 2014.

ELECTRIC DAISY CARNIVAL

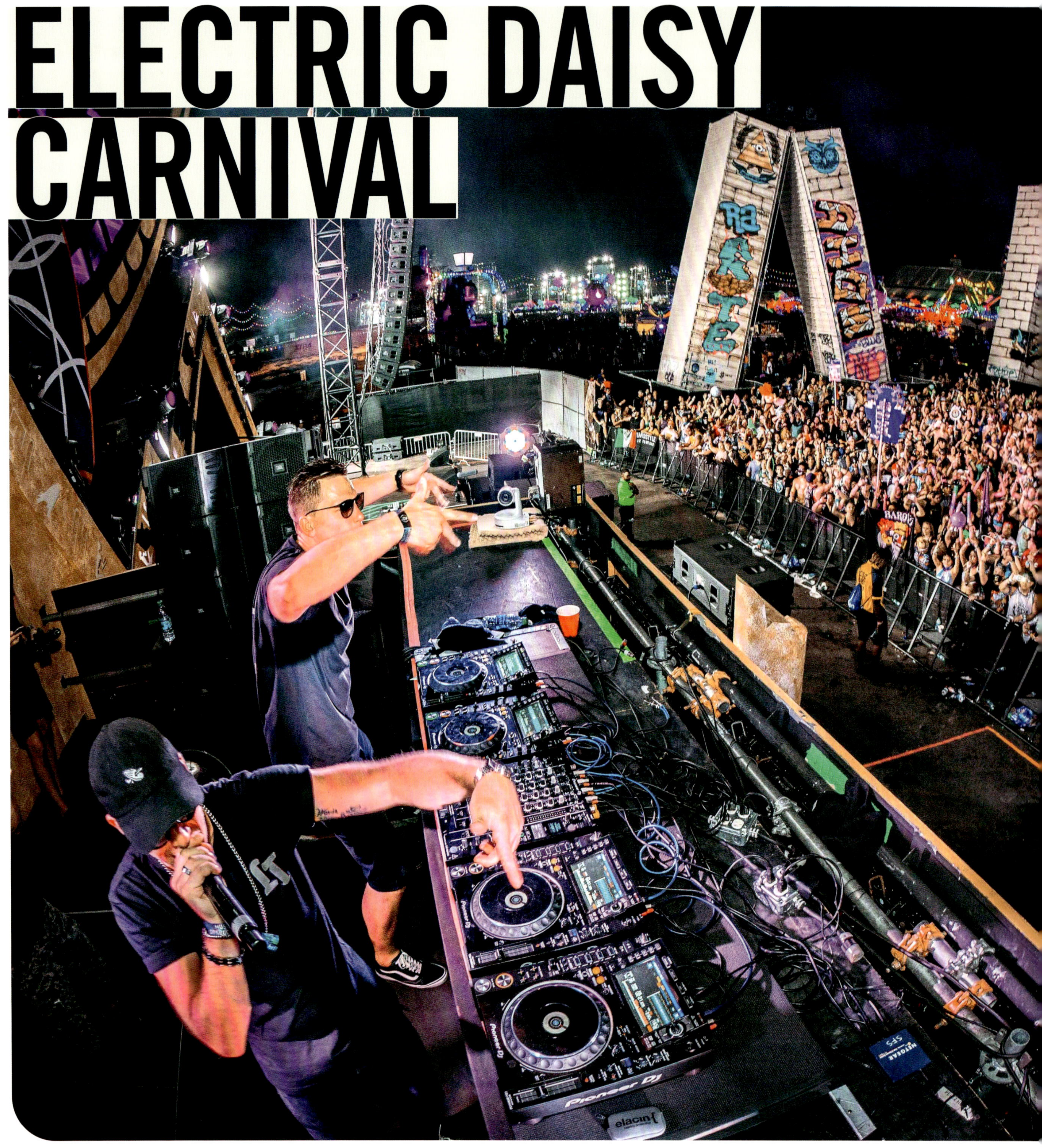

WHERE NIGHT BECOMES DAY

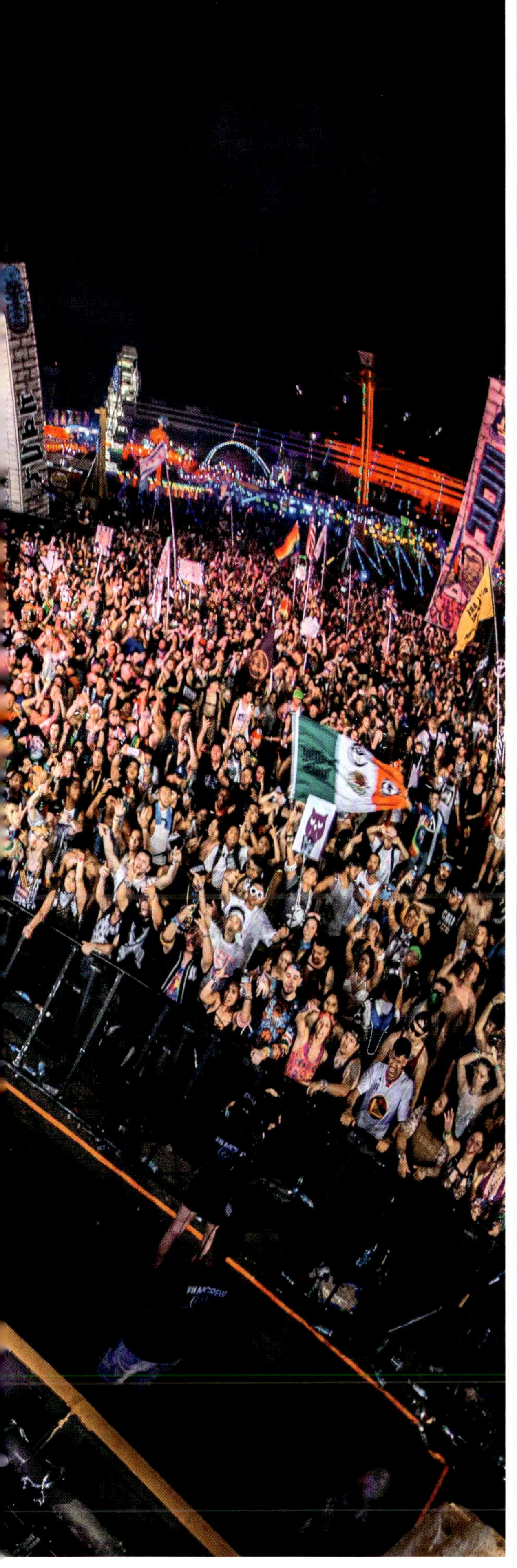

The best nights of your life can never be planned. They usually happen spontaneously, the result of a feeling, mood or special moment. But there are a few circumstances that increase the likelihood of such nights occurring: the right company (good friends), the right drink (Jägermeister) and the right setting. And there's one particular city that has the potential to turn any night into the best ever: Las Vegas. So it's no wonder that one of the world's most ostentatious music events has been held in this dazzling metropolis since 2011 – all night, from 8 pm to 6 am. After all, that's when neon lights and laser beams shine the brightest.

The Electric Daisy Carnival, EDC for short, premiered in Los Angeles in 1997, to a 5,000-strong crowd. Thereafter, the festival was held in various locations across southern California, growing in size every time. By 2010, 185,000 people were making the pilgrimage to the major EDM event, and today, with over 400,000 attendees, the Electric Daisy Carnival is the world's largest electronic music festival, with spin-offs right around the globe. But its home base is the oasis of pleasure in Nevada, where it is held at the 'Blue Oval' – the Las Vegas Speedway. For three days a year, the grounds around the famous NASCAR race track, spanning nearly 5 sq km in area, are transformed into a lavish fun fair, with a sparkling, twinkling and generally fantastical vibe that gives even the Las Vegas Strip a run for its money. With fire-

works, dancers, stilt walkers, three-dimensional illuminated art installations, action games, parachutists, circus acrobats and trapeze artists, silent discos, karaoke, carousels and rides, wedding chapels, bars and open-air theatres, the Electric Daisy Carnival has pleasure-seekers more than covered. And, most importantly of all, it offers a phenomenal line-up and a visual spectacle like no other – so very Vegas.

The crème de la crème of electronic music – from Tiësto to Steve Aoki, from Skrillex to A$ap Rocky – perform on ten different stages, each with their own theme. The Kinetic Field, for example, is reminiscent of a glittering crystal city, and the Cosmic Meadow appears as a vibrant fata morgana. The Neon Garden looks like a minimalist (at least by Vegas standards) deep-house and techno temple, while the Circuit Grounds are illuminated so brightly by giant LED walls that they probably even boost vitamin-D levels in many of these night owls. Art Cars, on the other hand, is not one single stage, but rather consists of multiple mobile sound systems akin to dazzling buggies or party boats, which cruise along the Las Vegas Speedway with various DJs on board. And wherever there are people living the best nights of their lives, Jägermeister is of course never too far away either.

Jägermeister first became active at live music events in the USA in the late 90s, making it the first ever alcoholic-beverage manufacturer to act as a

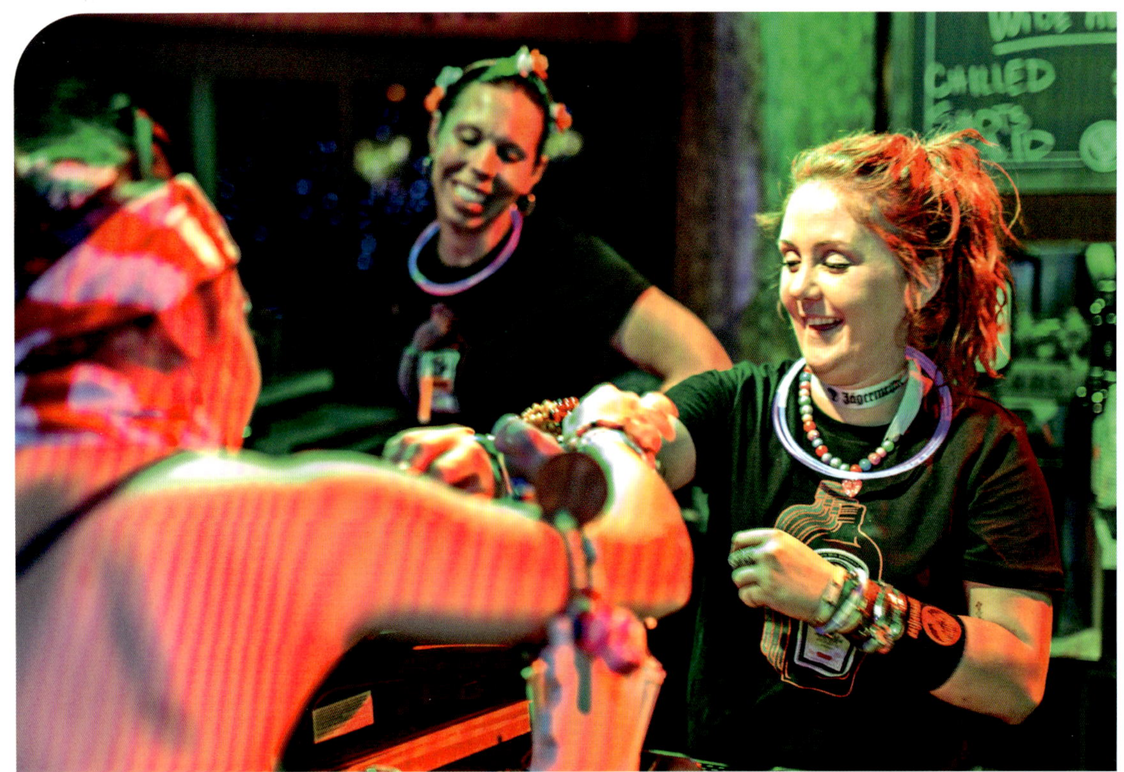

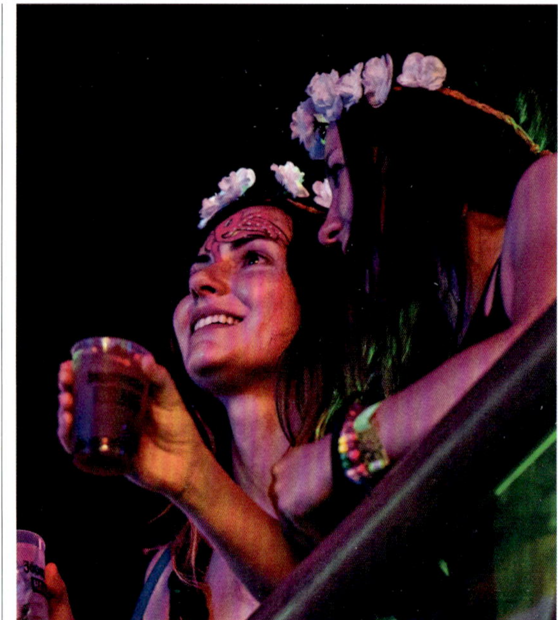

sponsor in this industry and on such a scale. The gigs included Ozzy Osbourne's famous Ozzfest, which toured through the USA and parts of Europe from 1996 to 2018. As the popularity of these festival tours grew, Jägermeister similarly expanded its activities and supported tour events such as the Mayhem Festival (2008–2015) and the World's Loudest Month, a fusion of several national rock and metal festivals. Over the following 15 years the brand's festival presence continued to grow, with 2009 finally seeing the debut of the Jägermeister Mobile Stage, an innovative concert experience in which a trailer is transformed into a fully fledged live stage for up to 20,000 fans. The stage has already travelled nearly 1 million miles across the USA in that ten-year period, and more than 5 million fans have danced, partied and drunk hundreds of thousands of ice-cold Jägermeister shots. A brand partnership with concert agency Live Nation gave rise to another festival highlight in 2016: Haus 56. And it is with Haus 56 (56 being the number of herbs contained in Jägermeister) that Jägermeister is present at the largest of all EDM festivals – the Electric Daisy Carnival – in addition to four others.

The two-storey treehouse, with stages, bars and a gigantic pair of antlers at the top, fits perfectly in the bizarre festival setting. Haus 56 gives attendees the chance to try cocktails inspired by their favourite festival artists, participate in games, enjoy live performances and meet musicians. This is pretty unusual even for Las Vegas, because, on the Strip, whether it be at the Flamingo, Planet Hollywood, Caesars Palace or Park MGM, it would require at least a backstage pass. Jägermeister makes it available for free. Even in the middle of the night. Topped off with an ice-cold shot. ⚹

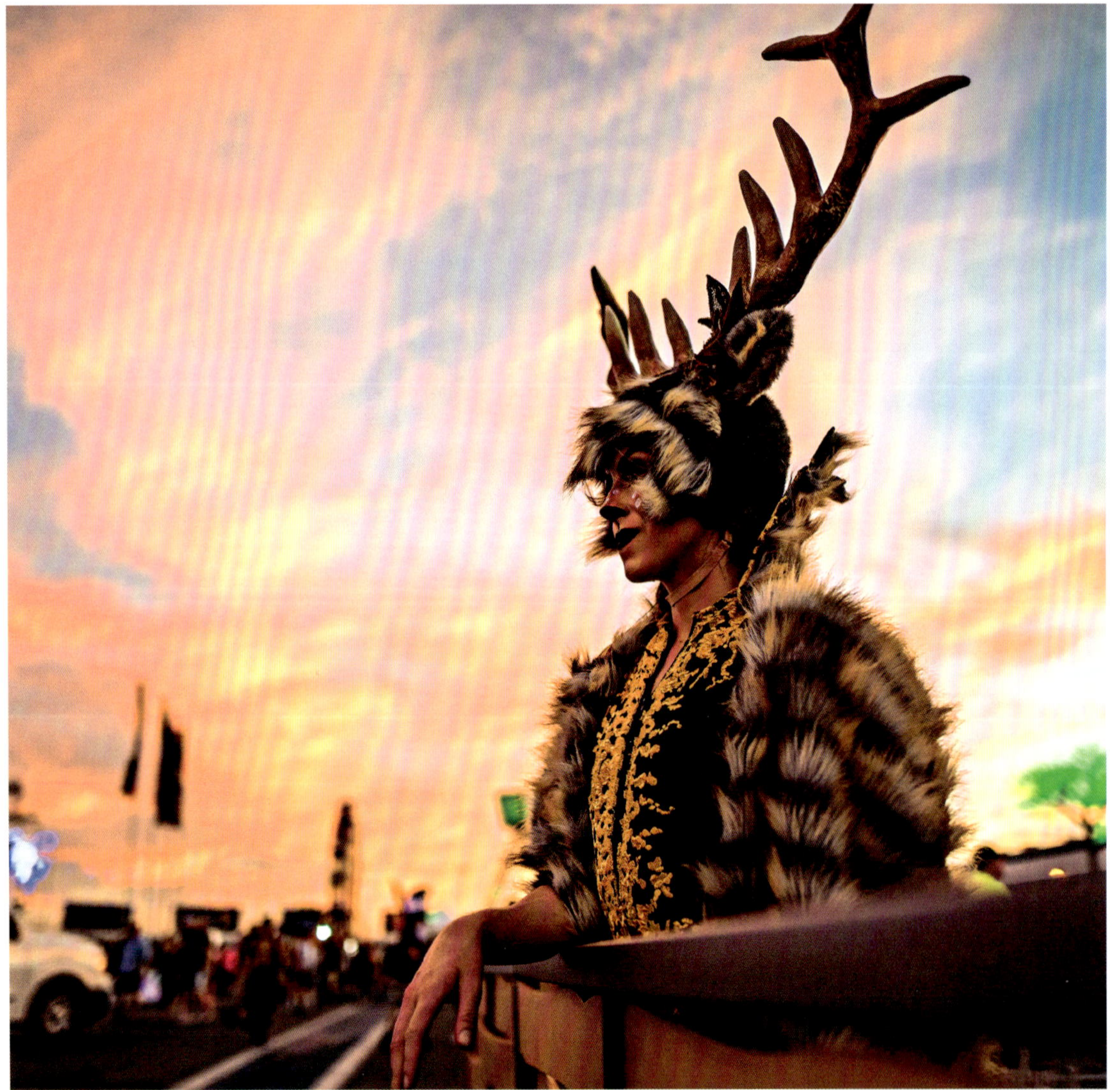

Bizarre backdrops, crazy parties, wild looks: An EDC attendee in a deer costume. <u>Top:</u> A popular festival highlight for Jägermeister fans is Haus 56, a two-storey tree house with stages, bars and a gigantic pair of antlers at the top.

WHAT A RIDE!

—More than **2 million people** have attended the Electric Daisy Carnival since it has been held in Las Vegas
—The festival grounds include **24 rides**
—**9,500 fireworks** light up the desert city's night sky over the festival weekend
—The giant **disco ball** in the Neon Garden has a diameter of 3 metres, and weighs **550 kilograms**
—The pyrotechnics on the Basspod Stage, one of the ten major festival stages, can generate a **flame 55 metres high** – the largest at the EDC

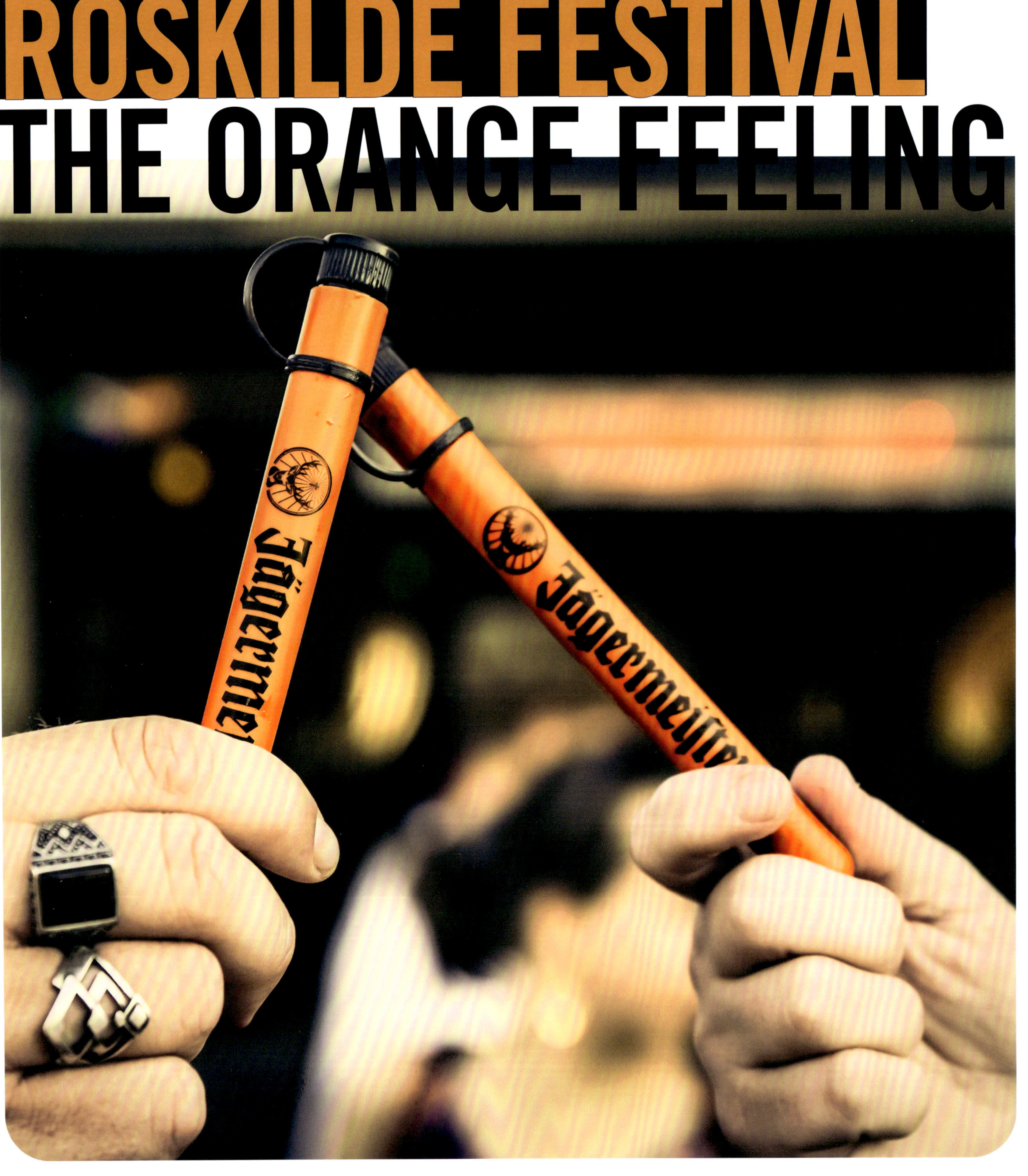

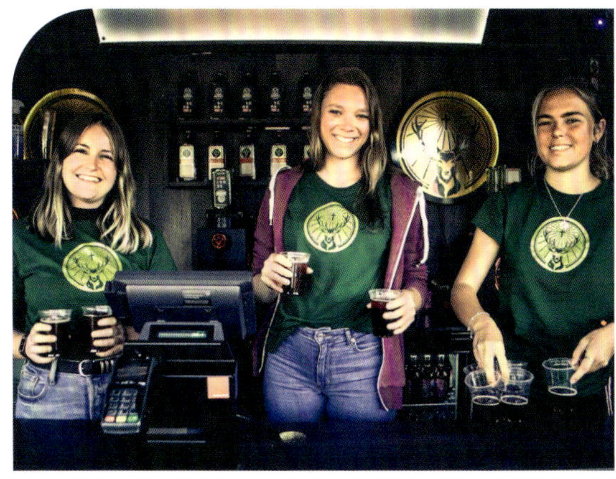

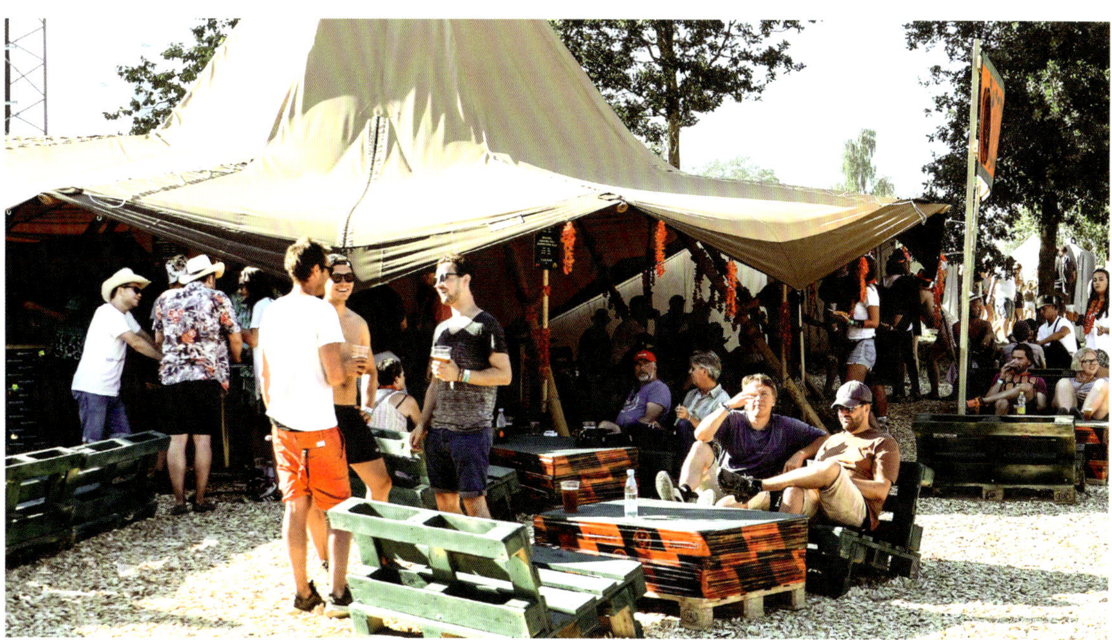

Jägermeister has been present at Roskilde since 2010, most recently with a signature bar housed in large tepees.

For some, it's enlightenment; for others, it's football: but there are many things people associate with the colour orange. At festivals, it has long been synonymous with one thing in particular: Jägermeister. For nearly ten years, the herbal-liqueur brand was both famous and infamous for its orange events at music festivals. These days, there's only one large open-air event that still decks out in orange: Denmark's Roskilde Festival. Not only does it have the Orange Stage (the orange-coloured main stage) and the *Orange Press* (the festival newspaper), it also has the fabled 'orange feeling' that so many Roskilde attendees rave about. What is it? What does it feel like? The orange feeling is apparently something that needs to be experienced live. And after that, there's no going back – which is the reason why some people have been making the pilgrimage to Roskilde for as many as 30 years.

The eight-day festival is one of the oldest and largest in Europe. Established in 1971, the open-air event has now become an institution in Denmark, with 130,000 attendees and a global appeal. Its programme transcends genre, scene and age boundaries, and proves that diversity isn't some kind of sorcery – the line-up includes more explicitly queer acts than anywhere else, and the total number of acts across the seven stages amounts to more than 180. While the first half of the festival revolves exclusively around young artists in the early stages of their careers, the headliner slots always feature big names, with the likes of Björk, Paul McCartney, Metallica, Bob Dylan, Foo Fighters, The Cure, Bruce Springsteen, The Rolling Stones, Red Hot Chili Peppers, Prince, Iron Maiden, Neil Young, Muse, Patti Smith, Leonard Cohen and many others all having performed in the past. It is unquestionably an honour, even for top stars, to play at Roskilde, which is why they often think up very special shows for the occasion. In 2013, Rihanna and Kraftwerk put on a spectacle that involved kitting out the 70,000 attendees with 3D glasses. In 2015, Damon

Albarn, frontman of Blur and the Gorillaz, teamed up with Africa Express to play a five-hour set. He eventually had to be carried off the stage because he refused to stop playing. All part of the orange feeling transmitted from the audience to the artists and vice versa.

After the euphoria of the concerts, many attendees keep the party going at the camp site, where an 8,000-watt music system provides the necessary sound. By day, there are street-dance workshops, a skate park, areas for beach volleyball, basketball and football, a cinema and a swimming lake. Another area, known as Dream City, enables selected fans to get creative 100 days before the festival even begins. The Nøgenløb, or naked race, is an additional activity that causes a great stir. As the name suggests, it is a race in which all participants are nude. It is organized by the festival's own radio station, Roskilde Festival Radio, which reports live from the festival over the entire nine days. And as if that were not enough, Roskilde is also the world's only open-air event to have its own daily gazette in the form of the *Orange Press*, containing festival news, live reviews and sharp critiques throughout Roskilde week. It is available free of charge from selected stands at the festival grounds. One particular name has regularly graced the cover over the last eight years, adding to Roskilde's distinct 'orange feeling': Jägermeister first began co-operating with the fes-

tival in 2010, and has been present as a sponsor ever since. It does this in the form of at least one large Jägermeister signature bar. In the early days, this bar was inside the Jägermeister bus, then in an old train carriage, which was converted into the Jäger Bodega. In recent years, however, it has been housed in large tepees. Roskilde is made for Jägermeister. The company is all about creating visionary ideas that make for unforgettable moments, and in the case of Roskilde, that means entertainment with attitude.

For years, the festival has been a pioneer – artistically, politically and socially. Roskilde is a non-profit association, and is organized and designed, with great fondness and care, by 32,000 volunteers. This volunteering concept strengthens the orange feeling; it's

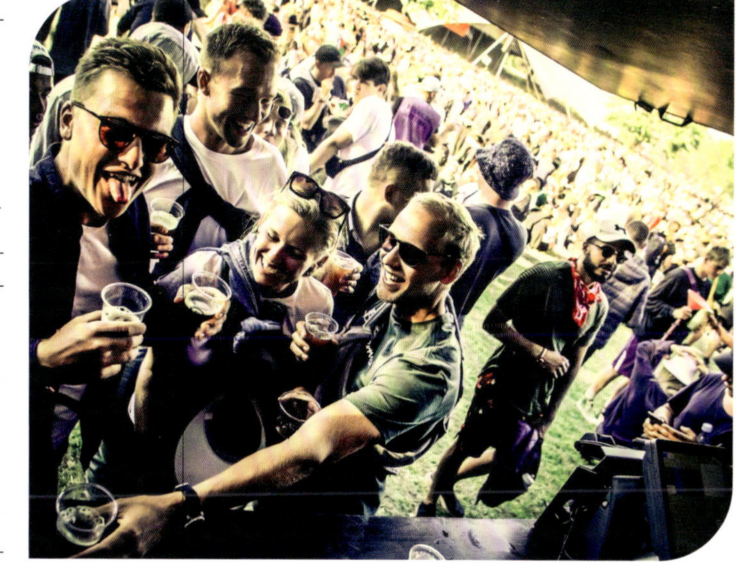

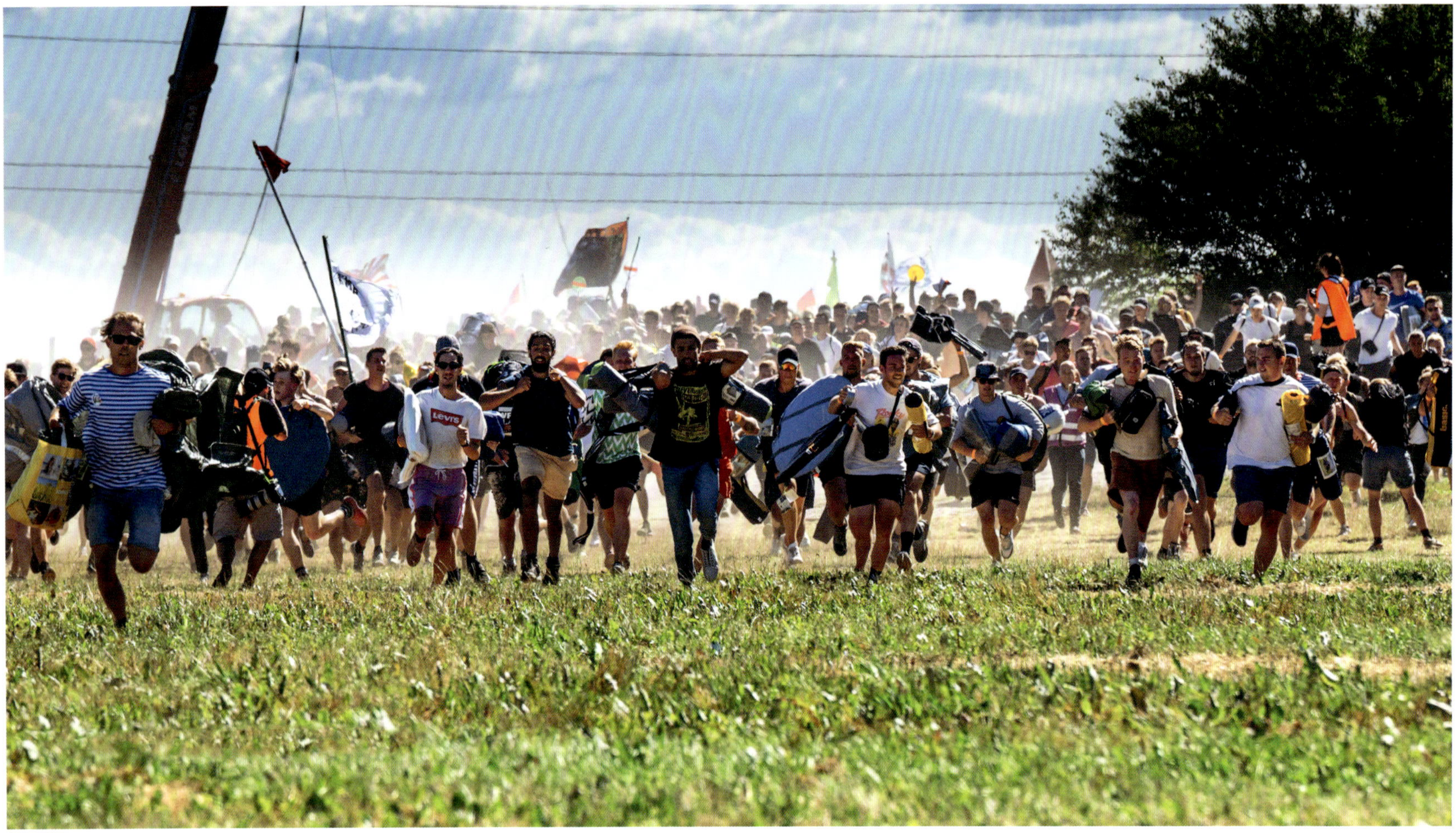

The eight-day Roskilde Festival, attracting a total of 130,000 attendees, is one of the largest open-air festivals in Europe. Below: Jägermeister's popular giveaways also add to the uniquely Roskilde orange feeling.

not about maximizing profit, but rather about maximizing the festival experience. Any proceeds left over (2.44 million euros in 2019) are donated to various humanitarian and cultural organizations. The organizers also try to raise awareness of issues such as social justice, equality, diversity and sustainability – and create precedents to boot. Beer at Roskilde, for example, is sold not only in returnable cans, but also at supermarket prices, thereby discouraging attendees from bringing their own beer with them and leaving the empty cans lying around on the grass. Roskilde takes a stance, including in its festival programme. There are panel discussions and workshops on political and social issues, with representatives from the likes of Occupy Wall Street and Fridays for Future. In 2014, the organizers invited activists from Russian punk group Pussy Riot, while the 2017 edition saw Edward Snowden beam in live from Moscow. When it became known that it was his birthday that day, thousands of Roskilde attendees sang 'Happy Birthday' to him at the top of their lungs.

There are many things people associate with the colour orange. A birthday serenade for the world's most hunted whistle-blower? Why not. It may be difficult to describe, but one thing's for sure – the orange feeling feels damn good! ♥

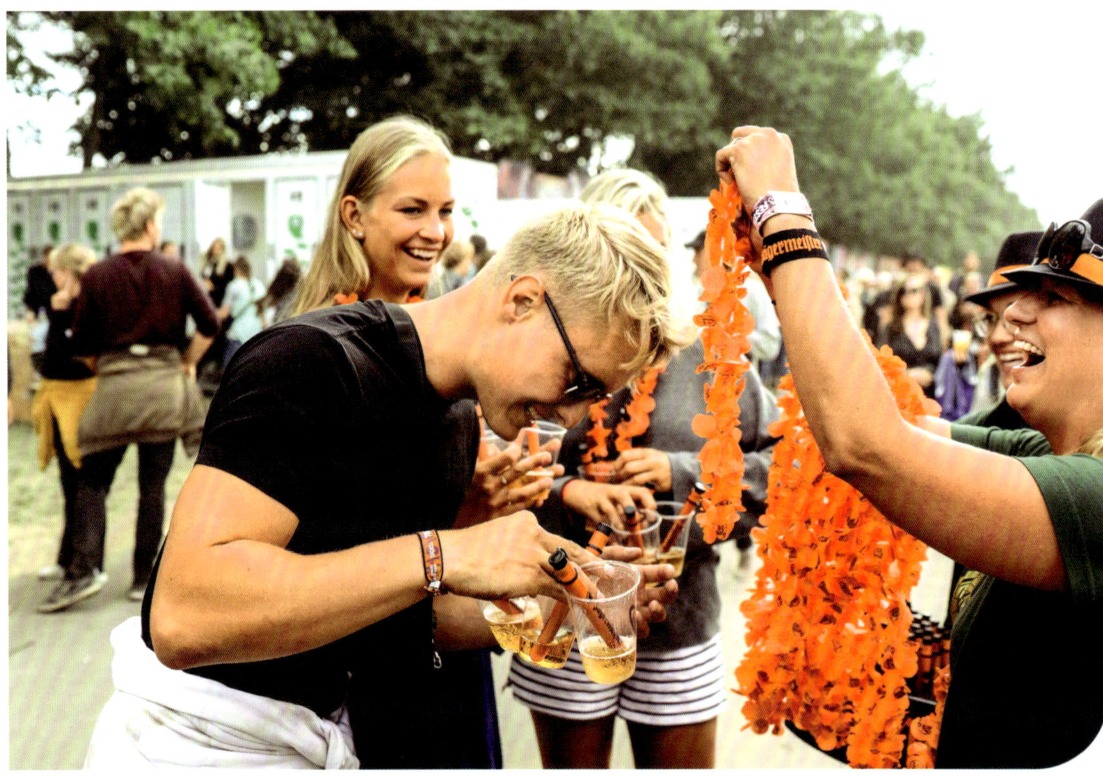

1
Ch-ch-ch-ch-changes

The stages are constantly changing – both visually and location-wise. The ground may be familiar, but there's a surprise every year. Sometimes the Apollo Stage gets a new design, sometimes old stages need to make way for new ones. But one thing's for certain: the Orange Stage isn't going anywhere.

2
Here comes the rain again

Just like the rest of Denmark, the weather in Roskilde is fickle. In other words: it rains often and a lot. One of the heaviest downpours came in 2007, after which waterproof gear for festival-goers became mandatory.

3
Evolution orange

The space in front of the Orange Stage, Roskilde's main stage, can accommodate approximately 60,000 people. The stage was imported from England in 1978, and was originally used by The Rolling Stones on one of their European tours.

4
Sound and vision

The first Roskilde Festival in 1971 saw 20 bands play over two days, though it was still known as the Sound Festival back then.

5
Rock your body

In addition to the famous naked race on the Saturday, there is a lot of bare skin to be seen on the other festival days too. Prudishness and free thinking do not go together. Dr Hook's Nude Show in 1976, which infuriated anti-Roskilde conservatives, has legendary status today.

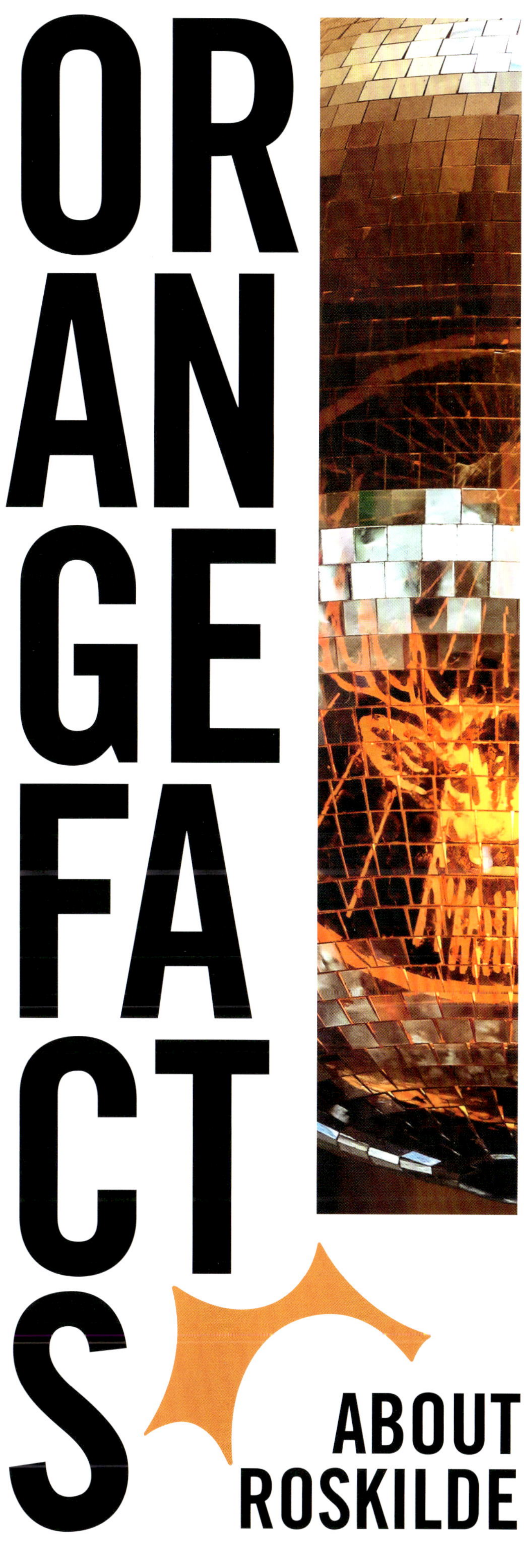

ORANGE FACTS
ABOUT ROSKILDE

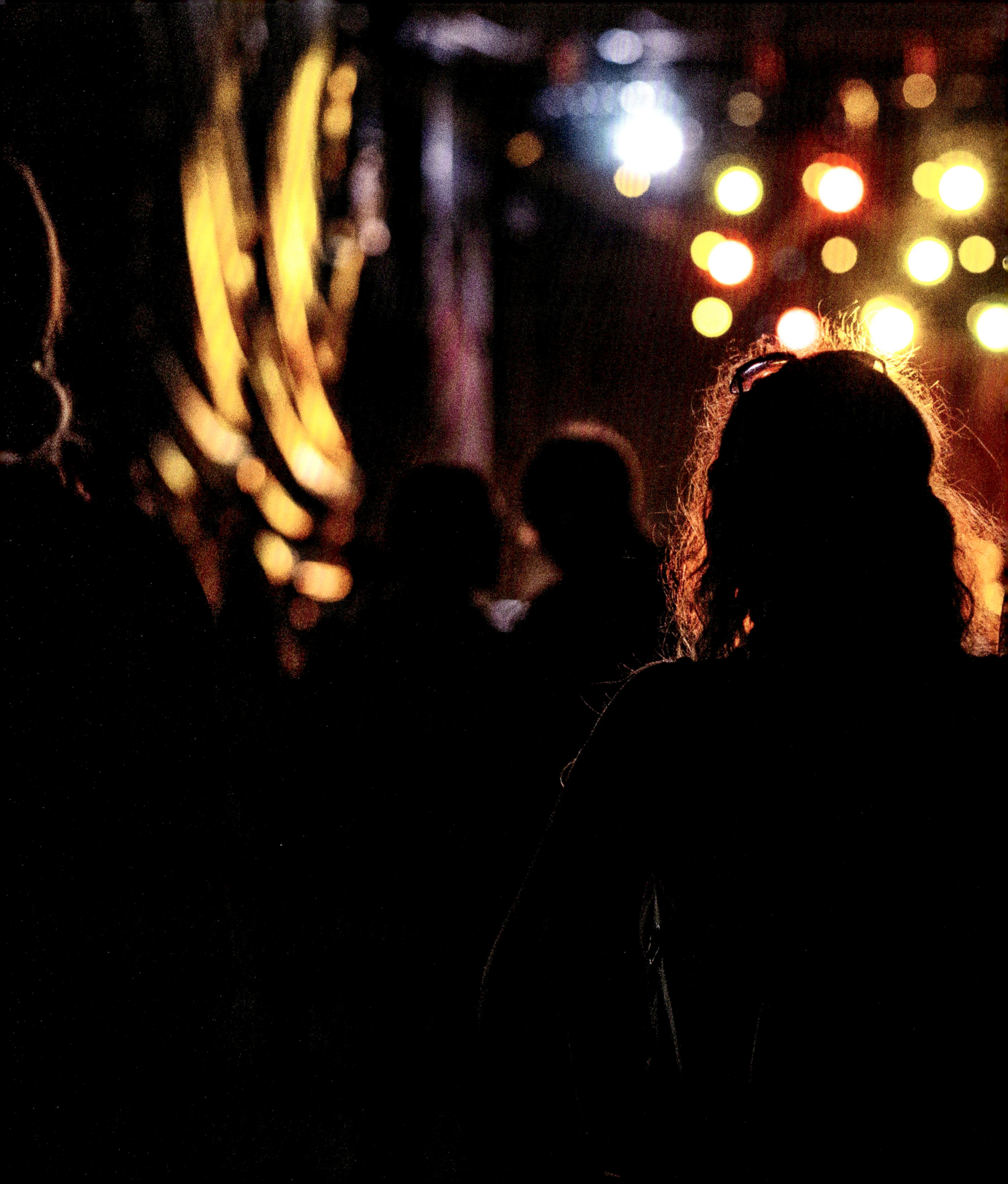

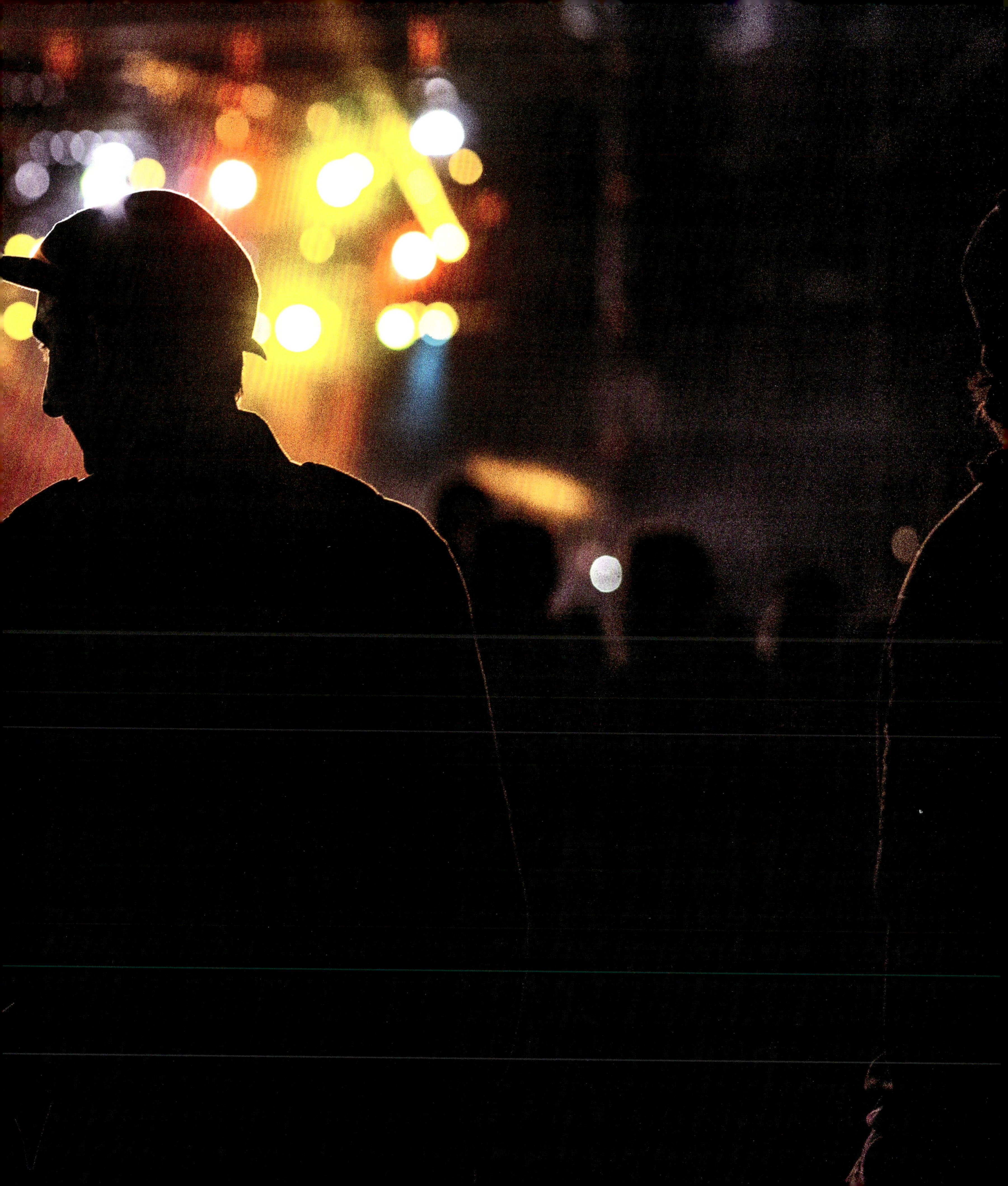

HALALI — THE HAPPY HUNTER'S CALL

The zeitgeist of the early 1950s in West Germany was all about the economic miracle, and the *joie de vivre* that came with it reached Wolfenbüttel too. While Conny Froboess, Heinz Rühmann, Peter Alexander and Heinz Ehrhardt delighted the people of Germany with cheer and merriment, the Mast company sent out a hunter's call (known as *Halali* in German, signifying the melodic horn tune played at the end of a hunt) for people to 'return to the happy days of old with a glass or two of Jägermeister', promising that 'Jägermeister always creates a good mood'. The brand's first advertising icon was Waidmanns Waldi, a gun dog whose loyal puppy-dog eyes told his master a hunt had been successful. With cutely designed advertising motifs, the little dachshund melted everyone's hearts in 1958.

He was followed by the brand's first television commercials, and it wouldn't be long before it scored its next big advertising coup. The new marketing strategy was largely the work of Günter Mast, nephew of Curt Mast. He had been the company's commercial director since 1952, and his unconventional advertis-

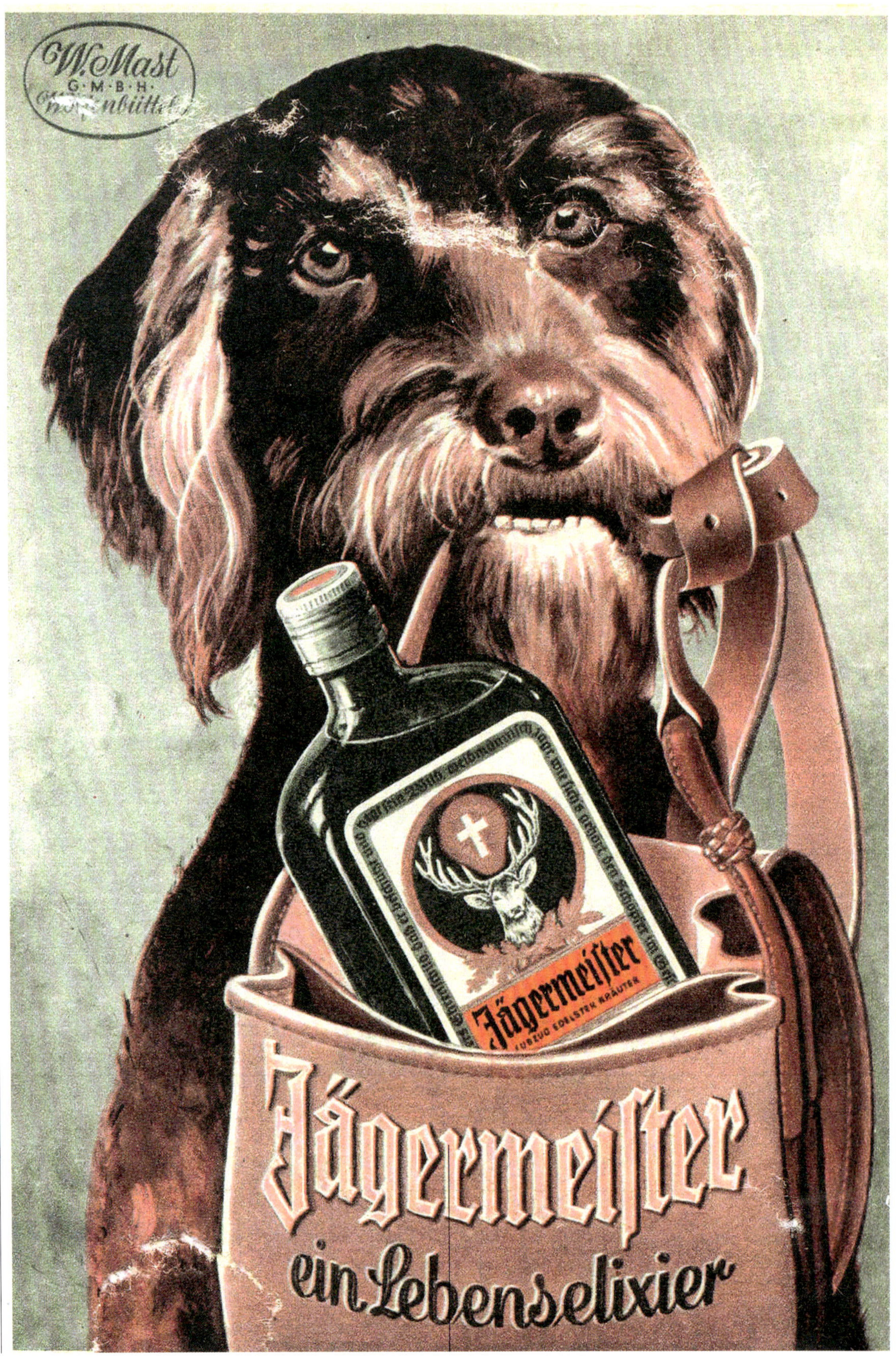

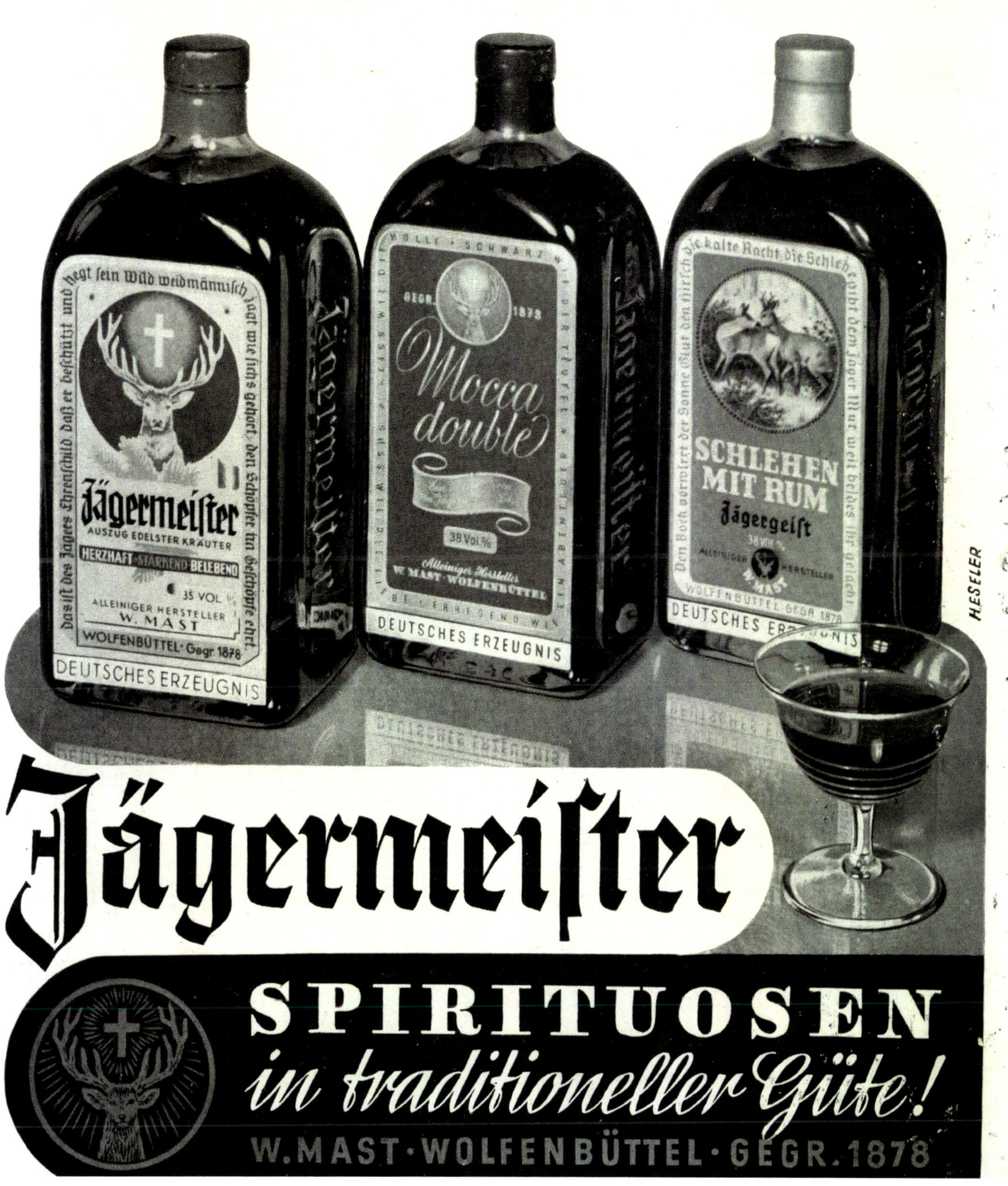

Left: Waidmanns Waldi, Jägermeister's first advertising icon, made its debut in 1958.
Right: In 1952, the product range also included other spirits, until Günter Mast began focusing all his activities on Jägermeister in the mid-60s.

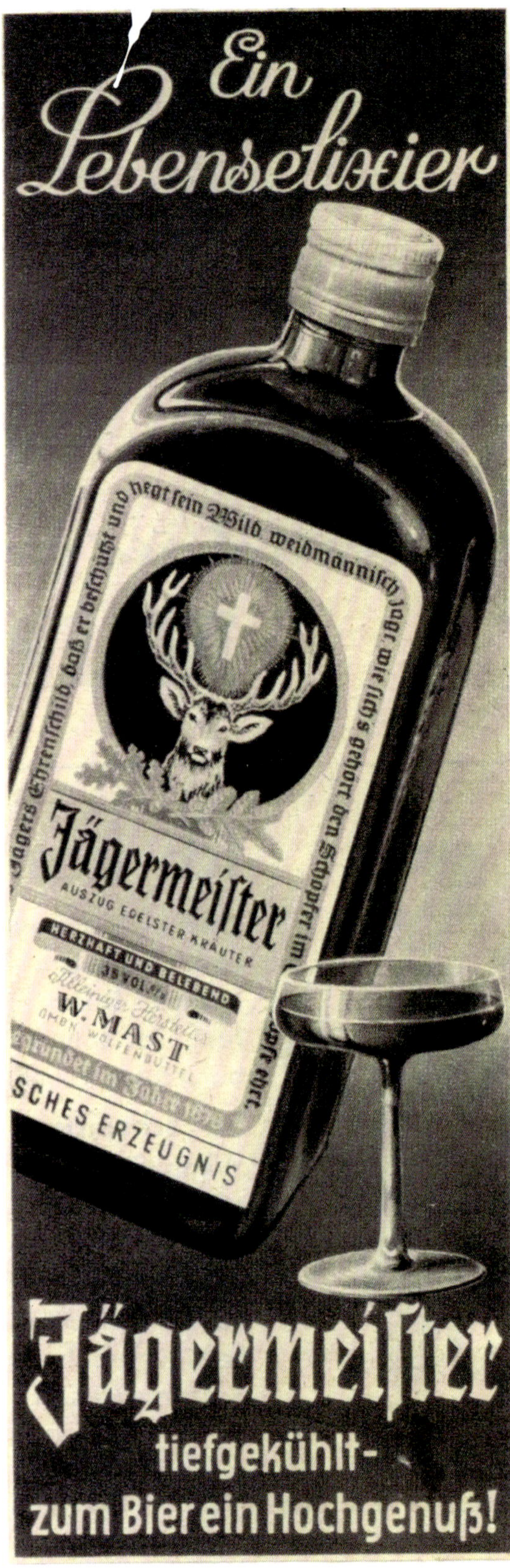

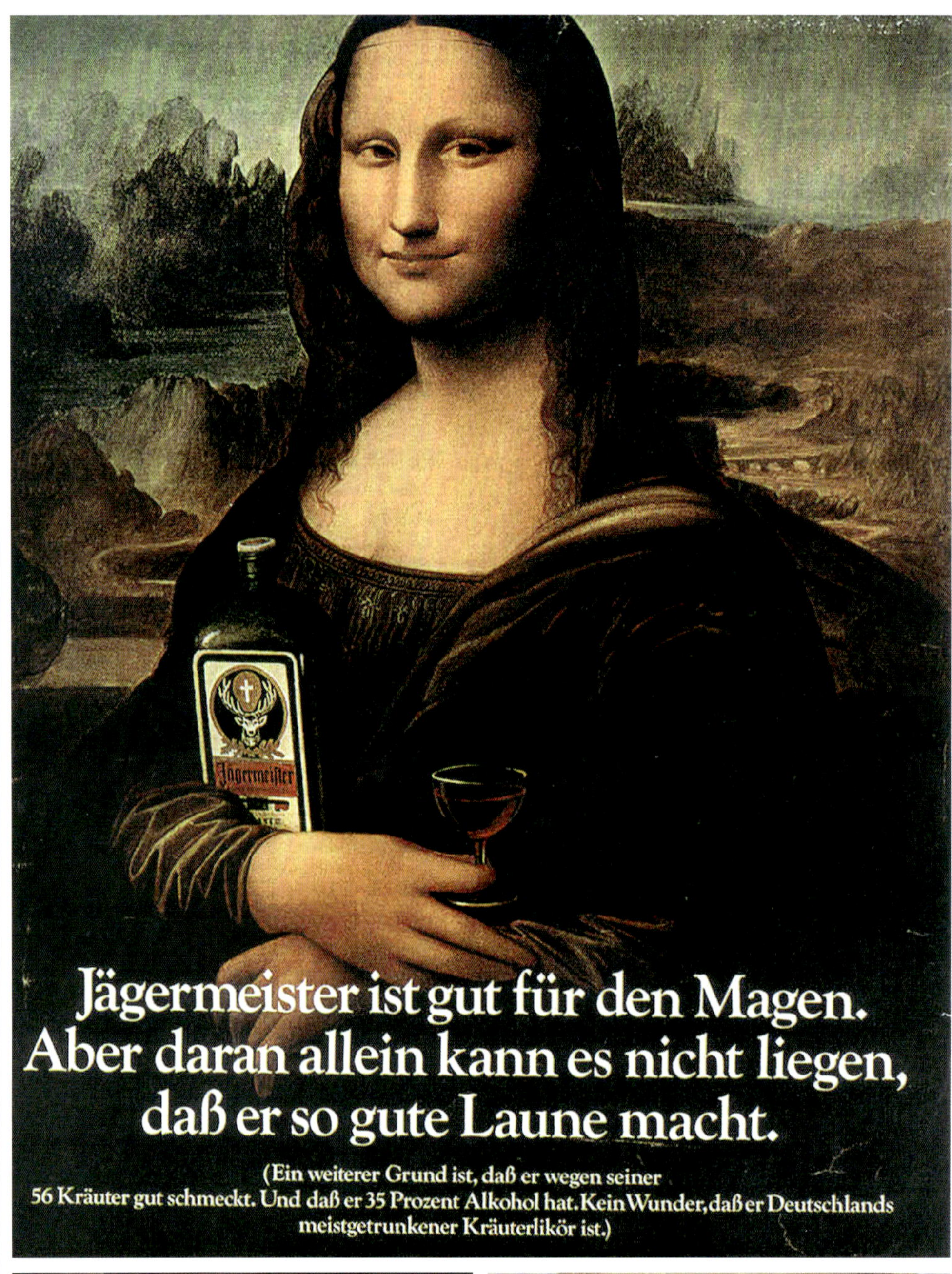

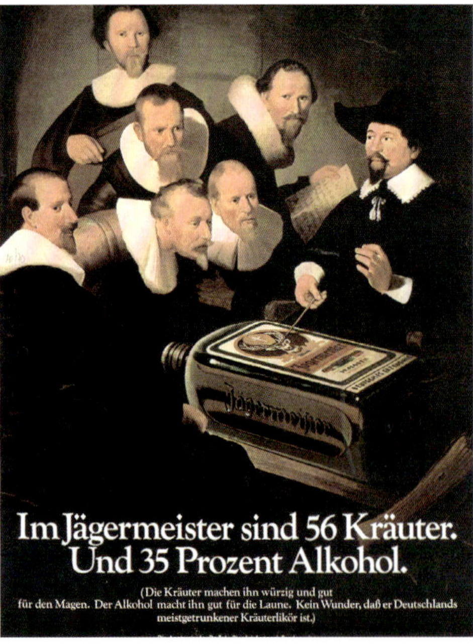

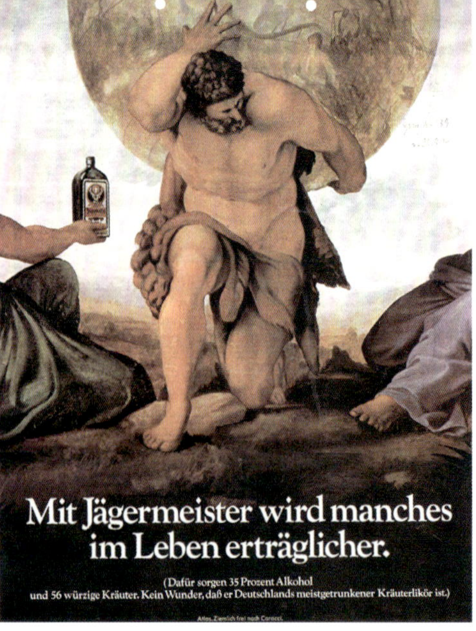

One of Jägermeister's early advertisements from 1957, recommending it be enjoyed chilled. Popular then as now: Jägermeister with beer, also known as 'Deer & Beer'. **Right:** Where art meets herbal liqueur: an advertising campaign from the early 70s, which saw Jägermeister featured in world-famous paintings

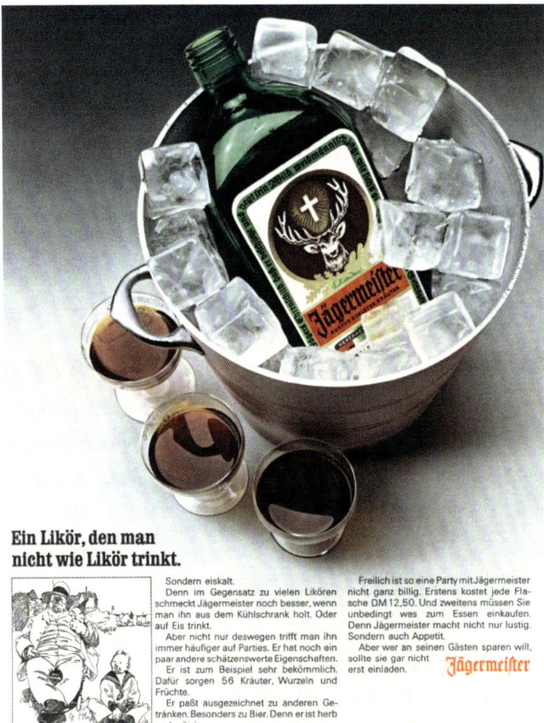

Typically Jägermeister: humour and self-depreciation have become an integral part of Jägermeister advertising, as this campaign from the late 1960s shows.

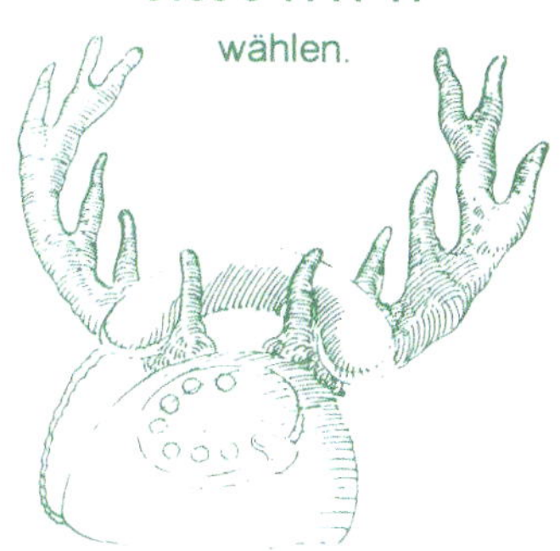

More than 17 million callers dialled the Rudifon hotline between 1980 and 1997, just to listen to the iconic bellowing of Rudi the Jägermeister stag.

ing strategies had helped Jägermeister gain new levels of popularity. Not only was he responsible for radically increasing the rather modest advertising budget in the early 1950s, but it was also thanks to him that, from then on, the product range focused solely on Jägermeister. The 1960s marked the start of many humorous and innovative campaigns alongside roadside hoardings and sports advertising. And self-depreciation was not out of bounds for Jägermeister either. 'By Ash Wednesday, you'll see that Jägermeister doesn't just tease,' said one late 1960s billboard, referring to the brand's clownery. Humour and self-depreciation were the much-appreciated attributes that would define Jägermeister's advertising for decades thereafter.

By the early 1970s, Jägermeister had gone from being a hunters' speciality to Germany's number one herbal liqueur, giving Düsseldorf advertising agency GGK an idea that was as simple as it was genius. The spectacular campaign it devised would mark a milestone in Jägermeister's public image. Each ad showed people drinking Jägermeister for all kinds of reasons, always with a totally logical explanation. One bartender's quote, for instance, translated to, 'I'm drinking Jägermeister because business picks up if you've had one too.' Each ad in this one-of-a-kind campaign would only appear once, hugely boosting its entertainment value. The resulting success was so great that many ideas for slogans were soon being submitted by the consumers themselves, who then ended up featuring in the respective ads. This campaign of unique-edition ads was a giant success, causing a sensation for several years. It ran from 1973 to 1988 (and once again between 1996 and 1998), producing over 3,500 different ads – and not just in Germany. The slogans were also adapted into foreign-language versions abroad.

The tremendous longevity and continuity of its advertising campaigns saw Jägermeister soon become one of the best-known brands in Germany – in part also thanks to other innovative measures that got many tongues wagging (or should that be ears buzzing) in the 1980s. By calling a hotline featured in ads and on coasters, fans could listen to the rutting call of a royal stag. And for those who didn't immediately get the point of the 'Rudifon', Jägermeister had an illuminating explanation at the ready: 'The best way to wash away the shivers down your spine is with a little Jägermeister,' said the ad. But even the company itself was surprised at how positive the response to

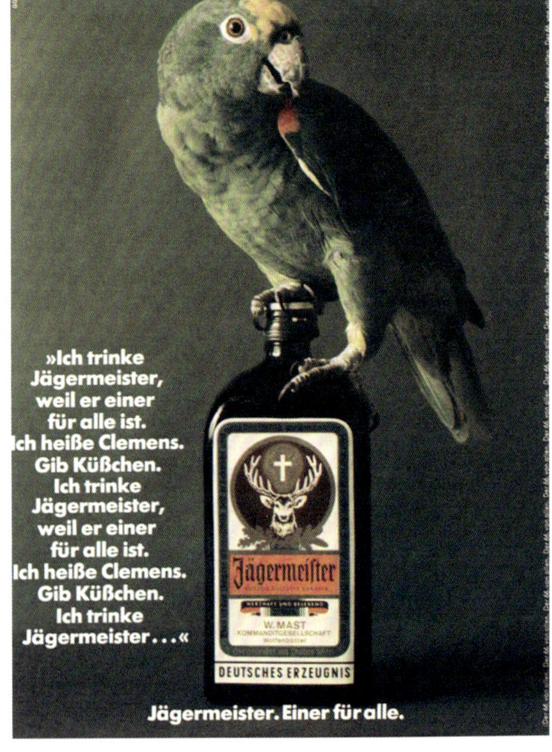

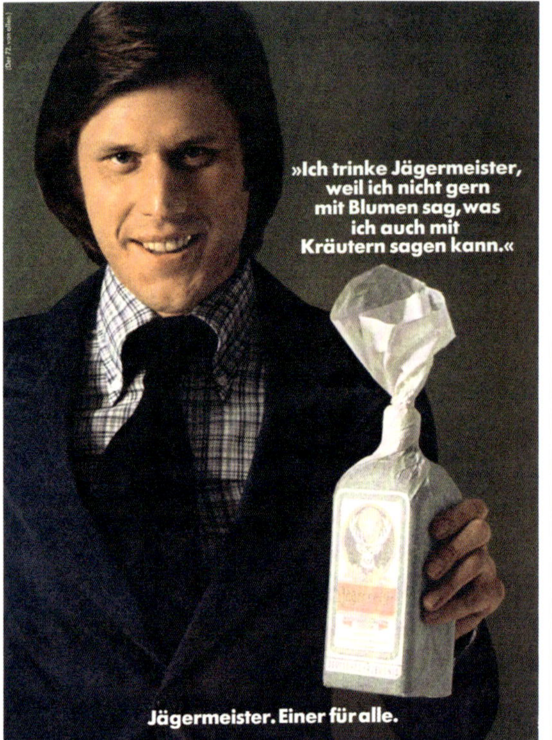

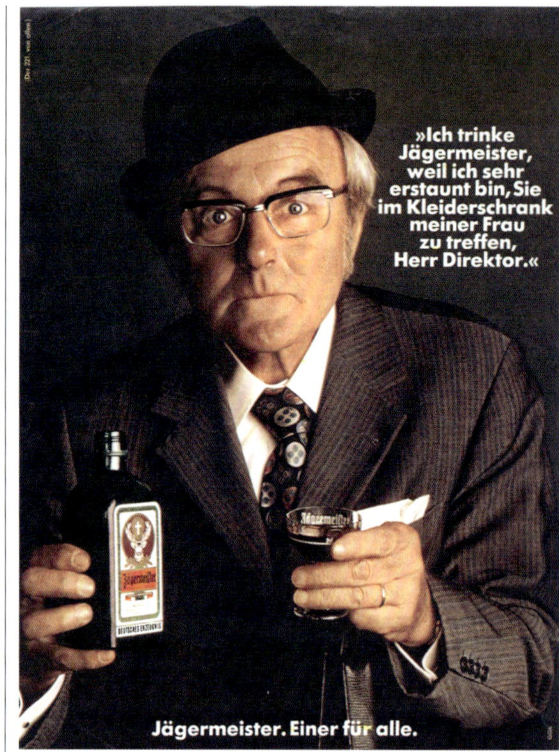

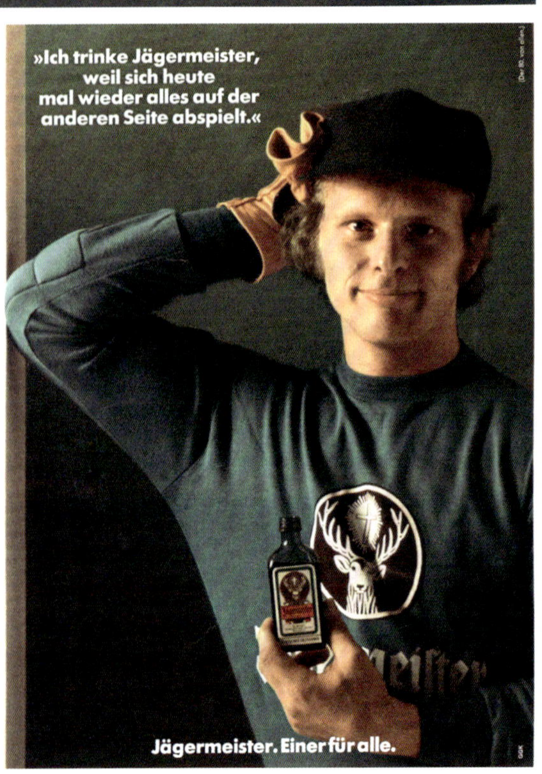

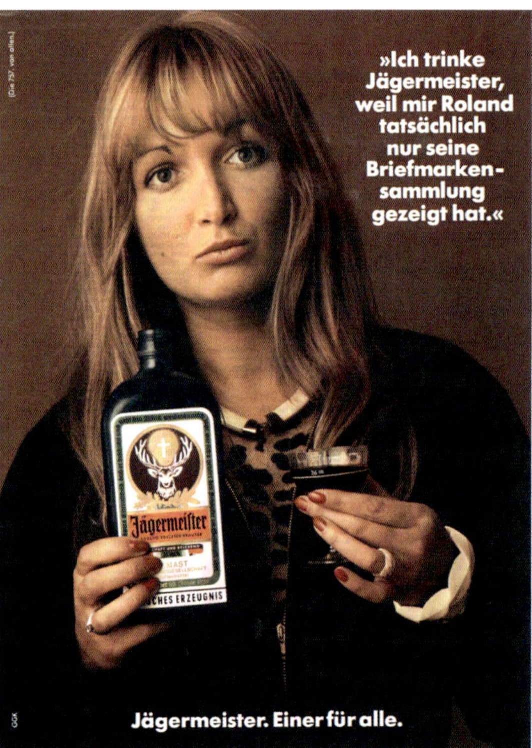

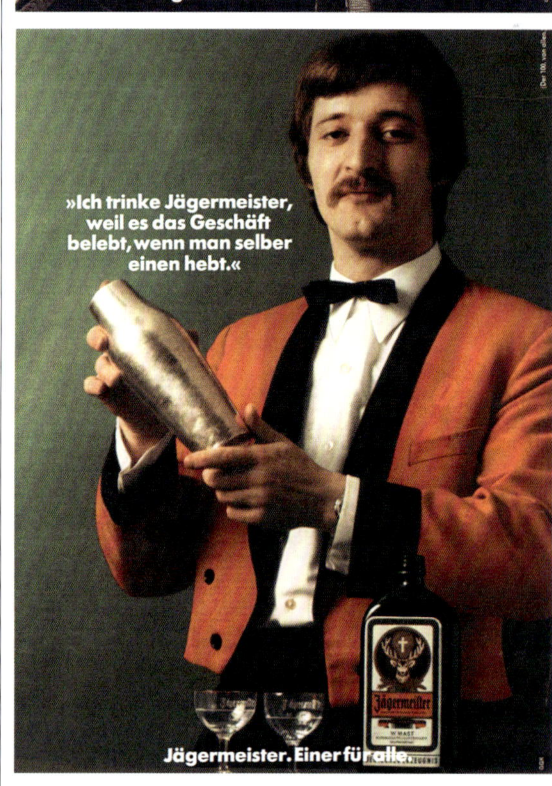

the unusual marketing measure was, with more than three million people calling Rudi in the space of just a few months. In 1984, this figure was over 12 million. A letter to sales-force staff in 1989 stated that 'in total, more than 15 million advertisement-driven contacts have been generated since the campaign began in 1980/81. What makes this number even more impressive is the fact that all of these contacts were at the caller's initiative.'

'As, pleasingly, a relatively large number of calls are still being made despite us not taking any action here ourselves,' as Günter Mast put it, Jägermeister simply kept the campaign going. The Rudifon was so successful that dozens of callers were dialling the stag's number every day until the mid-1990s, even in the United States. By the end of 1997, the company had recorded a grand total of 17,603,684 calls. And, indeed, the famous rutting call can still be heard today if you call Jägermeister and are put on hold while they are trying to connect you.

Jägermeister became a cult brand in the United States in the 1990s. The herbal liqueur's logo lettering – which no one there could read, let alone pronounce correctly – was fascinating enough. Local distribution partner Sidney Frank was a breath of fresh air here; his innovative marketing ideas helped

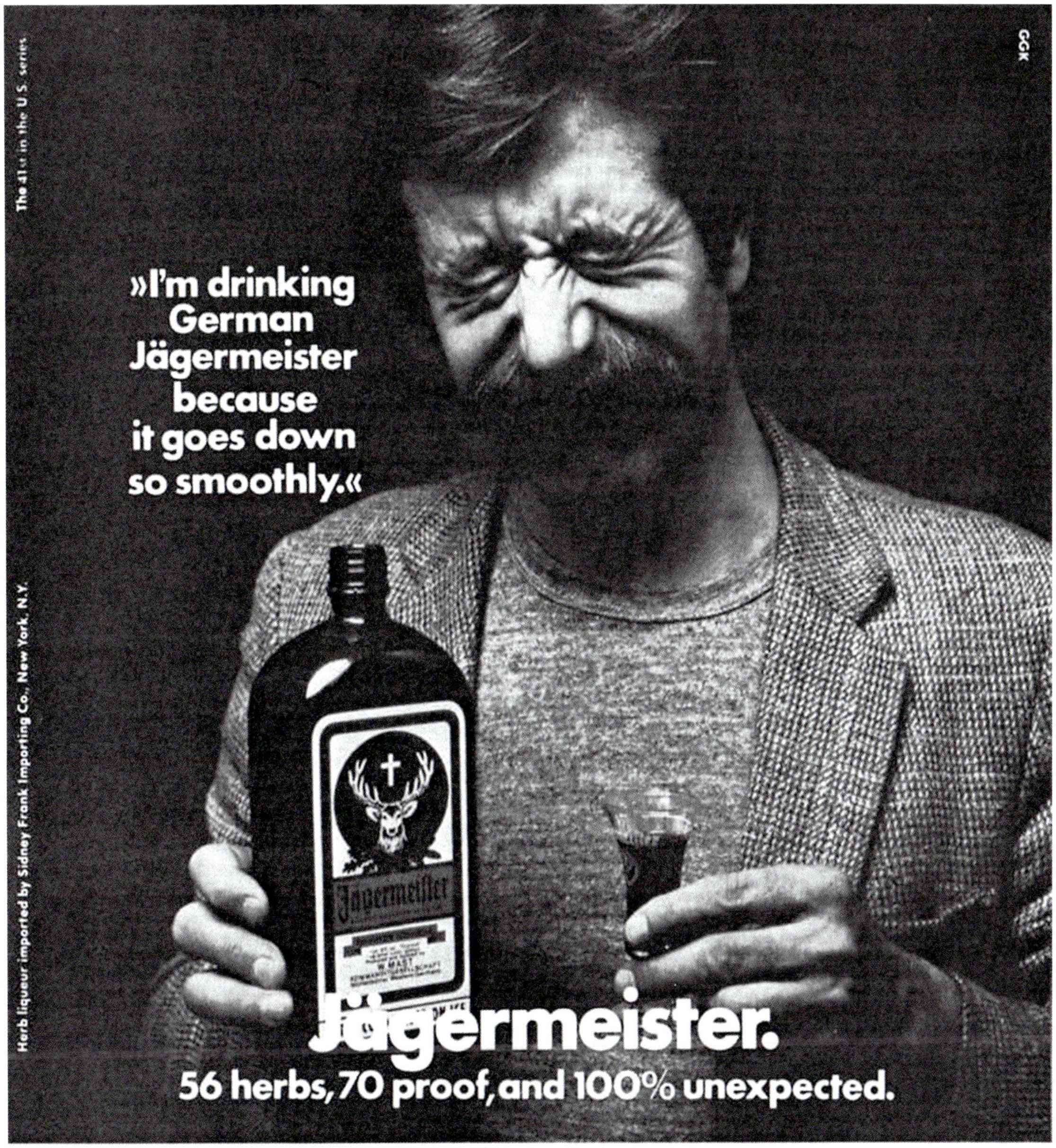

This is the 4,169th ad in the worldwide Jägermeister campaign:

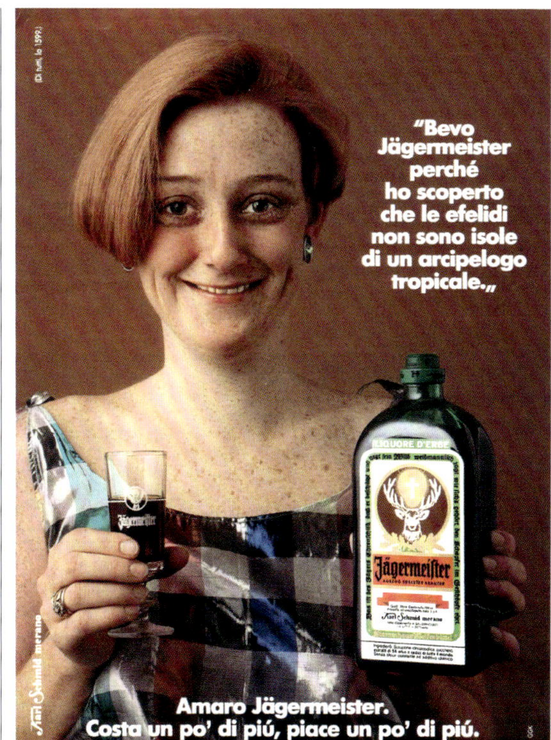

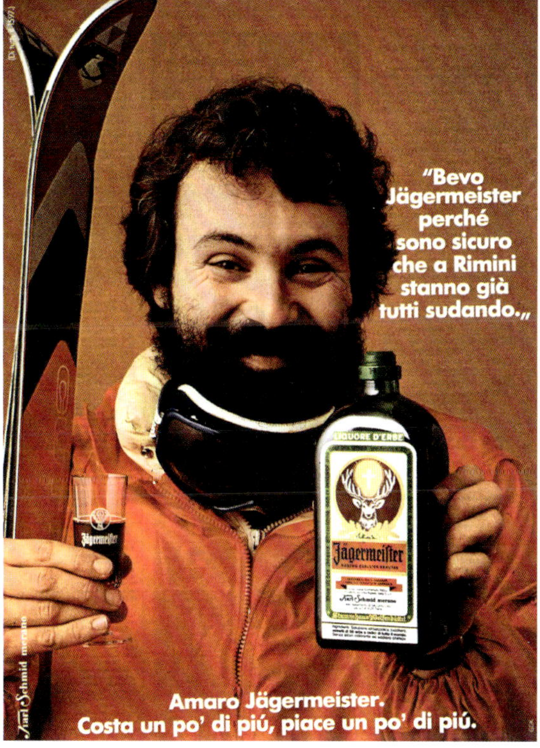

to make Jägermeister one of the most sensational spirit brands ever in the United States within a very short space of time. The cigar-smoking, tuxedo-wearing Frank, who oozed charm and idealism, convinced his fellow Americans that Jägermeister was *the* drink of the new age. It was a time of upheaval on the American spirits market, with trends shifting from highly alcoholic drinks to liqueurs. The secret recipe and the associated rumours of deer blood being one of the ingredients did their thing. Frank was responsible for introducing the now globally known tap machines in the United States, establishing the Jägerettes and Jägerdudes (the first hostesses and hosts in the history of spirits advertising), and setting up an intensive music-marketing campaign. A rather dramatic rejuvenation of the brand was a calculated by-product of the success, and, by 1990, sales figures in the United States were reaching almost dizzying heights. The

An advertising milestone: the legendary, one-of-a-kind 'I'm drinking Jägermeister because . . .' campaign ran for a total of 17 years, and was also adapted abroad, for example in the United States (top left) and Italy (below).

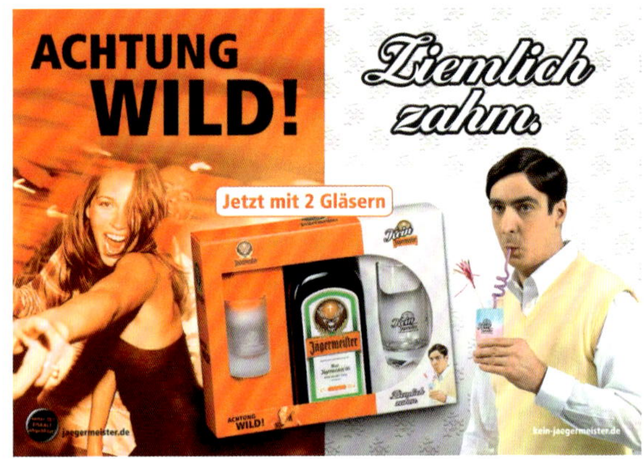

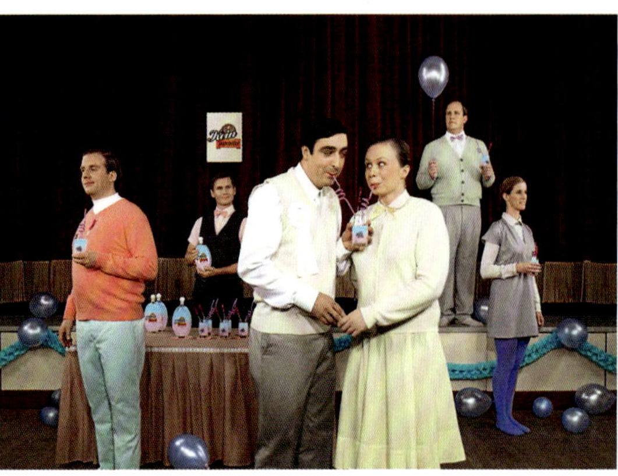

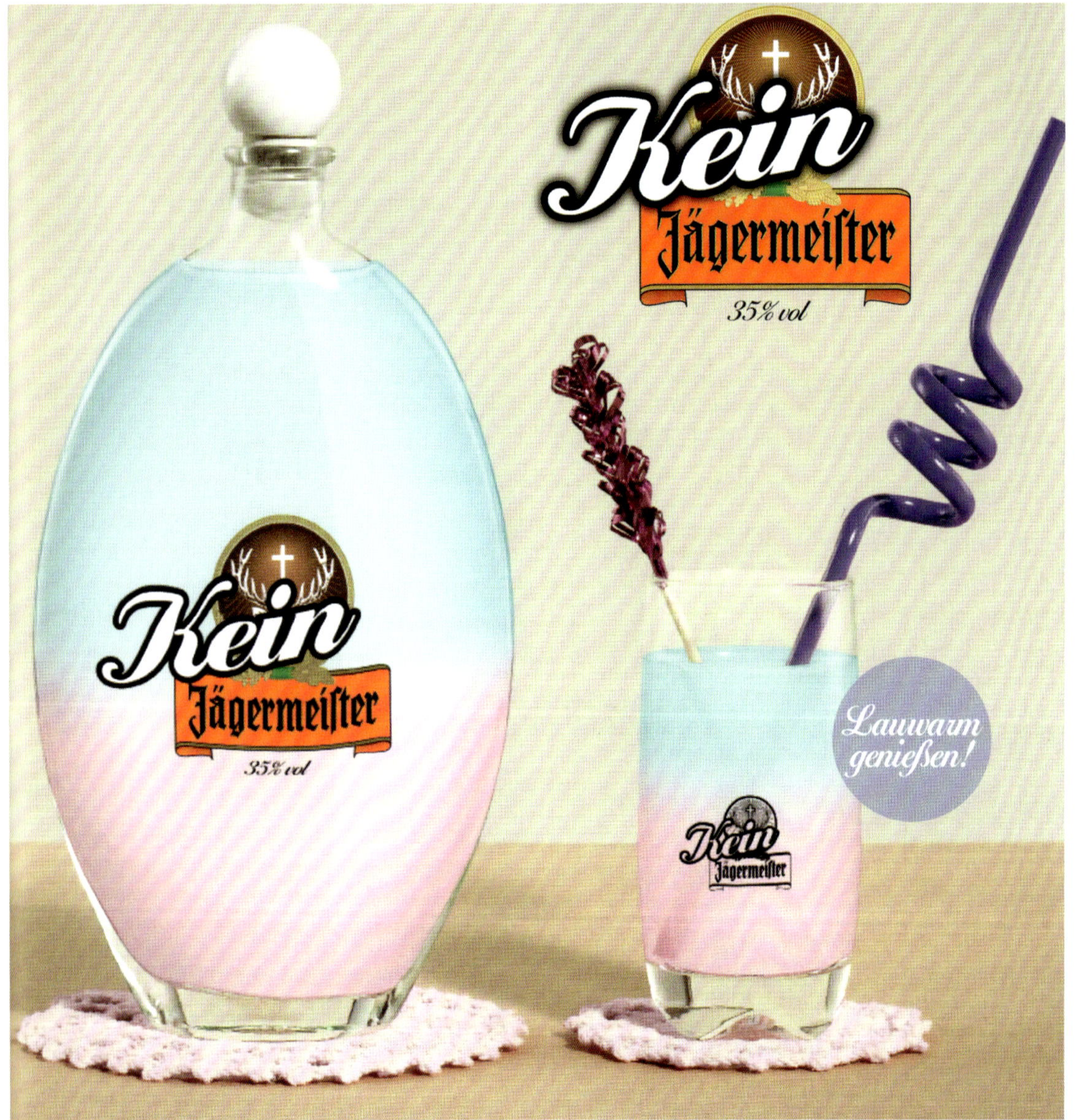

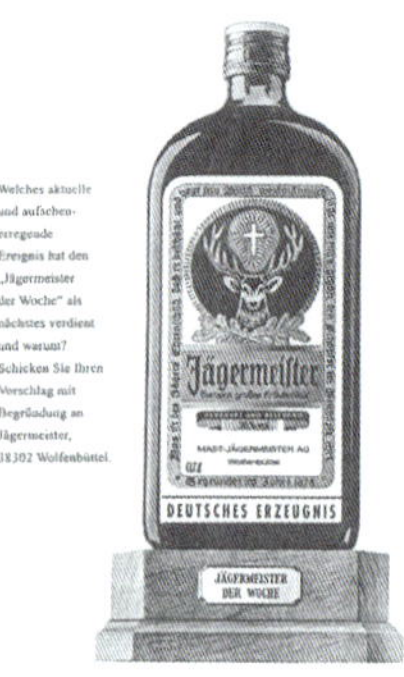

Der **Jägermeister der Woche** geht an den Reichstag, weil er nach seiner erfreulicherweise nur kurzen Zeit der Verhüllung für die Zukunft eine Glaskuppel tragen soll, die kein privater Bauherr für sein denkmalgeschütztes Gebäude genehmigt erhielte.

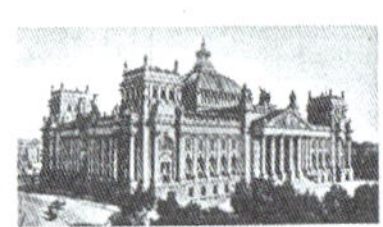

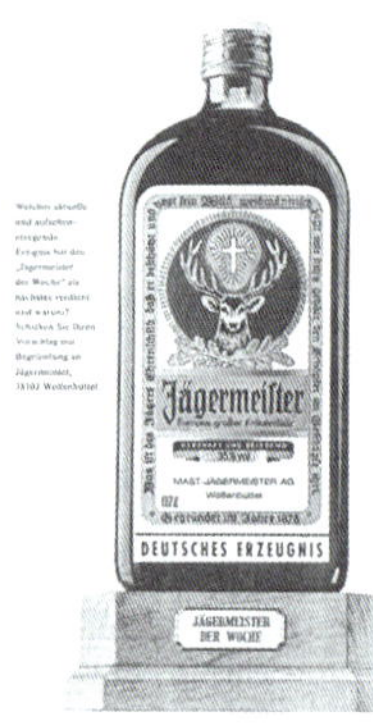

Der **Jägermeister der Woche** geht zumindest an unseren Rudi, weil man sich die Sympathien der Bundesbürger nicht nur im Sturm, sondern ebenso im stillen Wald erobern kann.

Neugierig? Wenn ja, dann wählen Sie Rudi unter der folgenden Nummer an:

030/3961034

Rudis Einschaltquote: Bisher 17 Millionen Anrufe.

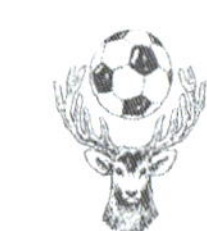

Jägermeister. Einer für alle.

scorecard? Over 3 million bottles sold – nearly double that of the previous year.

From the late 90s onwards, the marketing department in Germany also started taking advantage of Sidney Frank's successes and the findings obtained from the United States. Jägermeister's brand image became radically younger. 'Just how cold is a Jägermeister like this?' two talking deer heads suddenly asked television viewers in Germany in 1999. As part of the Rudi & Ralph television advertising campaign, the pair would make witty comments on bar life. With their pithy, sometimes even provocative, quips, they not only provided entertainment during scheduled advertising breaks, but sometimes they also surprised the audience by hosting short weather-forecasting segments. And Rudi and Ralph, too, very soon became iconic cult figures.

Left: What is the opposite of Jägermeister? That was the question posed – and promptly answered – by a 2008 campaign. It turns out things would be *ziemlich zahm* (quite tame)! Top: It happens even to cult brands: the 'Jägermeister of the week' advertising campaign failed to generate much hype in the mid-90s.

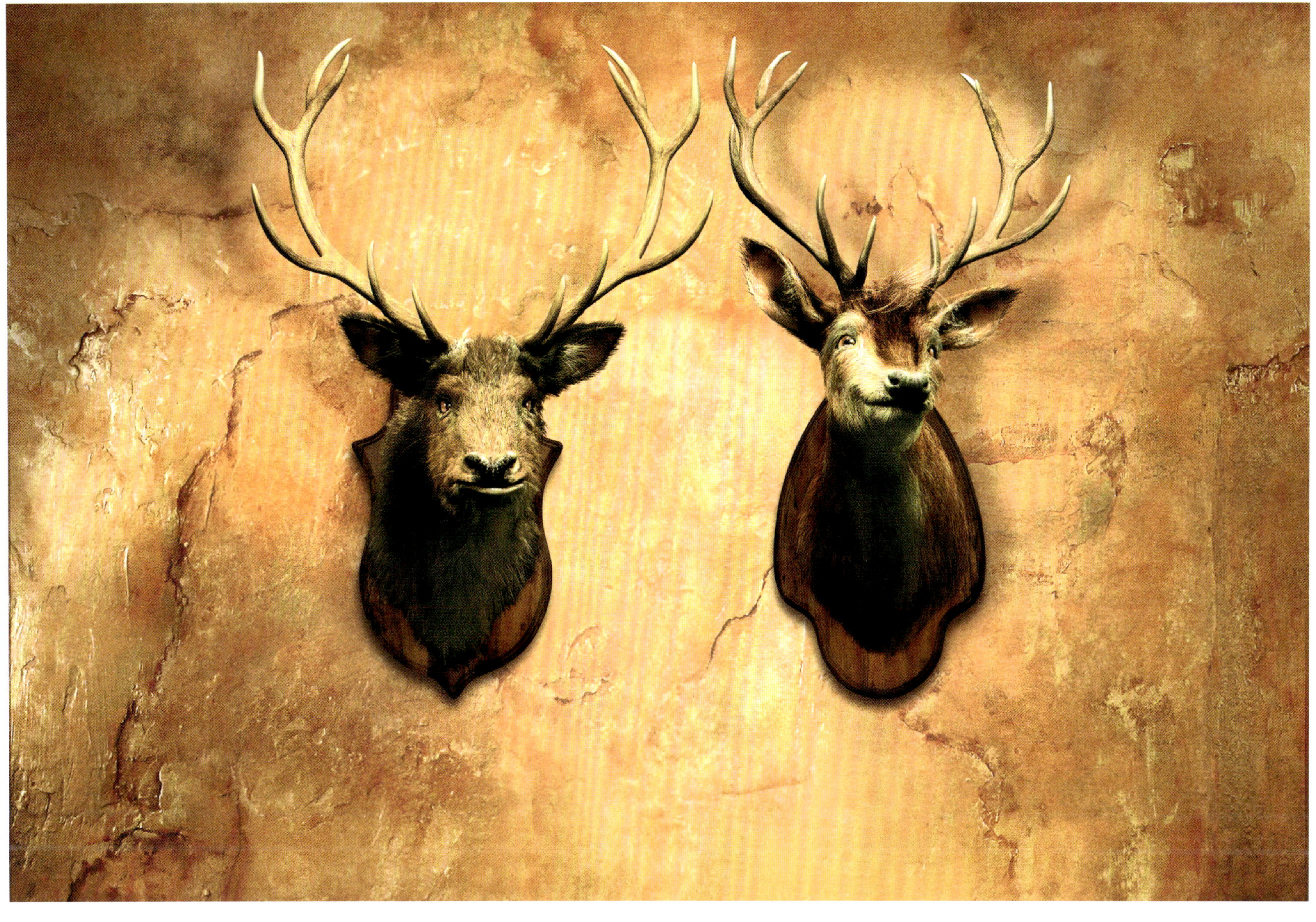

The noughties saw Jägermeister's branding turn flashy and loud; its promotions and campaigns mainly featured the colour orange, later increasingly shifting to a black theme. What would life be like without Jägermeister? Another original campaign provided an answer to this question in April 2008. A brand philosophy centred on a fictitious drink with a name that translated to 'No Jägermeister – quite tame' was cultivated in television and cinema ads, print ads, billboards and on a specially dedicated website. The bland drink and dull world were made out to be so boring that even homebodies wanted the wild Jägermeister party scene back – a clever move by the marketing pros, once again reminding consumers that, when it comes to getting wild, Jägermeister is simply a must.

The claim *Wer, wenn nicht wir* (Who, if not us), and other advertising spots such as Atrium, *Wofür Freunde brennen* (What friends are passionate about) and *Monument der Freundschaft* (Monument of friendship) are still fresh in the minds of German consumers to this day. But the media landscape was changing at the time; advertising was becoming faster-paced, needed to be more complex and sophisticated, and had to work on all channels. It had thus become all the more important to do things that would remain memorable, and the best way to achieve this was through experiences. Jägermeister took advantage of the super-hot summer of 2018 to launch a super-cool campaign, placing ice blocks containing bottles of Jägermeister at public squares in some of Germany's larger cities. Not only did the 850-kilogram blocks nail the idea of the perfect drinking temperature, they also served to directly advertise this. Super cool and so very Jägermeister.

In early 2020, however, the COVID-19 pandemic brought the world to a halt. Places of interaction and socializing, which had previously been so loud, colourful, diverse and inspiring, had to close their doors indefinitely. Under the motto 'Save The Night', Jägermeister launched an international initiative based on donations, microfunding and creative online entertainment in a bid to assist the nightlife community. A special Jägermeister bottle, the #SAVETHENIGHT, with a label designed by acclaimed German illustrator Max Löffler, was an added source of support; upon its launch, Jägermeister provided a further 1 million euros for artists, creatives, bartenders and club owners in the participating markets. After all, it's those masters of the night who have always made nights better, wilder, more creative and unique. ⩗

<u>Top:</u> Achtung wild!-campaign – the two talking stags, Rudi and Ralph, gained cult status between 1999 and 2009 with pithy sayings and humorous comments. <u>Overleaf:</u> An ad from the Superkühl campaign in Germany 2018.

BE THE
MEISTER

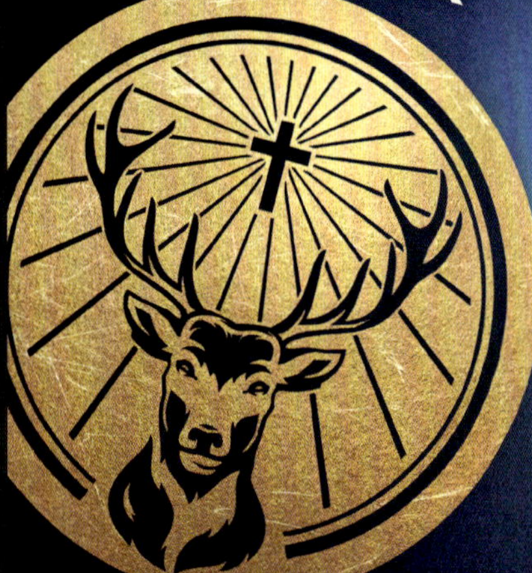

NOT
EVERYBOD
DARLING?

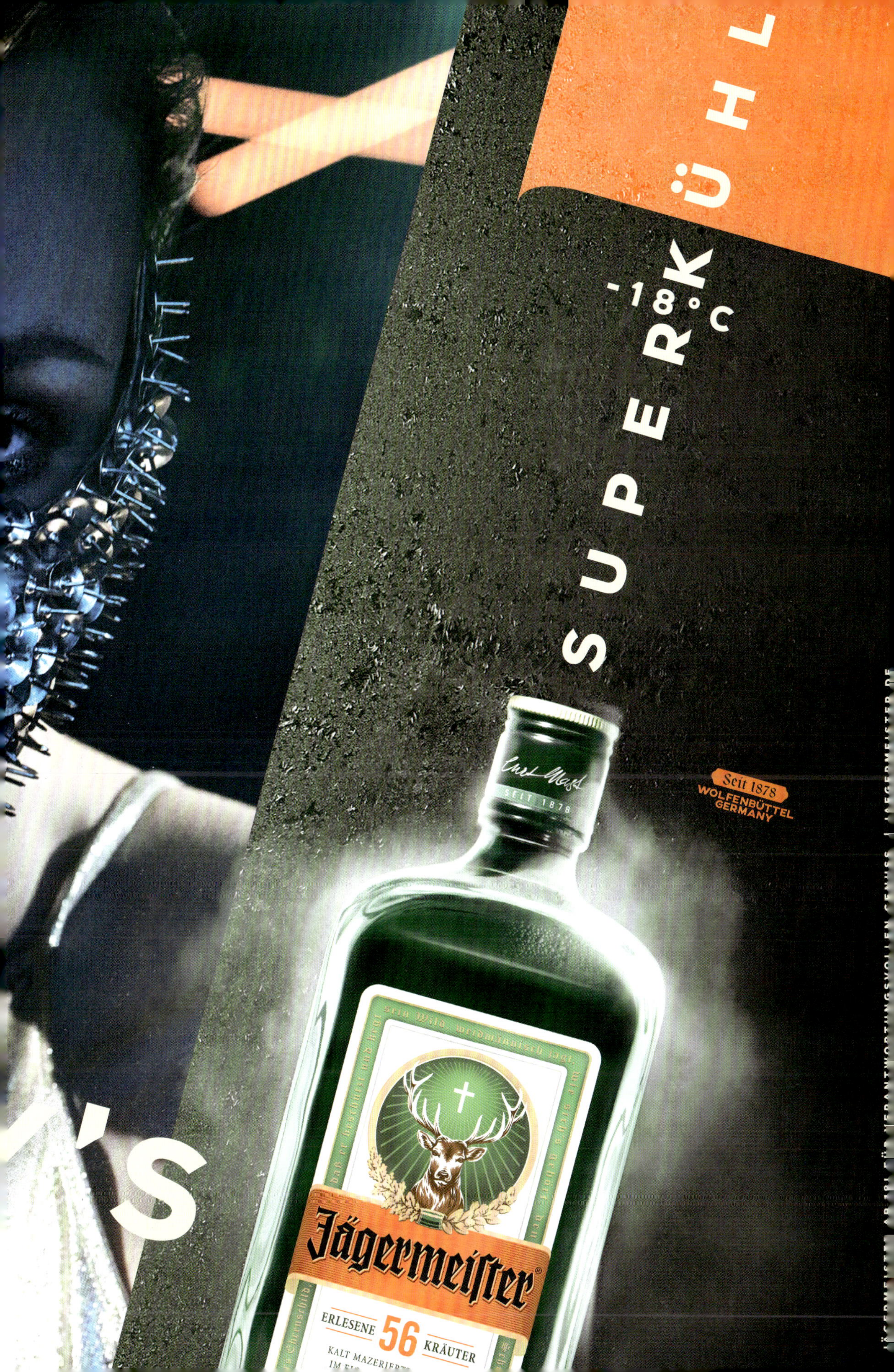
SUPERKÜHL
-18°C
's
Seit 1878
WOLFENBÜTTEL
GERMANY
Jägermeister
ERLESENE 56 KRÄUTER
KALT MAZERIERT
IM EIS
Seit 1878

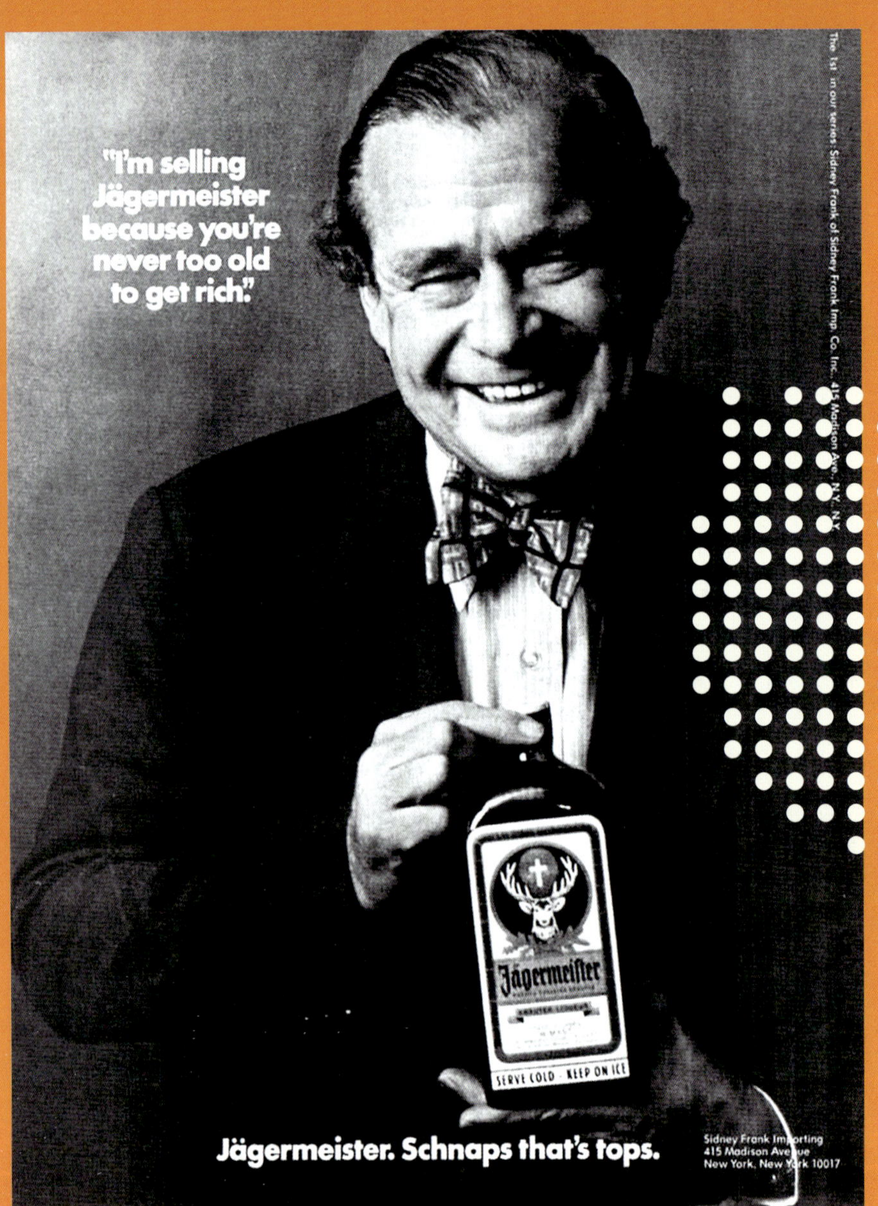

HOW JÄGERMEISTER CONQUERED THE USA

Mercedes Benz, Bayern Munich, Adidas, Rammstein: many German brands enjoy cult status in the United States. But there's only one German spirit brand known by everyone from Alaska to Louisiana: Jägermeister. Today, the German herbal liqueur is the number one party shot for many Americans, and its triumph in that country started about half a century ago.

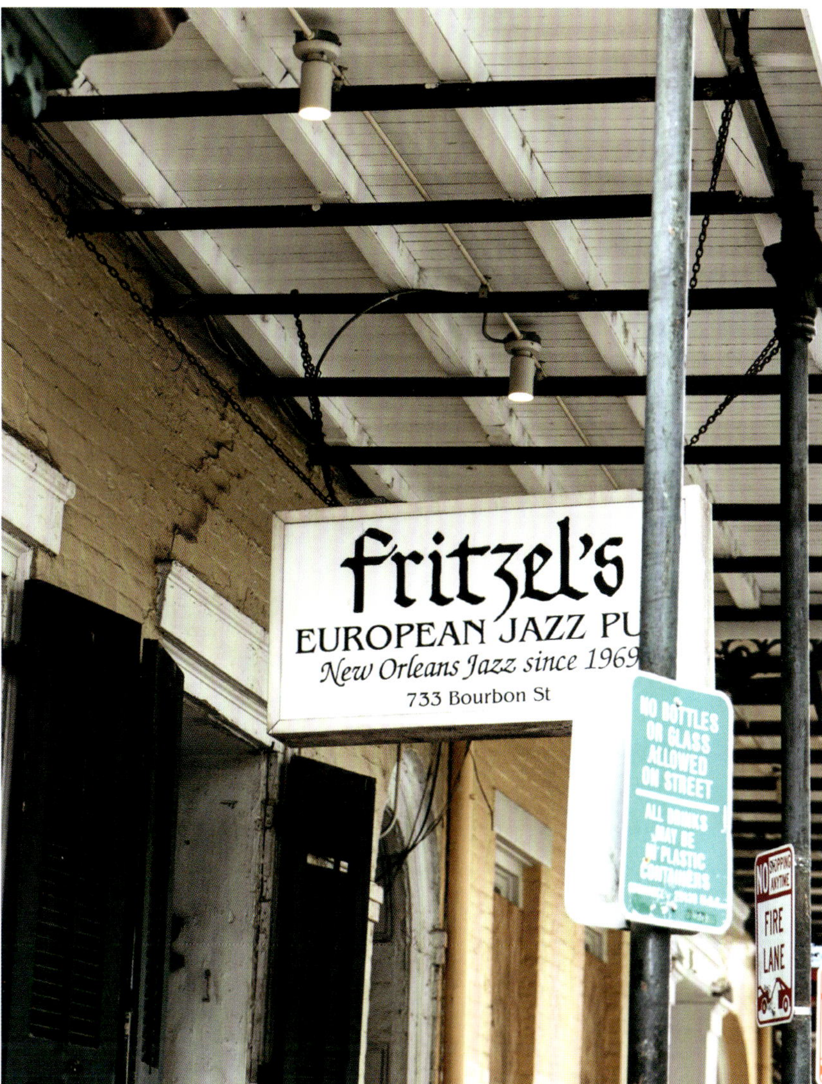

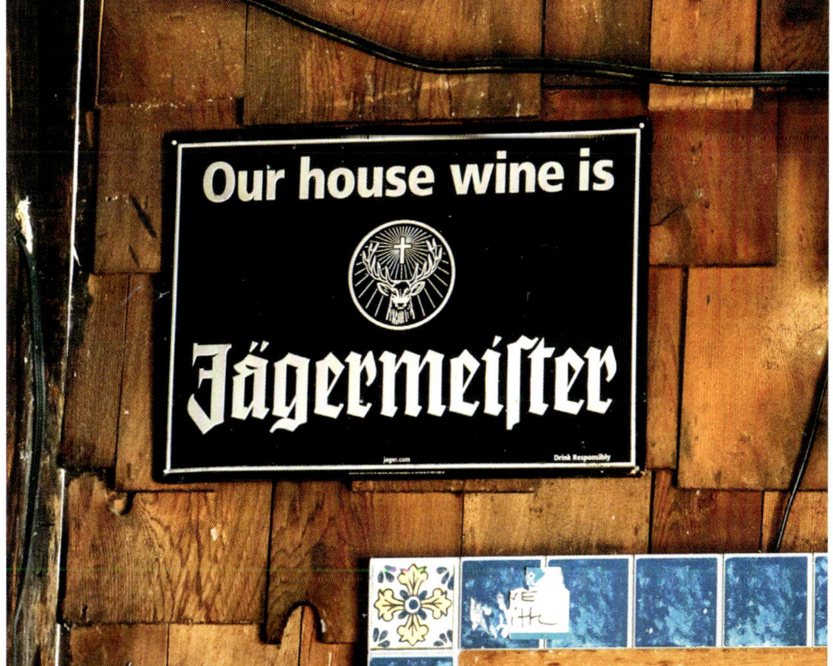

The Jägermeister cult took off in the United States at Fritzel's Bar in New Orleans. To this day, guests come specially to the jazz bar to enjoy an ice-cold shot of their favourite herbal liqueur at the long oak counter. <u>Left:</u> In the 1970s, Sidney Frank, Jägermeister's distribution partner, successfully adapted the one-of-a-kind 'I'm drinking Jägermeister because …' campaign for the US market – even featuring in it his own freely voiced admission: 'I'm selling Jägermeister because you're never too old to grow rich.'

Very German and mysterious

Convinced that the patrons at his Fritzel's European Jazz Club in New Orleans, a 1920s-style jazz bar, would love Jägermeister, the resourceful bar owner, Gunter 'Dutch' Seutter, a German immigrant, served the herbal drink primarily to university students, who came to party at his bar even on weekdays, turning Jägermeister into a popular drink with the in-crowd. It was considered 'very German', its lettering alone captivating consumers, who were unable to read it, let alone properly pronounce it. The carefully guarded secret recipe also sparked wild speculation that would only make the beverage even more popular. It was rumoured to contain cannabis, deer blood or even Valium, the last of which was due to an article in a regional daily newspaper that had described Jägermeister as 'liquid Valium'. But this only further fuelled curiosity Everyone wanted to try the mysterious brew from faraway Germany – and Jägermeister soon became legendary well beyond Louisiana's state borders.

The carefully guarded secret recipe also sparked wild speculation that would only make the beverage even more popular. It was rumoured to contain cannabis, deer blood or even Valium

Almost simultaneously, two businessmen set off to look for a new partner, on both sides of the ocean. As attractive as the American market was for German brands, it could also be very confusing. The distribution channels – importers, wholesalers, retailers – the state monopoly and the official requirements were a minefield for Europeans. This was a good enough reason to team up with a competent partner known for being a top dog in the American spirits industry: Sidney Frank. The distributor, who had just gone independent and whose trademark were oversize cigars and garish green, or sometimes even pink, suits, was considered a quick-witted and astute businessman and was on a quest for a new brand. With the support and financial backing of his company Sidney Frank Importing,

Jägermeister swiftly developed an advertising strategy for the US market, based on a long-established, tried-and-tested model from Germany: 'I'm drinking Jägermeister because . . .', adapted for American consumers and featuring the additional slogan of '56 herbs, 70% proof and 100% unexpected'. But after the campaign launch in Manhattan, it soon became clear that what had worked like a dream in distant Germany had proved to be a flop in the American East Coast metropolis of New York. There were no herbal liqueurs in the US and the Americans simply didn't know what Jägermeister actually was.

When was the last time you were Jägermeistered?

But Sidney Frank and his German partner firmly believed in Jägermeister's tremendous market potential. In the well-founded hope that going from west to east was a better way to introduce new brands in the United States, they dared to start afresh on the west coast of California. And lo and behold, the new 'When was the last time you were Jägermeistered' billboard campaign launched much more successfully, even though the big breakthrough took a while to happen. This only came after Jägermeister and Sidney Frank focused all their advertising and sales efforts on the southern states, where the way had already been cleared for Jägermeister by Gunter Seutter and the Fritzel's patrons. They had long turned the ice-cold herbal liqueur into a highly sought-after cult drink, so it's no wonder the large-scale campaigns were now catching on from here like wildfire.

The second half of the 1980s marked the start of a real triumph for Jägermeister, thanks to a couple of rather smart ideas and a few strokes of luck. While the now internationally renowned Jägerettes and Jägerdudes promoted tastings all over the place, distributed free merchandise bearing the Jägermeister logo and thus made the brand even more popular, the American spirits market was in a time of upheaval, with trends shifting away from hard liquor such as whiskey and brandy, and more towards lighter spirits and liqueurs. For people seeking new and surprising flavour experiences, the mysterious German herbal liqueur came right on cue. Unlike Germany and other countries of Europe, America has no tradition of herbal liqueurs, making the Jägermeister product unrivalled, unprecedented and unique. The consequence? A downright Jägermeister boom that had well and truly gripped America by the early 1990s at the latest.

Freshly tapped, ice-cold

Sidney Frank and Jägermeister had achieved the impossible. They had made an unknown German herbal liqueur one of the most popular spirits in the USA within the space of just a few years. Suddenly, all of America seemed to be 'Jäger crazy', and bartenders couldn't top up glasses fast enough. But how should you serve Jägermeister ice-cold without the bottles disappearing in the refrigerators of the bars and thus invisible to the guests? The solution: Sidney Frank swiftly initiated development of the Jägermeister tap machine, a high-performance tap system that not only enabled shots to be poured quickly and cleanly at the perfect ice-cold drinking temperature, but also helped Jägermeister gain a prominent profile right across the country in no fewer than 4,800 bars, clubs and pubs. To this day, Sidney Frank's ideas remain among the most innovative marketing campaigns in the American spirits industry, attracting considerable media attention even then. Reputable publications such as *The Wall Street Journal* and *Forbes Magazine* reported on the wild promotion concepts that had sparked a veritable boom in the country, which had seen Jägermeister receive the Barkeeper's Best Award of Excellence as early as 1994.

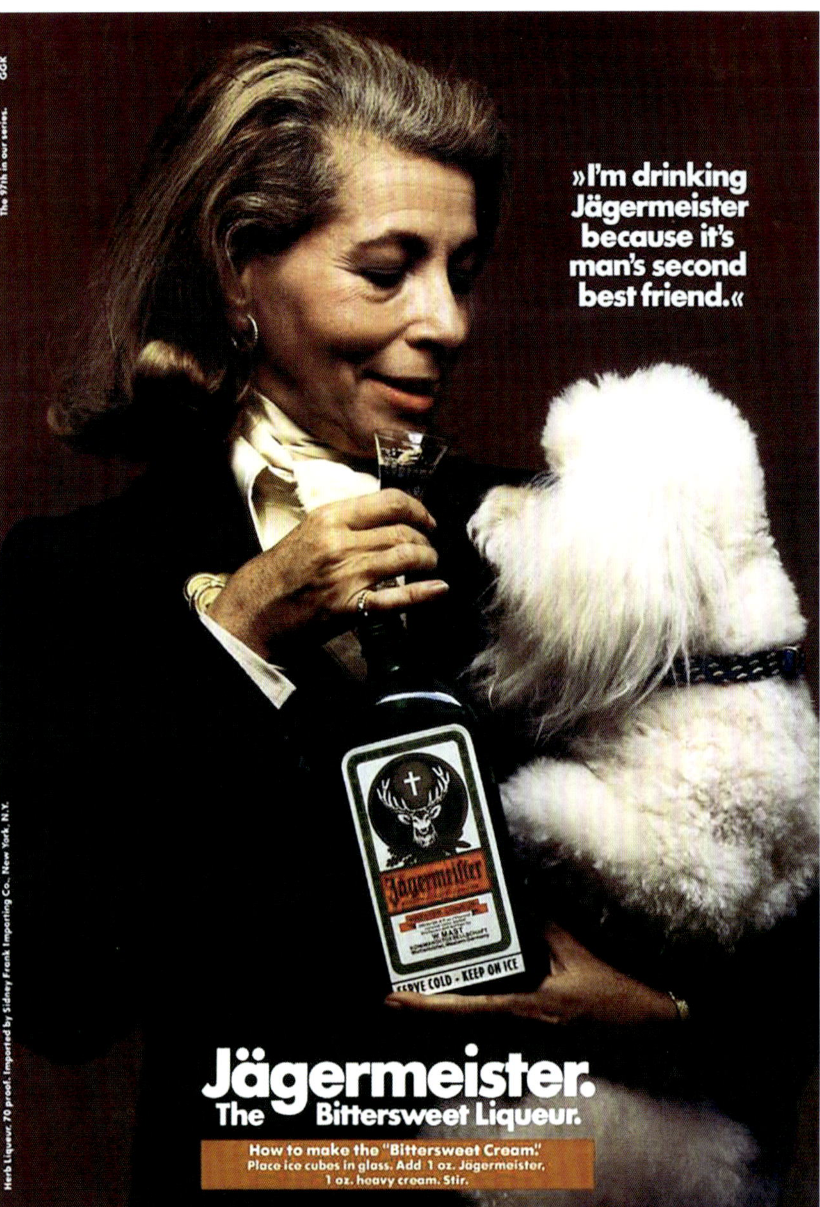

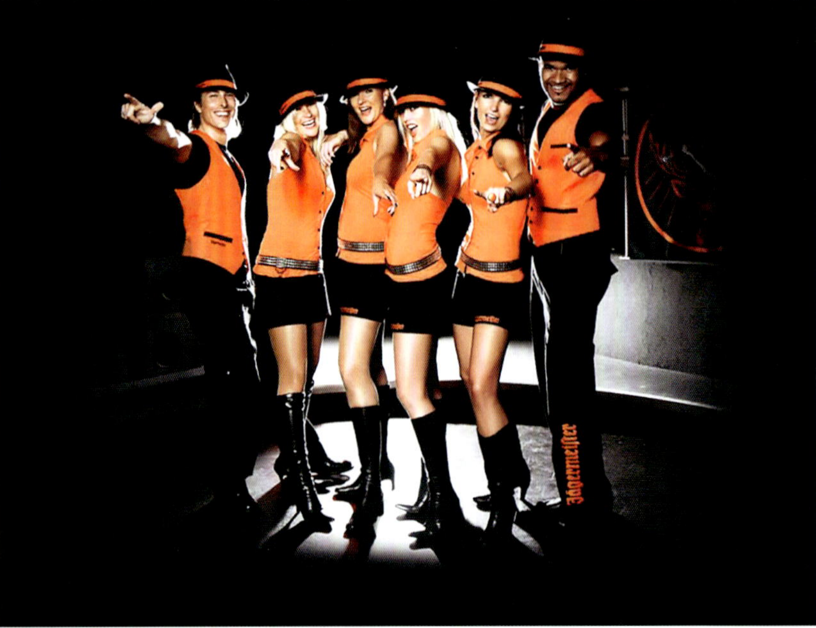

Jägerettes and Jägerdudes have been representing the cult brand since the 1980s – an innovation by Sidney Frank, which is now known around the world.

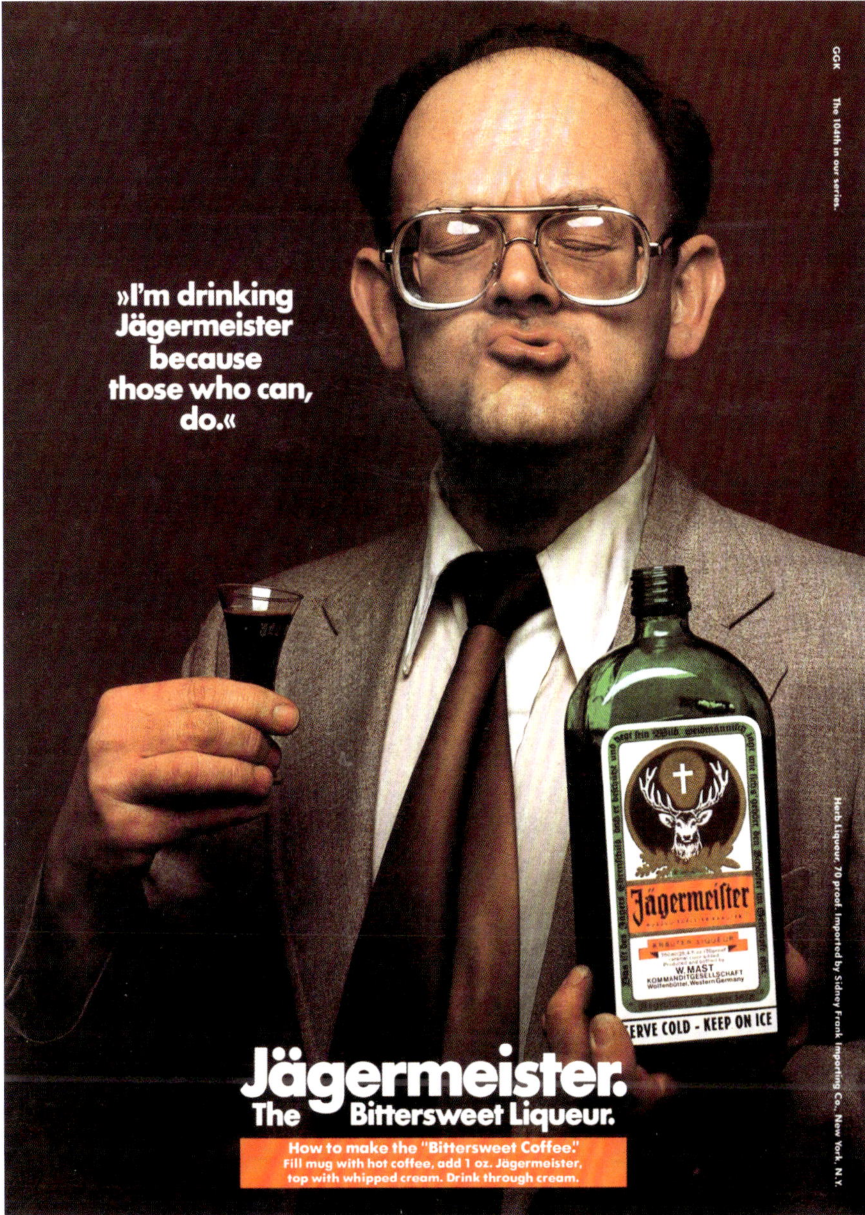

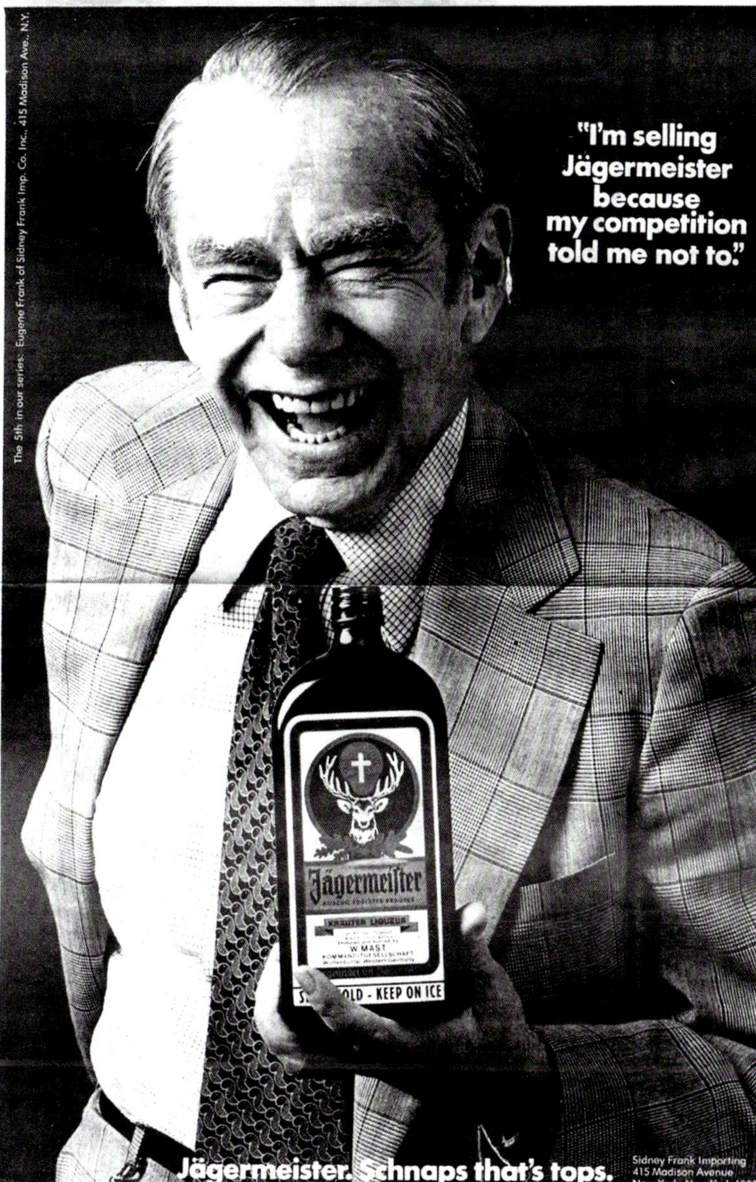

From left to right: Witty worldwide: the humour of the unique campaign was conveyed even beyond Germany's borders, such as on thus US version: 'I'm drinking Jägermeister because it's man's second-best friend.'; 'I'm drinking Jägermeister because those who can, do.'; 'I'm selling Jägermesiter because my competition told me not to' – so said an employee at Sidney Frank Importing.

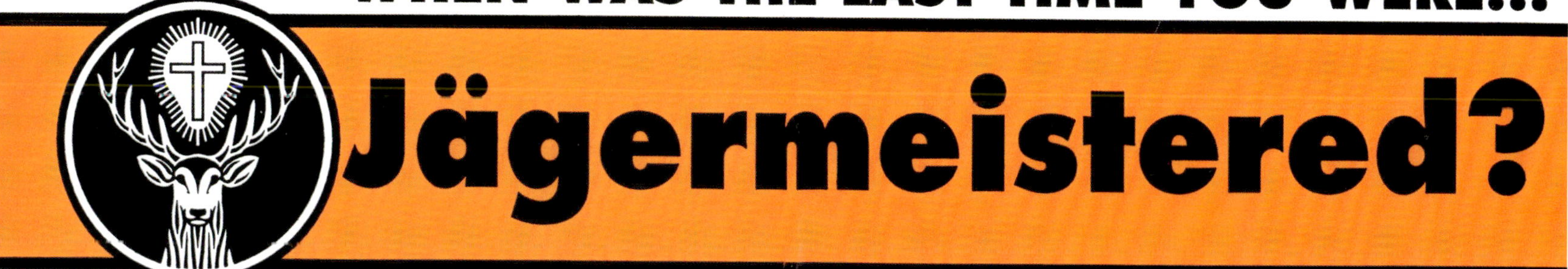

Jagermeister Liqueur. 35% Alc./Vol. Imported by Sidney Frank Importing Co., Inc. New Rochelle, NY 10801

The popular tap machines soon also started to be used outside the USA, and further successful American ideas were adapted internationally. In Wolfenbüttel, people had realized that the US concepts could also work well elsewhere, and had applied them virtually identically to Germany and other European markets. The

Reputable publications such as *The Wall Street Journal* and *Forbes* magazine reported on the wild promotion concepts that had sparked a veritable boom in the country

results spoke for themselves; from the late 1990s onwards, the Jägermeister brand was young, flashy and orange in the USA and throughout the rest of the world.

What really rocks? This is the question Jägermeister has been asking itself since the early 90s. Sidney Frank recognized music as being an emotive arena for making the brand more accessible to consumers, which is why Jägermeister has been involved in music sponsorship ever since. It set up a platform for all music enthusiasts in the form of the Jägermeister Music Tour. With well-known heavy-metal bands such as Slayer and Slipknot, as well as plentiful ice-cold shots, it created the perfect billing for a whole host of unforgettable concerts.

Jägermeister is now one of the best-known and most popular spirits ever, in Germany, the United States and worldwide. Yet very few people know there is actually an official day commemorating the German herbal liqueur. During a visit by the Jägermeister executive board to Philadelphia, the city's then mayor, John F. Street, honoured the company through a proclamation ceremony. He highlighted the role Jägermeister was playing in the local restaurant and dining sector and praised the economic contribution the company was making in the region, before officially declaring 14 November 2006 the first Jägermeister Day. Since then, however, people continue to wonder why this day of honour still doesn't exist in Germany. ♥

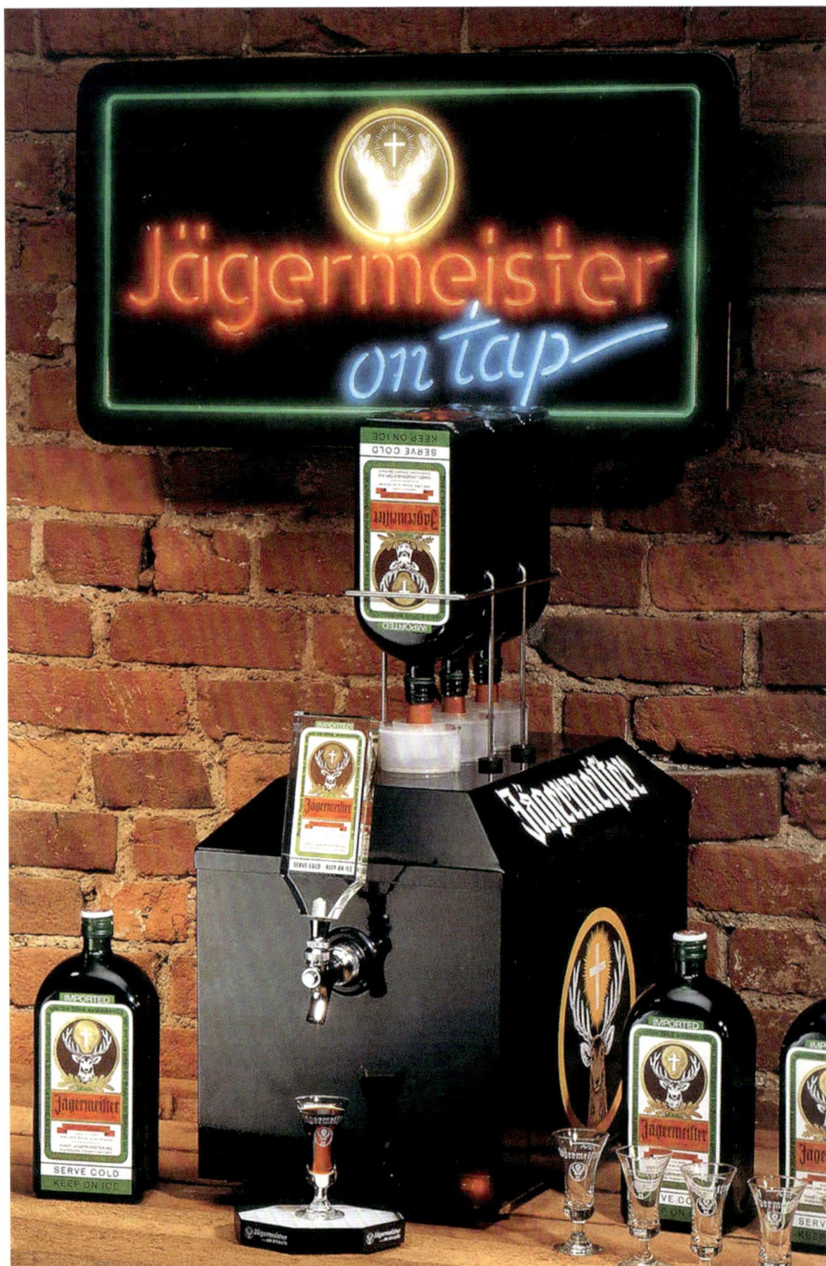

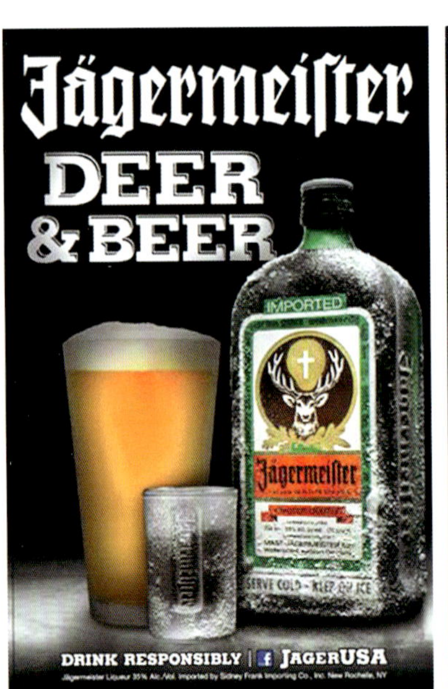

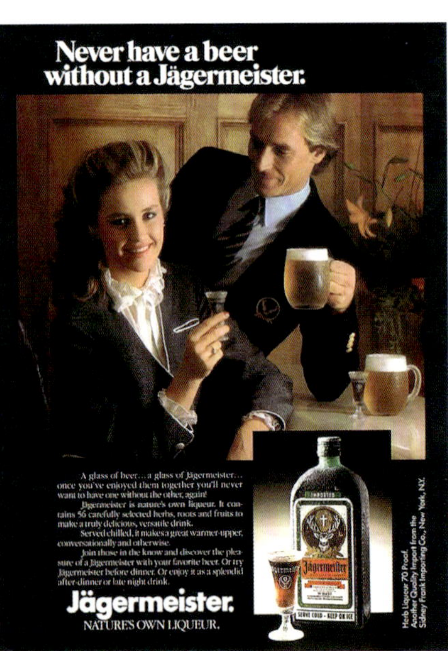

Left: Cool drinking recommendation, cool campaign: Deer & Beer. Right: A forerunner of the Deer & Beer campaign? At least the ad recommended even then: 'Never drink a beer without a Jägermeister.'

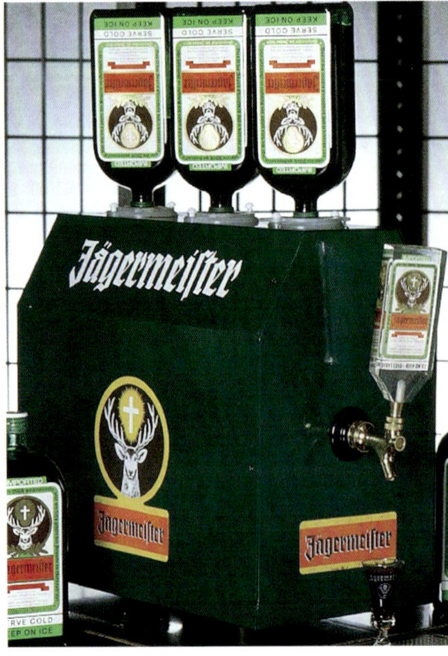

Sidney Frank's invention of the Jägermeister tap machine remains a stroke of marketing genius to this day. Apart from fast, chilled shot-drinking pleasure, they also helped the brand become one of the most prominent at bars, clubs and pubs. The first tap machine, the legendary J81, was followed by many others, such as the Multiplex Tap Machine (above) and the Perfection Tap Machine (bottom right).

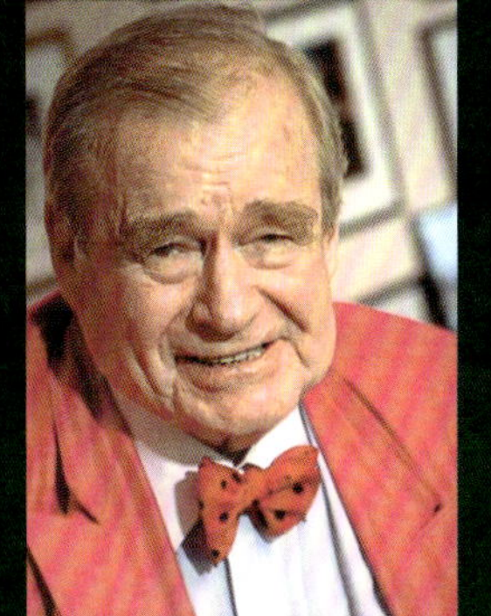

TEAMWORK: SIDNEY FRANK AND JÄGERMEISTER

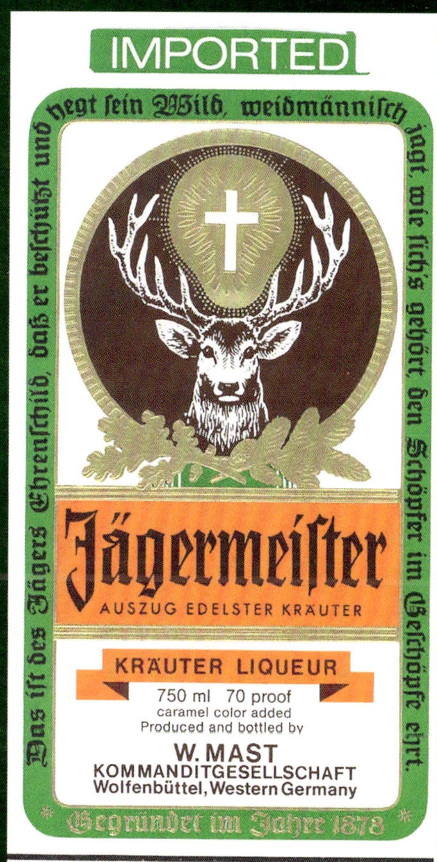

The label on American Jägermeister bottles in the 1970s and 80s.

When Sidney Frank told his friends he wanted to bring a German herbal liqueur to the USA and distribute it there, they thought he was crazy. At that time Jägermeister was looking for a new partner in the US, after all, they sold only a few tens of thousands of bottles in the States in the early 1970s (today that number is nearly 20 million), and this potential tantalized Frank. 'I was looking for a niche product, and that's exactly what Jägermeister was at the time,' he later explained in an interview, adding, 'It was mostly drunk by college students and Germans — after all, we had a lot of them in the country.' His strident, confident manner immediately convinced the Germans, and it wasn't long before Frank was responsible for USA-wide distribution of Jägermeister — with the exception of two states, for which another importer was contracted. This problem would soon resolve itself, during an excursion to Disneyland. Jägermeister chief of distribution Walter Sandvoss had asked the importer to accompany him there, but on the way, the latter got so badly lost that Sandvoss became sceptical; was he not familiar enough with his own distribution area? Sandvoss gave him short shrift, retracting his distribution licence and giving it to Sidney Frank. The rest is history. With his bold and innovative campaigns, Frank not only made the Jägermeister brand popular right across America, but also invented the concept of guerrilla marketing to boot.

His simple yet impressive business card? Sidney Frank always had a small Jägermeister bottle in his pocket. When asked for his business card, he handed out a Jägermeister bottle with the words: 'On the front you'll find the name of the producer — on the back you'll find the name of the importer: Sidney Frank.'

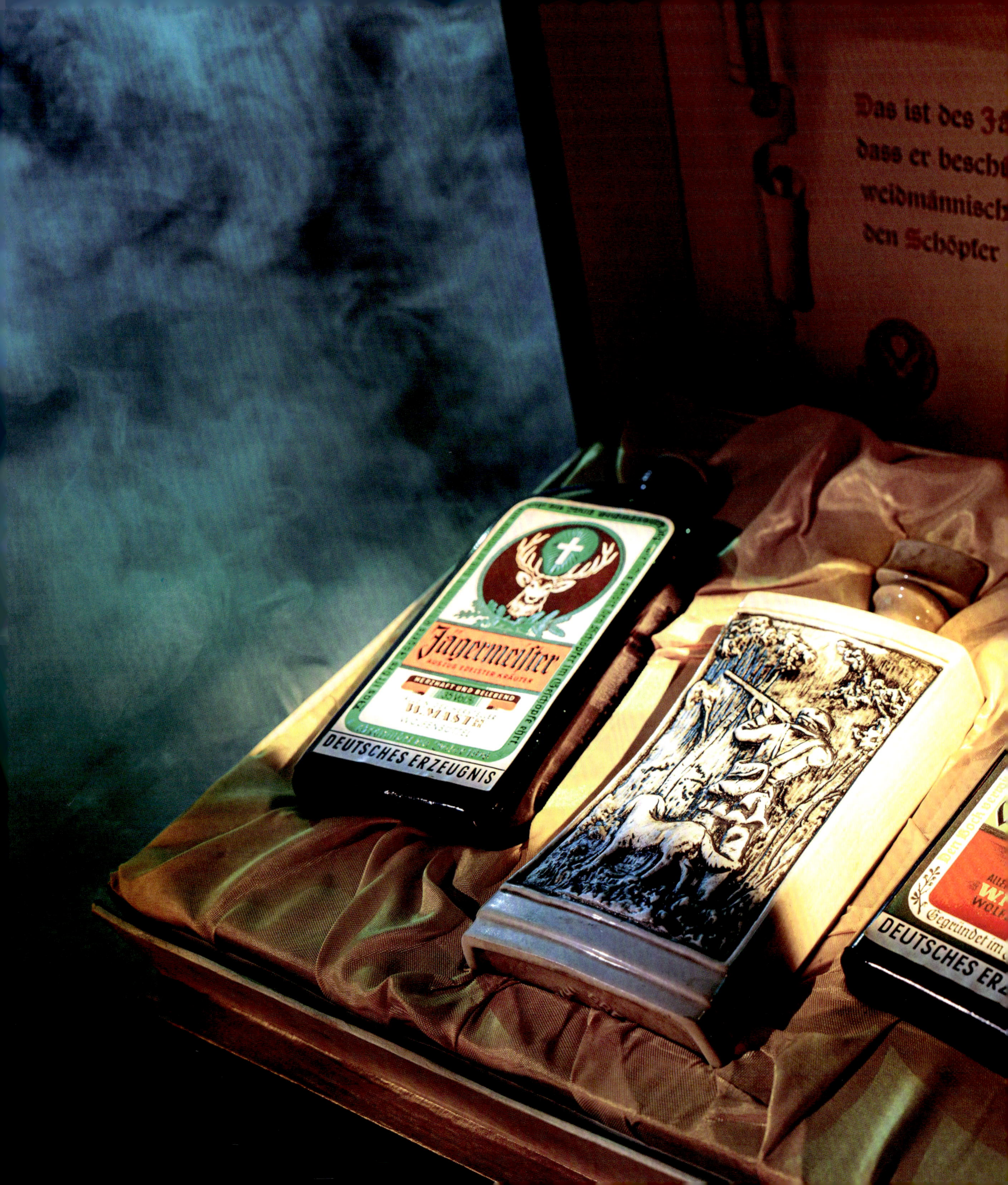

Das ist des Jä...
dass er besch...
weidmännisch...
den Schöpfer
Jägermeister
AUSZUG EDELSTER KRÄUTER
HERZHAFT UND BELEBEND
JM. MEISTER
DEUTSCHES ERZEUGNIS
Gegründet im
DEUTSCHES ERZ...

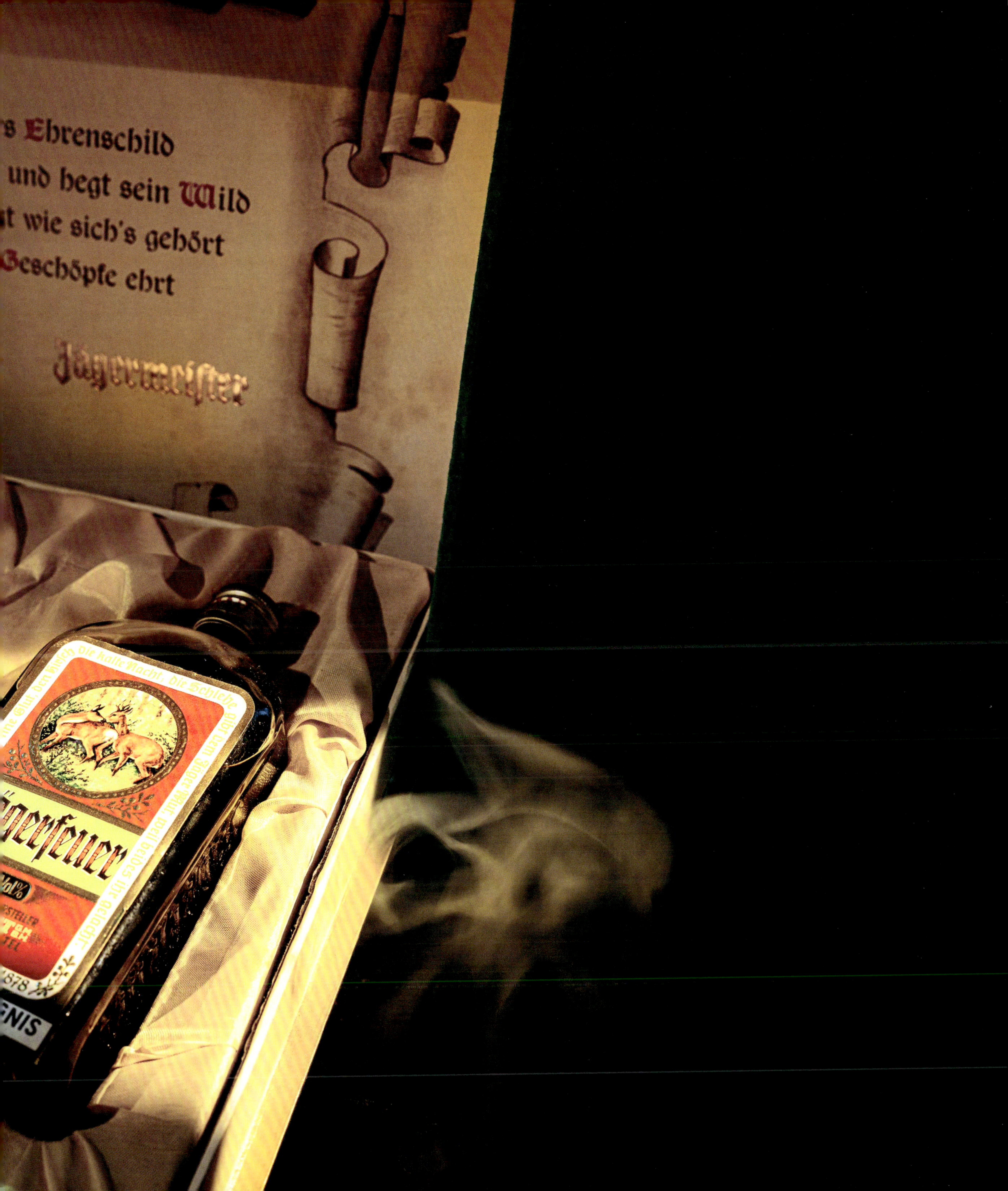
s Ehrenschild
und hegt sein Wild
t wie sich's gehört
Geschöpfe ehrt

Jägermeister

MUSIC TO EVERYONE'S EARS

Always up with the times, always right amid it all – Jägermeister is known for its flamboyant, innovative and, in particular, striking marketing campaigns. While the company played a pioneering role in sport sponsorship in the 1970s, the focus since the 2000s has been on music, event and festival sponsorship. It was a change in image that made the herbal liqueur the hottest – and also the coolest! – drink on the party scene.

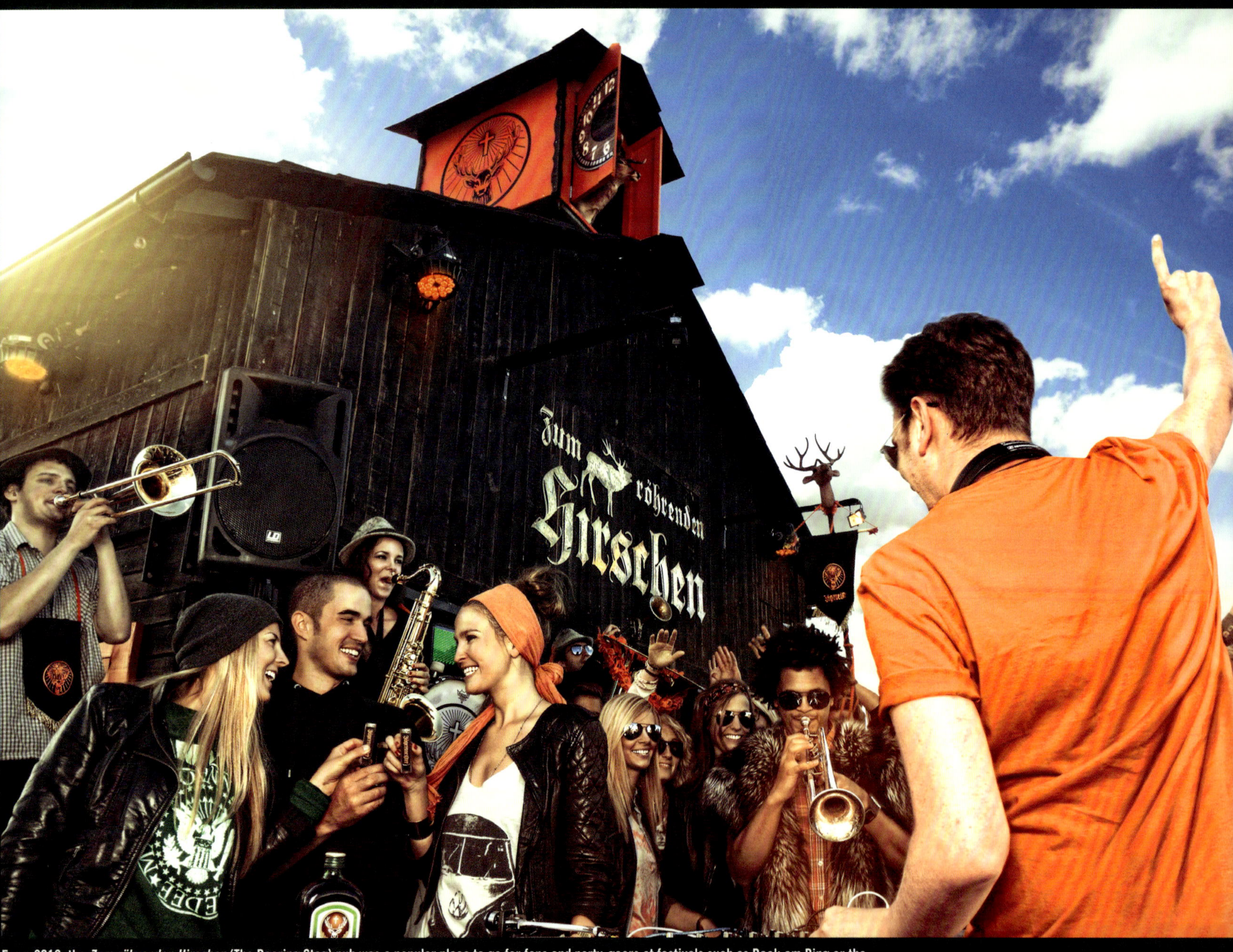

From 2012, the *Zum röhrenden Hirschen* (The Roaring Stag) pub was a popular place to go for fans and party-goers at festivals such as Rock am Ring or the Wacken Open Air for four years. DJs and bands (including the Jägermeister brass band) regularly played living room gigs here in an intimate setting. How did the inn get its name? A loud roar from the huge cuckoo clock on the gable announced every day: the spectacle begins now!

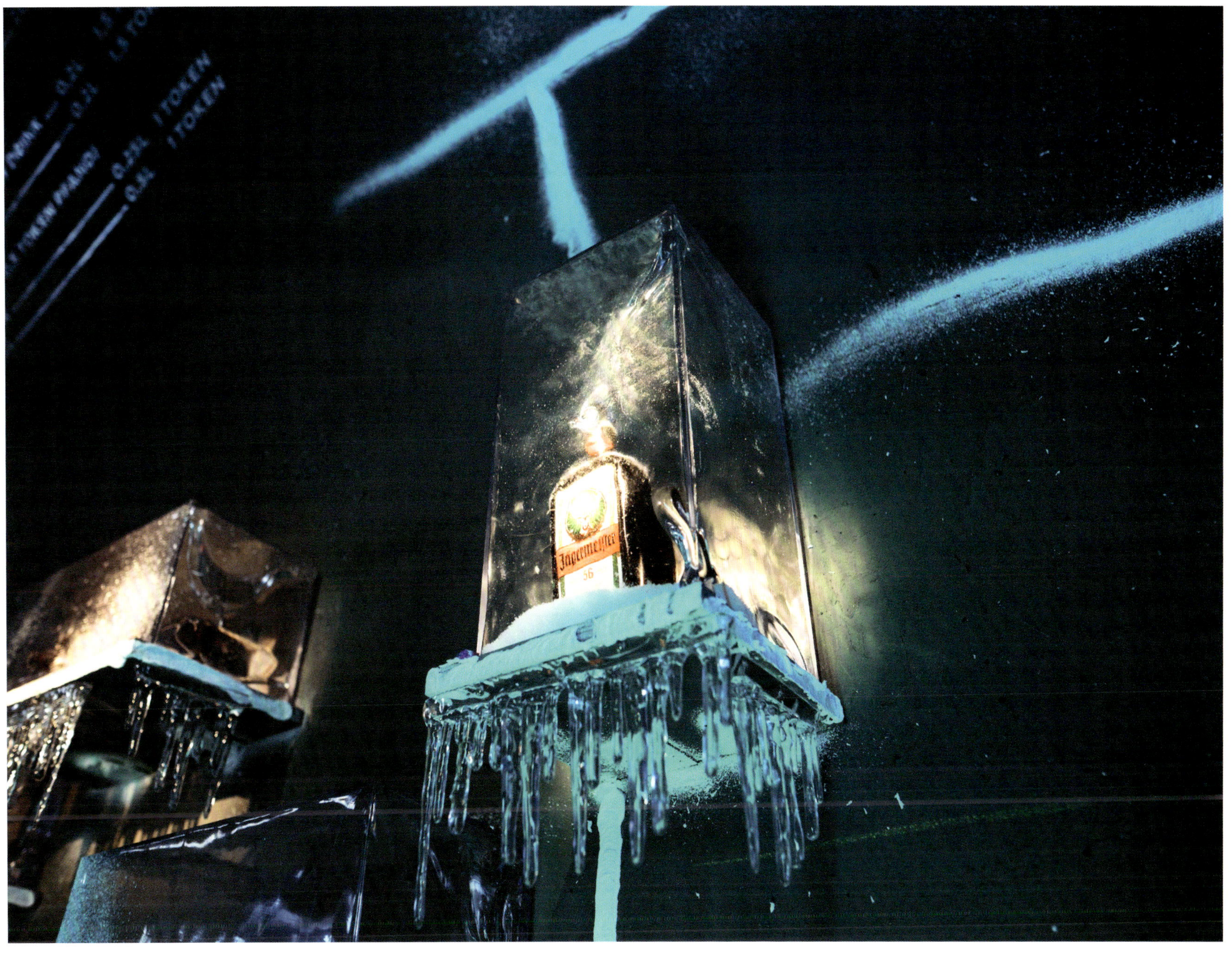

The start of the new millennium also marked a new era for Jägermeister. The long-established brand transformed into a young party brand with loud music and club sounds, shifting its focus to music marketing, festivals and music events. The three-level Jäger station was first used at Jägermeister-Schlagermove in Hamburg in 2001, its bar level at the festival grounds having been created by fashion designer Rudolph Mooshammer. What was the Schlagermove? Twenty-nine lorries with DJ sound systems and 300,000 partygoers heading down the streets, led by a Jägermeister truck. Thanks to the giveaways that were being thrown out of the lorries en masse, the St Pauli district was soon a sea of orange. Promotions like these increased the brand's emotional charge, enabling a new and younger audience to identify with it.

Party like there's no tomorrow

The first 'Orange Events' were held a year later. Jägermeister was present at various festivals with a mobile bar, giveaways (usually orange hats) were handed out to attendees en masse, and several of the sets were also bright orange in colour. A number of events, such as G-Move and Christopher Street Days, saw Jägermeister sold in test tubes, which soon became a popular collectors' item. Many fans at the time had six to twelve of these at home, kept in homemade wooden blocks. Jägermeister pursued this strategy so fiercely (and overtly) that the festivals' main sponsors eventually lodged a complaint, prompting organizers to declare they could no longer allow Jägermeister to look like a main sponsor with all its orange giveaways while only paying minimal fees.

So, from 2005 onwards, the company devised new, larger-scale sponsorship tools for Germany's major festivals. The *Hochsitz*, the *Zum röhrenden Hirschen* pub and the Platzhirsch provided even more opportunities for attendees to experience every aspect of the brand up close and personal. The Jägermeister perch was first used in 2009, affixed to a crane nearly 50 metres up. For three whole years, those fans with a head for heights could sit at an open-air bar and enjoy seeing the main stage and festival grounds directly beneath them, with ice-cold shots directly in front of them. At the 2011 Highfield Festival, Dave Grohl from the Foo Fighters took several minutes out on stage to celebrate Jägermeister, spontaneously dedicating the song *These Days* to 'all the crazy ones up there on the crane drinking Jägermeister'. Meanwhile, at festivals such as Rock am Ring, Southside, Deichbrand, the Helene Beach Festival and the Wacken Open Air, the *Zum röhrenden Hirschen* (The Roaring Stag) pub created a Jägermeister-style sense of home for four whole years from 2012 onwards. And the name said it all. The booming roar of a stag from the 15-metre high cuckoo clock signalled that the pub was open! Particularly popular were the lounge gigs by top DJs and bands, including the Jägermeister Brass Band. In addition to cool drinks and good music, the verandah at the front also served up a spectacular view of the festival grounds.

Large, loud and fire-breathing, the Jägermeister Platzhirsch was *the* attraction at many festivals from 2016 to 2019. With its three-level bar, terrace and lounge areas, the 17-metre stag, built out of several shipping containers and clad in wood, was a popular hang-out spot, because it was also a place to relax and watch everything happening in front of and on the main stage. The highlight, however, was the stag's roar, which was accompanied by fire and smoke, and signalled a round of ice-cold Jägermeister shots on the house. After a three-year festival tour – from Deichbrand and Wacken to SonneMondSterne to Sputnik – Germany bade farewell to the colossus as it set off to explore other festivals in Europe (and even in China).

The Platzhirsch stag never needed branding during its festival appearances. As the emblematic animal of the Jägermeister brand, its allegiances were very clear. It even kept up appearances in the most extreme of situations. When a fierce storm forced the infield and camping grounds to be evacuated after three hours at the 2016 Southside Festival, the Platzhirsch stood proudly on its own, solid as a rock. Completely undamaged, it provided festivalgoers with protection from the driving rain.

Bands, burlesque and bar-hopping

Jägermeister had entered the music-sponsorship game in Germany as early as 2002, initially with the talent sessions and later with Jägermeister Band Support. The idea was for established artists and bands to take newcomers with them on tour, with Jägermeister providing branded sleeper buses and the necessary advertising. This campaign would pave the way for Jägermeister to enter the music industry. Authentic, harmonious, tailored – Jägermeister Band Support was not just about branding, but rather about customized (and highly sought-after) concepts for the participating bands. For example, Jägermeister spent four weeks touring through Germany with Finnish gothic

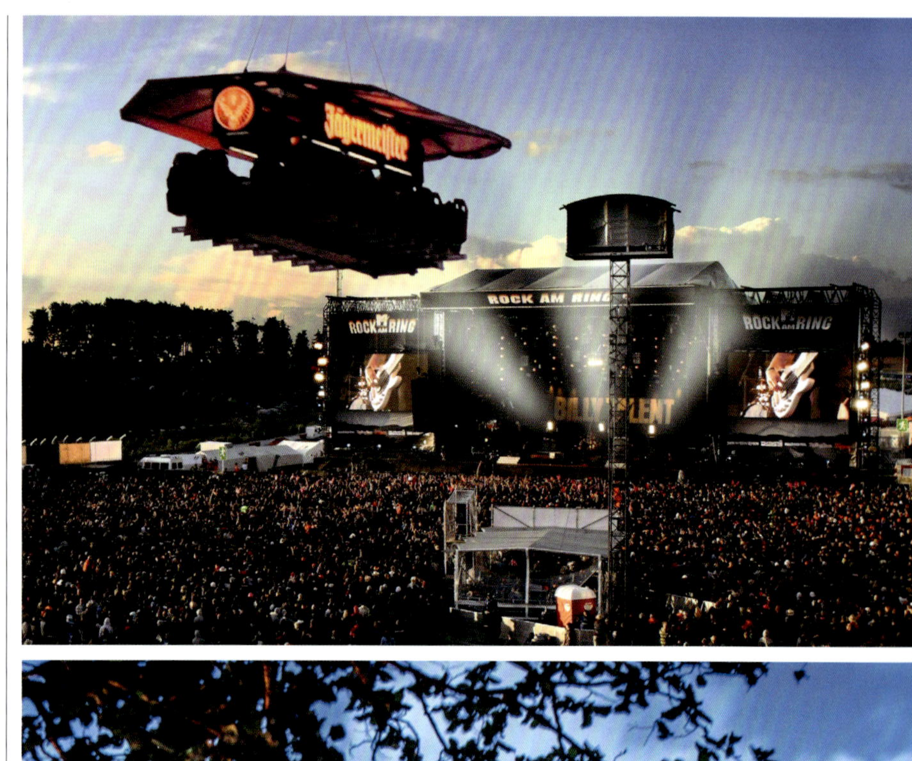

On the Jägermeister perch, fans can enjoy ice-cold shots and a phenomenal view over the festival area at a height of almost 50 metres.

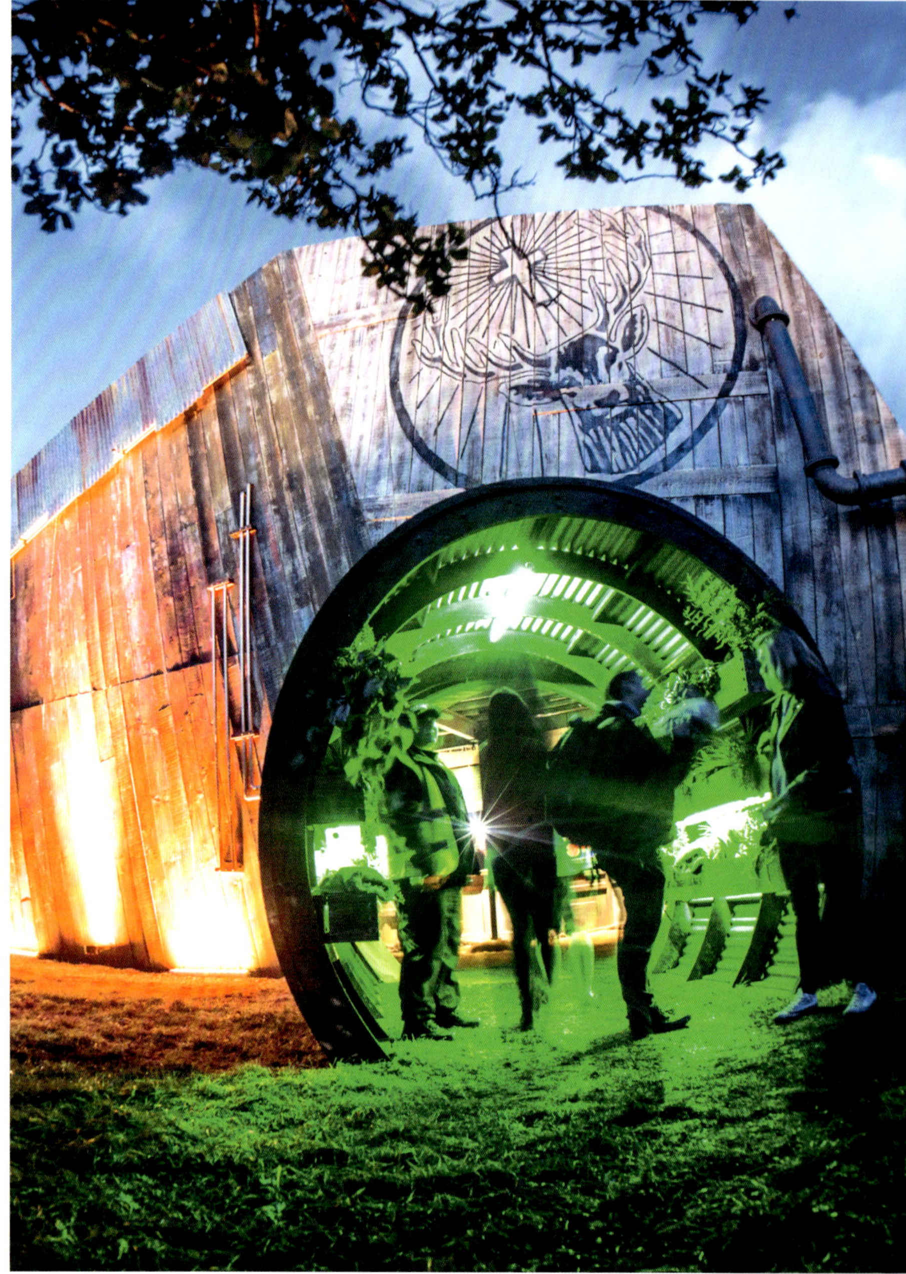

The innovative Jägermeister festival meeting places at open-air events are popular eye-catchers around the world, such as here in Great Britain.

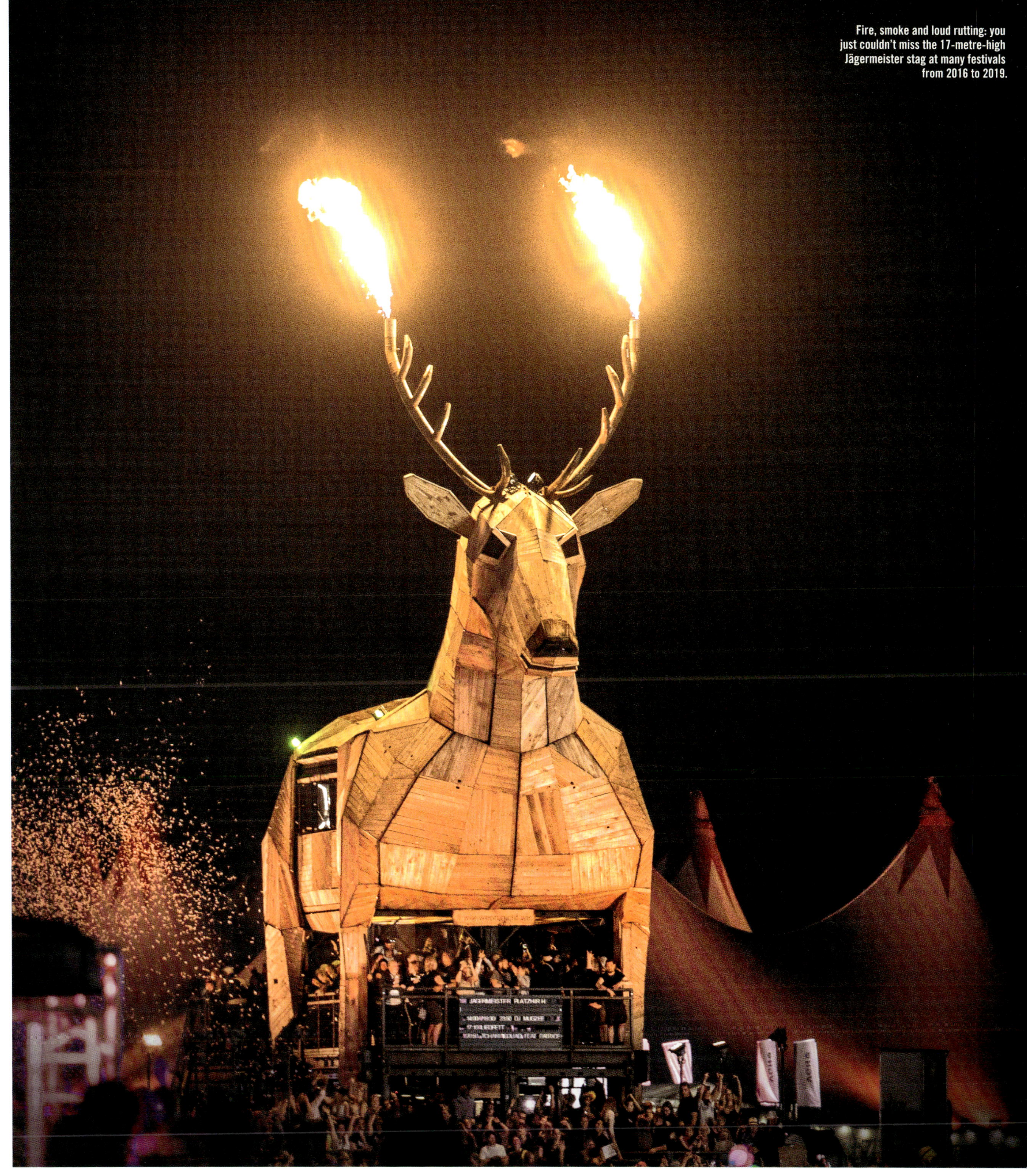

Fire, smoke and loud rutting: you just couldn't miss the 17-metre-high Jägermeister stag at many festivals from 2016 to 2019.

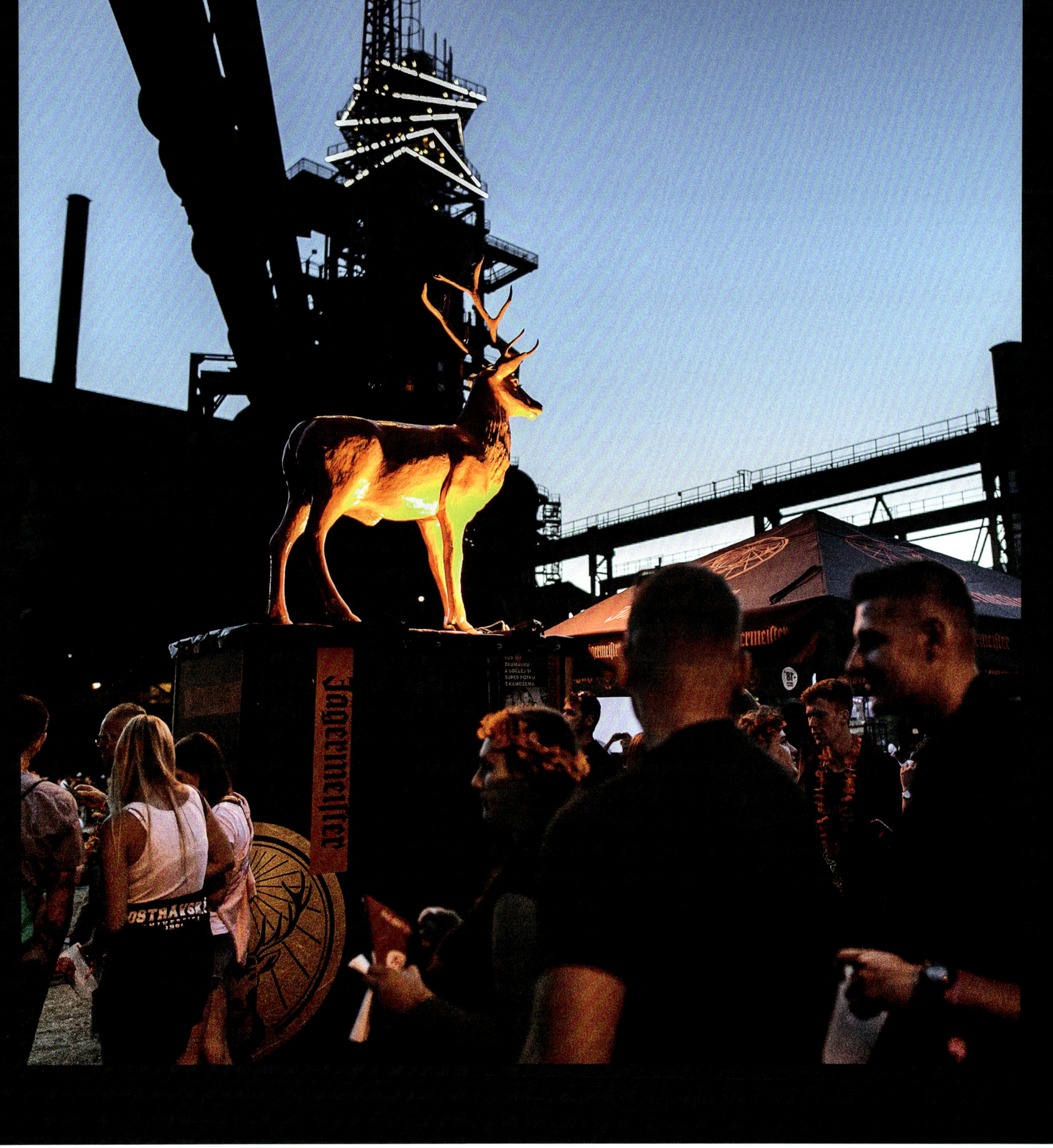
Jägermeifter
OSTRAVSKI

rock band HIM – and, at a secret gig in Hamburg, every attendee aged over 18 was given a small bottle of Jägermeister labelled as a HIM edition. The brand also held two larger events of its own in 2003: the Jägermeister Band Support Festival in Cologne, headlined by Fury in the Slaughterhouse and the Jägermeister Party Akademie with Lexy & K-Paul. The three-storey perch and lots of orange Jägermeister hats also featured prominently, of course.

Finally, Jägermeister Rock:Liga made its debut in August 2004, in a large Berlin disco, complete with a *Miss Arschgeweih* (Miss Trampstamp) contest. Liqueur meet antlers, antlers meet arse – an approach that saw Jägermeister hit just the right note in the early 2000s. The 50 female contestants would sit on a Jägermeister director's chair, where the gap between seat and back provided the perfect frame for tailbone tattoos. 'Anja, the hottest policewoman in Berlin', as *BILD* newspaper put it, was named the winner. The Rock:Liga itself was based on the Bundesliga football concept, with 18 bands from 18 cities competing against each other for the championship. The concept was fine-tuned over the years that followed; national and international bands such as the Bloodhound Gang, Deichkind, Bosse and many others played in various cities in a bid to make it to the final in Berlin. After six years of Rock:Liga, the brand's involvement and message had become well known nationally, and Jägermeister had finally established itself as a heavyweight on the music scene.

New event series were starting up in clubs across the country: Wild Girls was a burlesque show with an avant-garde touch, while Dance Box Hero in 2009 saw Jägermeister create the first simultaneously digital concept. In the box – a cross between a dance studio and recording studio with mirrored walls, LED panels and a mad sound system – club goers could film themselves dancing, and then upload the clips directly to YouTube.

The Jägermeister Wirtshaustour also made its debut that same year, with traditional corner pubs transforming into nightlife hotspots. DJ sounds provided a backdrop to pub sports such as bowling, darts, football and billiards, along with hearty food and ice-cold Jägermeister shots. The trendy urban Wolfenbütteler Festspiele was similarly designed as an event tour, with a unique twist: each night could be organized by the performing artists themselves. The event at Bochum's Club Matrix, for example, had the motto of 'Light in the shaft', with Ruhr-based rappers 257ers performing in a mine-themed setting under neon lights. Meanwhile, Heisskalt auf Hirschkurs saw rock band Heisskalt take attendees on a musical scavenger hunt through their hometown of Stuttgart, with the band performing 30-minute concerts at their favourite clubs, accompanied by acts from various genres.

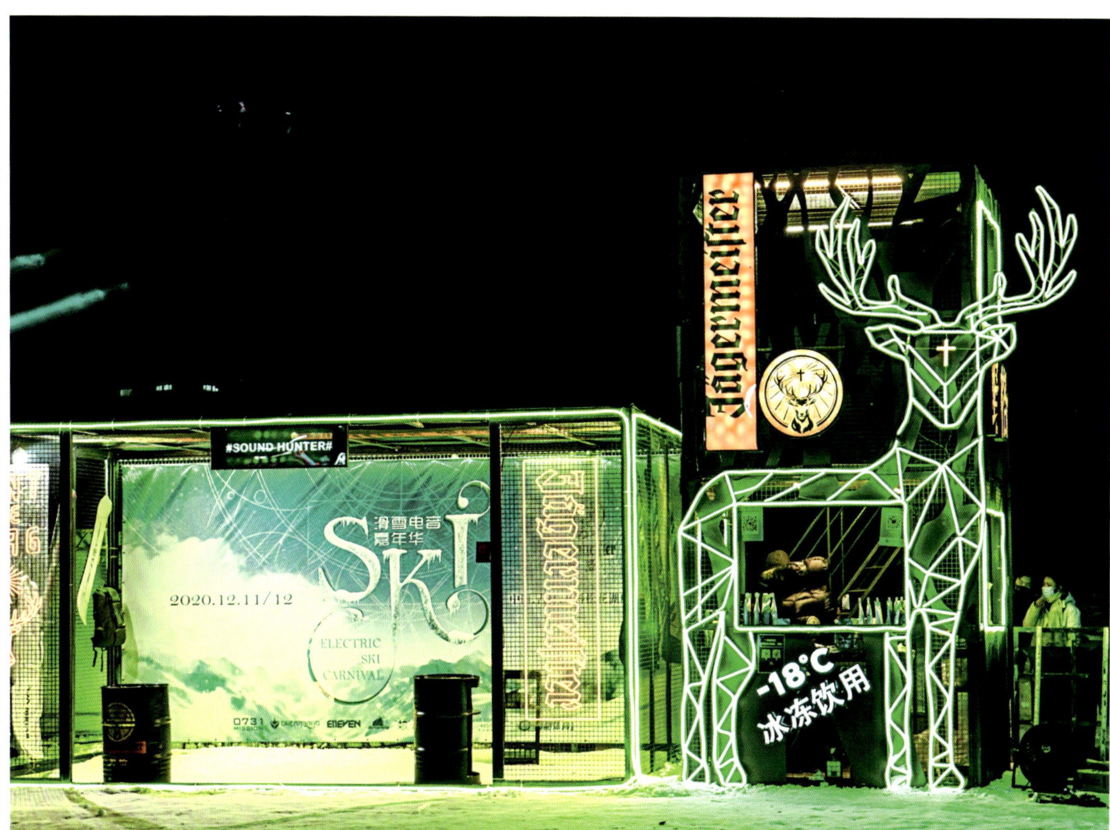

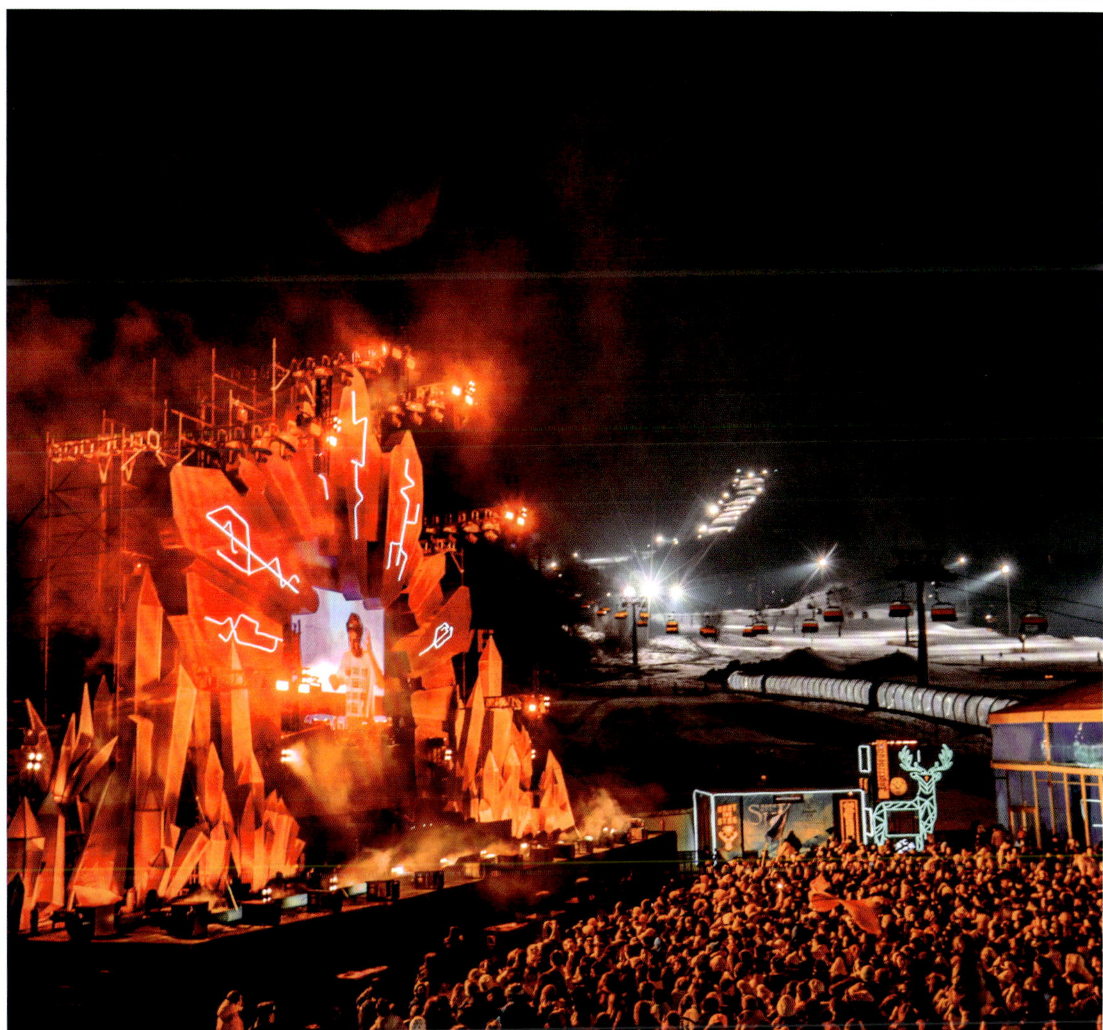

Today Jägermeister is present at festivals all over the world – like here in China.

We need brass! Brass!

The now famous Jägermeister Brass Band was formed in 2012, and it quickly became clear that brass interpretations of the latest hits and party classics were wildly popular. The group of 16 professional tubists, saxophonists and trumpeters soon gave rise to a show concept involving dancers and break-dancers, which could work both in large clubs and on larger stages – as evidenced by successes such as performances at Rock am Ring and Wacken. With its energy-charged belter shows, the Jägermeister Brass Band swiftly became a sneaky headline act at many festivals. The crazy troupe was booked by FC Bayern Munich for its Champions League celebrations, and by tennis icon Boris Becker for his birthday party – and it even went on tour with bands like Scooter and The Boss Hoss.

The Jägermeister Brass Band effectively invented a new genre with its unique sound. It released its first album, *Move Your Brass*, in 2014, promptly breaking into the official sampler charts. Renowned artists such as Cro, Sido, Marteria, Das Bo, Alexander Markus, Jennifer Rostock and Haddaway teamed up with the band to record some of their greatest hits, while the 2019 Remade project saw three songs by German rappers Kool Savas, Jamule and Prinz Pi mixed with timpani and trumpets. Opposites have rarely attracted more harmoniously.

And there's still no other drink that does harmonious opposites better than Jägermeister.

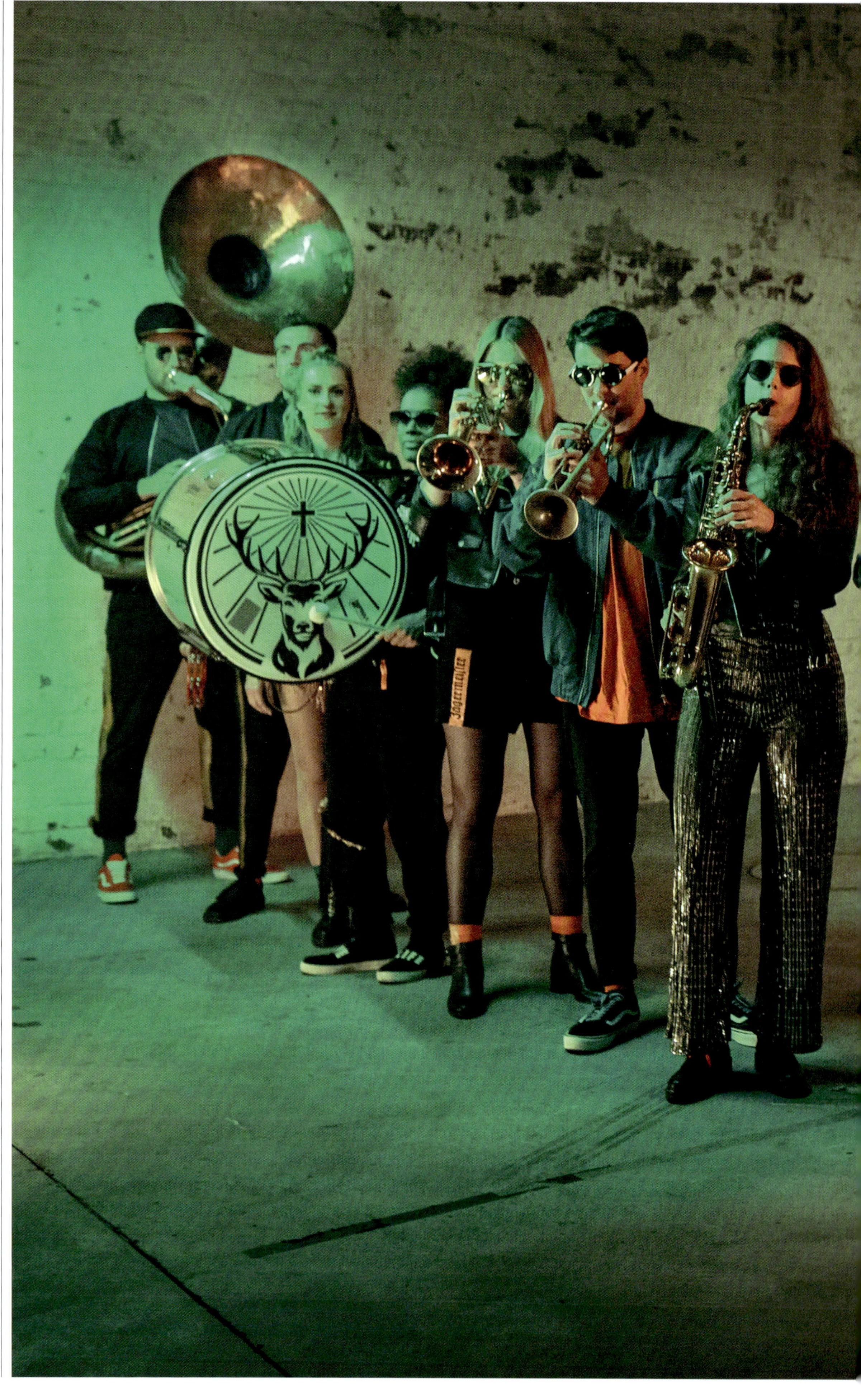

The trademark of the Jägermeister brass band is an incomparable mix of pop, hip hop and party hits – all reinterpreted in brass band style

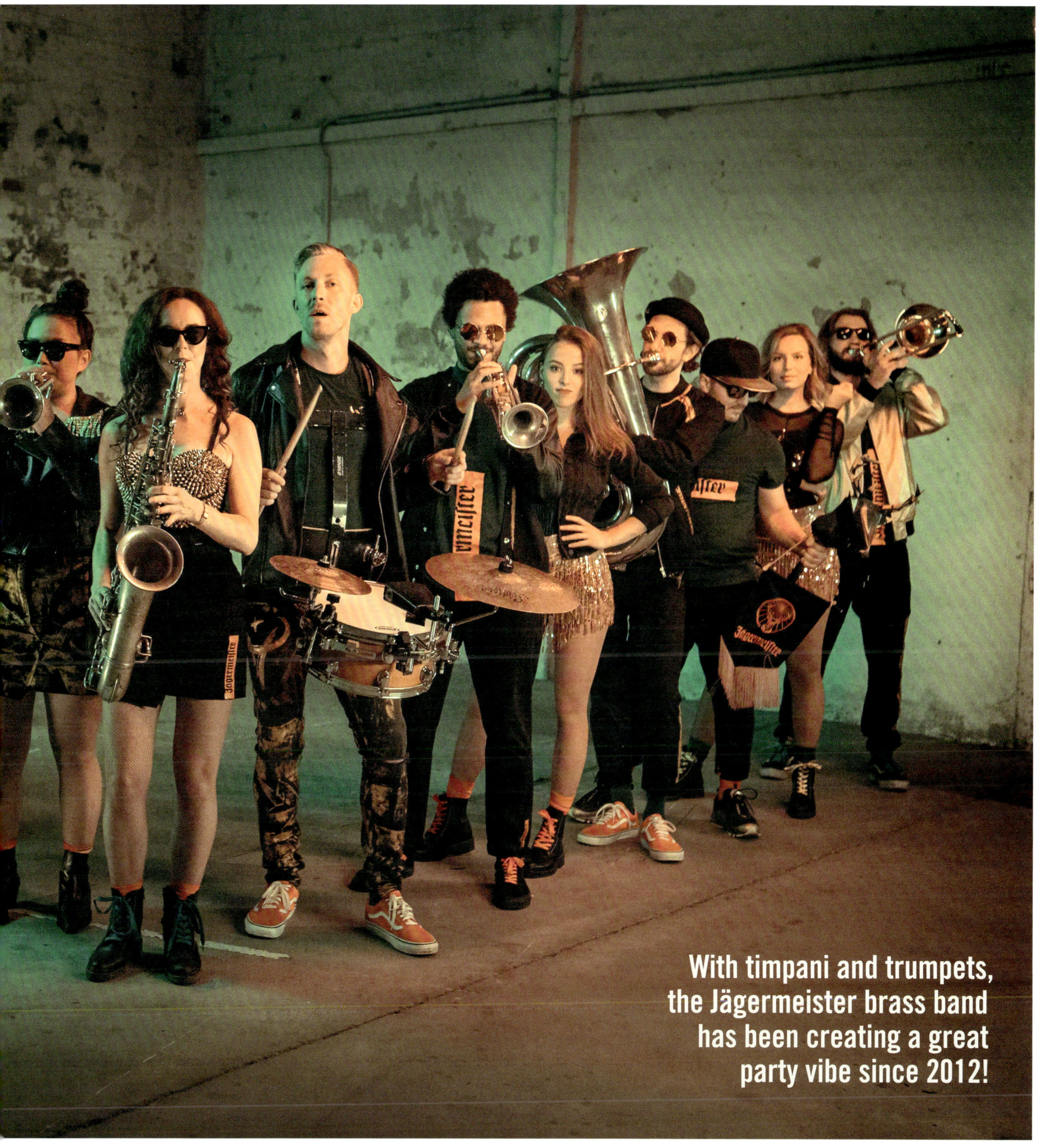
With timpani and trumpets,
the Jägermeister brass band
has been creating a great
party vibe since 2012!

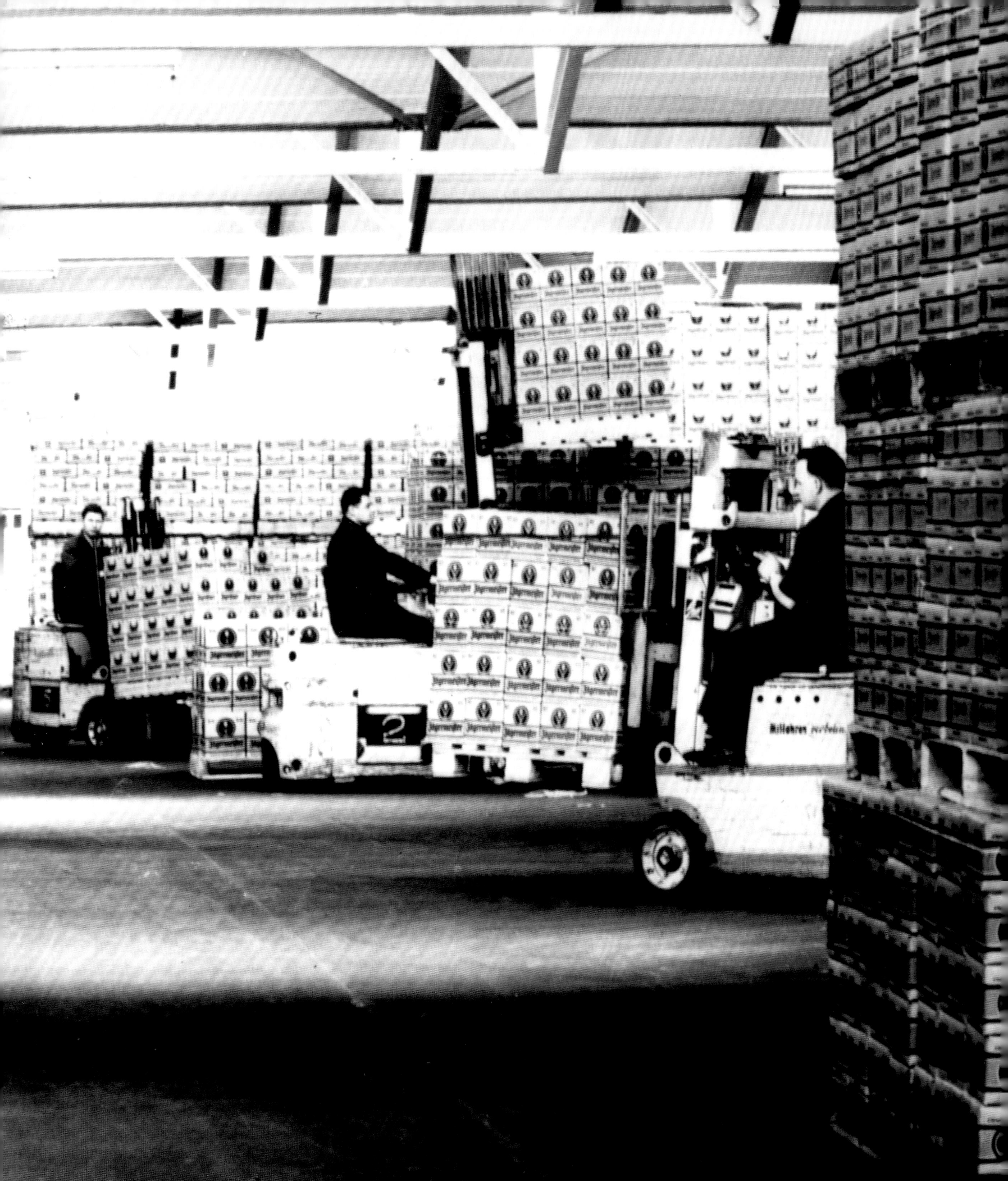

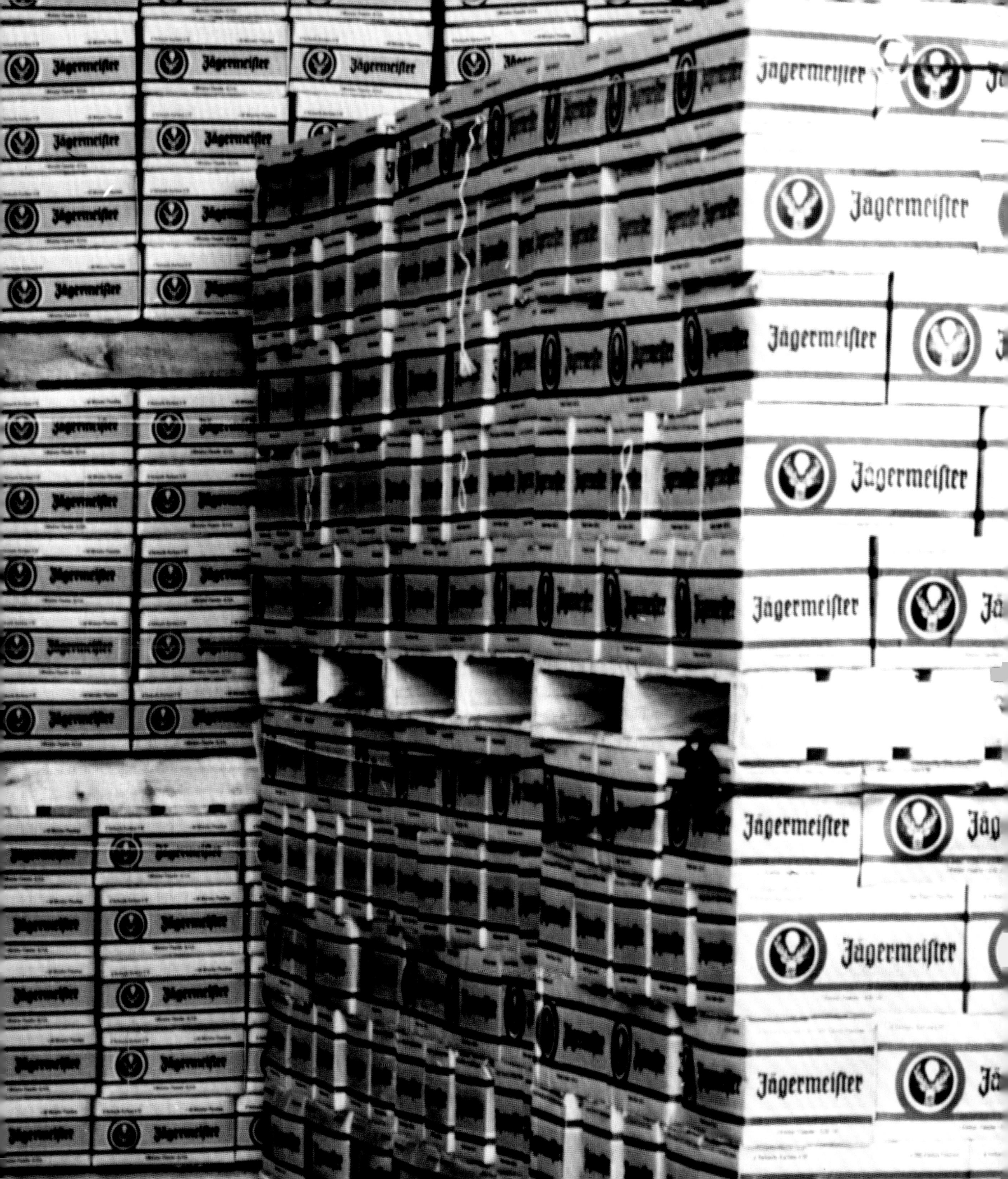

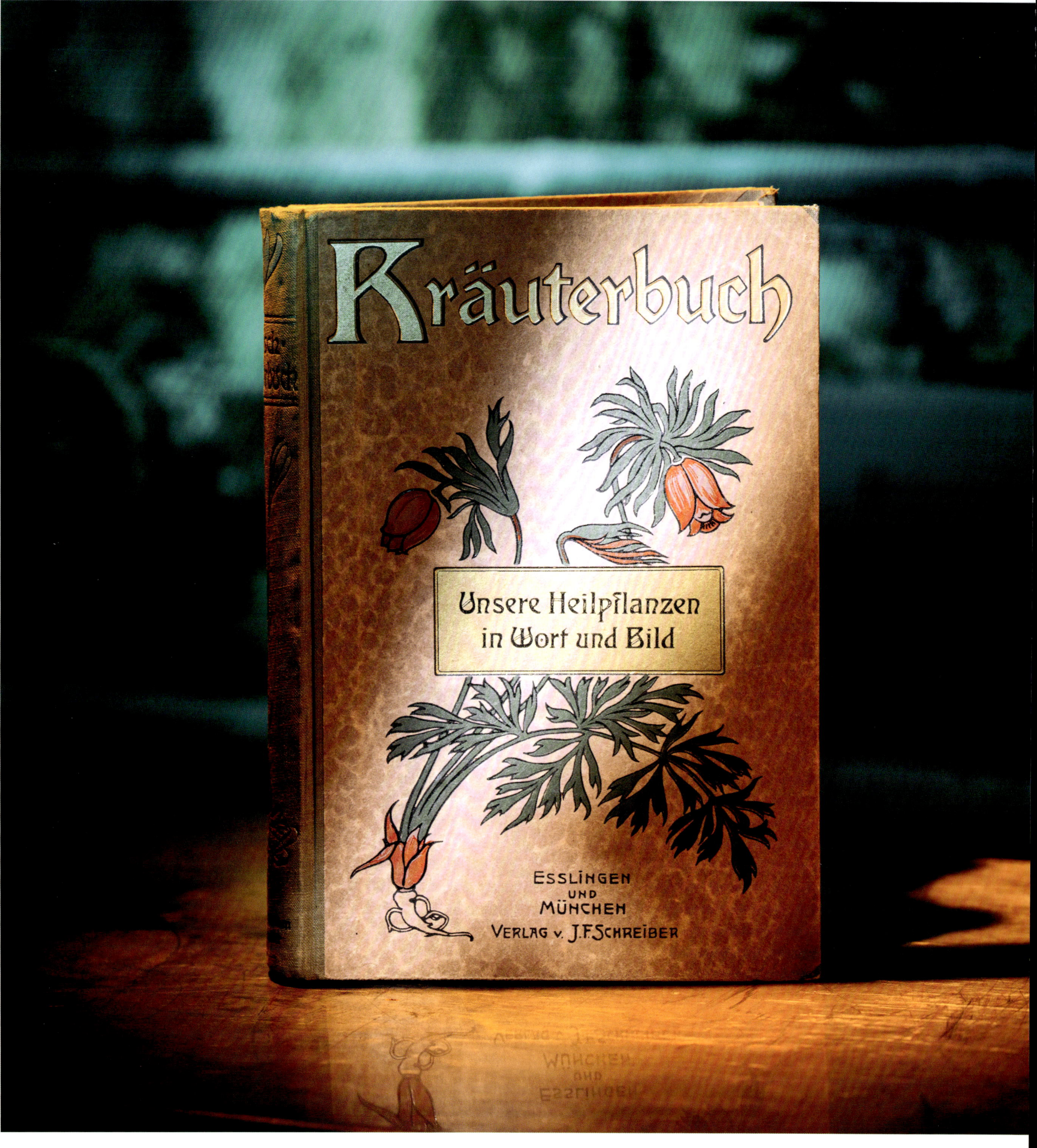

Kräuterbuch
Unsere Heilpflanzen
in Wort und Bild
Esslingen
und
München
Verlag v. J.F.Schreiber

PRODUCT AND LIFESTYLE IN HARMONY

Jägermeister is so much more than the sum of its ingredients, and it famously has plenty of those; indeed, the world's most successful herbal liqueur is a harmonious fusion of 56 herbs, blossoms, roots and fruits.

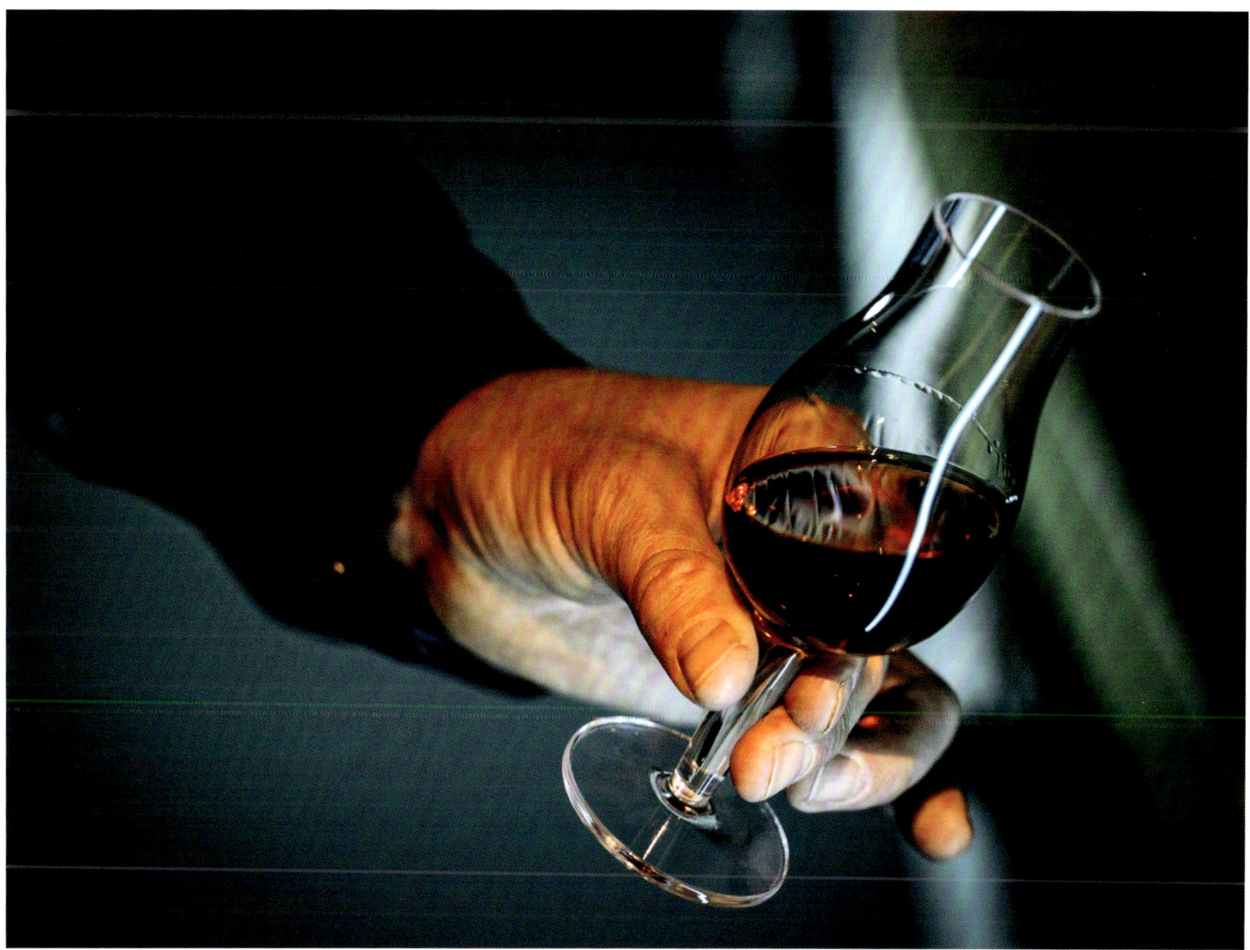

The recipe has remained unchanged since it was first invented by Curt Mast – and it's top secret, locked away in a giant black safe in Wolfenbüttel. Only a handful of people on the planet know the exact formula that makes Jägermeister what it is: a liqueur and lifestyle in one.

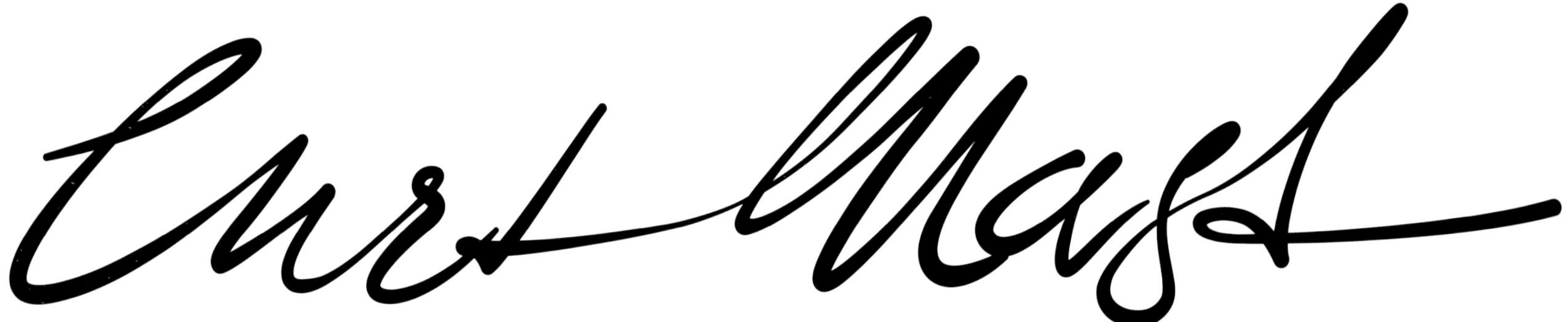

Along with the recipe, it is the perfect unison between the unique product and its iconic brand that is the secret to success. Because Jägermeister is the embodiment of its own brand: made for hunters (*Jäger* in German) by masters (*Meister* in German) – a commitment to solidarity in the world and to the German cliché of perfection. The ingredients come from all corners of the globe, making Jägermeister a liqueur that celebrates the world's diversity – at 35% alcohol by volume.

Without the Jägermeister product, the Jägermeister brand would of course not have been possible, so both have gone hand-in-hand for what will soon be a century.

The effort that goes into making Jägermeister is huge. Before the ice-cold shot is gulped down by revellers in seconds, it goes through a sophisticated maceration process, 383 quality controls and a year sitting in giant oak vats. It is based on botanical raw materials of a quality that always needs to be outstanding. Jägermeister also places extreme emphasis on eco-friendliness and socially and economically sustainable farming. The master distillers avoid extracts or mix-

tures of various plant components, instead only using unprocessed raw materials. It is important to work as closely and for as long-term as possible with the local farmers or dealers to ensure only top-quality ingredients ever make their way to Wolfenbüttel.

The first step involves precisely weighing the botanical raw materials sourced from all over the world, and dividing them into four different dry mixtures. Harmony is once again crucial here, for not all raw materials release their ingredients at the same speed – and this is particularly important for the next step: maceration. This process sees the herbs pickled in a mixture of alcohol and water for several weeks in order to bring out flavour components such as essential oils. The principle is comparable to that of tea preparation, and here, too, the leaves' ingredients are released into a liquid as gently as possible. But the four macerates produced during this step have a much higher alcohol percentage.

To make a harmonious base substance out of these four preparations once again takes patience and care, because the base substance is the heart and soul of Jägermeister's flavour. The four macerates are combined, and then spend an entire year sitting in mighty oak vats with capacities of up to 24,000 litres. The particularly fine-pored vats are hand-crafted from wood sourced from Germany's Palatinate Forest. They are then opened, before being resealed in Jägermeister's cellars. Some of the oak vats in the cellars in Wolfenbüttel are indeed as old as the herbal liqueur itself. If new vats are used, they must each first be neutralized, because, unlike whisky storage, where the aim is for the product to absorb

some of the woody flavour, the wood's flavour here must not permeate the liquid. Jägermeister relies on the pure taste of the herbs. The pores in the vats enable the base substance to mix with the external environment, enabling optimum fusion of the ingredients without any extra aids or additives. Anyone who has the chance to take a tour through the Jägermeister factory will be able to experience the process with all their senses, because the entire room is filled with the aromas of the 56 herbs.

The last step sees the base substance finally enriched with liquid sugar, alcohol, pure water and caramel. The result is the Jägermeister that eventually gets poured into the distinctive green bottles, before helping to bring together people all over the world.

Just as the Jägermeister brand creates a sense of solidarity and community out of human diversity, so the Jägermeister beverage combines the world's diversity in a single shot. It is the one constant in a lifestyle that the Jägermeister brand consistently reinvents and translates into all the world's languages.

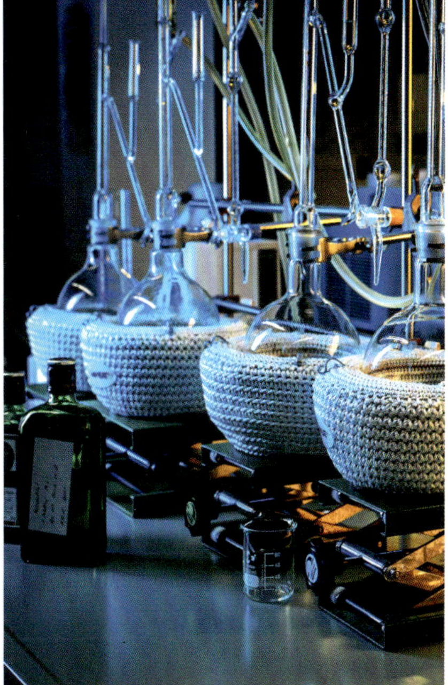 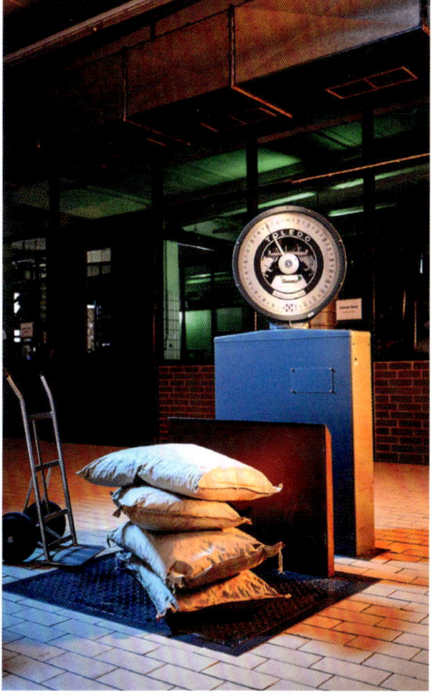

1 The botanical raw materials initially undergo quality checks in the laboratory before being weighed and split into four different dry mixtures.

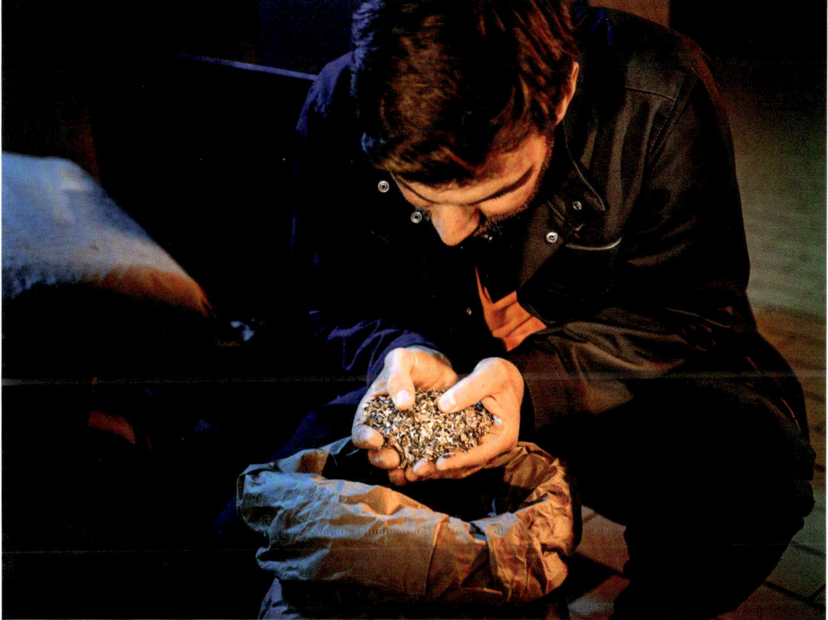

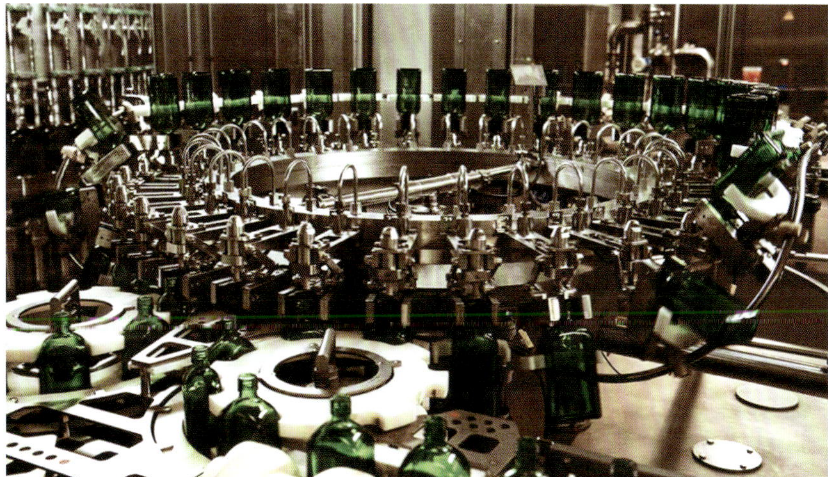

5 The herbal liqueur is finally poured into the distinct green bottles at the bottling plant.

2 The next step is maceration, during which the herbs are pickled in a mixture of alcohol and water for several weeks to release flavour components.

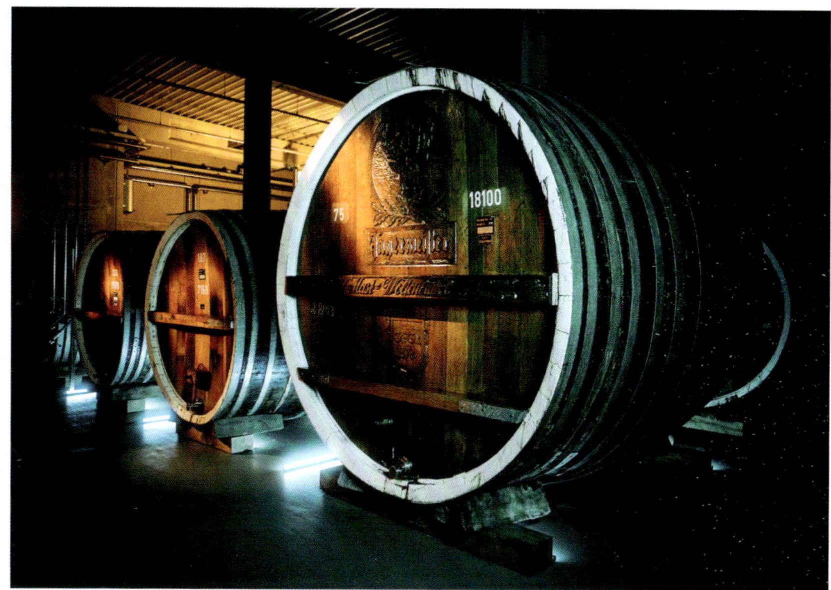

3 Four macerates are mixed together and spend a year sitting in oak barrels. The aroma of the wood itself must not be transferred to the liquid, so if new barrels are used they always need to be neutralized first. <u>Right:</u> A master distiller (probably in the 1960s) takes a taste test.

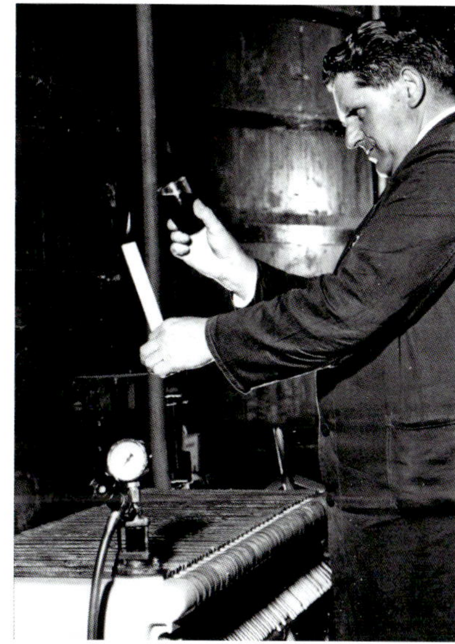

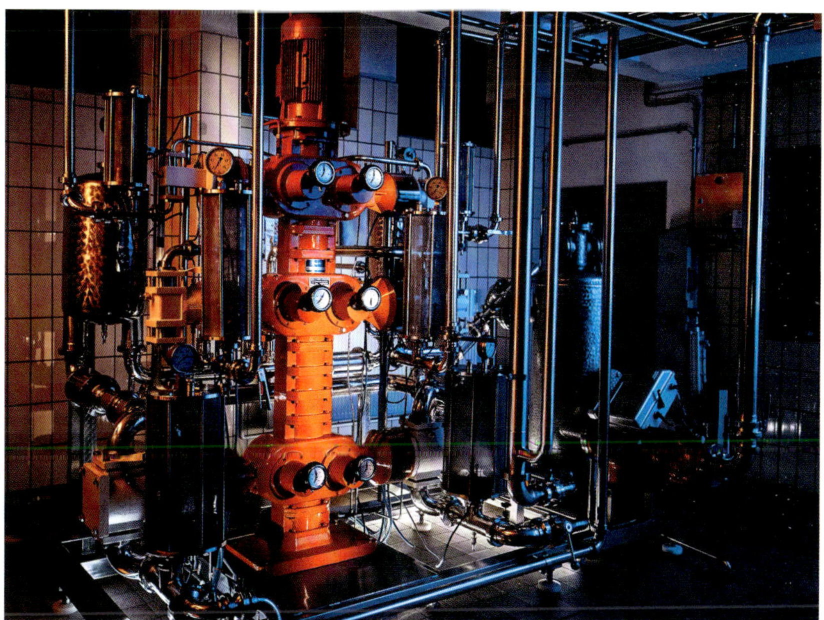

4 Once the base substance has spent enough time in the barrels, it is enriched with liquid sugar, alcohol, pure water and caramel.

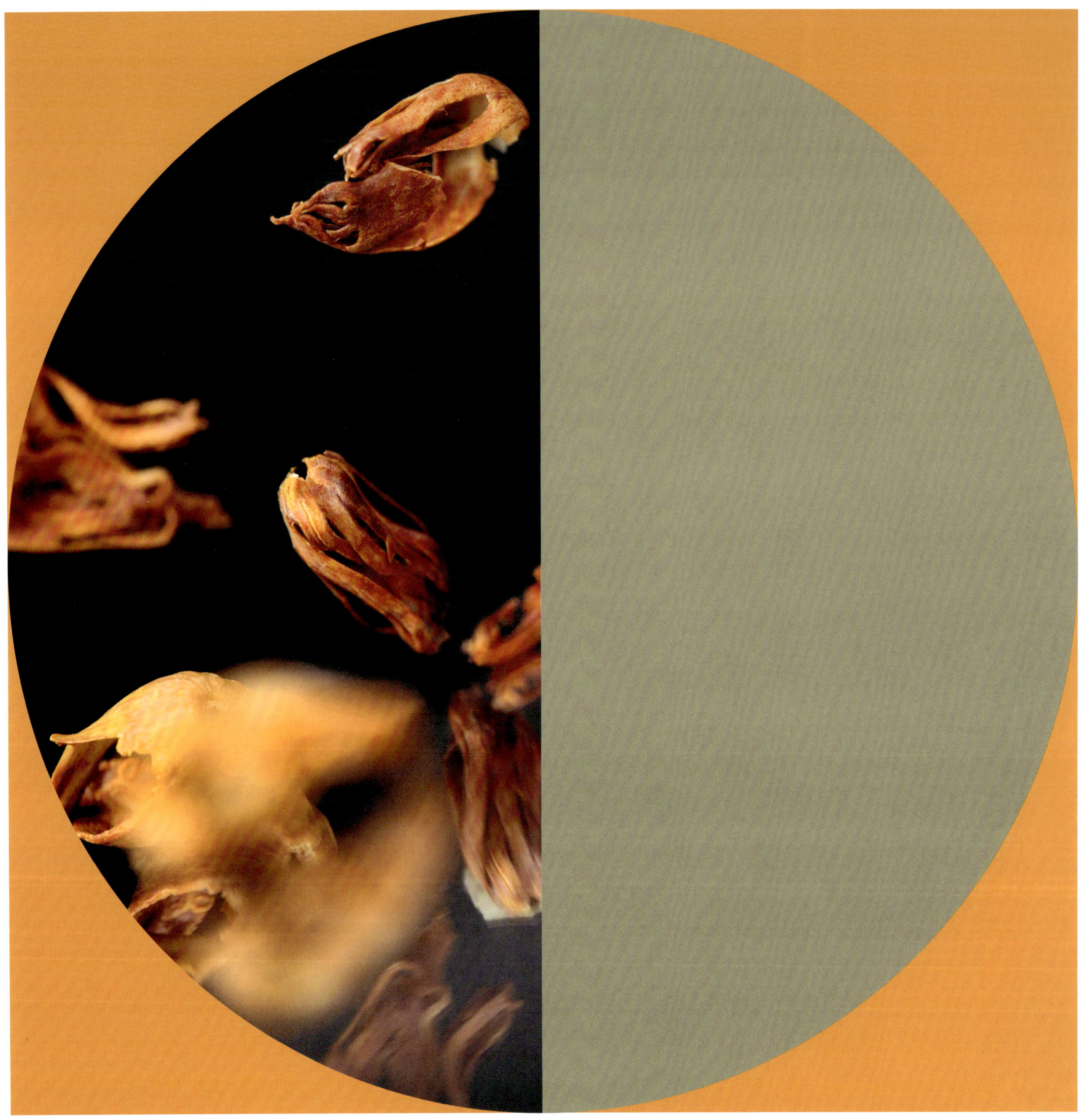

MACE

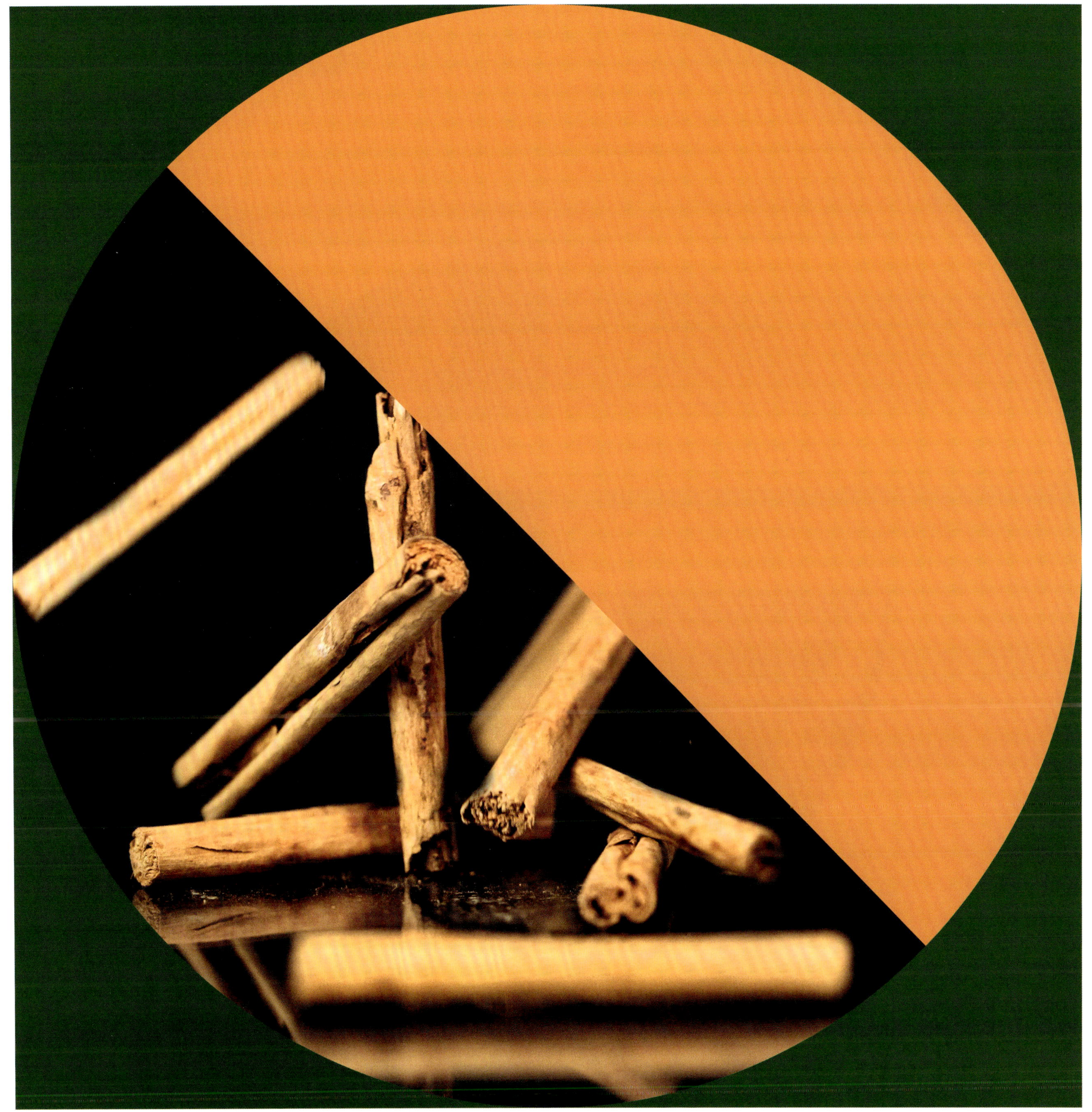

CINNAMON

GALANGAL ROOT

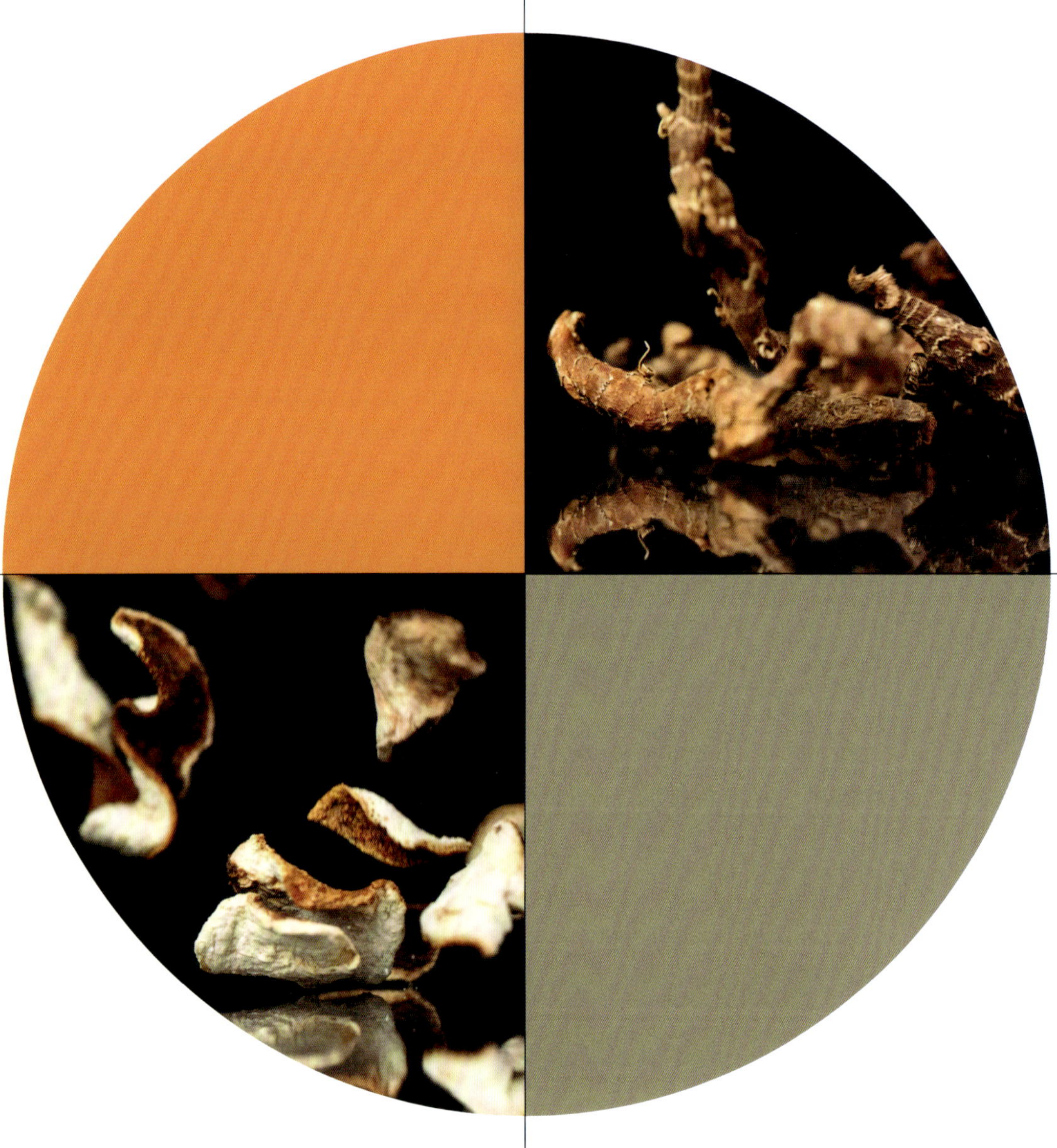

BITTER ORANGE PEEL

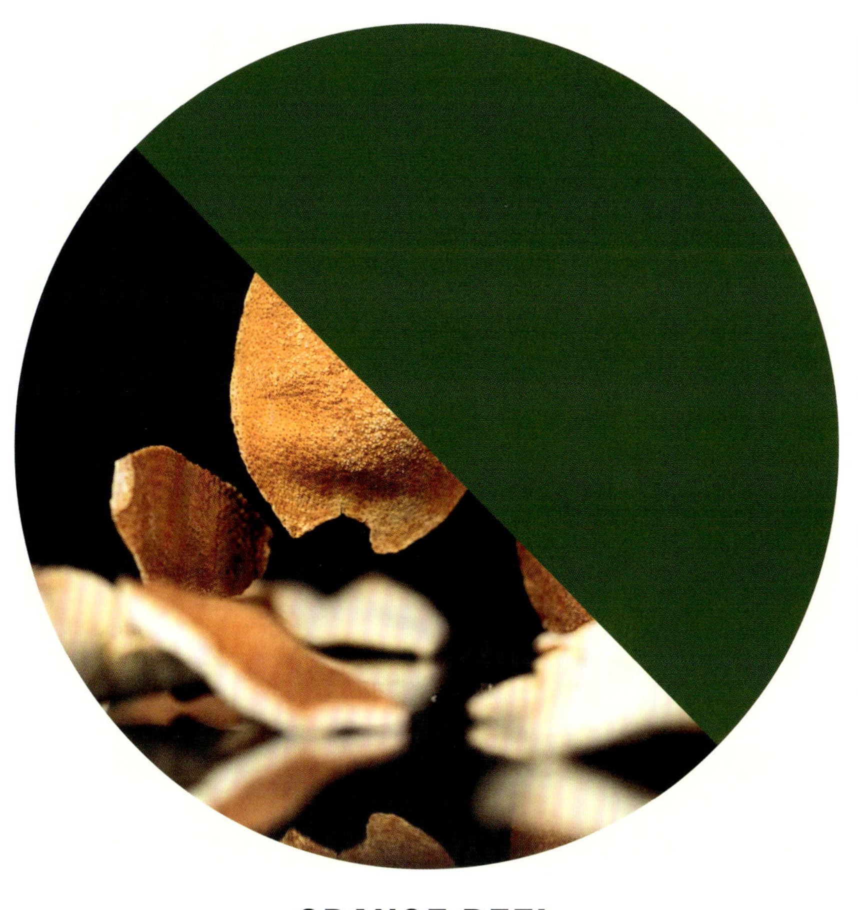

ORANGE PEEL

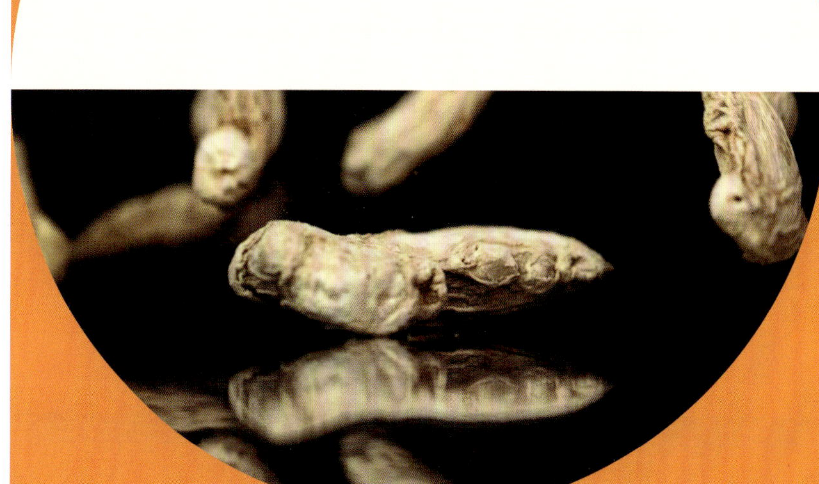

GINGER ROOT

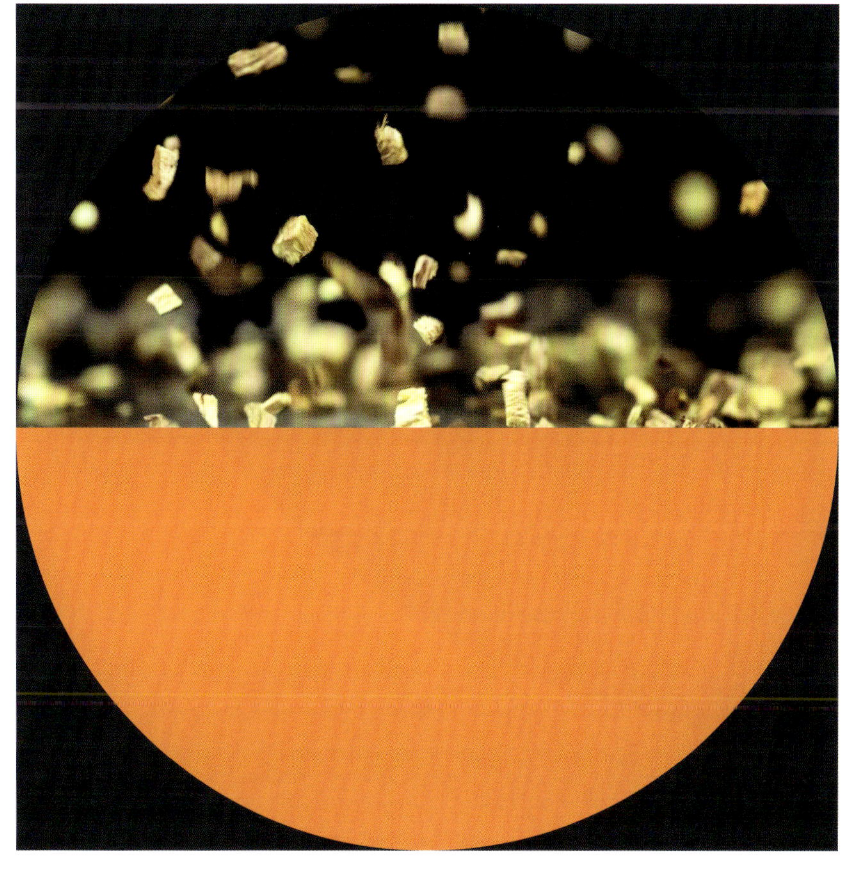

LIQUORICE ROOT

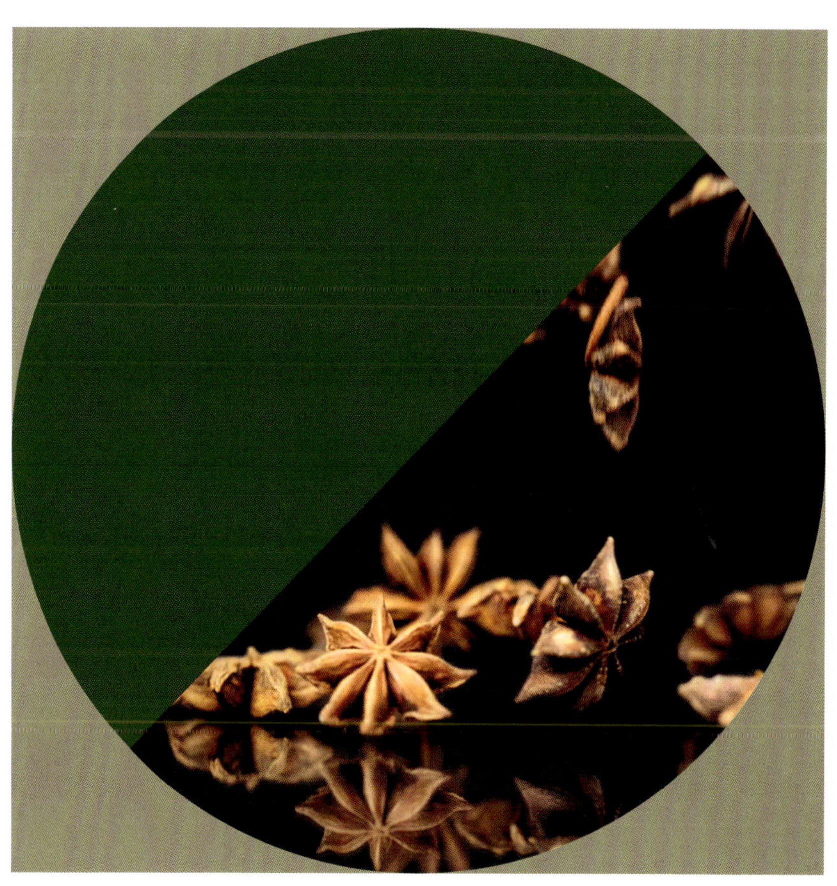

STAR ANISE

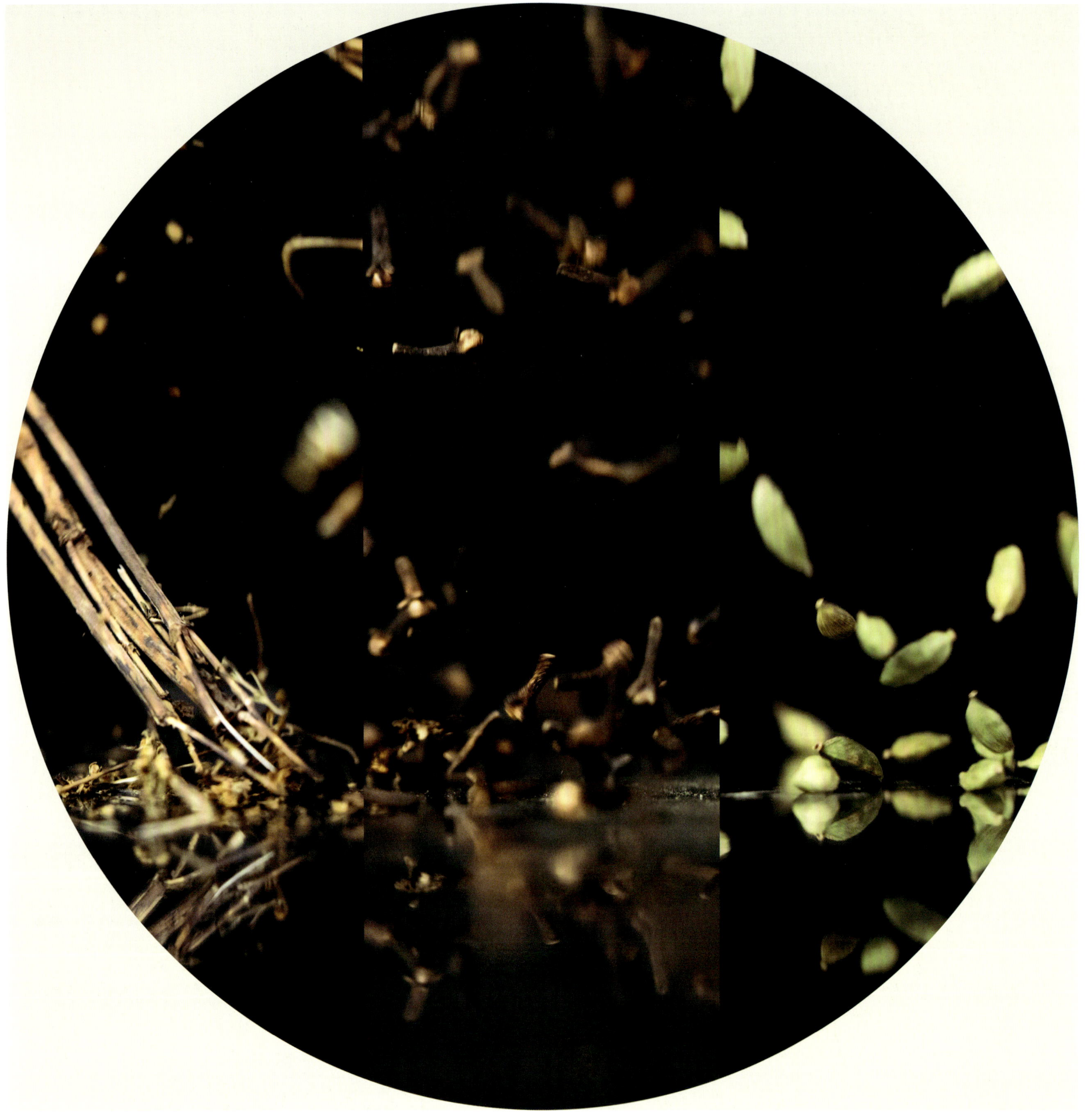

CHIRETTA
CLOVES
CARDAMOM

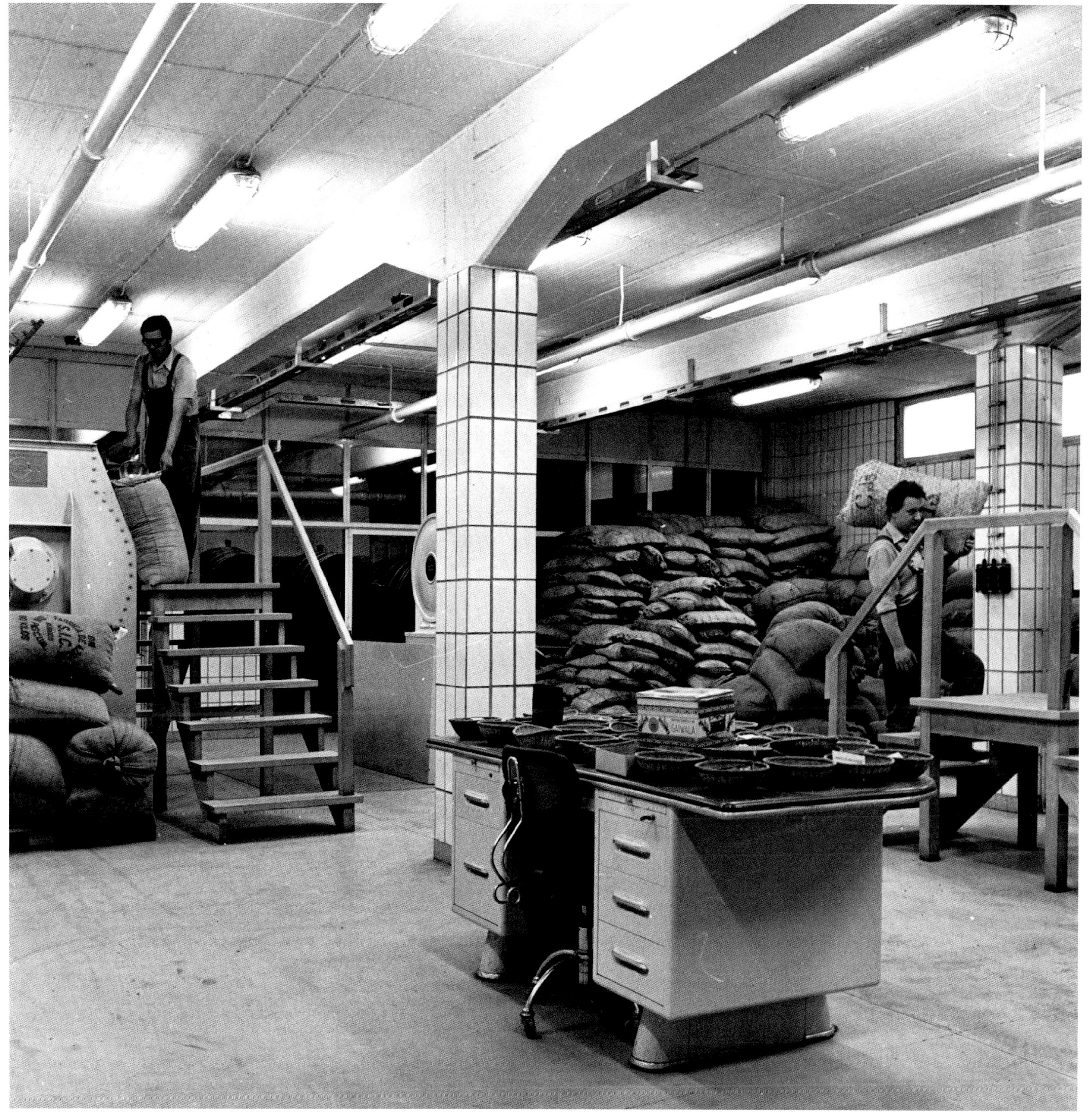

The botanical raw materials sourced from all over the world are delivered in sacks, weighed, ground in the large machine on the left, and mixed — and this same process from the 1960s continues to be applied to this day.

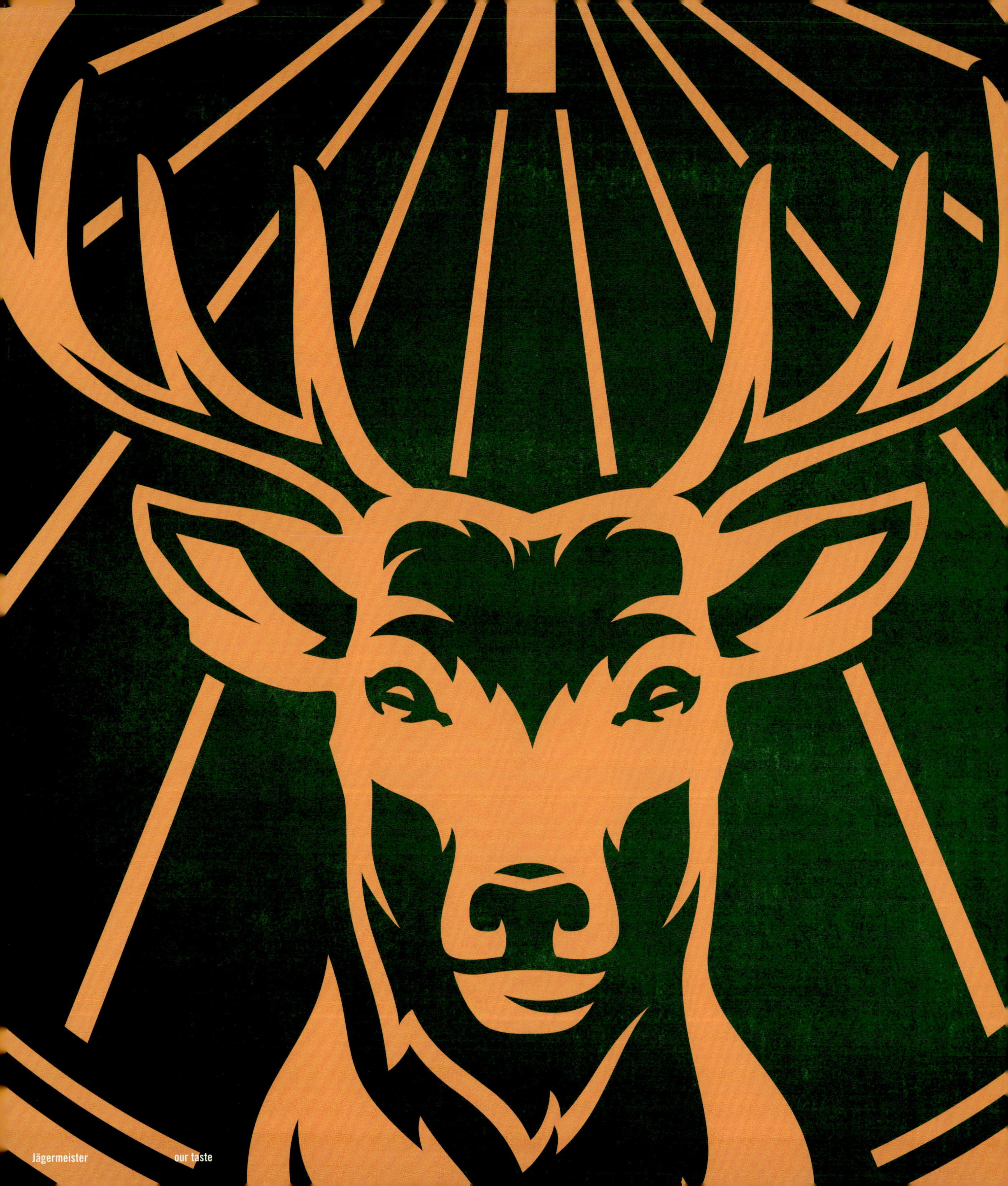

Jägermeister
our taste

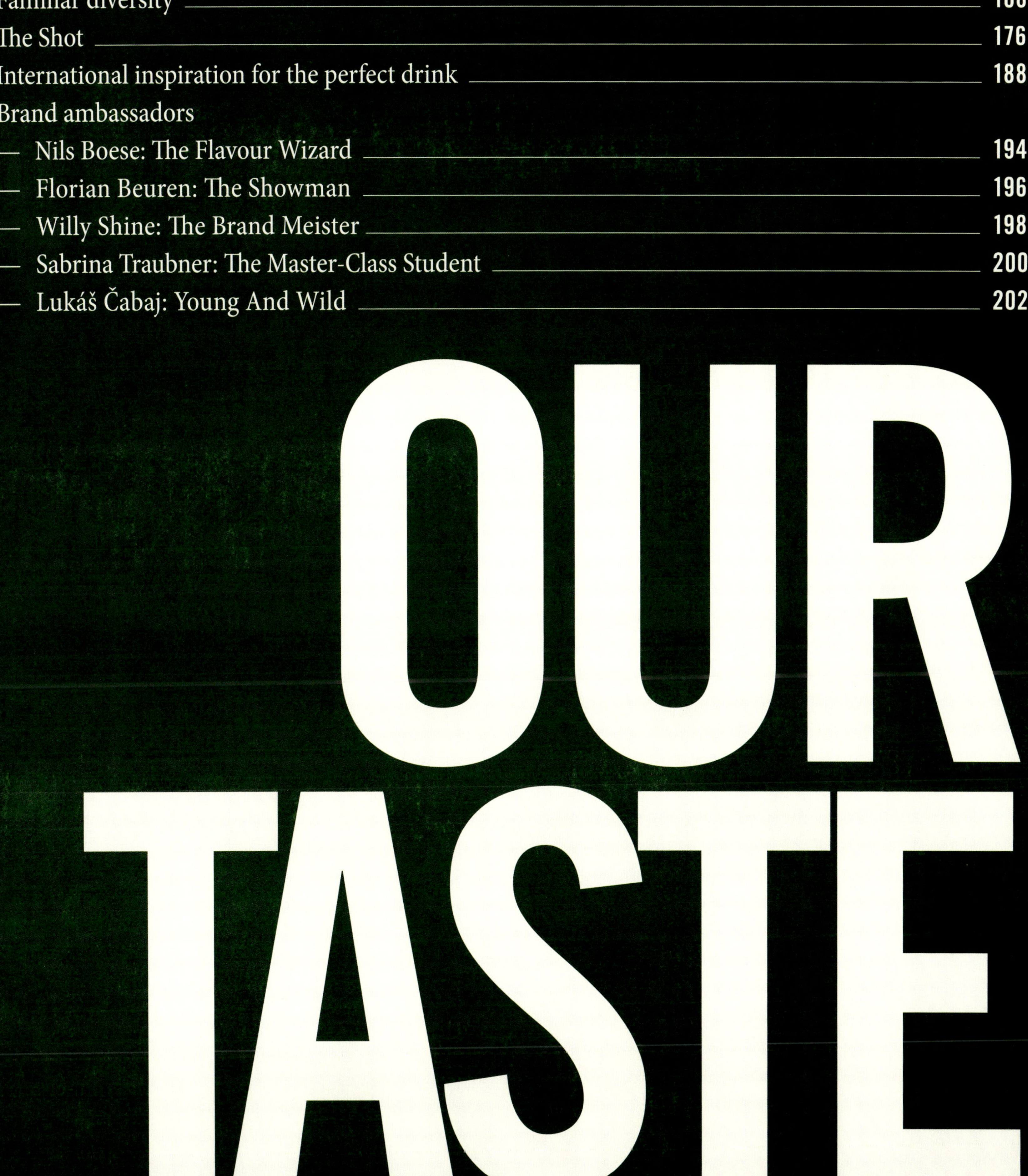

Familiar diversity — 168
The Shot — 176
International inspiration for the perfect drink — 188
Brand ambassadors
— Nils Boese: The Flavour Wizard — 194
— Florian Beuren: The Showman — 196
— Willy Shine: The Brand Meister — 198
— Sabrina Traubner: The Master-Class Student — 200
— Lukáš Čabaj: Young And Wild — 202

FAMILIAR DIVERSITY

Daring to try new things while also staying true to itself is the Jägermeister way. Innovative, different and exciting, yet always traditional and distinctive: few brands embody these attributes as perfectly as Jägermeister – attributes that have long been reflected in more than just creative, contemporary marketing campaigns. The Jägermeister brand now encompasses a diverse range of products appealing to completely new target audiences through innovative flavours. But there's one thing all products have in common: they're based on the original classic.

PORTFOLIO

FOCUSING ON THE CLASSIC

Jägermeister is the original. The planet's most successful herbal liqueur has been thrilling its fans for generations – and now does so all over the globe. At the German town of Wolfenbüttel, Jägermeister's master distillers apply the greatest of care to mix the 56 herbs, blossoms, roots and fruits from right around the world into a harmonious composition containing 35% alcohol by volume. The recipe has remained unchanged since Jägermeister was first invented by Curt Mast in 1934 – and remains top secret to this day. The diversity of its ingredients and the months spent stored in a barrel define the character of this legendary liqueur, whose complexity and harmonious nature become most apparent when the beverage is consumed as an ice-cold shot at -18°C. Five distinct flavour profiles shape the unique Jägermeister experience: sweet, bitter, fruity (citrus), spicy and aromatic (herbs). Given this vibrant and harmonious variety, Jägermeister is also ideal as a spirit in fancy cocktails and long drinks – at least for anyone who understands what really matters. The classic original is the epicentre of all Jägermeister product innovations, and is the basis for the entire range.

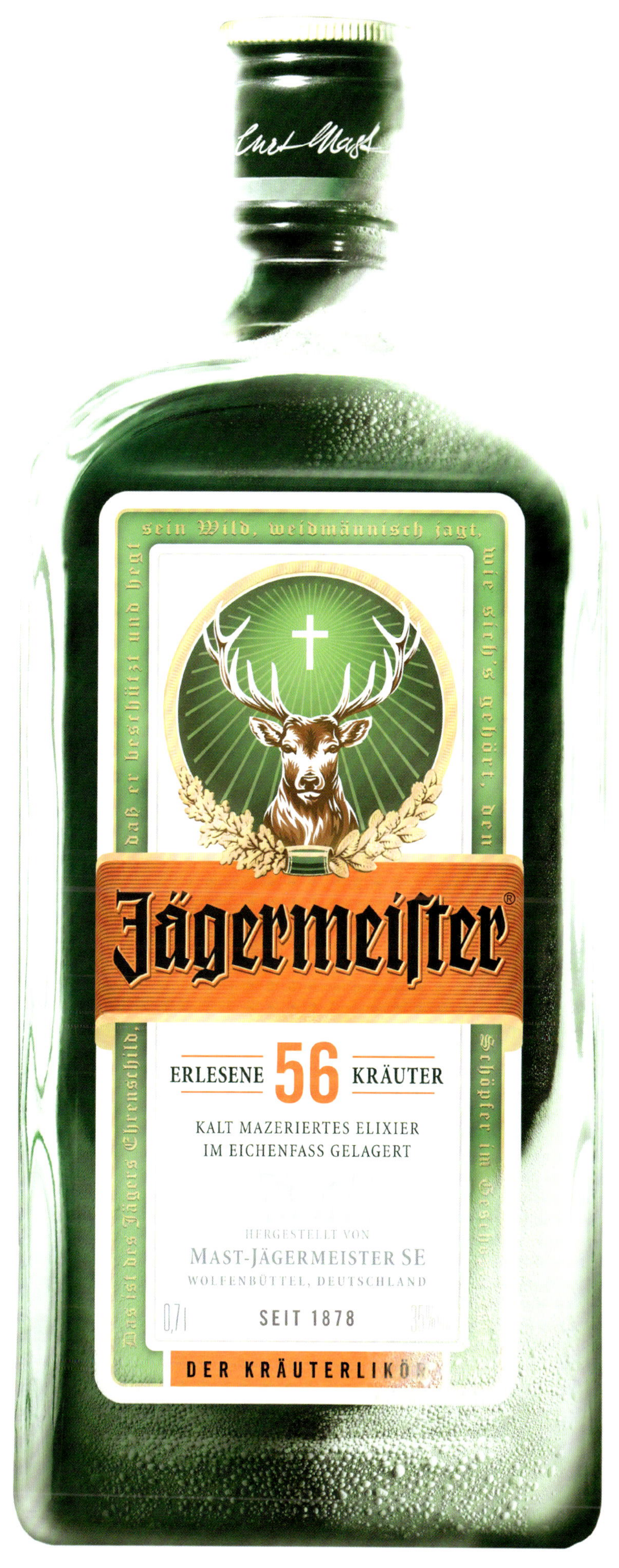

THE HIGHEST OF STANDARDS — WITH A HERB BASE

A manifest of perfection: Jägermeister Manifest is the world's first super-premium herbal liqueur, and its label bears the statement that has defined Jägermeister's master distillers since day one: 'The things we dare to do. The rules we rewrite. The offbeat spirit we embrace. The truth we stand for. The taste we savour.' Jägermeister Manifest is based on the original recipe, which has been further developed by Jägermeister's experienced distillers by adding extra exquisite ingredients to the 56 traditional Jägermeister herbs and extracting their essential natures through an elaborate five-fold maceration process. At 38% alcohol by volume, Manifest's alcohol content is somewhat higher than that of the original Jägermeister, creating a complex new consumption experience without tasting too strong. A fine wheat distillate obtained from a whole wheat grain serves as the base alcohol for Jägermeister Manifest. It is stored in small, toasted oak barrels for around 15 months, which gives the Jägermeister Manifest woody, vanilla-like notes and hints of smokiness. The result? Liquid perfection best enjoyed at one's leisure with a group of good friends – pure and slightly chilled. And Jägermeister has invented a special ritual for that: the sipping shot.

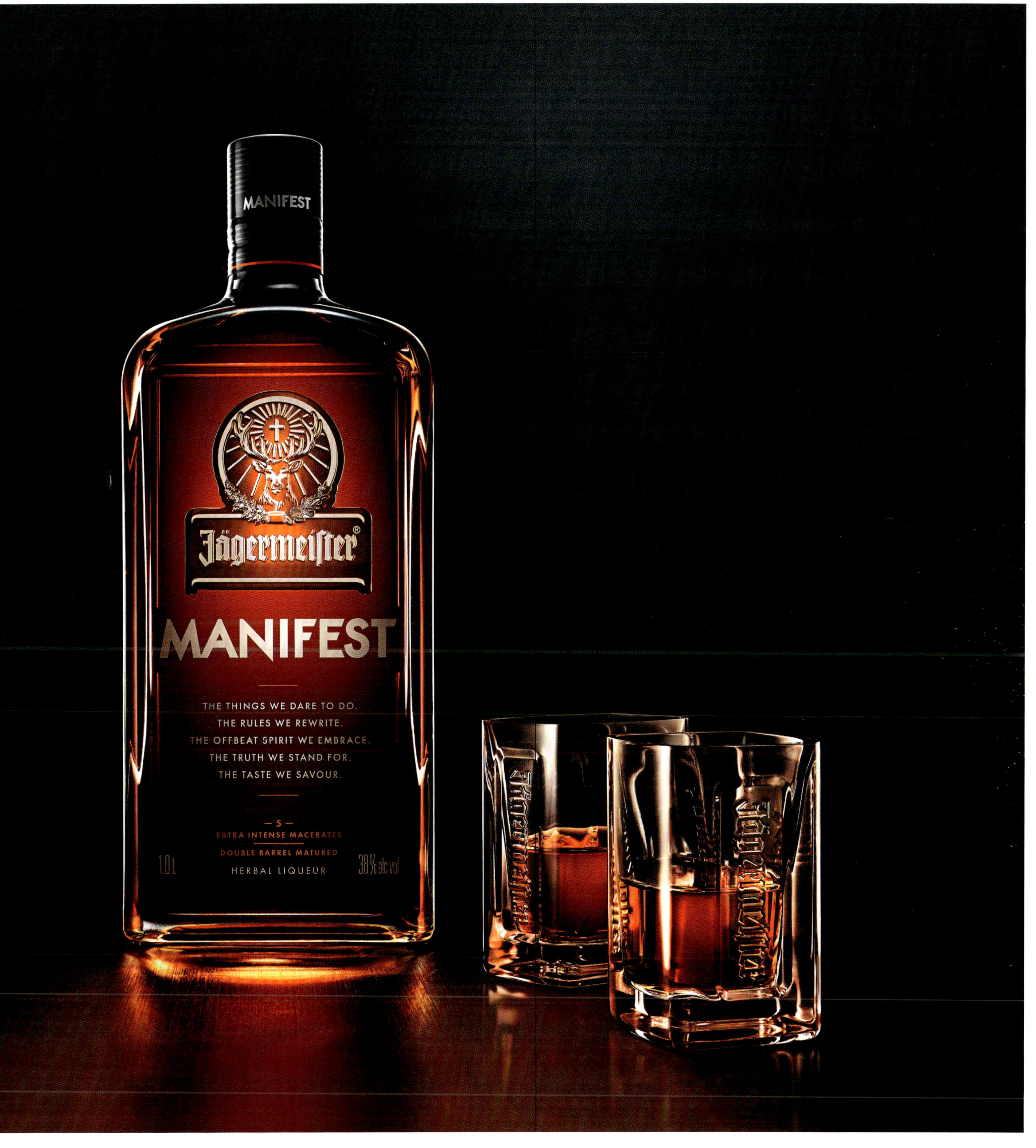
MANIFEST
Jägermeister®
MANIFEST
THE THINGS WE DARE TO DO.
THE RULES WE REWRITE.
THE OFFBEAT SPIRIT WE EMBRACE.
THE TRUTH WE STAND FOR.
THE TASTE WE SAVOUR.
— 5 —
EXTRA INTENSE MACERATES
DOUBLE BARREL MATURED
1.0 L
HERBAL LIQUEUR
38% alc vol

WINTER VIBES, JÄGERMEISTER STYLE
— 2013–2017

JÄGERMEISTER SCHARF: THE HOT SHOT
— since 2019

JÄGERMEISTER AND COLD BREW COFFEE IN PERFECT HARMONY
— since 2019

Jägermeister Spice *(Winterkräuter)* was an extraordinary liqueur in which the typical Jägermeister flavour was enhanced with the winter ingredients of vanilla, cinnamon, saffron and, of course, cloves. But bitters and alcohol content were reduced to 25% by volume, making for a pleasant mildness. Jägermeister Spice was a liqueur for frosty days and cosy nights, and not just at Christmas markets. It was originally designed to be a single season product, but remained in the Jägermeister range for over four years due to its worldwide success.

Headstrong, independent and a real character: the hottest Jägermeister of all time packs a punch! Its profile is both fierce and fine, with a harmonious sweetness, zest and perfectly balanced hotness. And this hotness aligns closely with the Jägermeister tradition – hot *(scharf)* being one of the five top notes of the original elixir! Flashes of galangal and ginger make it a version of Jägermeister that stimulates the senses, and cranks up the heat on unforgettable nights.

Jägermeister Cold Brew Coffee is the perfect fusion of the legendary premium liqueur and intense Arabica coffee. Available in the USA, UK and at duty-free shops, Jägermeister Cold Brew Coffee is a harmonious shot rich in coffee notes, with echoes of chocolate and an alcohol content of 33%. It is produced by roasting and grinding Fair Trade coffee beans from regional coffee companies, after which the Jägermeister distillers infuse the coffee with clear, cold water in a specially built extractor system as part of the cold-brew process. Shaking this mixture well is the perfect way to release all the natural ingredients. And at an ice-cold -18°C, Jägermeister Cold Brew Coffee truly demonstrates its headstrong but harmonious nature.

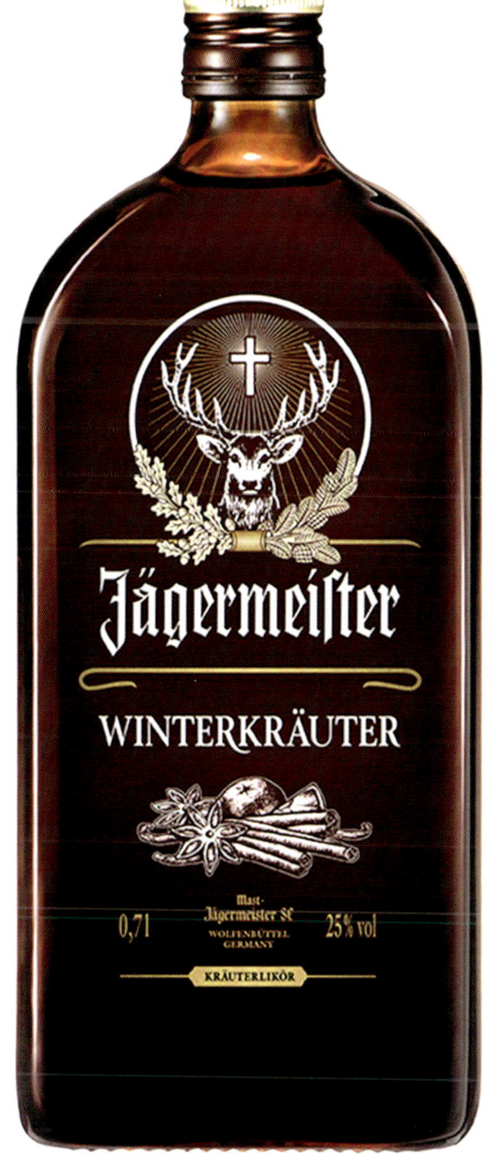

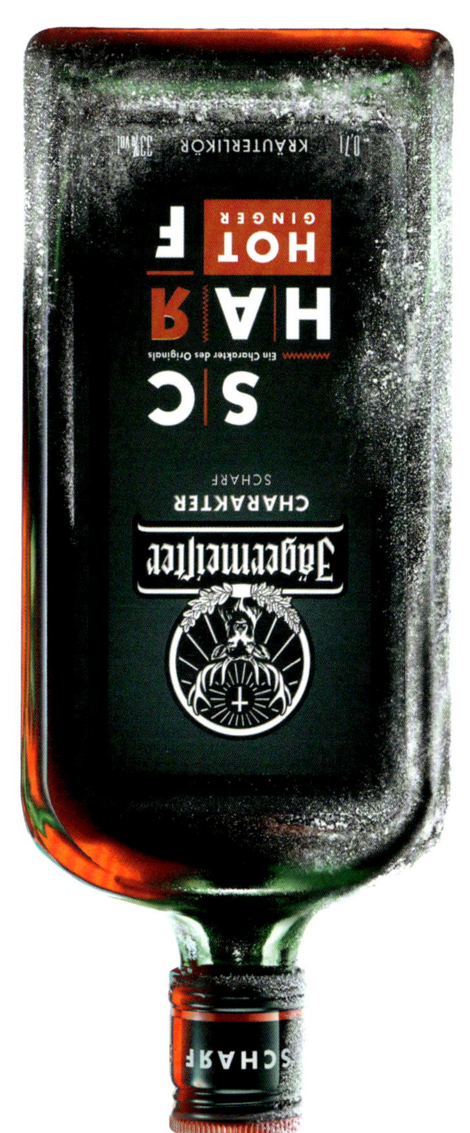
KRÄUTERLIKÖR
0,7l 33% vol
SCHARF | HOT GINGER
HART
Ein Charakter des Originals
CHARAKTER
SCHARF
Jägermeister
SCHARF

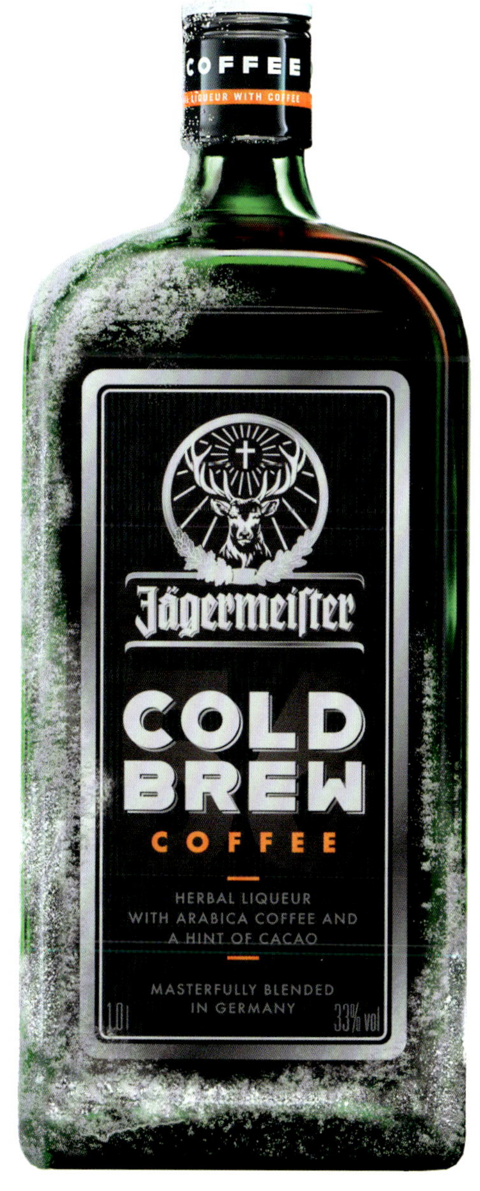
COFFEE
LIQUEUR WITH COFFEE
Jägermeister
COLD
BREW
COFFEE
HERBAL LIQUEUR
WITH ARABICA COFFEE AND
A HINT OF CACAO
MASTERFULLY BLENDED
IN GERMANY
1,0l 33% vol

Special moments require special drinks. Whether shot glass or bottle – here are some of the rare treasures from the historical Jägermeister archive that will make any collector's heart beat faster.

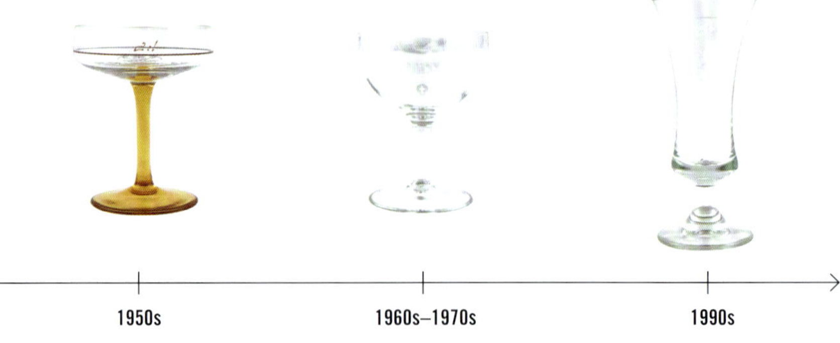

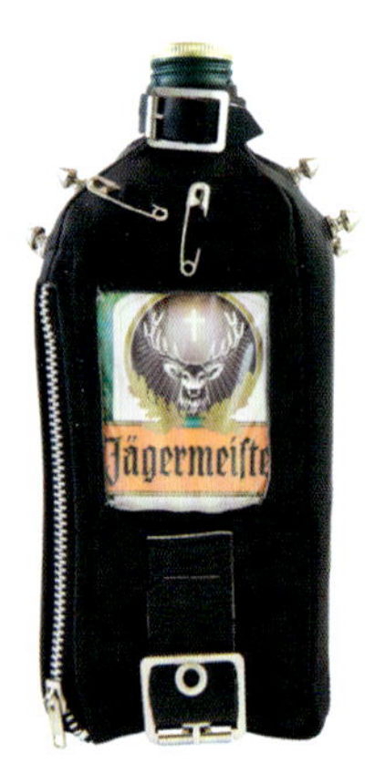

This punk bottle is a custom-made gift for rock bands in Great Britain from the year 2007.

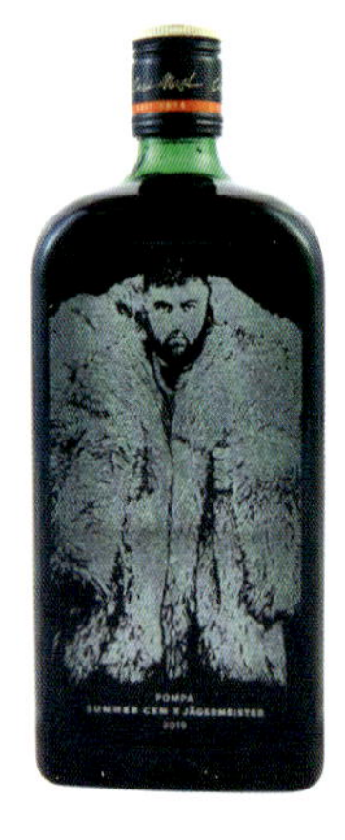

A flip book made of Jägermeister bottles. Exactly 756 frames from Summer Cem's video for the song *Pompa* were immortalized on the green Jägermeister bottles in 2019, resulting in the so-called bottle track. All 756 bottles also appeared in the video and were sold in the Jägermeister shop after the clip's premiere!

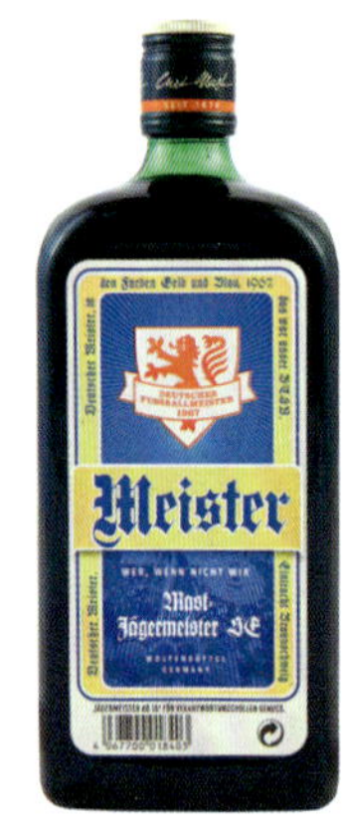

A special relationship between Jägermeister and Eintracht Braunschweig: to celebrate the 50th anniversary of Eintracht Braunschweig's championship title in the Bundesliga, Jägermeister launched this special Meister bottle in 2017.

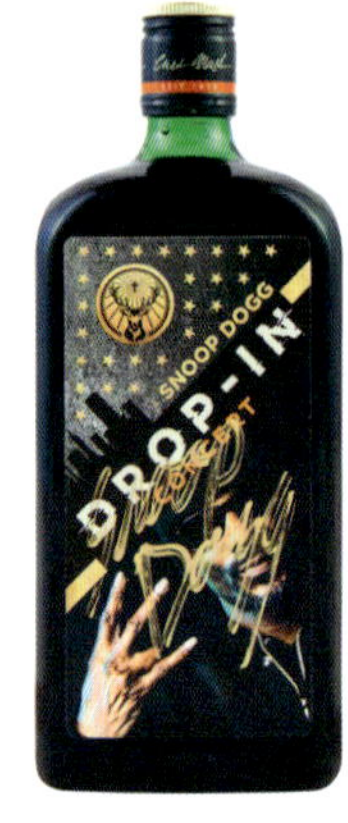

A very special competition from 2018: the winner was offered an exclusive concert by Snoop Dogg, who, with a bit of luck, could appear in the winner's living room with the help of Alexa. The bottle was personally signed by Snoop Dogg on the occasion of this drop-in concert and is now being kept in the Jägermeister archive.

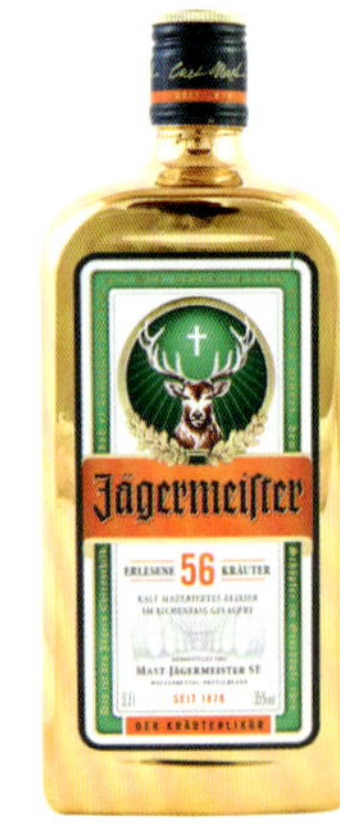

At the beginning of 2019, everyone was talking about Franck Ribéry's gold-covered steak. Jägermeister was also able to do that – and brought out this gold-plated special bottle within a very short time: only 56 bottles were produced at a price of € 560.56.

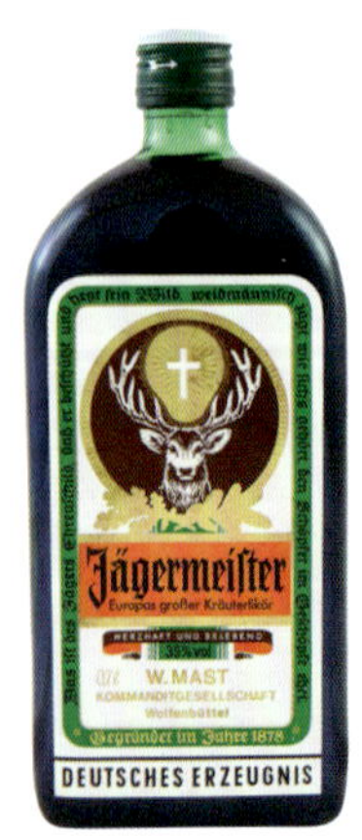

Only experts can correctly date historic Jägermeister bottles: This 0.7 litre bottle, for example, dates from the period between 1984 and 1987. Would you have known it?

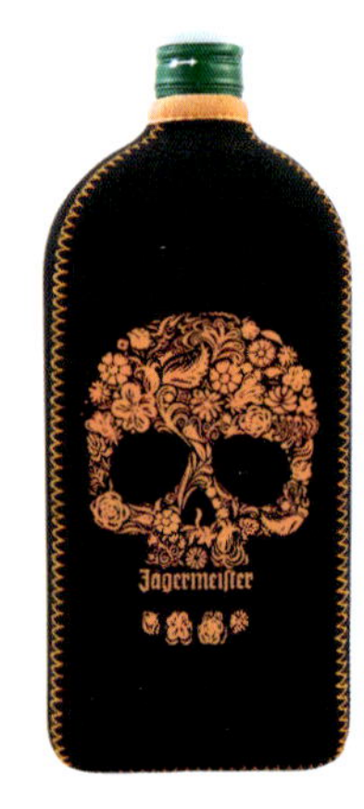

Trick or treat with Jägermeister: for Halloween, a Jägermeister bottle was wrapped in a creepy, opaque neoprene cover.

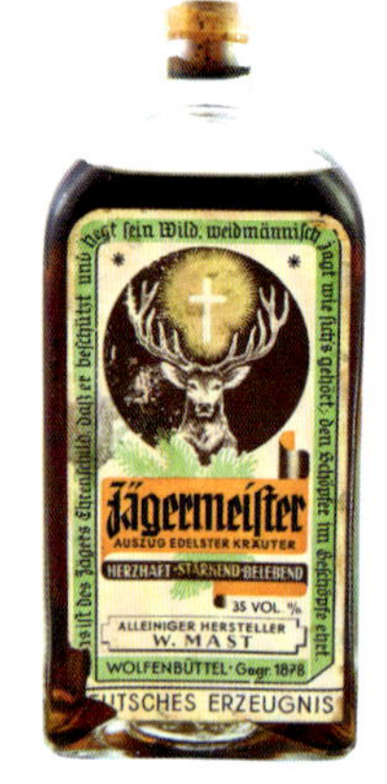

A special treasure and one of the oldest preserved, unopened Jägermeister bottles – this originally filled 0.7-litre Jägermeister bottle dates from the period before 1954.

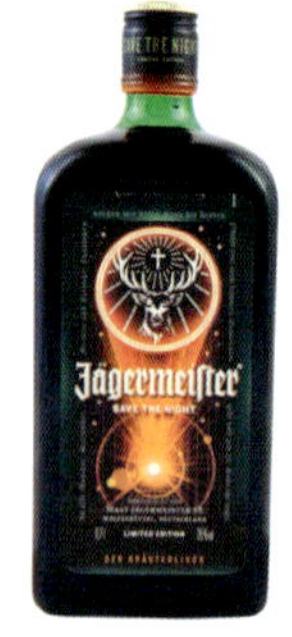

During the 2020 Covid pandemic, Jägermeister started its SAVE THE NIGHT initiative to support the nightlife community. The label of this #SAVETHENIGHT Limited Edition Bottle was designed by the German illustrator and graphic designer Max Löffler. With the launch of the limited edition, Mast-Jägermeister SE made an additional 1 million euros available, which benefited artists, creative people, club owners and bartenders in the participating markets.

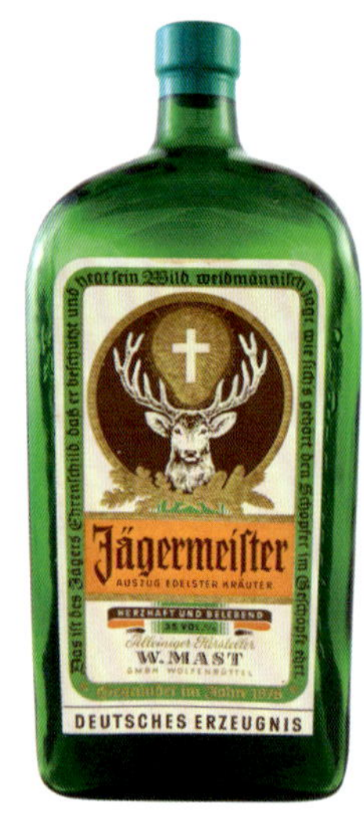

One of the largest-ever Jägermeister bottles! This 3-litre magnum bottle was sold between 1951 and 1961; this particular bottle dates from the period between October 1954 and November 1961.

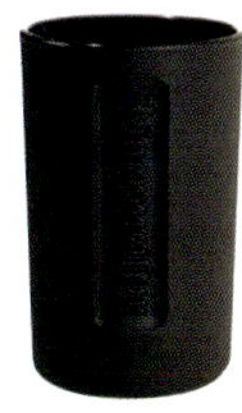

2000s	2021	2021 – black shot glass

Jägermeister is best drunk ice cold at -18°C. In order to prevent the distinctive Jägermeister bottle from disappearing somewhere deep in a dark icebox, the American partner Sidney Frank came up with something special at the beginning of the 1990s: his very own dispensing system for the herbal liqueur from Wolfenbüttel.

The tap machines quickly became a permanent fixture in American bars and made their way out into the rest of the world from there.

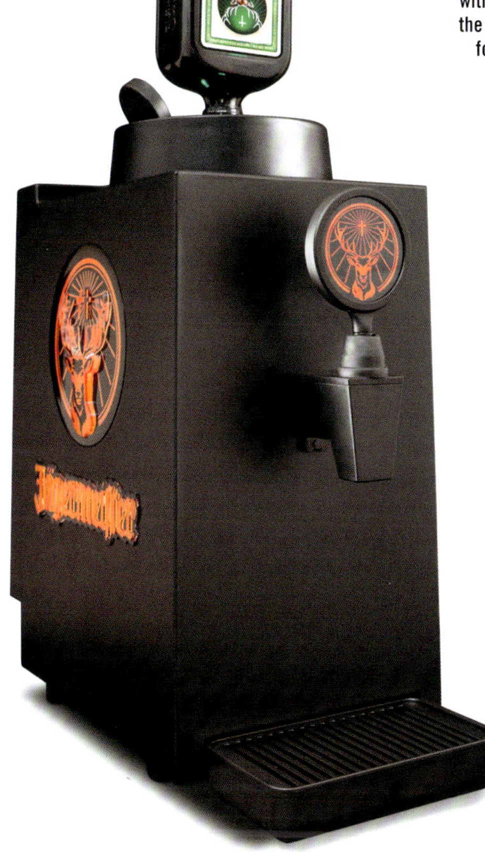

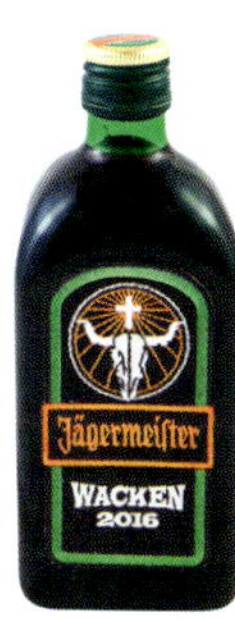

Wacken and Jägermeister – they go together. This special bottle brought out on the occasion of the 2016 festival even features a peel-off, iron-on patch as its label!

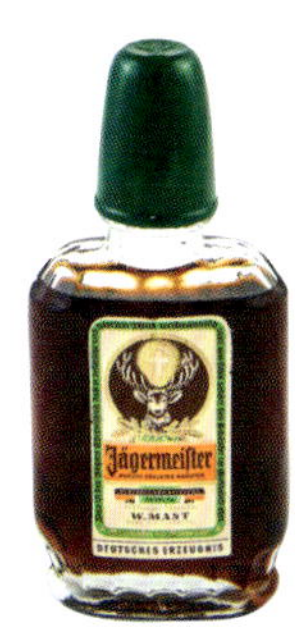

The Jägermeister pocket bottle was, so to speak, the multi-functional tool in the bottle portfolio. The lid doubles as a shot cup – guaranteeing enjoyment anywhere. The pocket bottle was available from 1948 to 1969, and this beautiful specimen designed to hold 150ml dates from before 1959.

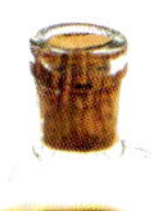

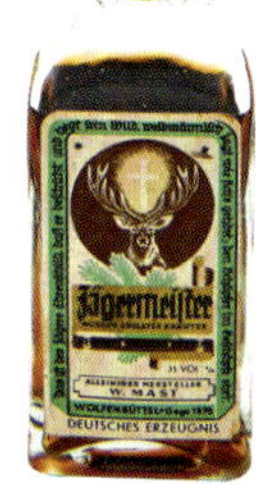

One of the oldest and best-preserved bottles from the Jägermeister archive, this 0.1 litre bottle made of white glass and with a cork stopper dates from the 1930s.

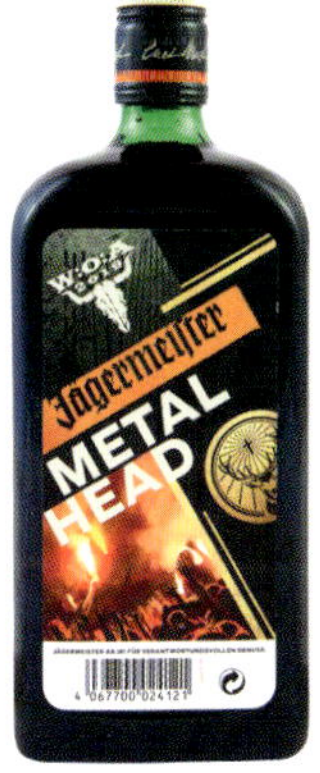

When bull and stag work together: Wacken and Jägermeister – a perfect match, as this bottle created on the occasion of the 2019 Wacken Open Air proves.

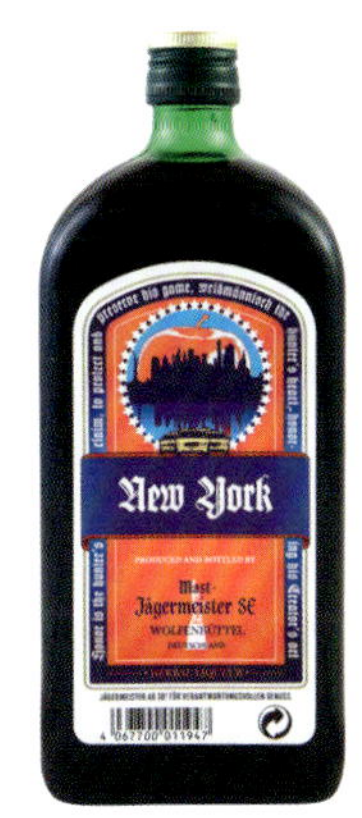

Jägermeister is now sold in 150 countries, and so quite a number of beautiful regional promotional labels can be found, such as this New York bottle from the early 2010s.

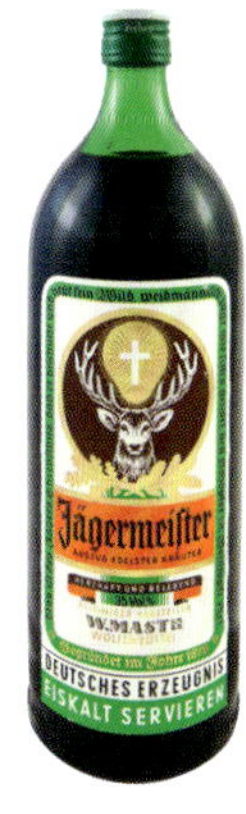

And at some point there you could even find this: round 1-litre bottles of Jägermeister, which were produced especially for the catering industry because they were a perfect fit for their drinks coolers. This bottle was made between 1954 and 1970.

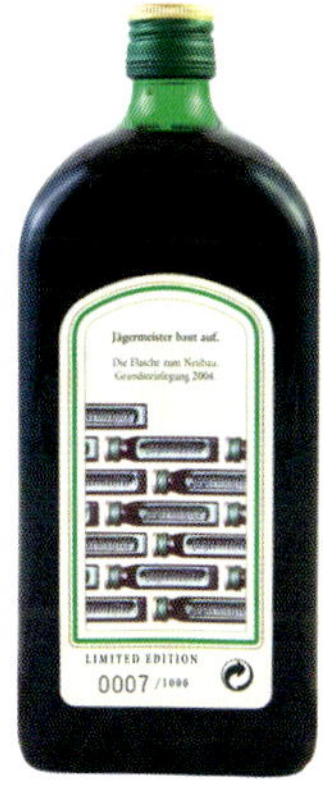

In the early 2000s, Jägermeister treated itself to a new building for the company headquarters. This special bottle was produced on the occasion of the laying of the foundation stone in 2004.

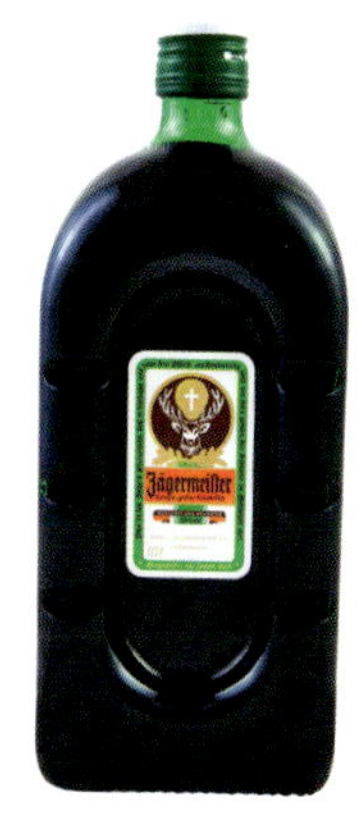

The Jägermeister bottle ashtray: this campaign bottle from 2001 could simultaneously serve as an ashtray when placed on its back.

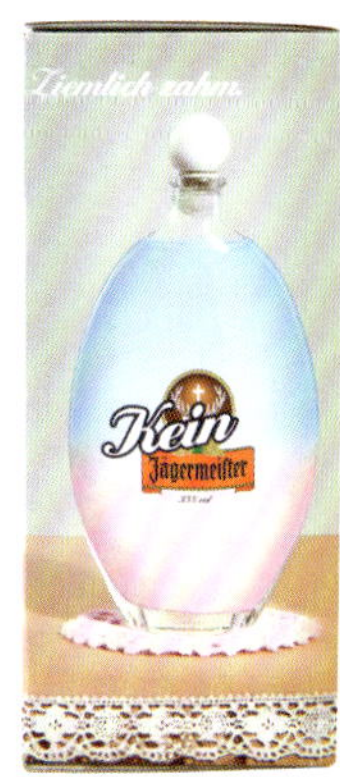

The fictional brand Kein Jägermeister (no Jägermeister) was the exact opposite of what Jägermeister 2008 stood for. The imaginary product was never sold. How does Kein Jägermeister taste? The Jägermeister archivist is probably the only one who knows.

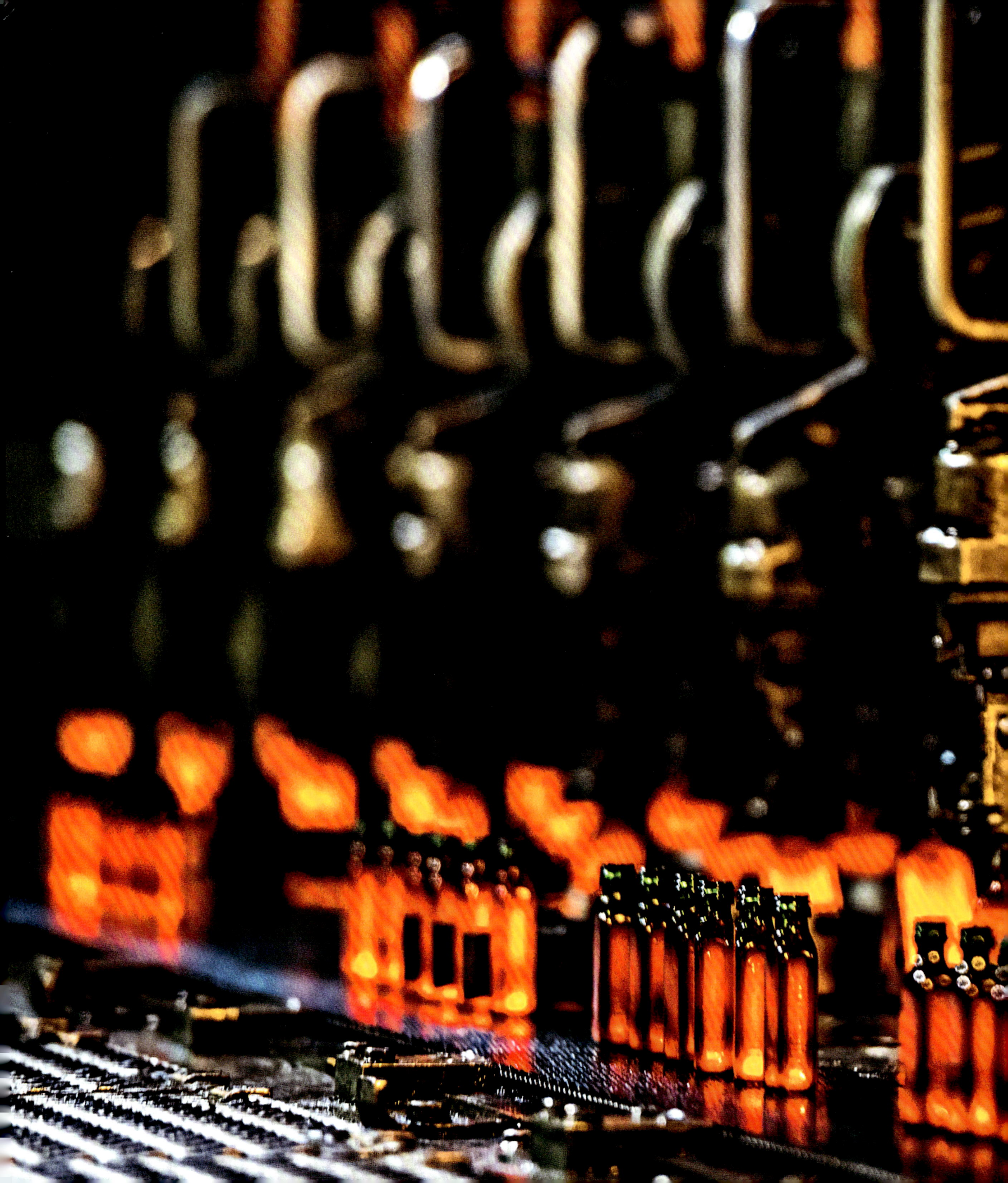

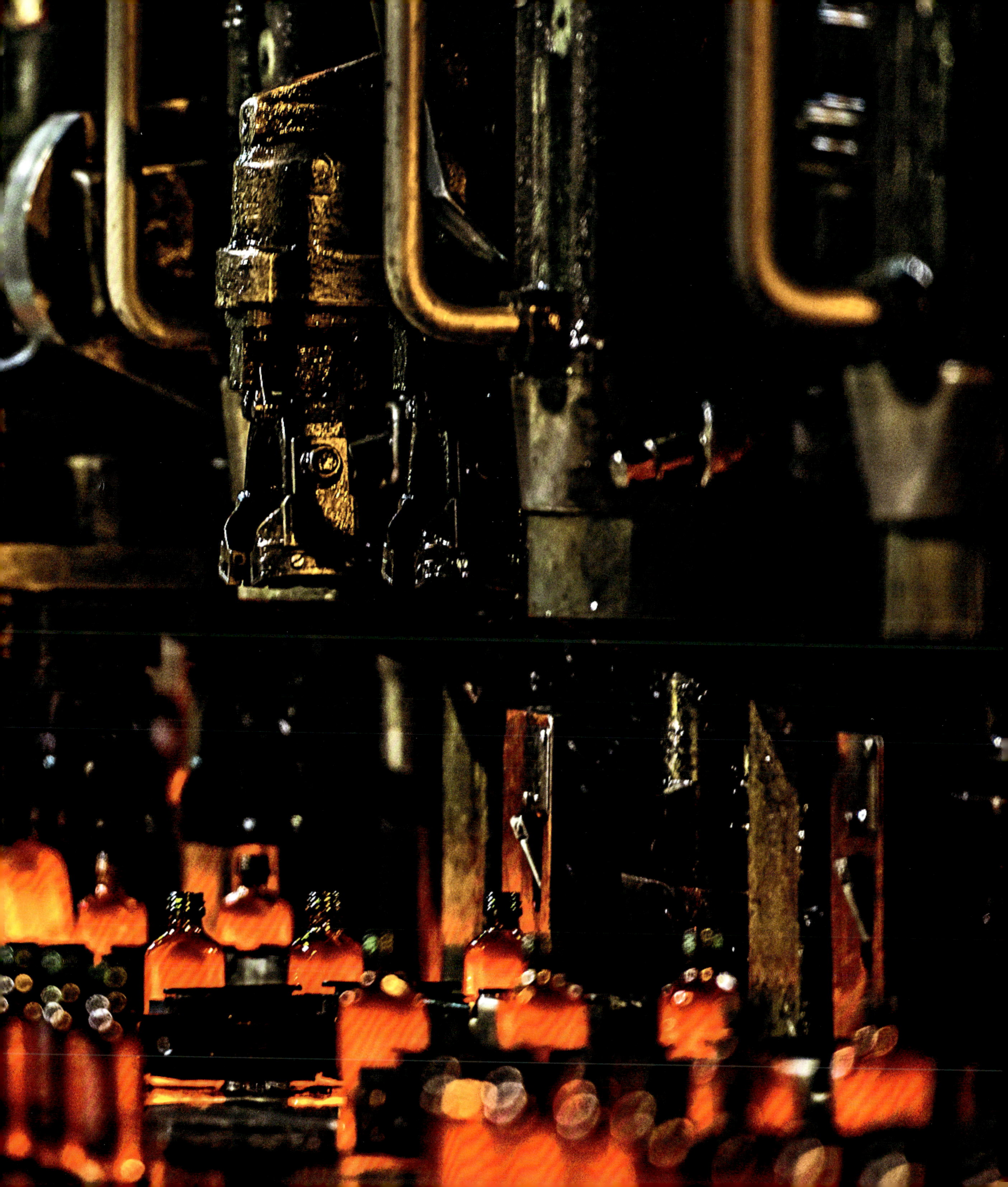

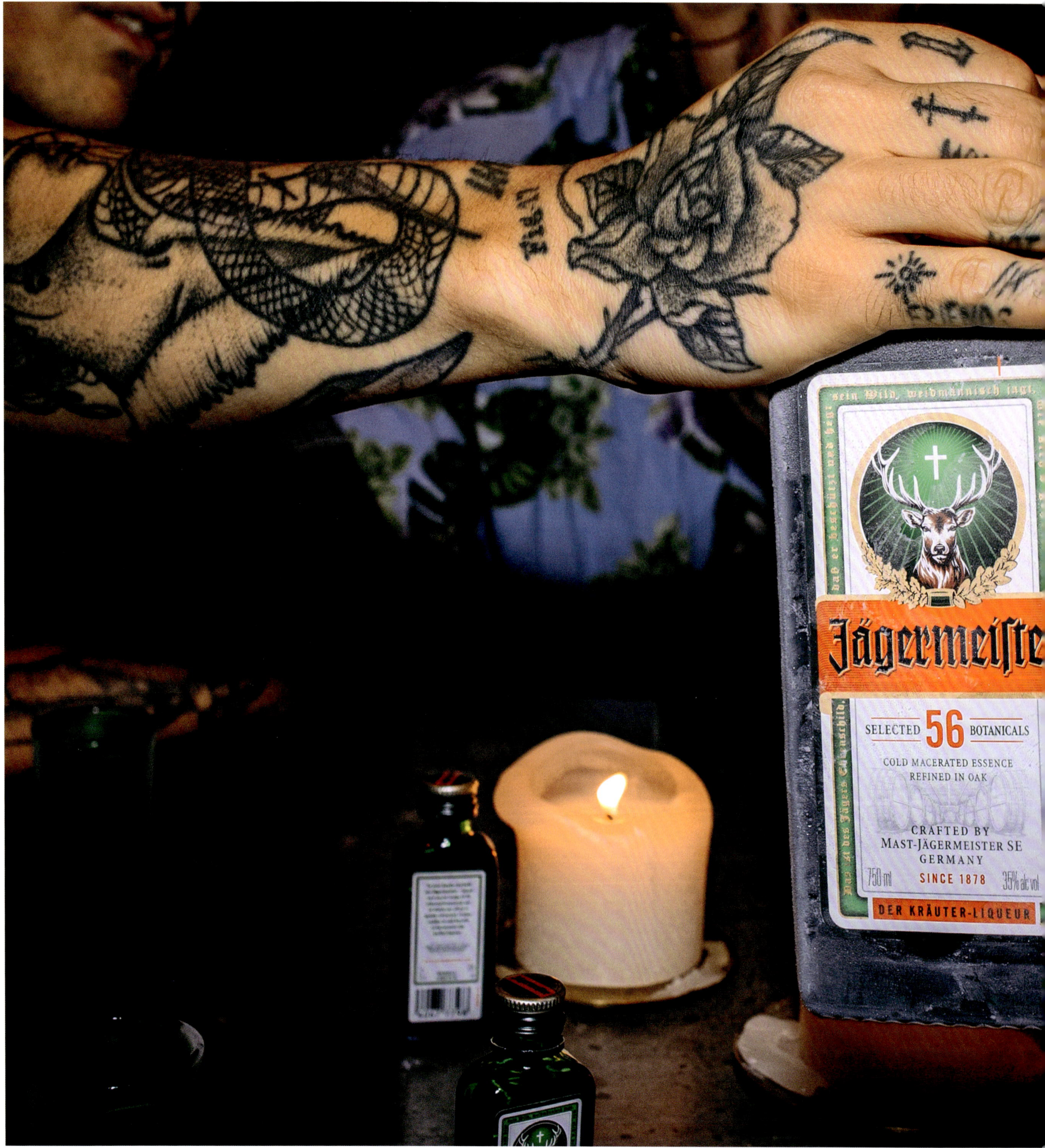
Sein Wild, weidmännisch jagt,
Jägermeister
SELECTED 56 BOTANICALS
COLD MACERATED ESSENCE
REFINED IN OAK
CRAFTED BY
MAST-JÄGERMEISTER SE
GERMANY
750 ml
SINCE 1878
35% alc vol
DER KRÄUTER-LIQUEUR

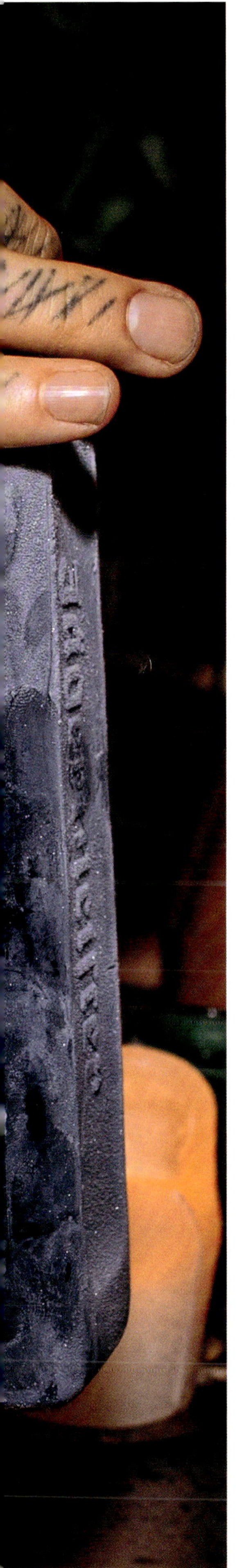

THE SHOT

The shot is the most direct way of appreciating a beverage, but it's much more than just a short.

...With a shot, what you see is what you get – there are no compromises, and nor should you make any when choosing your shot. The ice-cold Jägermeister shot combines all these attributes. It will accompany you and your friends through the night.

The shot is the most direct way of appreciating a beverage, but it's much more than just a short. It's both a ritual and a statement. When enjoying a shot with others, few words are required. That's when everyone knows they're savouring a pure moment of friendship – the shot is indeed often what kick-starts one of the best nights of your life. There is no more emotional way of drinking than from a 20 ml iced shot glass. Nothing can be hidden in a drink so neat and pure. No juice, no ingredient, no preparation method can mask this drink's character or quality.

THE CLASSIC

With 56 ingredients, a Jägermeister shot doesn't need anything more added to it – but there are a few tricks worth knowing if you want to unleash the ultimate Jägermeister flavour experience.

Ingredients:
20 ml Jägermeister
Chilled shot glass

Method:
A chilled glass is one of the secrets. It's already been mentioned, but the best way to enjoy a Jägermeister is when it's at -18°C.

And an extra tip: ideally, always keep a few glasses ready in the freezer, so you're always perfectly prepared for your next Jägermeister moment!

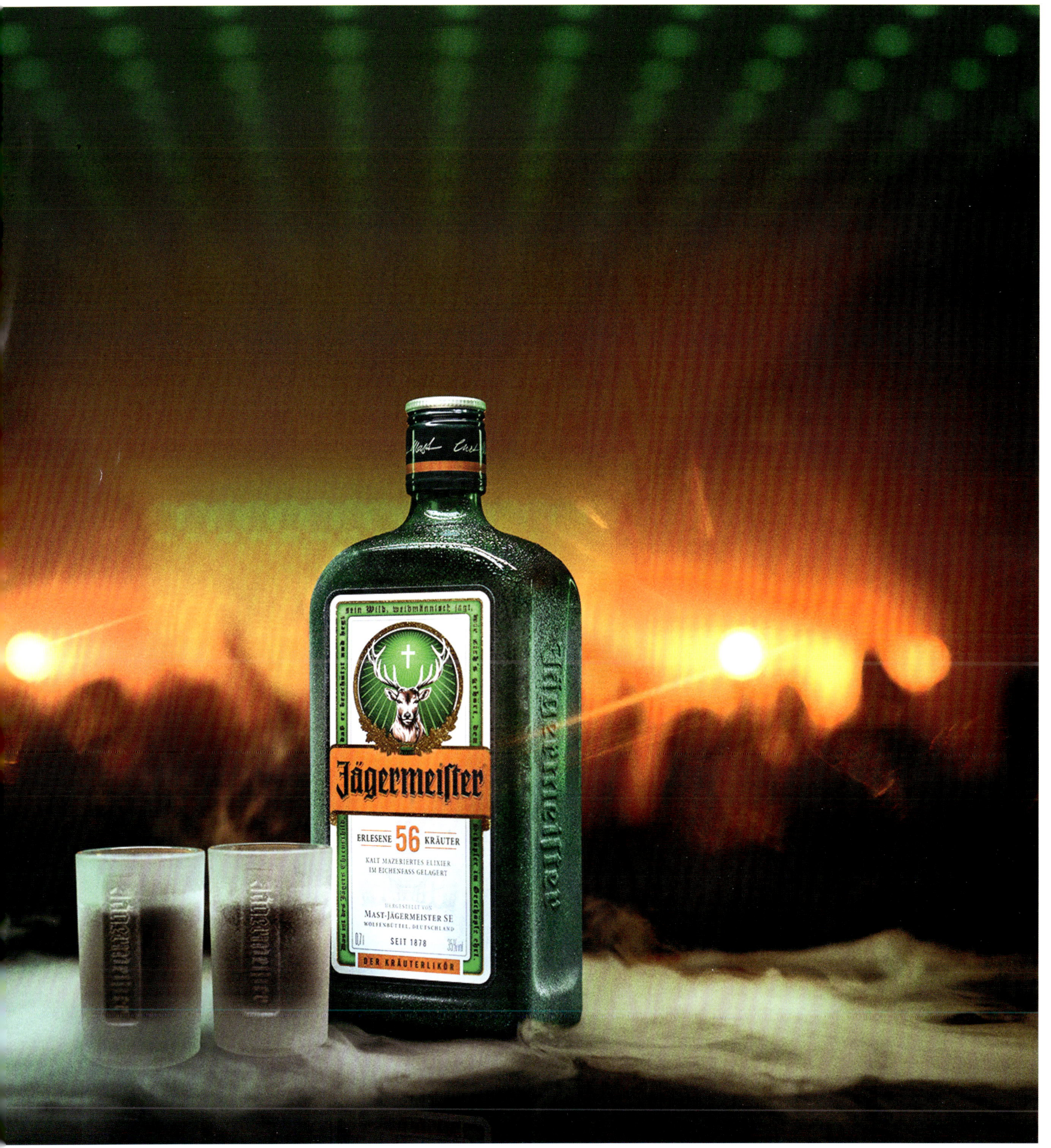

Jägermeister
ERLESENE 56 KRÄUTER
KALT MAZERIERTES ELIXIER
IM EICHENFASS GELAGERT
MAST-JÄGERMEISTER SE
WOLFENBÜTTEL, DEUTSCHLAND
SEIT 1878
DER KRÄUTERLIKÖR

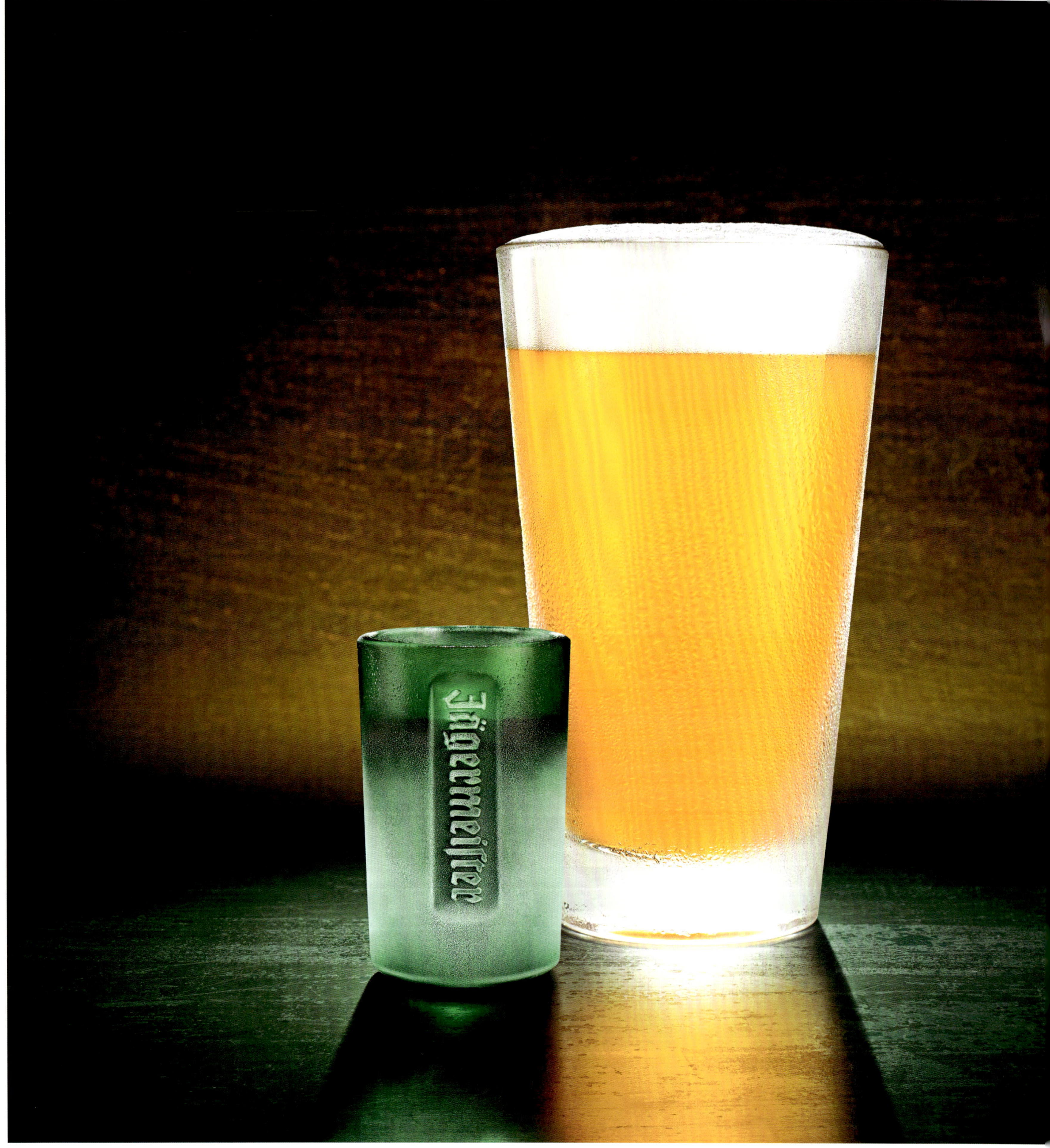

Jägermeiſter

DEER & BEER

Lager loves Jäger: a shot of Jägermeister, chilled of course, with fresh draught beer. No frills, no spiel, just straight-up honest goodness. Cheers!

Ingredients:
40 ml Jägermeister
Chilled shot glass
330 ml lager
Beer glass

Method:
Pour a double shot of ice-cold Jägermeister into a chilled shot glass. Pour your beer of choice into a chilled glass. Have a sip of the shot, drink the beer along with it, enjoy, repeat!

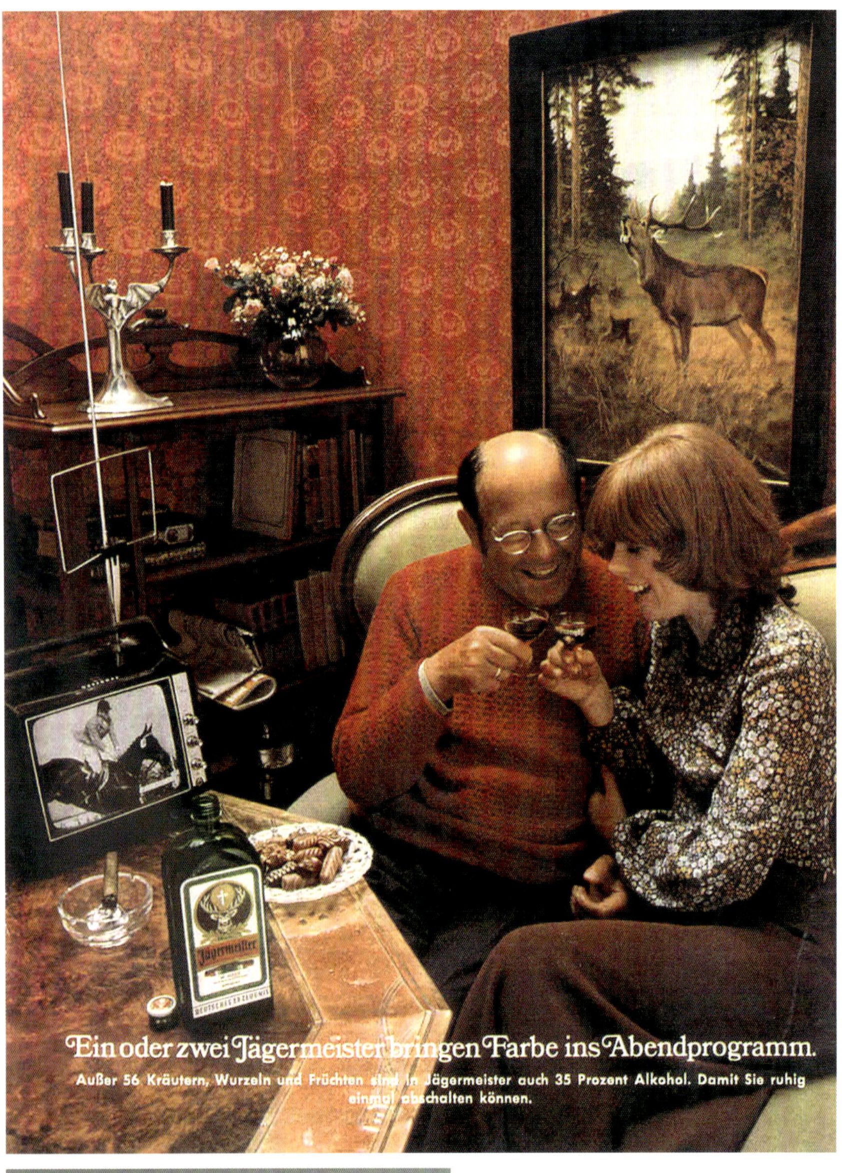

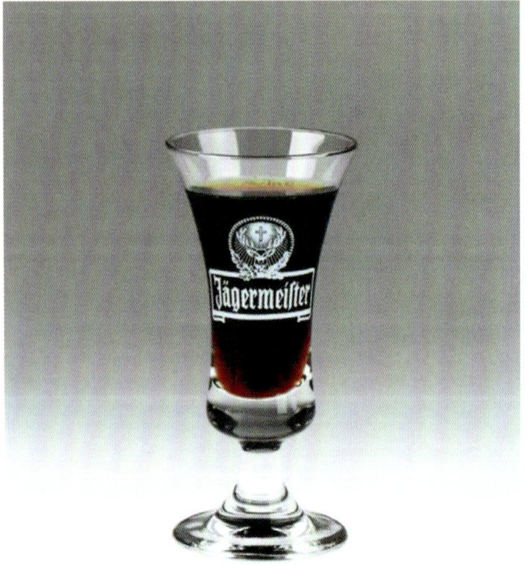

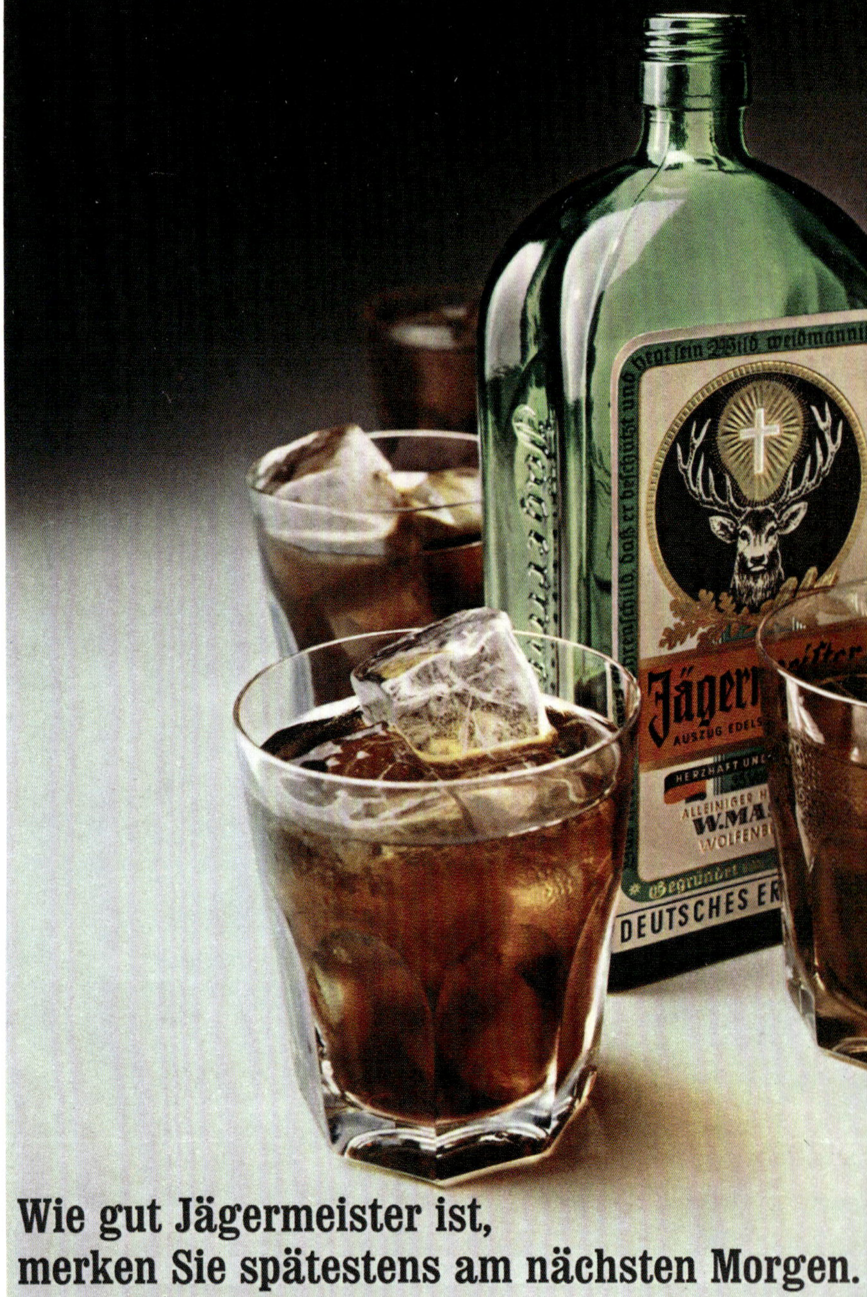

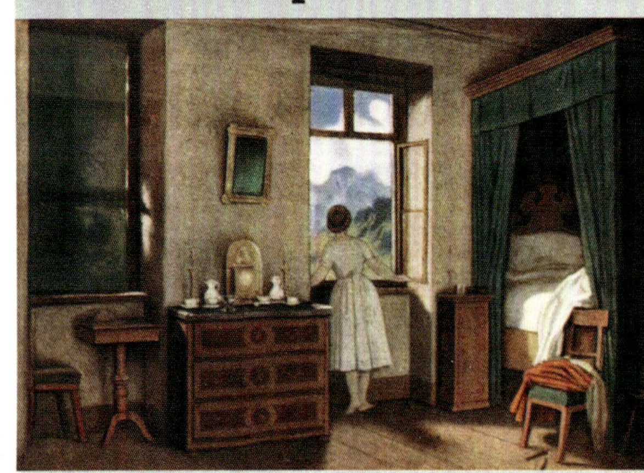

56 herbs, 1 shot, 100 % Jägermeister

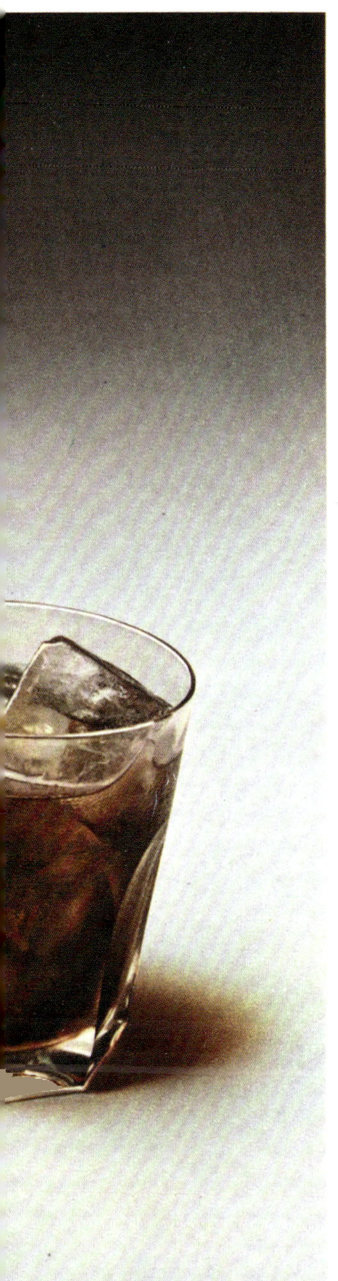

Wenn eine Party so ausgelassen zu werden droht wie das Jahrestreffen der Sanskritforscher. Oder einfach so.

Man kann Jägermeister auch mixen. Mit Cola zum Beispiel. Einige Avantgardisten trinken ihn sogar mit Sekt. Aber das sollten Sie nur nachmachen, wenn Sie einen guten im Haus haben.

Denn sonst haben Sie garantiert spätestens am nächsten Morgen einen Kater.

Jägermeister

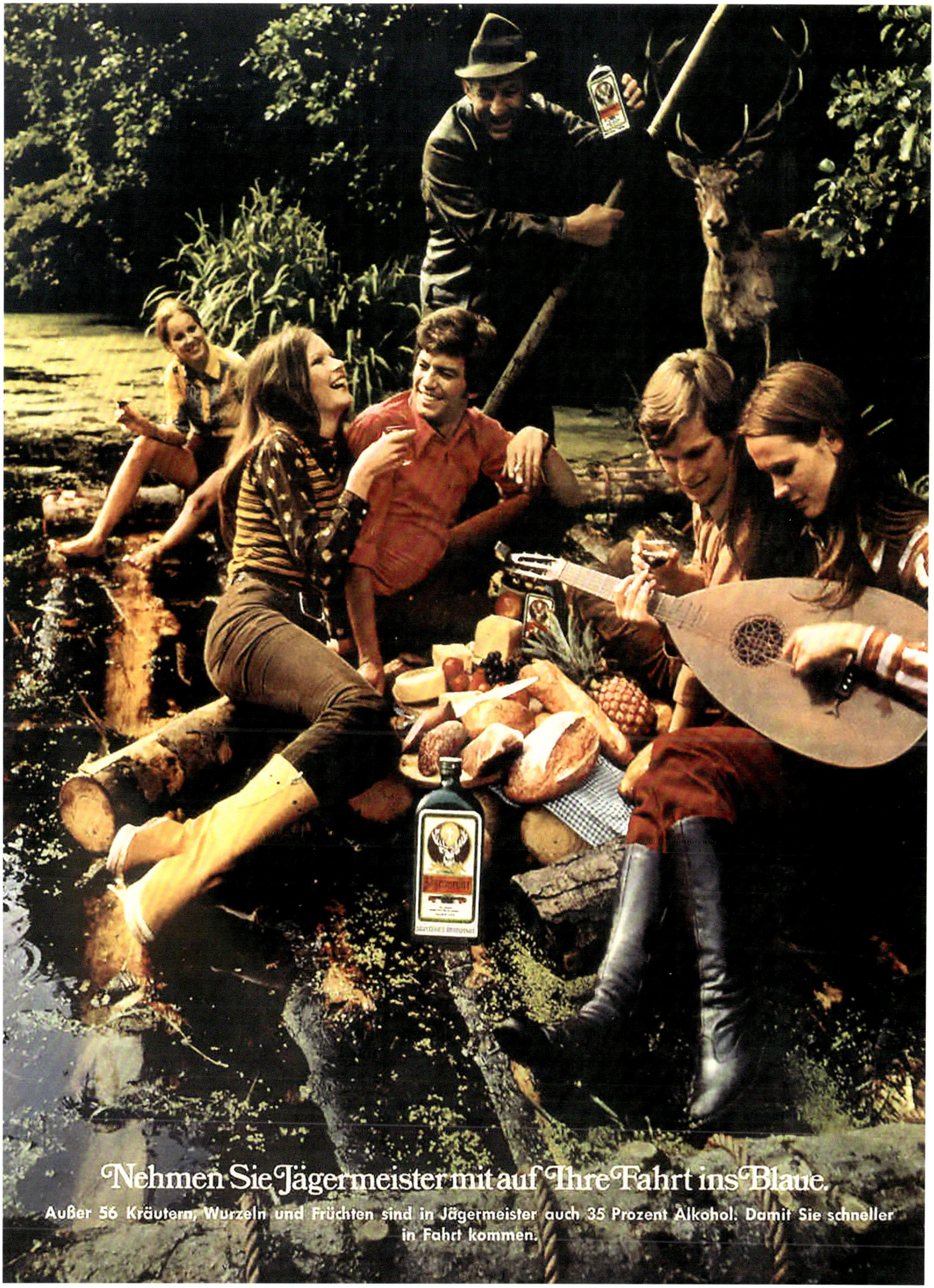

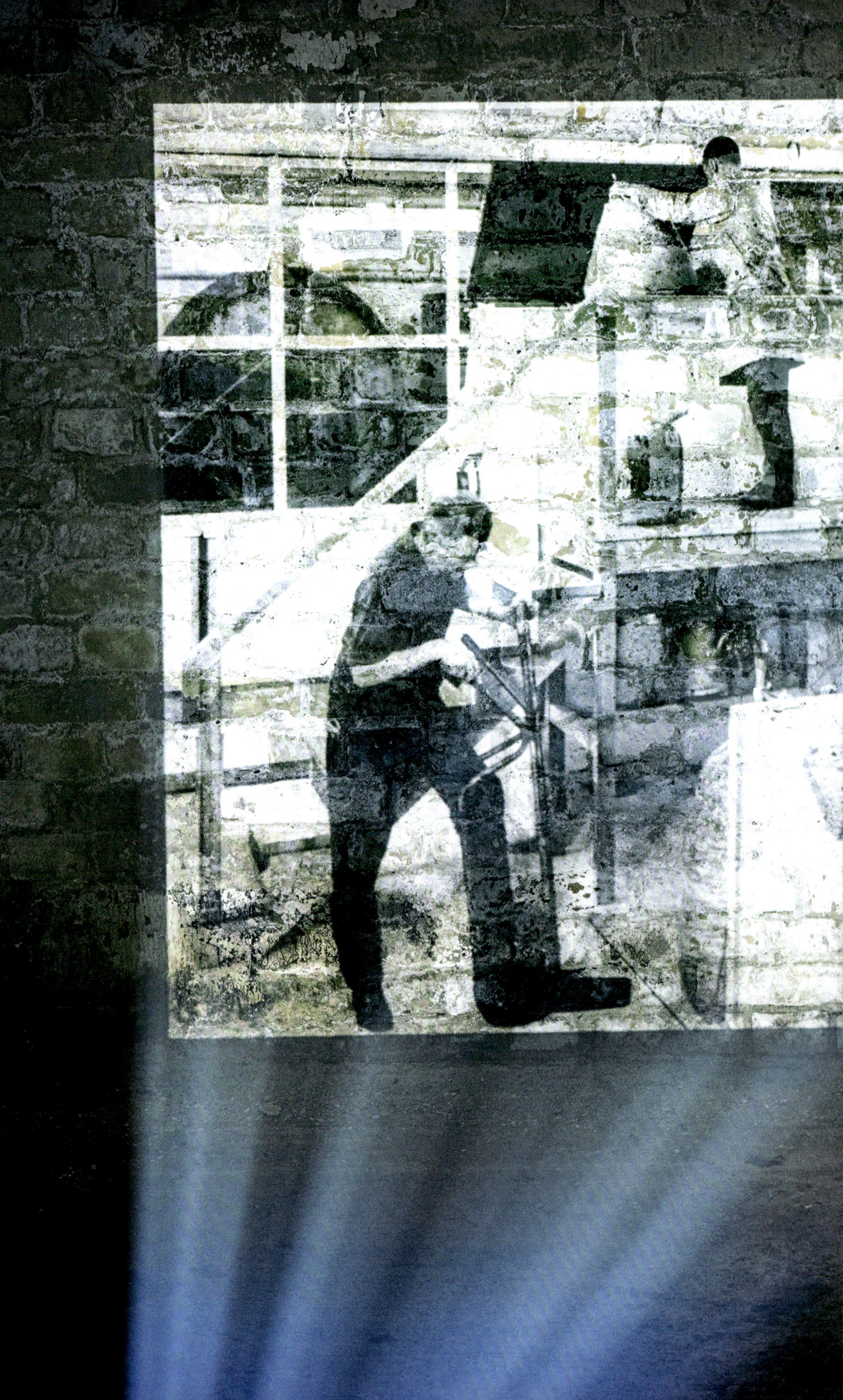

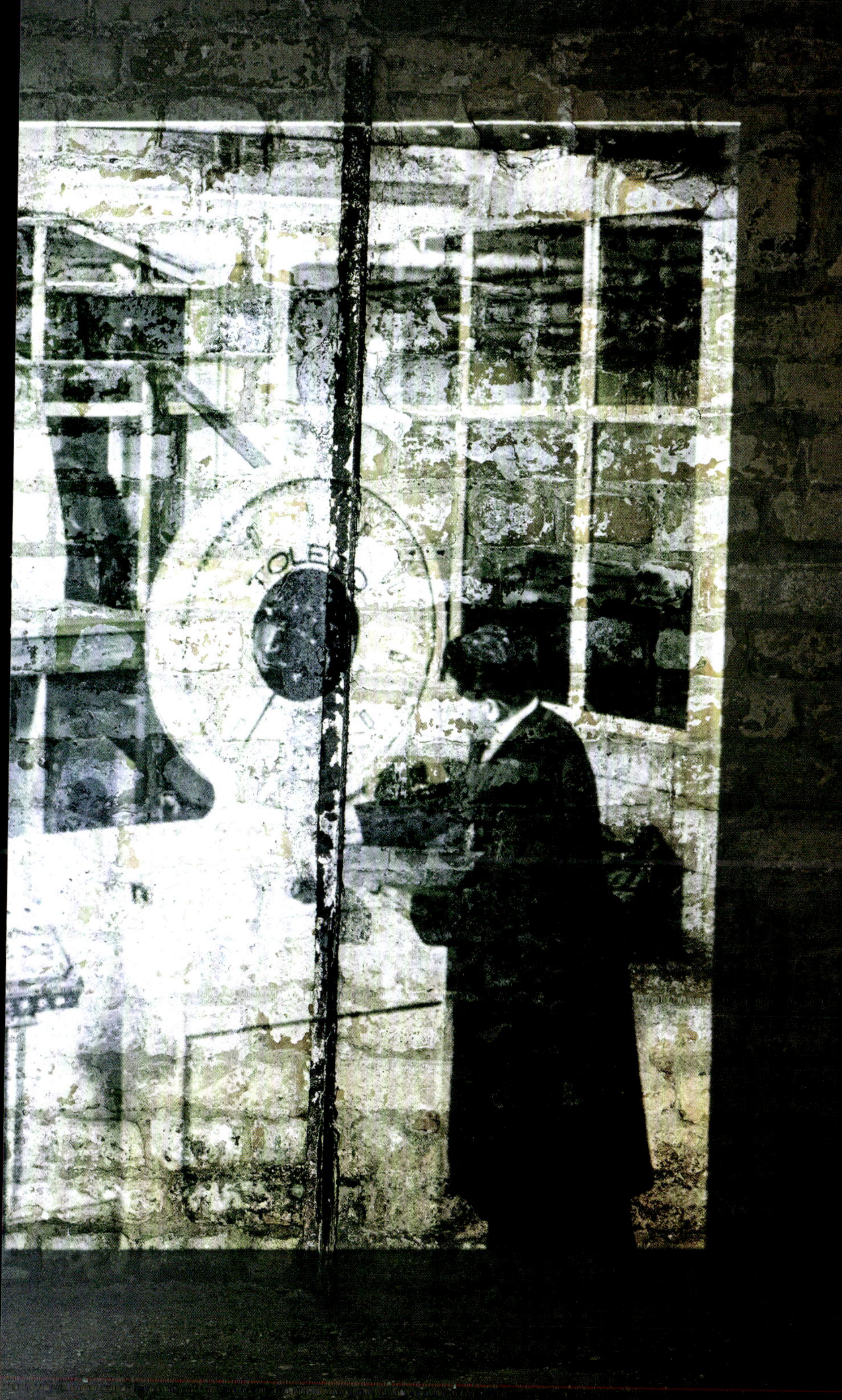

TOLEDO

INTERNATIONAL INSPIRATION FOR THE PERFECT DRINK

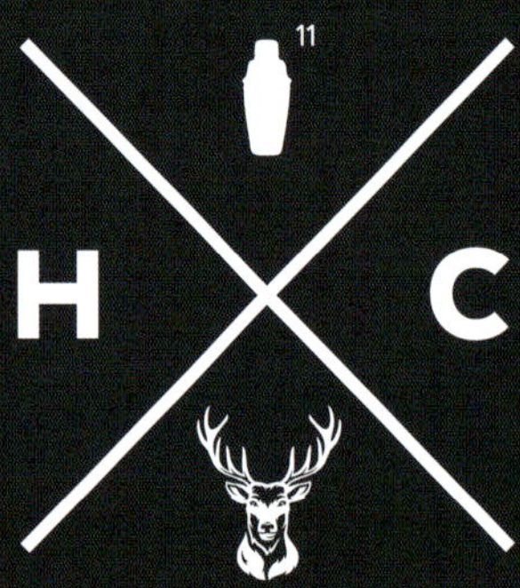

Jägermeister has been one of the most active supporters of the bar scene for many years, using its bartender network, the Hubertus Circle, to assist both established and junior bartenders. Founded in Germany in 2011, it now has a total of 18 international branches, including in Australia, Norway, Russia, the United States and South Africa. The participating bartenders have grown into a strong community, sharing ideas, collectively advancing their skills in mixing the perfect drinks and thus gaining inspiration and experience. Jägermeister also organizes masterclasses at which these international bartenders learn the latest tricks and secrets. Fancy a taste? The exclusive creations by Sabrina Traubner, Lukáš Čabaj, Florian Beuren, Willy Shine and Nils Boese, who are all part of the Hubertus Circle, can be found right here in the book.

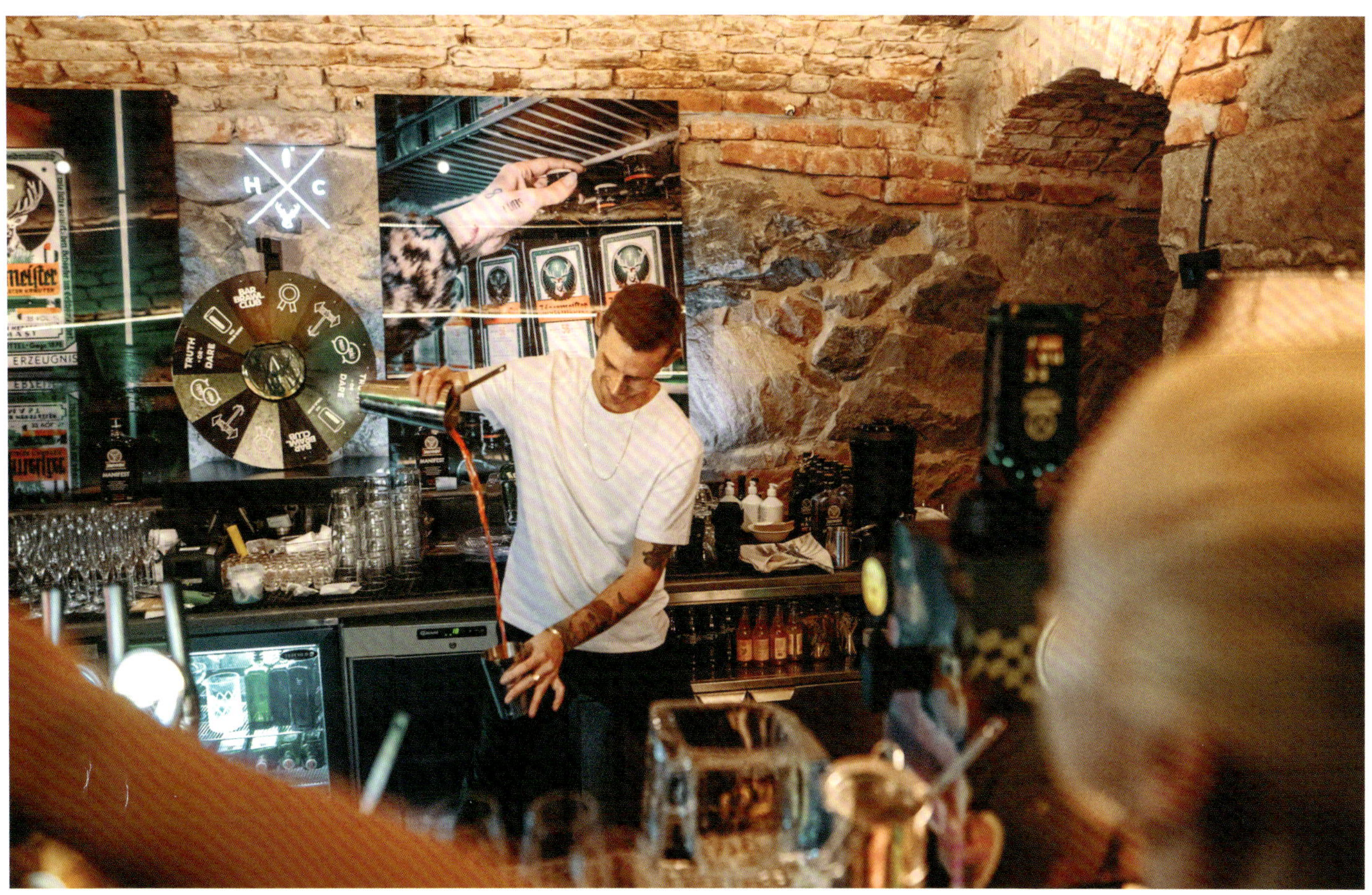

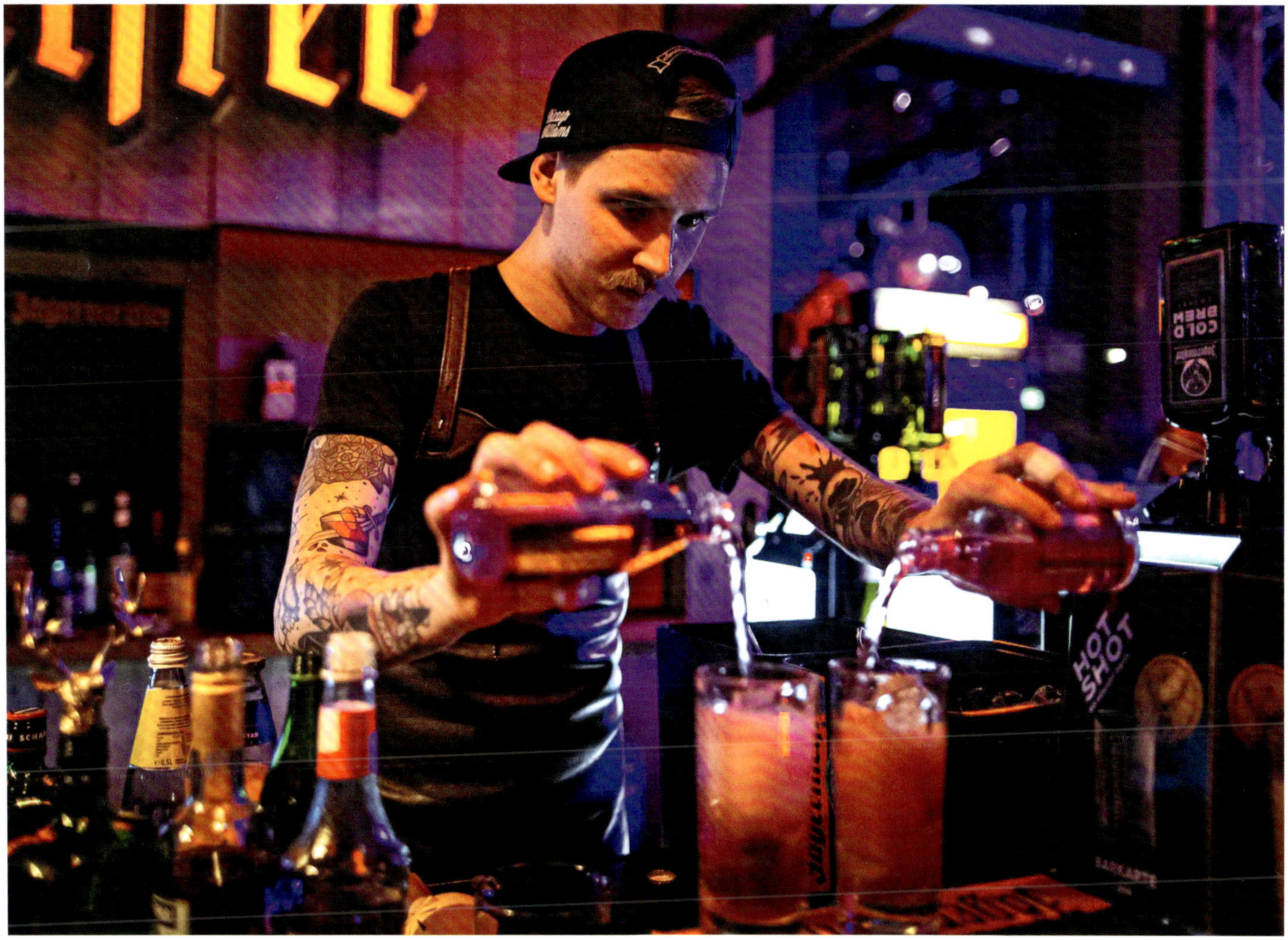

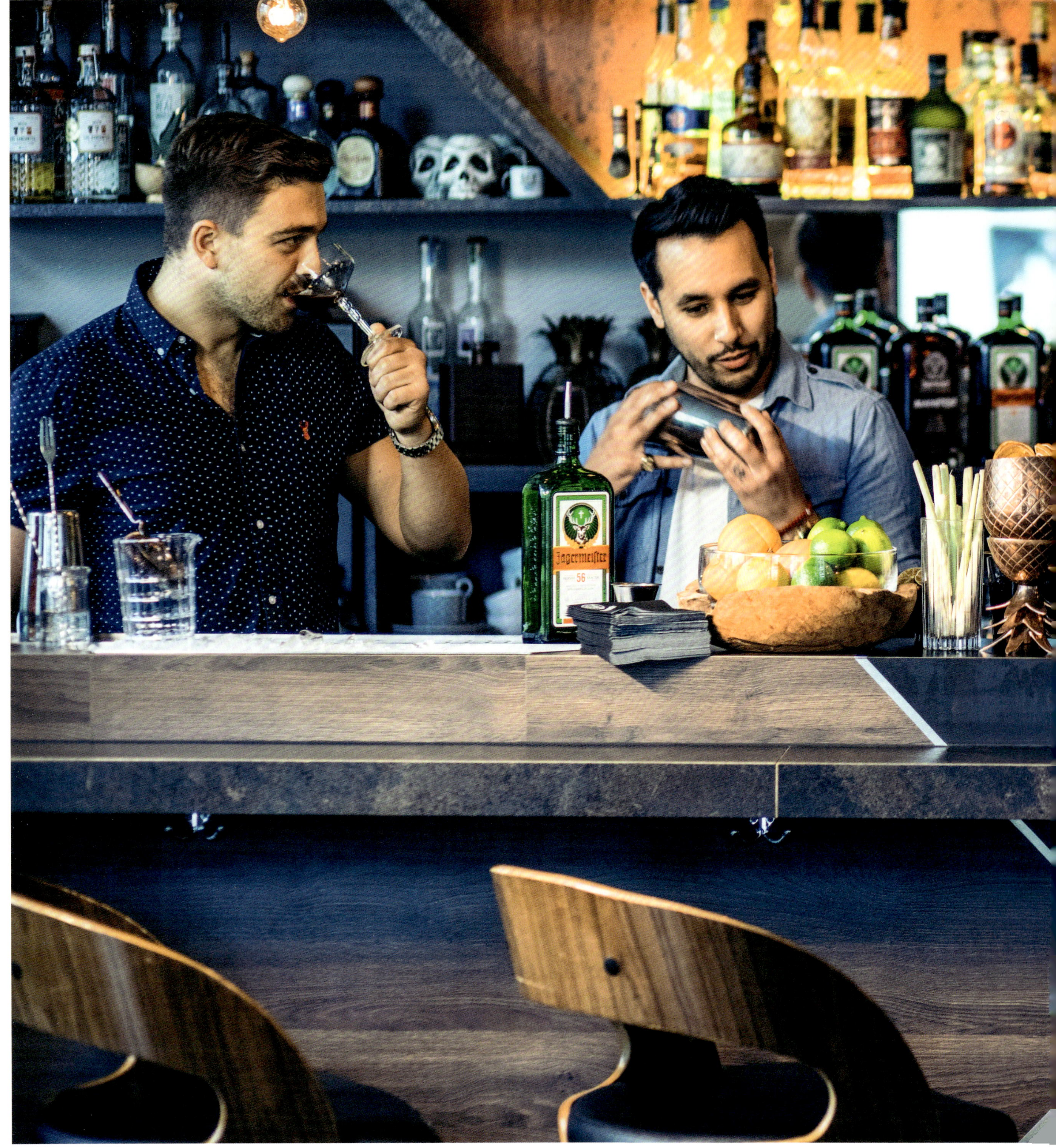

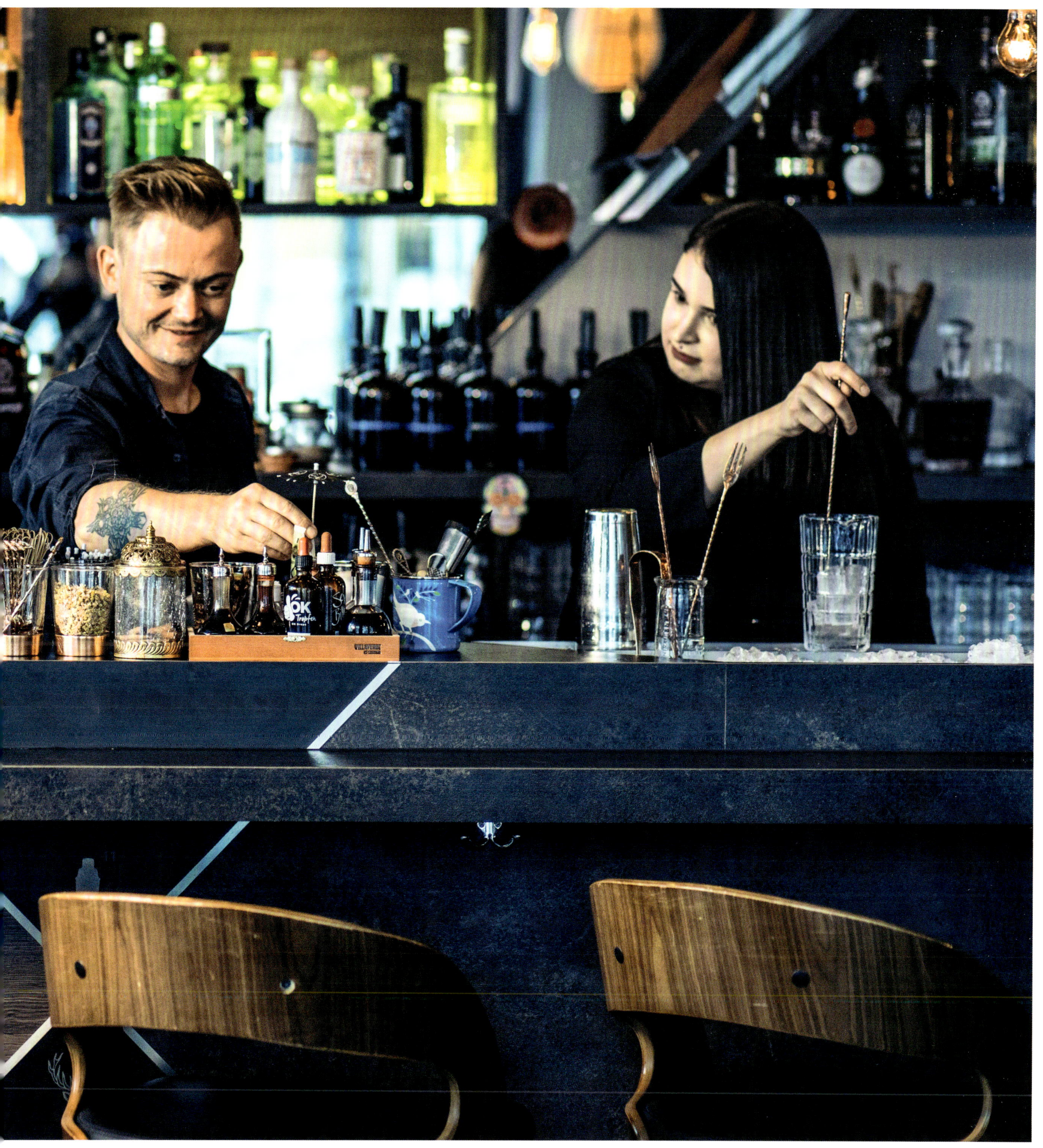

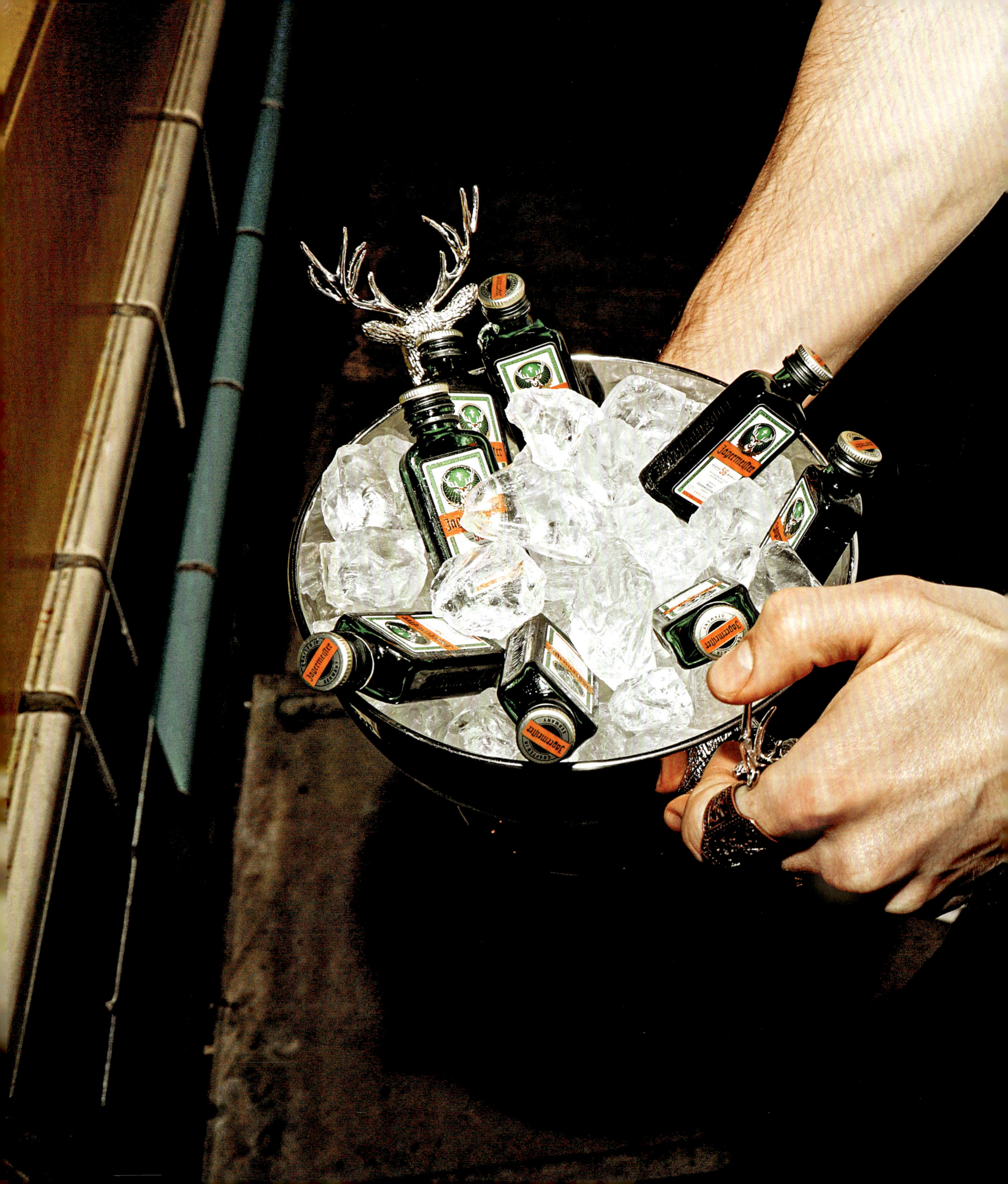

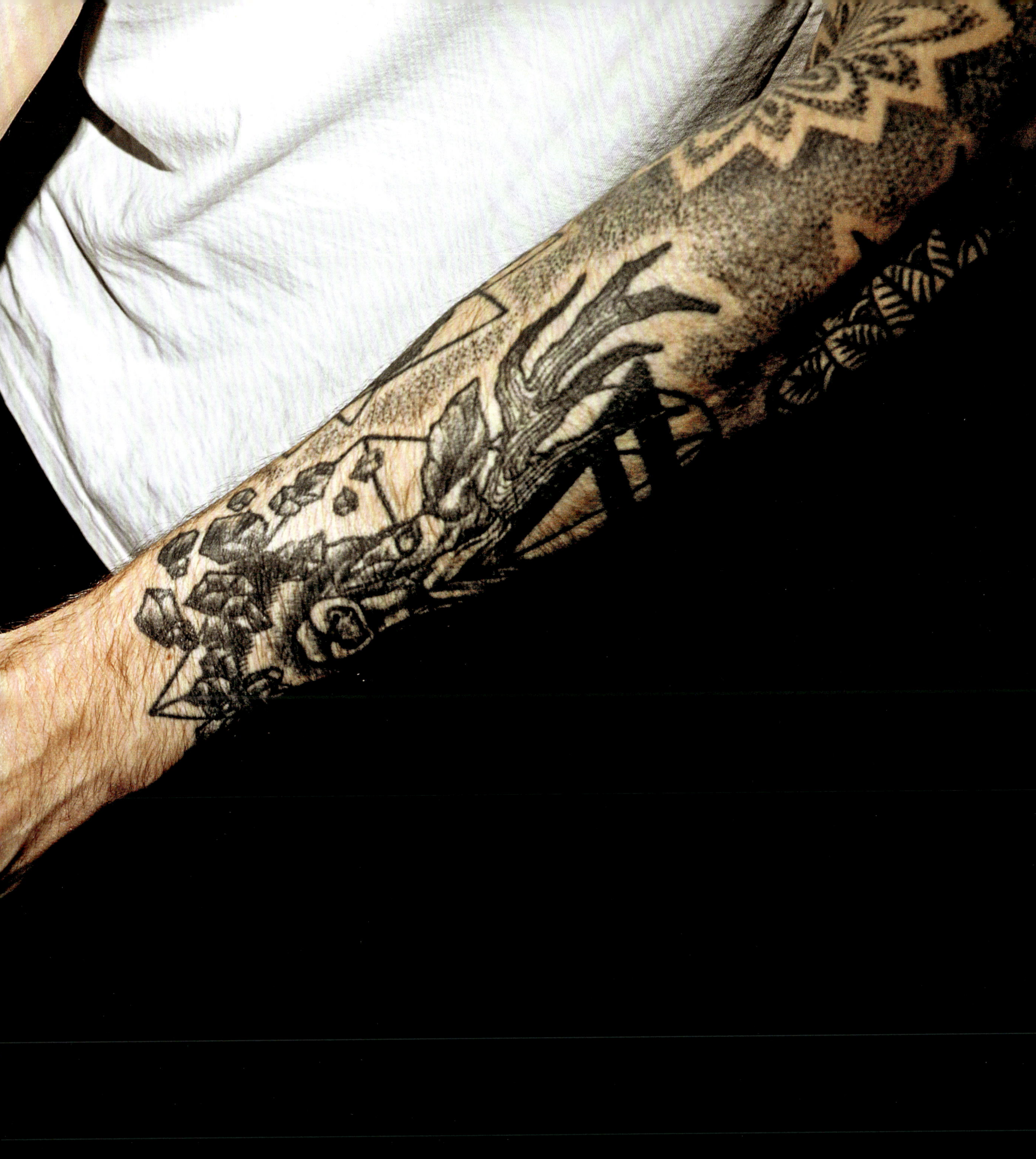

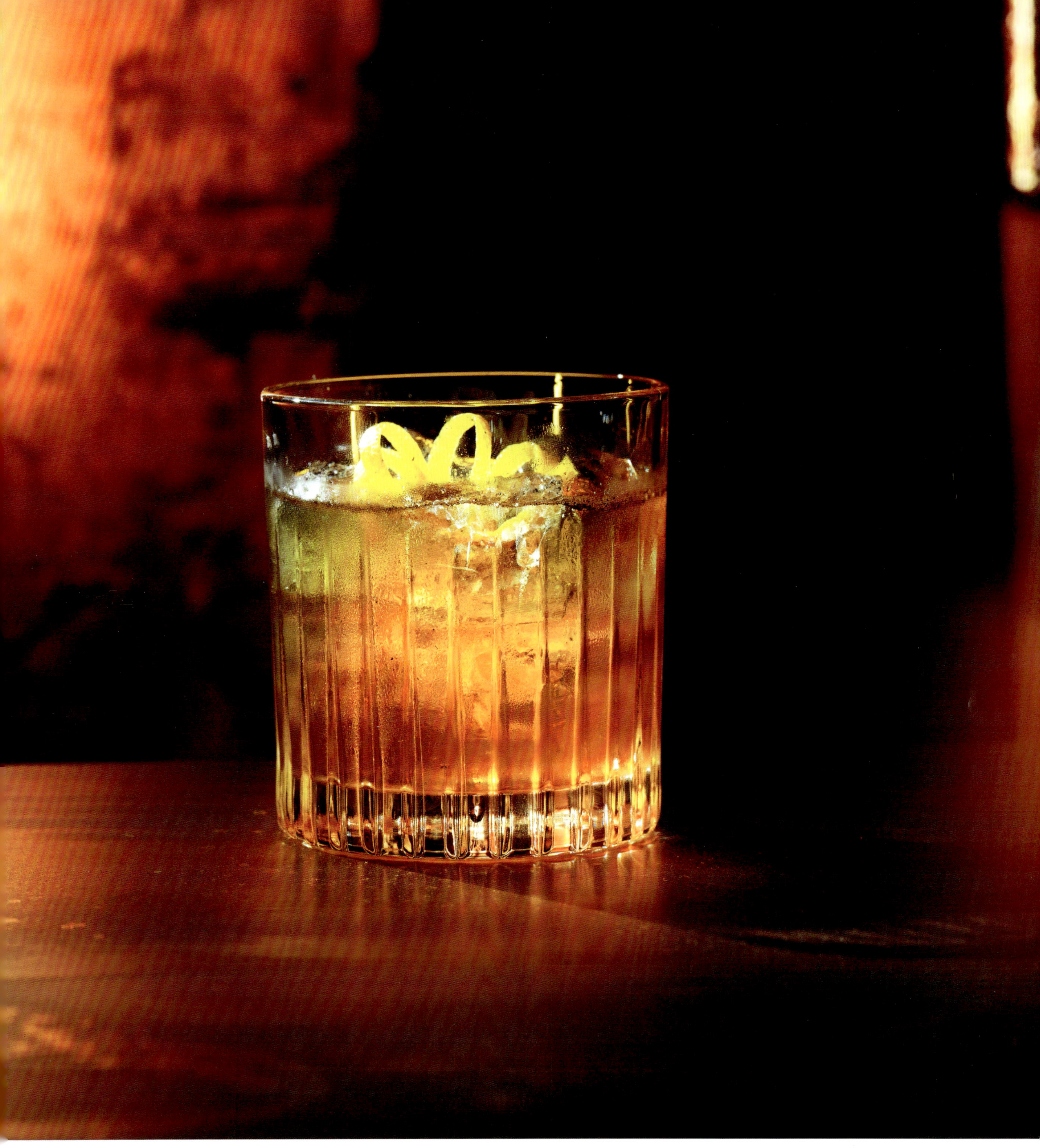

His drinks are explosions of flavour and he is considered the magician of mixers, an indispensable figure on the German cocktail scene. Though it was by accident that Nils Boese ended up in the dining industry, and ultimately behind the bar, he soon discovered his passion for mixing. He invented his own cocktails and experimented with unusual ingredients – including at some point with Jägermeister. 'I didn't find it a contradiction that Jägermeister was mainly known as a shot; I saw it more as an invitation,' explains the self-taught mixer. His mission is to 'mix Jägermeister cocktails that people who don't like Jägermeister shots not only consider acceptable, but magical'. This man from Germany's Lower Saxony region creates countless drinks that not only captivate drinkers but also convey the versatility of Jägermeister's flavour. In a bid to educate other bartenders and amateur mixers about this versatility, he helped to develop the Jägermeister Aroma Advisor, travels the world as a brand ambassador, and introduces new talent to the magic of mixing. What makes a good cocktail magical? This is what he reveals right here.

NILS BOESE
THE FLAVOUR WIZARD

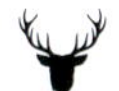

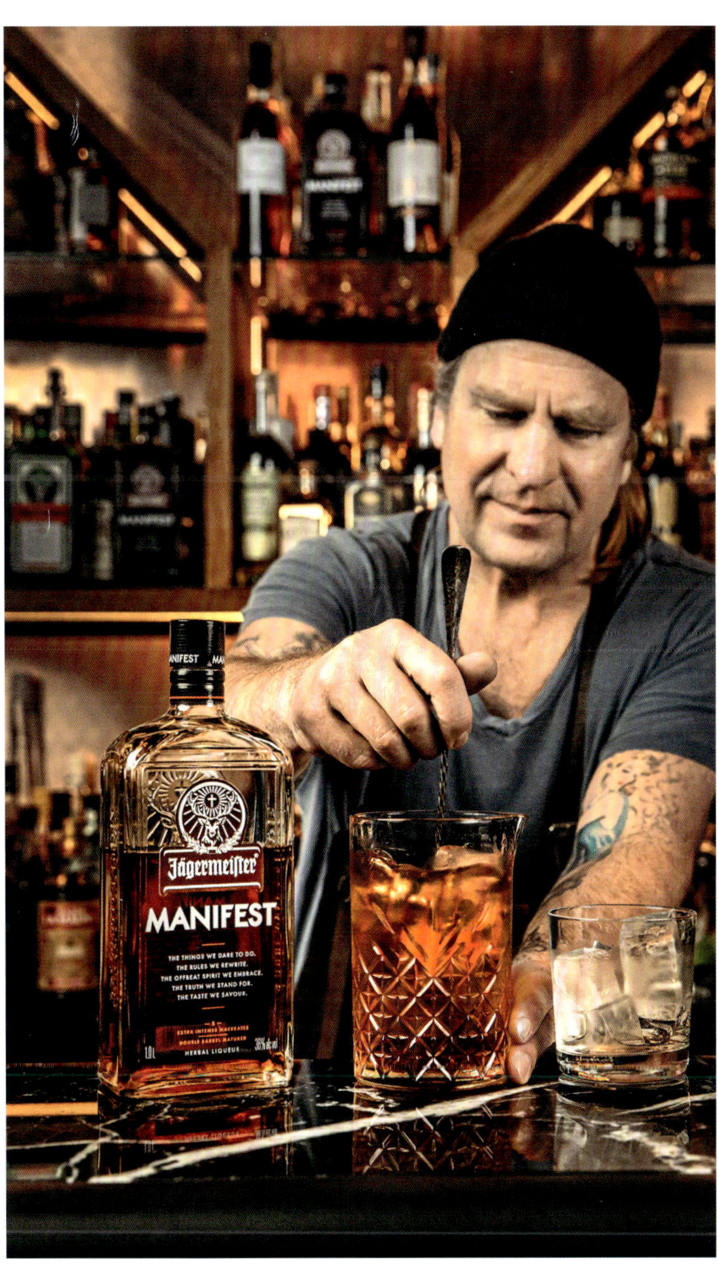

DEAR JÄGERMEISTER CONNOISSEUR,

To introduce you to the world of mixing with Jägermeister, there's only one thing I want to pass on to you: a drink is always more than its ingredients. This simplified message unlocks all the magic that cocktails can offer. You won't necessarily see a drink's depth just by reading a recipe. The flavour of a well-made drink is not simply a case of ingredient A plus B and C; it's about creating something new.

I would also like you to understand that reactions to taste are involuntary, so we can't control them. In other words, it's not that the drink is bad – it's just that we haven't (yet) acquired a taste for it. Perhaps it would be ideal in another situation or at a later date?!

Mixed drinks don't represent the perfect Jägermeister taste – that's what we have the ice-cold shot for. But they can accentuate or suppress some of Jägermeister's flavours. They are an invitation to try Jägermeister in a form and scenario other than as a merry round of shots, and they highlight Jägermeister's versatility.

To chill or not to chill? Jägermeister behaves similarly to some white wines, with both developing a different level of intensity when well chilled. At room temperature, Jägermeister is incredibly dynamic, acquiring a different degree of sweetness in its taste, and citrus notes in its nose – and so much flavour may be a challenge, particularly for the uninitiated. When chilled, however, everything is focused, leaving a great freshness and a certain lightness.

YOURS,
NILS BOESE

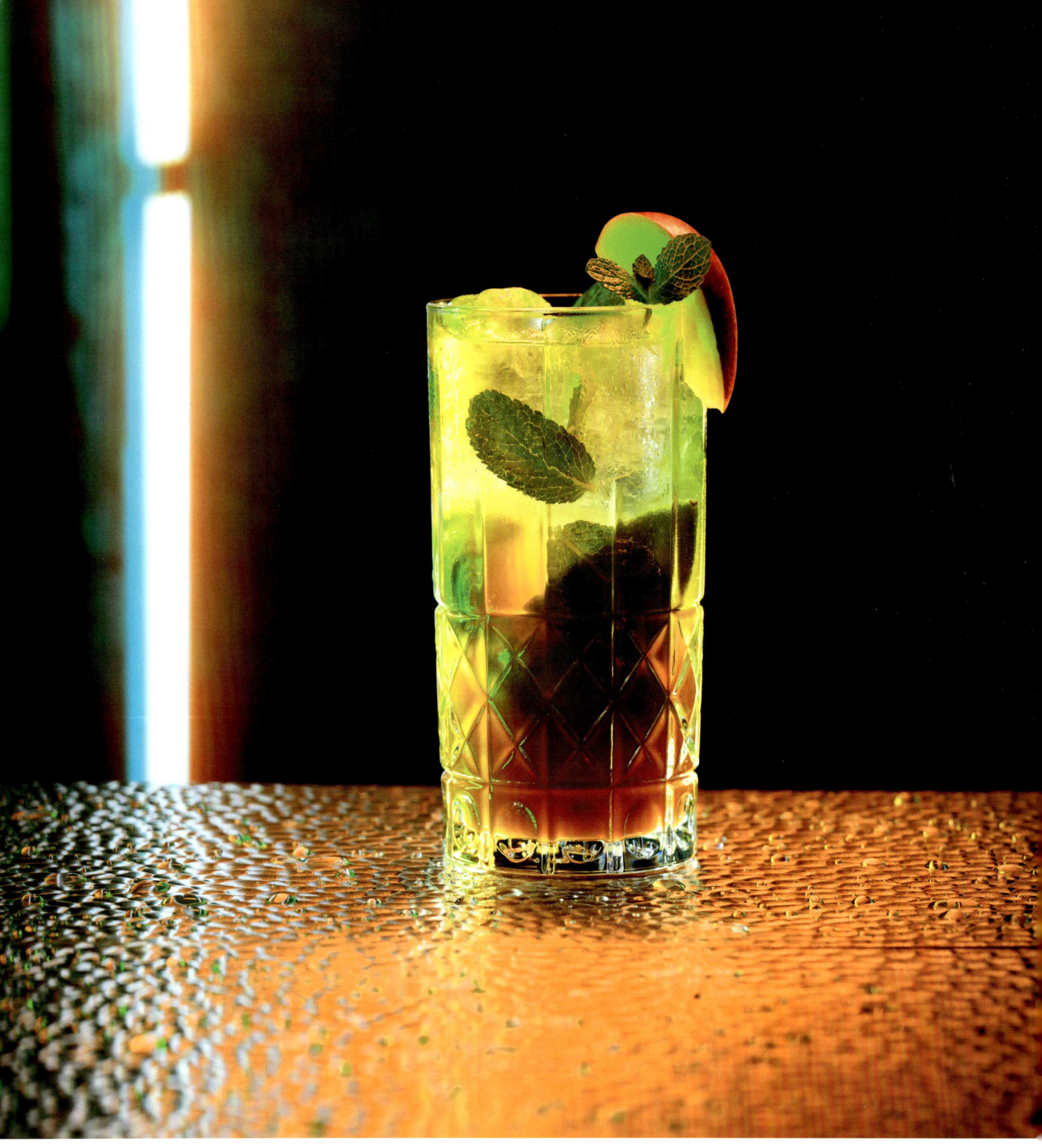

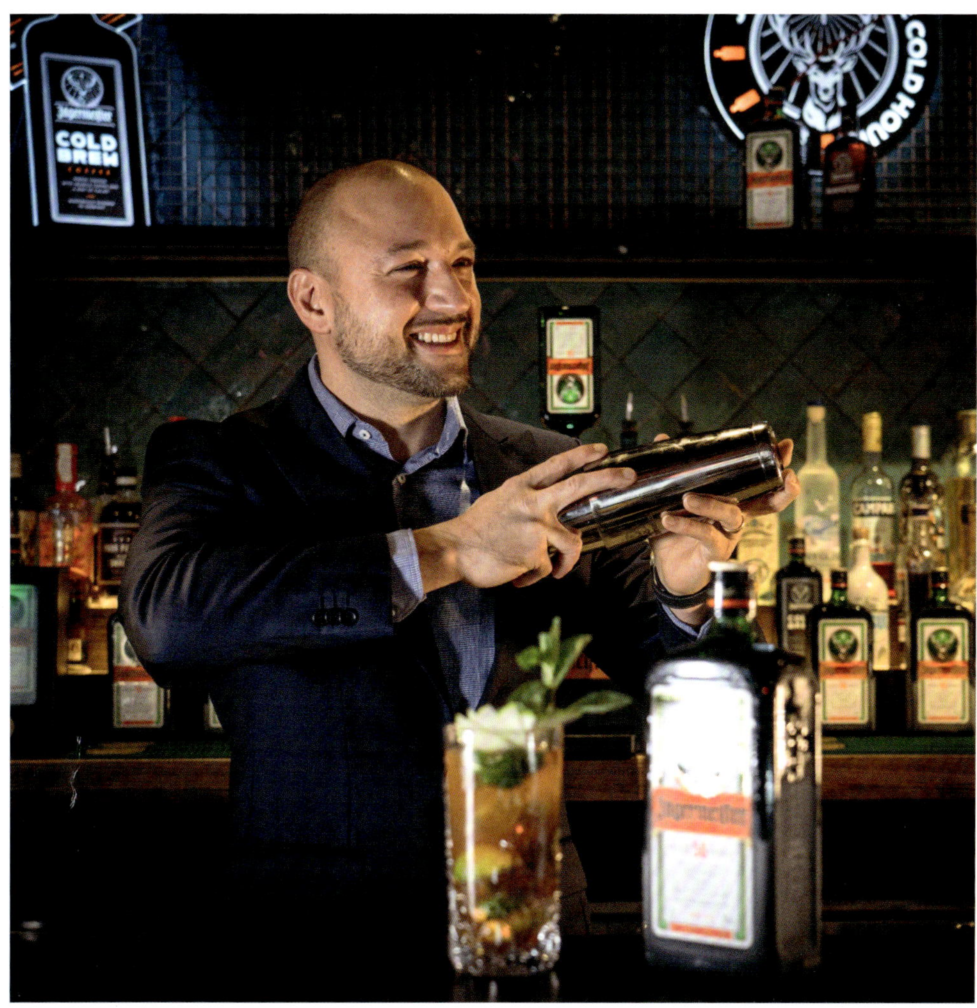

FLORIAN BEUREN
THE SHOWMAN

Playing to the gallery behind the bar, mixing delicious drinks, putting on cool shows, juggling shakers like Tom Cruise in *Cocktail* – these were all things that greatly appealed to Florian Beuren. To learn the trade of 'flair' bartending, the then-21-year-old gave up his job as head bartender for Alfons Schubeck and moved to London – to the very restaurant chain famous for performative drink-mixing, and the set of some of the *Cocktail* scenes. And the inevitable happened: the Darmstadt-born Beuren fell in love with flair bartending, won a load of prestigious competitions, 'flaired' at major film premiers and sporting events, taught other bartenders, and travelled the world in the process. Florian Beuren has been a brand ambassador for Jägermeister UK since July 2014, and admits that, 'as a German in Britain, I am of course highly motivated when it comes to our delicious German herbal liqueur'. Because, even though he loves to put on a show when mixing his drinks, he always chooses his ingredients with utmost care.

WHAT DO YOU LIKE ABOUT JÄGERMEISTER?

It had always been my shot of choice, even though I am quite a lightweight when it comes to drinking. But, unlike rum or vodka, Jägermeister is one of the shots I can handle very well. Only very few people know, however, just how exquisite, complex and diverse Jägermeister is. My mission is to change this perception and get people to see Jägermeister in a different light, not just as the world's best shot but also as a fantastic ingredient for other drinks.

OR FOR CHOCOLATE?

Yes, ha ha, we did actually make a Jägermeister chocolate. And Jägermeister salami, barbecue sauce, pizza … and of course Jägermeister coffee. These are all ways of showing people what's in Jägermeister. Its many herbs and flavours can be brought out by mixing it with other things. Jägermeister mixes excellently, and isn't overbearing. Its individual flavours also combine very well with others.

IS THERE SUCH AS THING AS THE PERFECT JÄGERMEISTER COCKTAIL??

For me, it's the Jägermeister Sour, a delicious beverage you can drink all night long. Lemon is one of the most important notes in Jägermeister, which is why this drink in particular accentuates the taste of the herbal liqueur in a wonderfully balanced manner.

WHICH COCKTAILS ARE TRENDING RIGHT NOW?

Less is more – that's the main motto at the moment. Many bars in London infuse various flavours in spirits, such as vodka. They then mix it with just one other ingredient – and that's it! So you no longer need four or five ingredients, but rather just two. The advantage of Jägermeister, however, is that it already contains 56 different ingredients. And that means it's super cool to mix with.

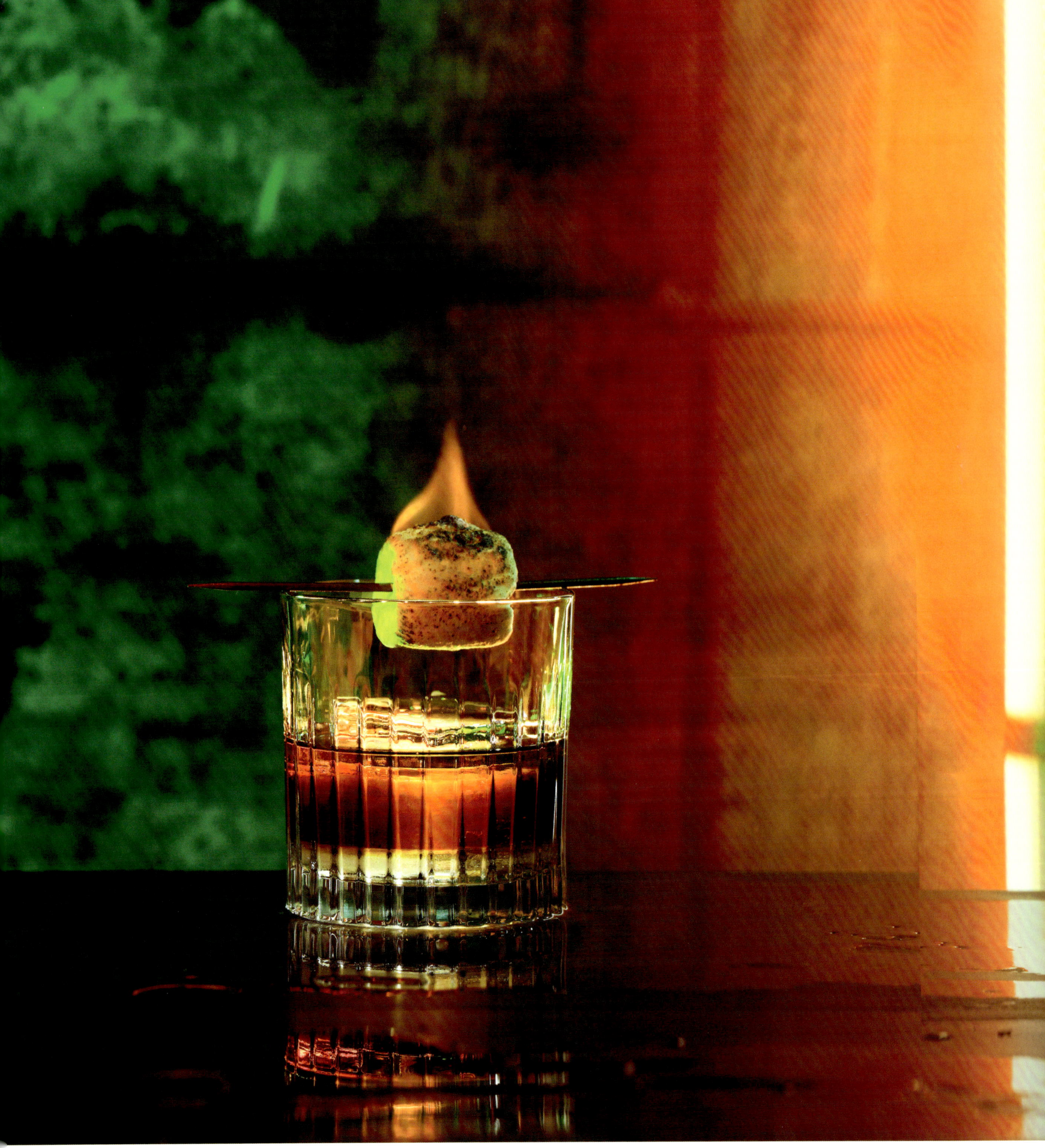

WILLY SHINE
THE BRAND MEISTER

He looks a bit like Ernest Hemingway and loves surfing, a good barbecue, good music and, most importantly, good drinks. Willy Shine is the face of Jägermeister USA, particularly within the bartending community, to which he has always maintained close and passionate ties – for it was here that his successful career began in the early 1990s. Willy Shine is now an award-winning restaurant and bar owner, and a high-flying industry consultant for bars, hotels and major events. Since 2015, he has also been contributing his unconventional ideas, his network and his creative pioneering spirit to Jägermeister as the 'brand meister'.

WHAT EXACTLY DOES A BRAND MEISTER DO?

I'm kind of like a brand ambassador; I appear in front of the camera a lot, promote advanced in-house training, take care of innovations, develop new cocktails . . . The term 'brand ambassador' didn't fit for all of that, so we called the role 'brand meister'.

WHAT'S YOUR FAVOURITE WAY TO QUENCH YOUR THIRST?

I generally prefer to enjoy what I'm drinking rather than just quenching my thirst. There's something celebratory about raising a glass and having a toast with friends. That's why I love our Deer & Beer campaign, in which special craft beers are paired with Jägermeister. I sip at the shot, have a gulp of beer along with it, and enjoy the interplay of flavours.

HAS THAT ALWAYS BEEN THE CASE?

When I first started doing bar work in 1992, Jägermeister was mostly known as a shot. It was called the 'bartender's handshake', because it was sometimes drunk as a welcome drink among bartenders. And I loved it even then.

WHAT WAS THE FIRST DRINK YOU EVER MIXED WITH JÄGERMEISTER?

It was called Surfer on Acid and consists of Jägermeister, pineapple juice and coconut rum. Eric Tecosky, one of my good friends from L.A., invented it.

HOW DO YOU GO ABOUT CREATING A JÄGERMEISTER COCKTAIL?

Over the years, I have learned all there is to know about bartending and cocktails. When you put so much time, effort and passion into something, you end up creating a kind of register in your brain of different flavour profiles, and how they develop and interact with each other. That's how it all starts. Eventually, you think about which direction you want to go in – such as whether you want a sour or spritz . . . Every drink has a story, a recipe, a special technique, and a specific glass for a specific reason. You can manipulate all of that by using different flavours, different ice, different glasses, different base spirits, and by adding Jägermeister to the mix or using it to replace an ingredient. Everything has a meaning or stems from a specific experience or inspiration.

WHAT INSPIRES YOU?

I like being inspired by recipes from other cultures, and sometimes also by music, a painting, a scent or a feeling. It's more of an advanced form of creation, almost like an art form, perhaps comparable to composing music.

DO YOU HAVE A FAVOURITE COCKTAIL?

One of my favourites is an old-fashioned variant with whisky and Jägermeister, a shot of bitter and some maple syrup. It's a great drink for sipping in a rocking chair after a meal. In any case, it's always about the occasion, the situation, the moment. It's never just about the drink.

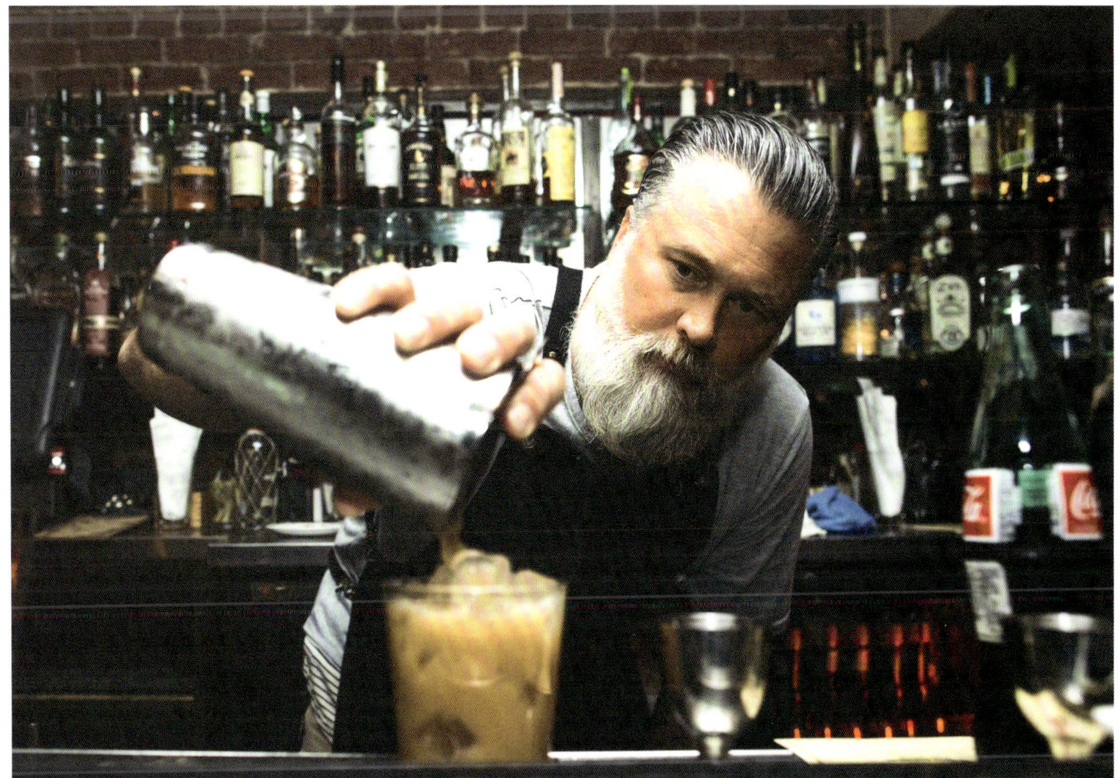

Willy Shine is the national brand ambassador and Head of Brand Education, Trade Advocacy & Mixology at Jägermeister USA. Or, for short, brand meister.

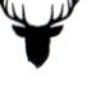

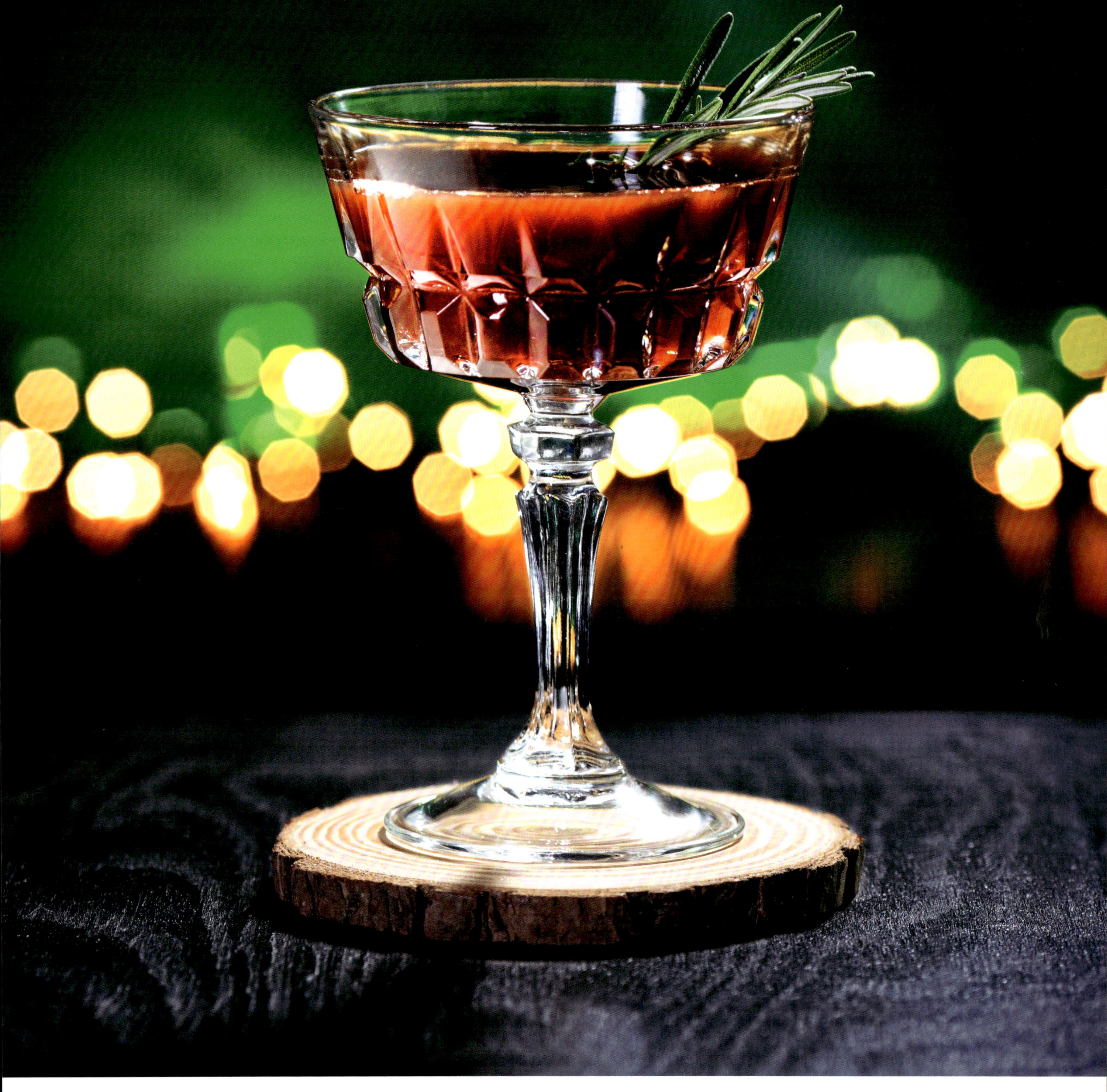

SABRINA TRAUBNER

THE MASTER-CLASS STUDENT

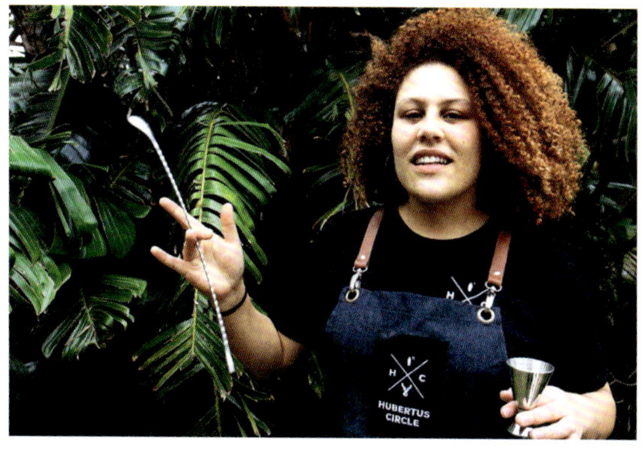

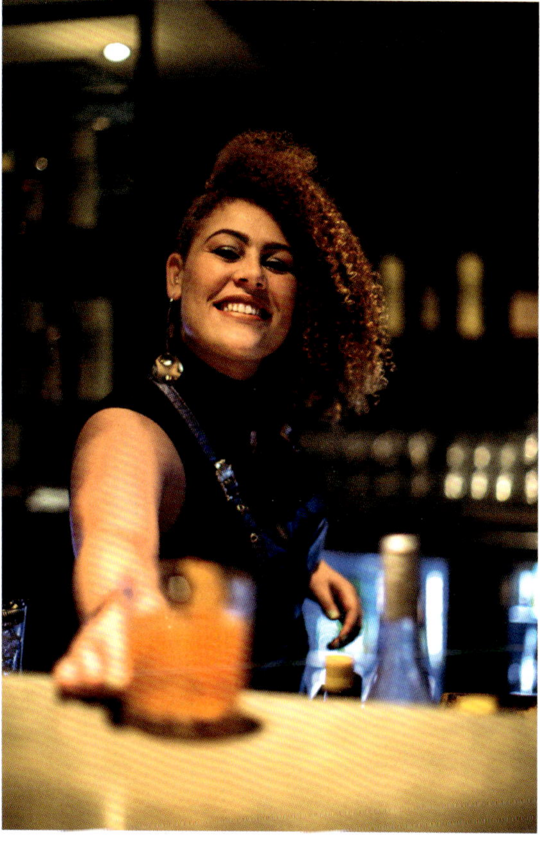

People say you need to have a plan in order to be spontaneous. The career of young South African bartender Sabrina Traubner is a perfect example of how determination and spontaneity can complement each other marvellously. At just 24 years of age, she is head bartender at The Athletic Club & Social in Cape Town. Before that, she worked as an events barmaid, a manager for bartending events and bar services, and then as a cocktail designer and head bartending coach. However, what sounds like a thoroughly planned CV began with a spontaneous idea. 'A friend suggested doing a bartending course to earn a bit of extra money after finishing school,' says Sabrina. Then 19 years old, she was soon hooked. 'I immediately fell in love with the fast-paced working environment that enabled me to interact with different people.'

Sabrina has now become a permanent fixture on the bartending scene thanks to her charismatically cheerful manner, her ambition and her unbridled passion for anything to do with great spirits and cocktail-crafting – including Jägermeister. That, too, is thanks to a spontaneous idea. 'I still remember the first time I mixed with it. I had a few friends over and was playing around a bit, using Jägermeister as a bitter.' The result? 'The drinks were great!' Since then, she has loved the herbal liqueur and everything the brand represents. 'The diversity of the drink harmonizes with the diversity of the Jägermeister brand, which is involved in so many creative events and brings together all kinds of creative minds. It's incredible how positive, cheery and authentic this brand is,' she enthuses. 'I feel honoured and privileged to have been able to get to know Jägermeister so well.'

Because, in 2020, the South African took part in the Jägermeister Scholarship × Virtual Edition. The three-month scholarship supports new talent in the Hubertus Circle, Jägermeister's renowned international bartender community. 'I love engaging with others and learning from them,' says Sabrina. 'It's important to get different perspectives on your life and career; it makes you more agile and open to change.' The young bartender is convinced her personality is the right fit for the Jägermeister brand, and proves it perfectly with the two drinks she has created for Jägermeister: The Hunter's Vacation and Hunter in the Woods. 'They reflect my personality, because they're both full of flavours but still very different', says Sabrina Traubner. 'That represents my versatility and multi-faceted nature.'

Behind the bar is her favourite place to let this shine – ideally forever. 'It makes you incredibly happy to have found a career you really enjoy.' Her next goals? 'To win one or two international competitions and become a mentor for other bartenders in my community.' The newcomer is ambitious and loves new challenges that drive her to become better and learn more about herself – at both a personal and creative level. But she still always leaves room for spontaneity. 'I'm inspired by the process. There's no greater feeling than designing a drink you've been unsure about for so long, and then finally trying it and discovering it's just magic.'

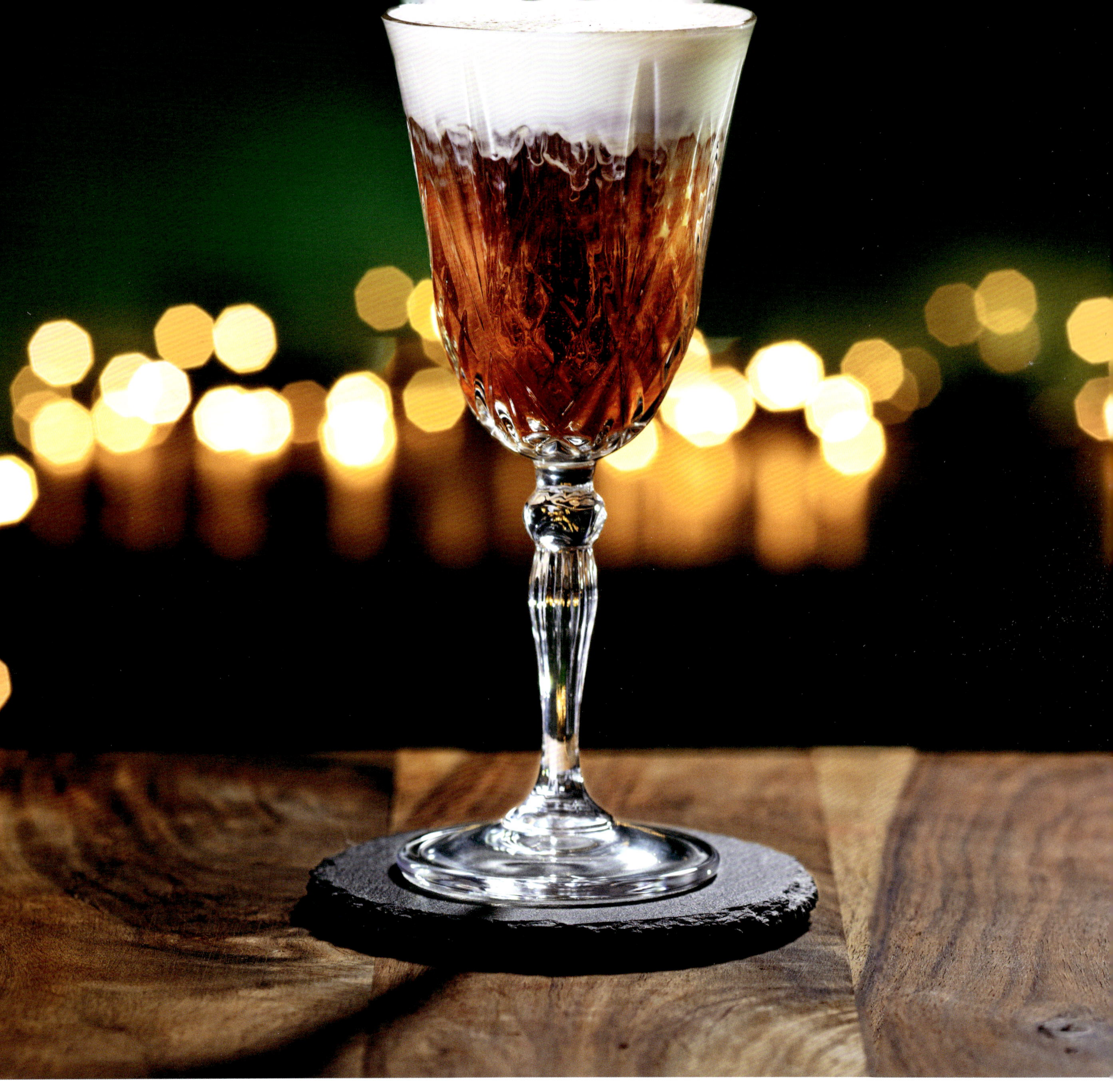

LUKÁŠ ČABAJ YOUNG AND WILD

When he's not behind the bar, Lukáš Čabaj prefers to spend his time in nature. 'That's where I can recharge my batteries,' he says. 'It helps me concentrate on what I want in life.' And the young Slovak knows precisely what that is. By the time he had completed his first bartending course at the Hotel Academy in his hometown of Brezno, he had realized that 'bartending gives me the chance to be creative and be myself'. So it's no wonder that much of his personality is manifested in his cocktails – sometimes sophisticated, sometimes funky, such as his two creations Jägermeister Wild Coffee and Jägermeister Jamming. 'One reflects my passion for coffee and minimalism, and the other the fun and creativity required when behind the bar,' he says. And Jägermeister is always within reach when Lukáš Čabaj is giving classic drinks an exciting twist or inventing new cocktails at Bratislava's Sky Bar & Restaurant. 'You can create any kind of drink with Jägermeister – from a refreshing sour, highball or fizz to a heavy old-fashioned,' he says, 'because it gives you countless ways to combine flavours.'

Lukáš Čabaj was able to further advance his mixing skills in 2020 during a three-month Jägermeister scholarship. Established to foster new talent, the scholarship has been running since the founding of the Hubertus Circle in 2011. This international bartenders' network combines the quality and tra-

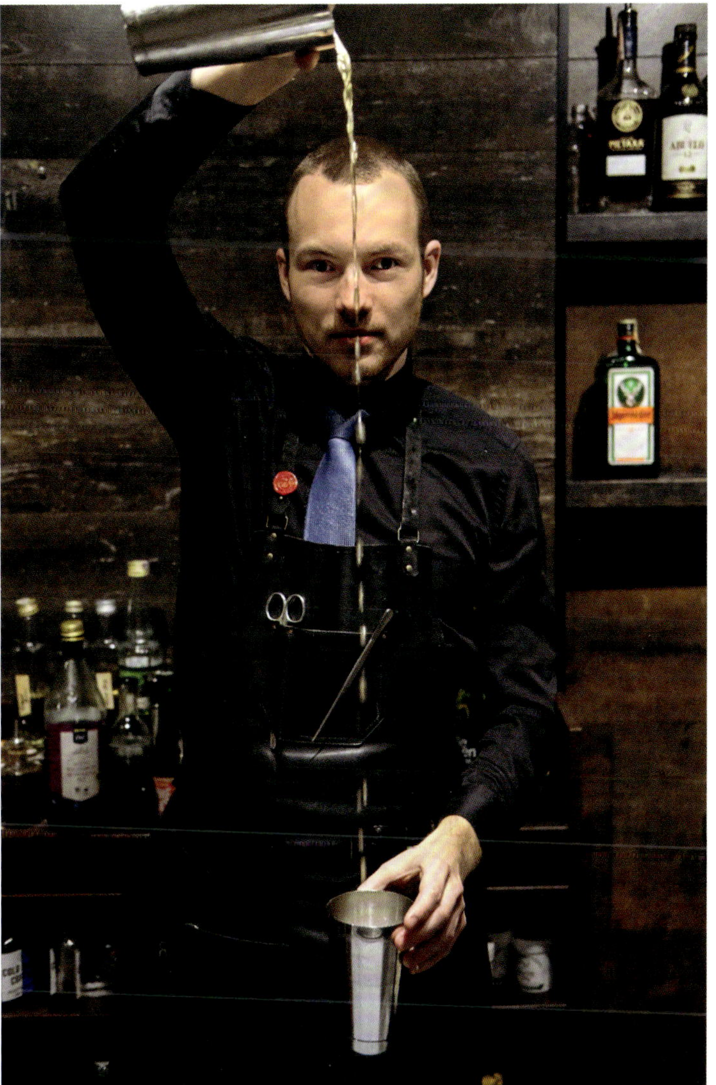

dition of the Jägermeister brand with the requirements of modern bartending as well as promoting exchange, conveying knowledge, and providing a source of fun. The Covid pandemic meant the 2020 events had to be held via video conference. 'And we sure had a lot of fun, even though I'm not a big fan of virtual meetings,' laughs Lukáš. 'But surprisingly I really liked it, and it was great to see how virtual teamwork can operate. Although we weren't physically together, we spent the whole time working together.' And these experiences continue to benefit the 22-year-old. 'I'm currently helping to put together our new bar menu and preparing everything for our guests,' says Lukáš. 'I can't wait to be back behind the bar, mixing drinks and creating wonderful moments for my customers.'

One of his own favourite moments also has a link to drinking. 'My grandmother says, "A little sip of a good drop every day will keep you chipper." To this day, she always has one or two bottles in her fridge just in case. When I visit her, we sit together and chat over a cup of tea, before having a little nip to our health.' Tea with a shot is something Lukáš Čabaj also has handy when he's back out in nature. 'I mix my hiking tea with Jägermeister. It keeps me and my friends nice and warm when on a long trek.'

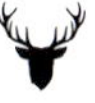

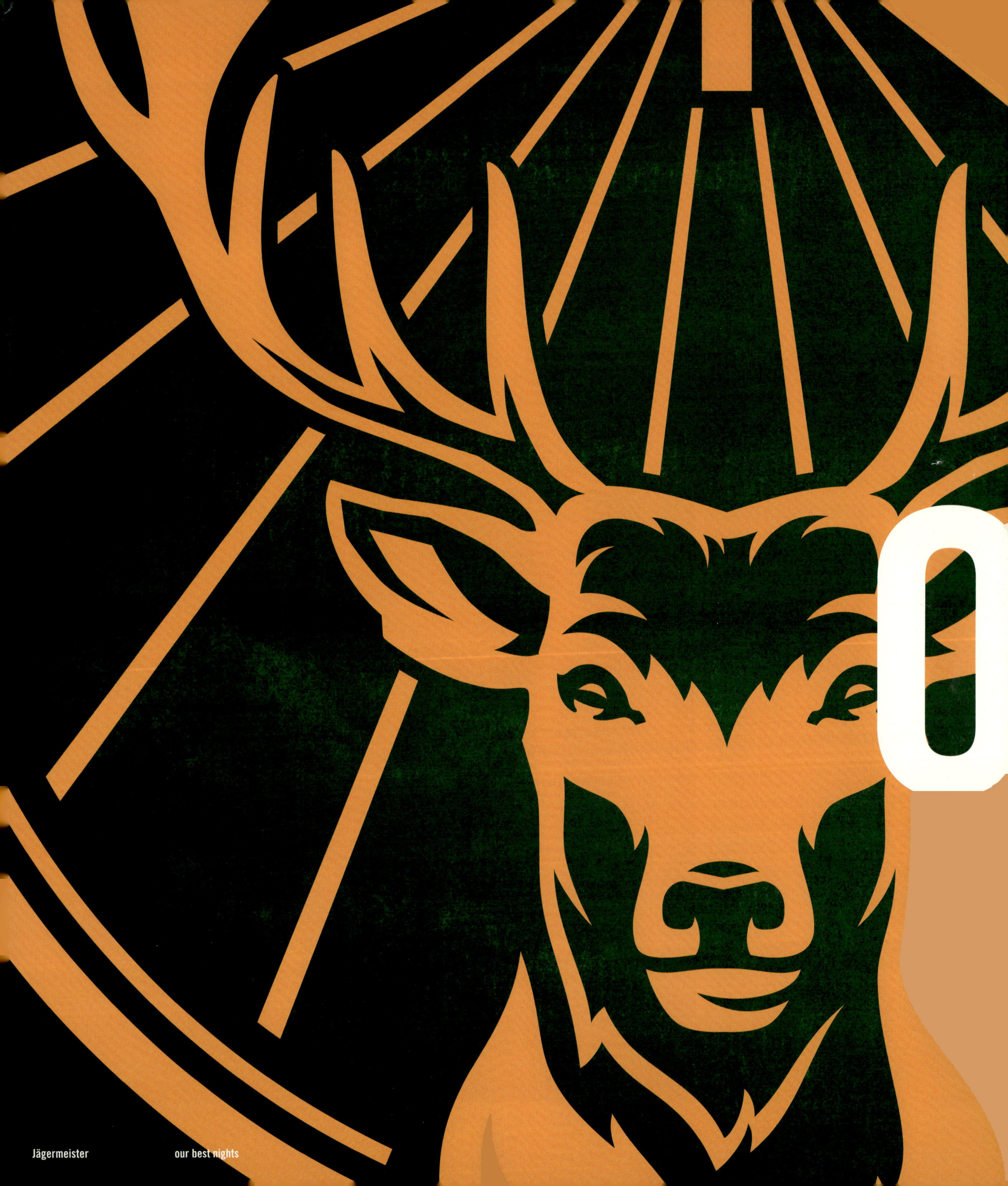
Jägermeister
our best nights

Wolfenbüttel: for a Jäger at Theo's — 208
Berlin: mastering the art of partying — 212
New Orleans: liqueur shots on Bourbon Street — 216
Dubai: the height of ice-cold pleasure — 218
Spain: ultimate house party — 220
Buenos Aires: deer crossing — 222
Shanghai: "Ye Ge" in Shanghai's underground scene — 226
Switzerland: ice-cold heat — 230
Czech Republic: ice-cold apparition — 234
Netherlands: Rudi and Ralph, the pert Dutch stags — 236
Sibiria: the coldest gig of all time — 238
Hollywood: shot moments on the big screen — 240
South Africa: loud and vibrant — 242
India: the 'ice cold tour' — 248
Israel: memories for eternity — 250

UR BEST NIGHTS

UNFORGETTABLY & UNMISTAKABLY

...

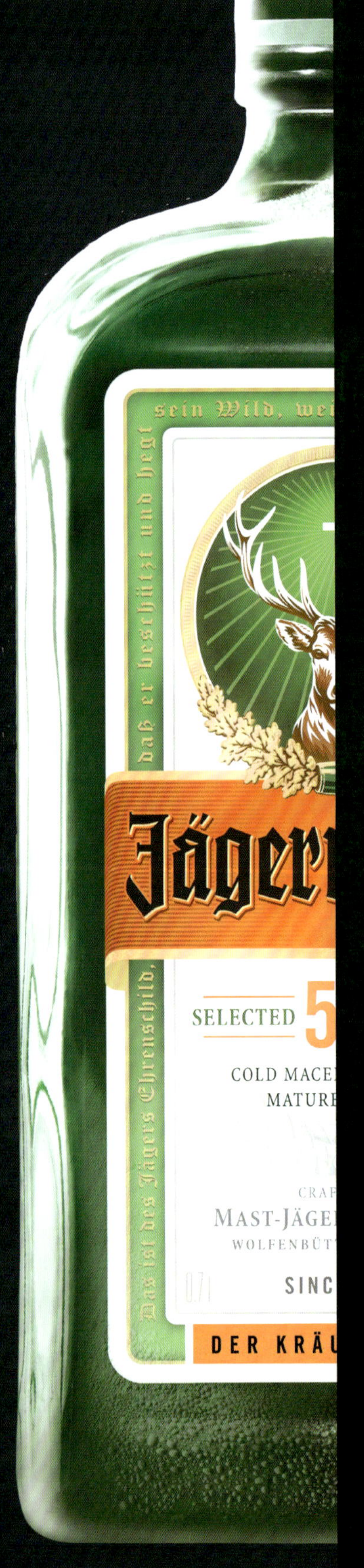

Special moments call for especially good drinks, and clinking glasses with your nearest and dearest is what Jägermeister is all about. The company's founder, Curt Mast, created his herbal liqueur for the hunting community, and for anyone he enjoyed raising his glass with. And this is precisely the idea that continues to live on today: people drink Jägermeister together, for special occasions, in precious moments, at festivals and concerts, during the best nights of their lives — whenever a normal night becomes an unforgettable one, leaving lifelong memories.

These days, you don't necessarily have to be out deerstalking in the woods, for since at least the time of Rudi and Ralph, it has been clear that a cool evening with friends at the bar can quickly turn into a legendary night when Jägermeister is involved – whether it be at the corner pub, at the highest bar in the world, or in a jazz club in New Orleans. From Wolfenbüttel to the world, the German herbal liqueur is a hit on every continent. And no matter where you are on the planet, one thing's for sure: special shot moments will stay with you forever.

Jägermeister is creative, edgy, daring, exceptional and so much more than just a liqueur. The proof? Unique moments, in keeping with the times, while still committed to tradition. What unites Jägermeister fans worldwide are the most exciting of adventures, the craziest of nights and the wildest of parties. And although it may be hard to believe, all of these stories did actually happen!

...JÄGERMEISTER

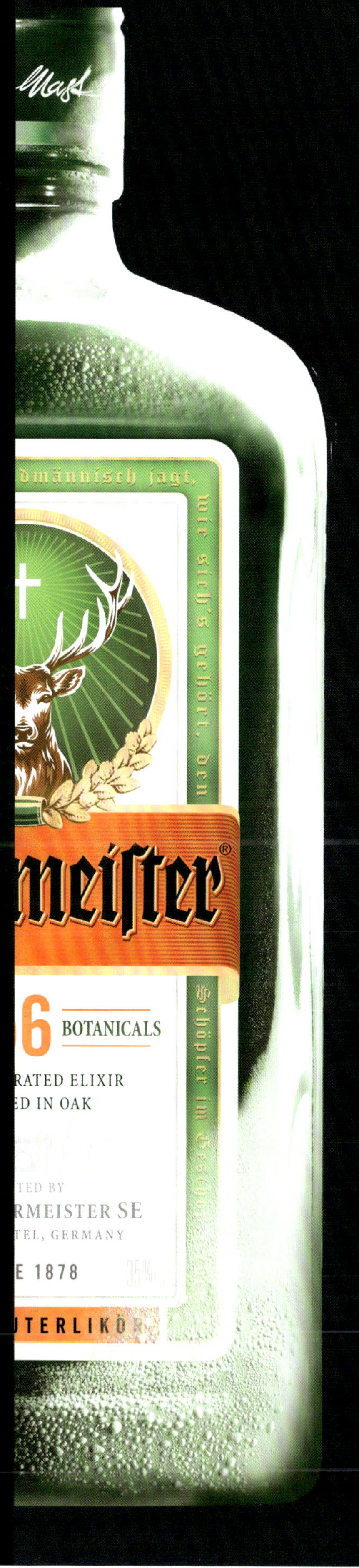

WOLFENBÜTTEL: FOR A JÄGER AT THEO'S

Shots have always needed moments, atmosphere and context. Traditional pubs are places for socializing and meeting people; they are often cultural sites, sometimes even places of refuge, all over the world, including in Wolfenbüttel. Here, near the Krambuden in the heart of the old town, is the city's oldest tavern, the Alt-Wolfenbüttel, known simply as Theo's by locals. The quaint half-timbered building has been operating as a public house since 1660. For centuries, people have been sitting together here, playing cards, partying and drinking – and for many years, their absolute favourite beverage has been the Jägermeister shot. The world-famous herbal liqueur and historic pub are inextricably linked: both epitomize a sense of harmony between tradition and modernity, and anyone seeking to trace Jägermeister's history through the streets of Wolfenbüttel will sooner or later end up at Theo's. The pub is a meeting place for many visitors to the Jägermeister factory, located just a few minutes' walk away. And it is timeless – the fact that its patrons range in age from 18 to 100 speaks for itself.

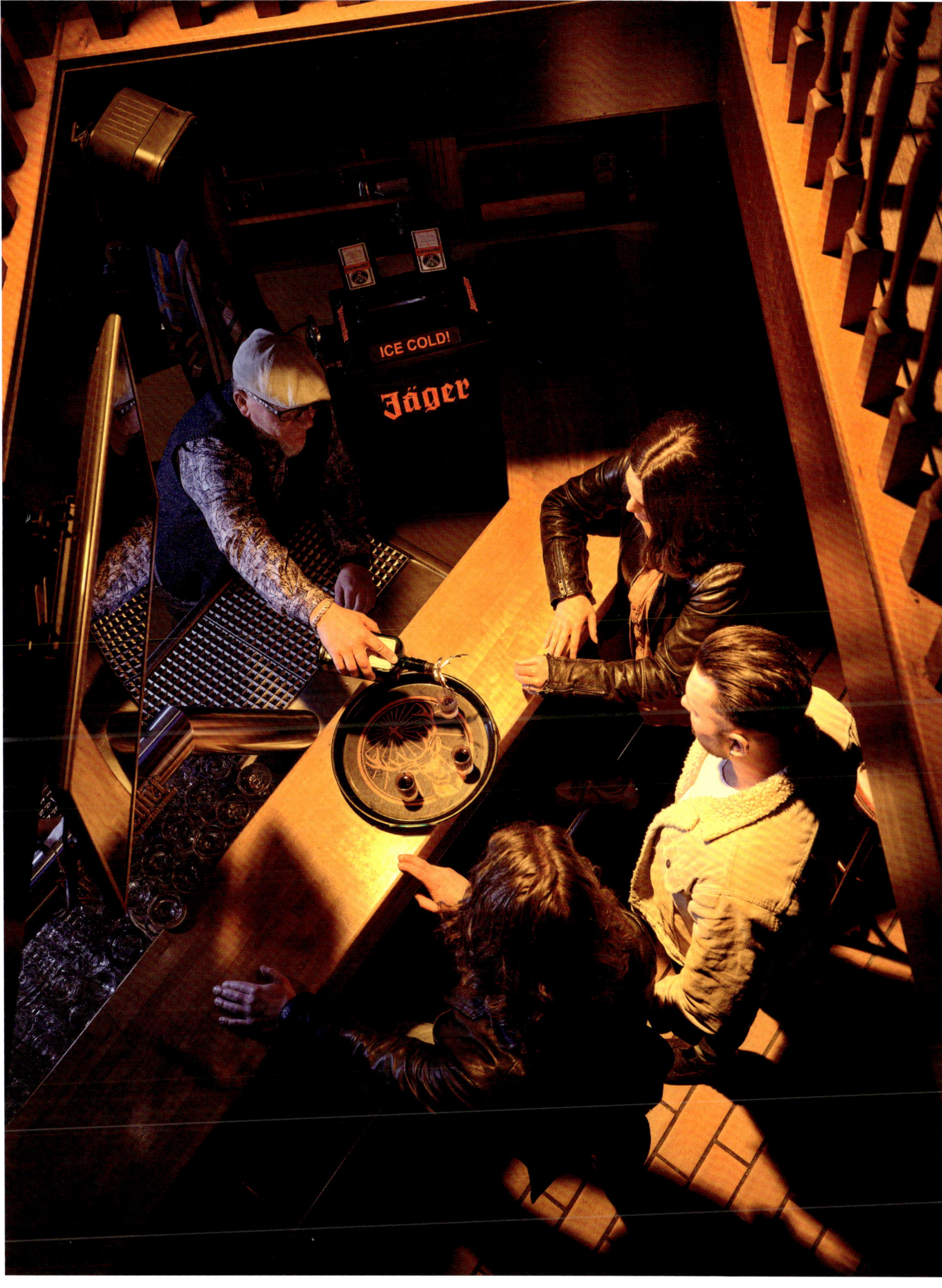

In the 1980s, Theo's was owned by a publican by the name of Ferdinand Fricke, who would regularly imitate actor Theo Lingen for his patrons, hence the pub's nickname. Today, the Jägermeister watering hole is run successfully by Robert Glowacki.

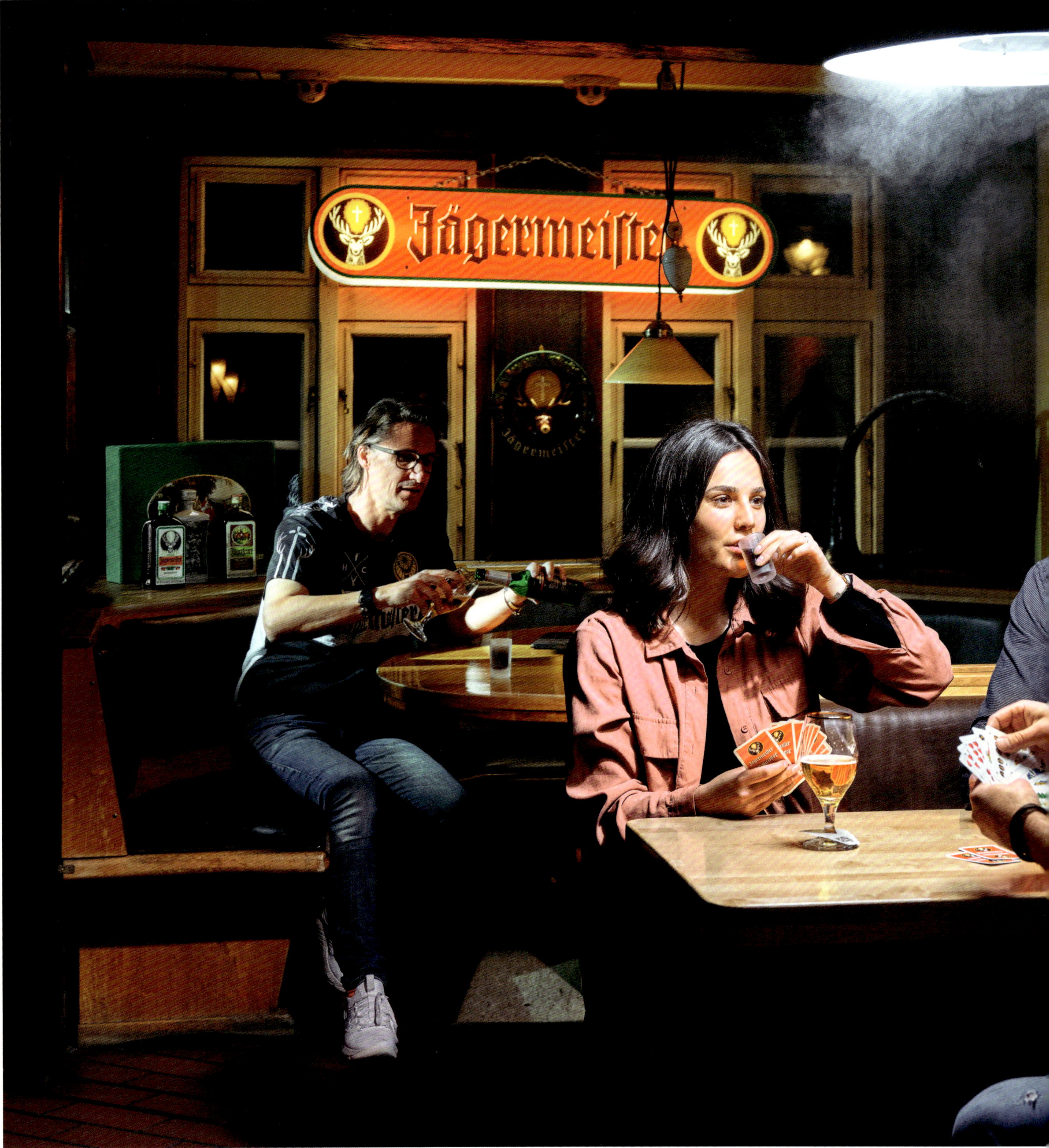

Jägermeiſter

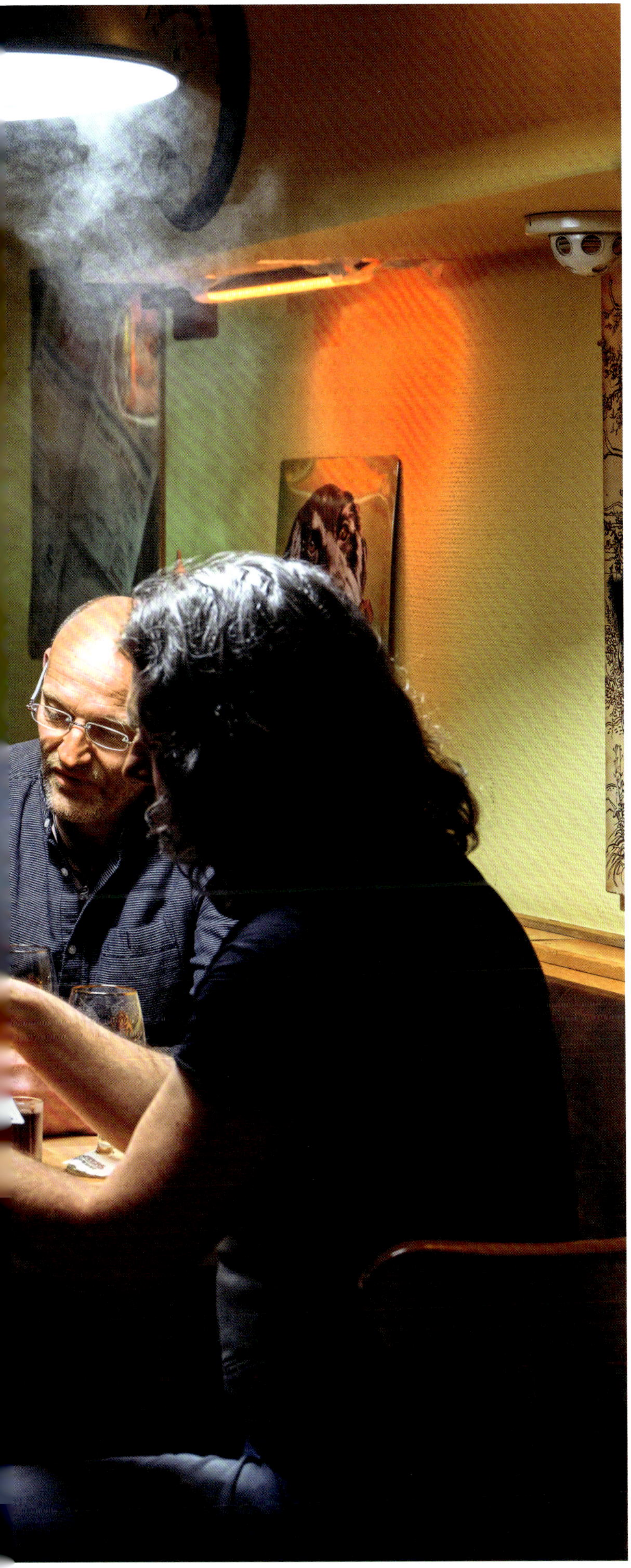

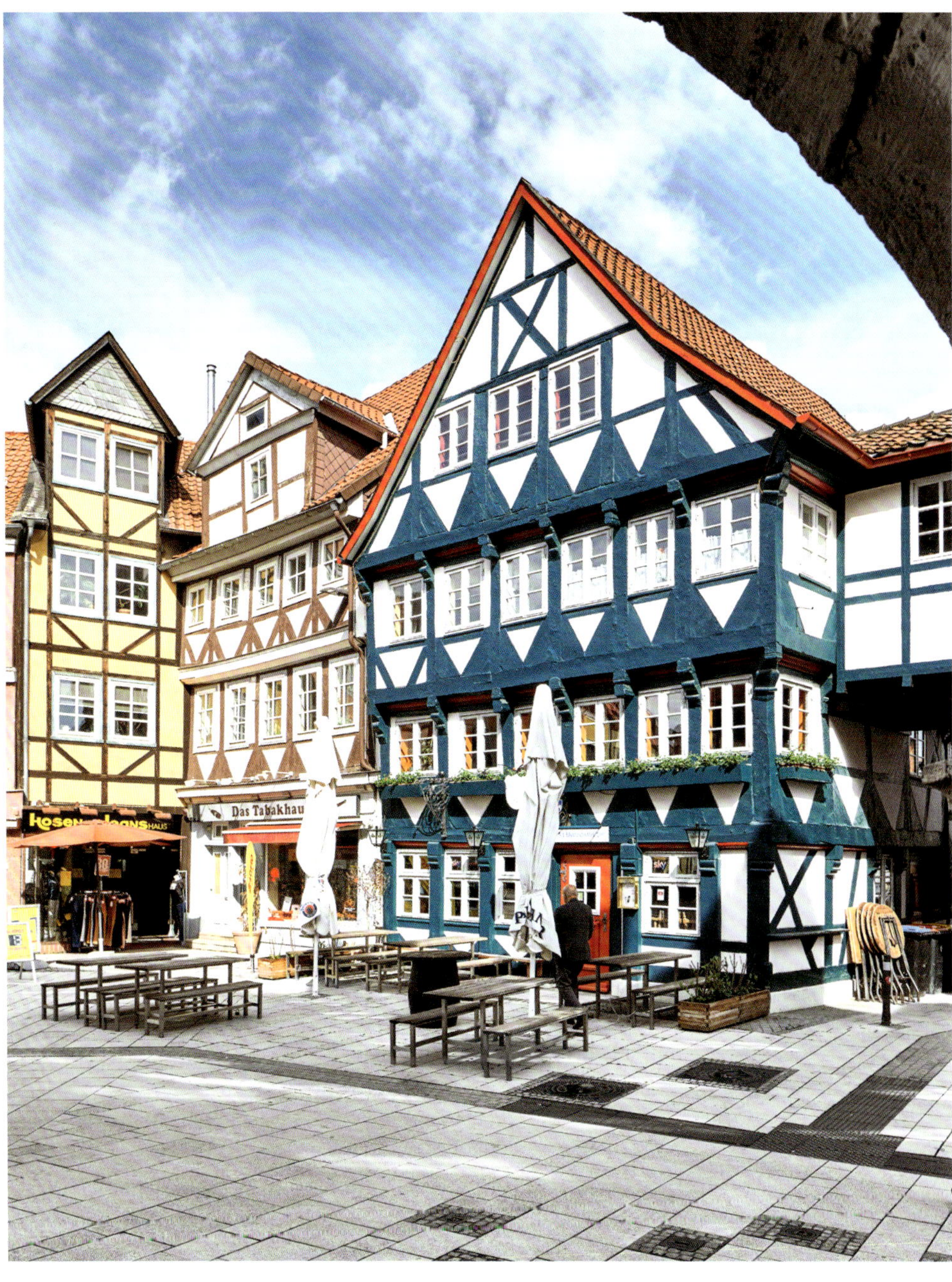

The Alt-Wolfenbüttel (or Theo's, as it is more often called) is only a few minutes' walk from the Jägermeister factory and is a popular meeting point for local and visiting Jägermeister friends.

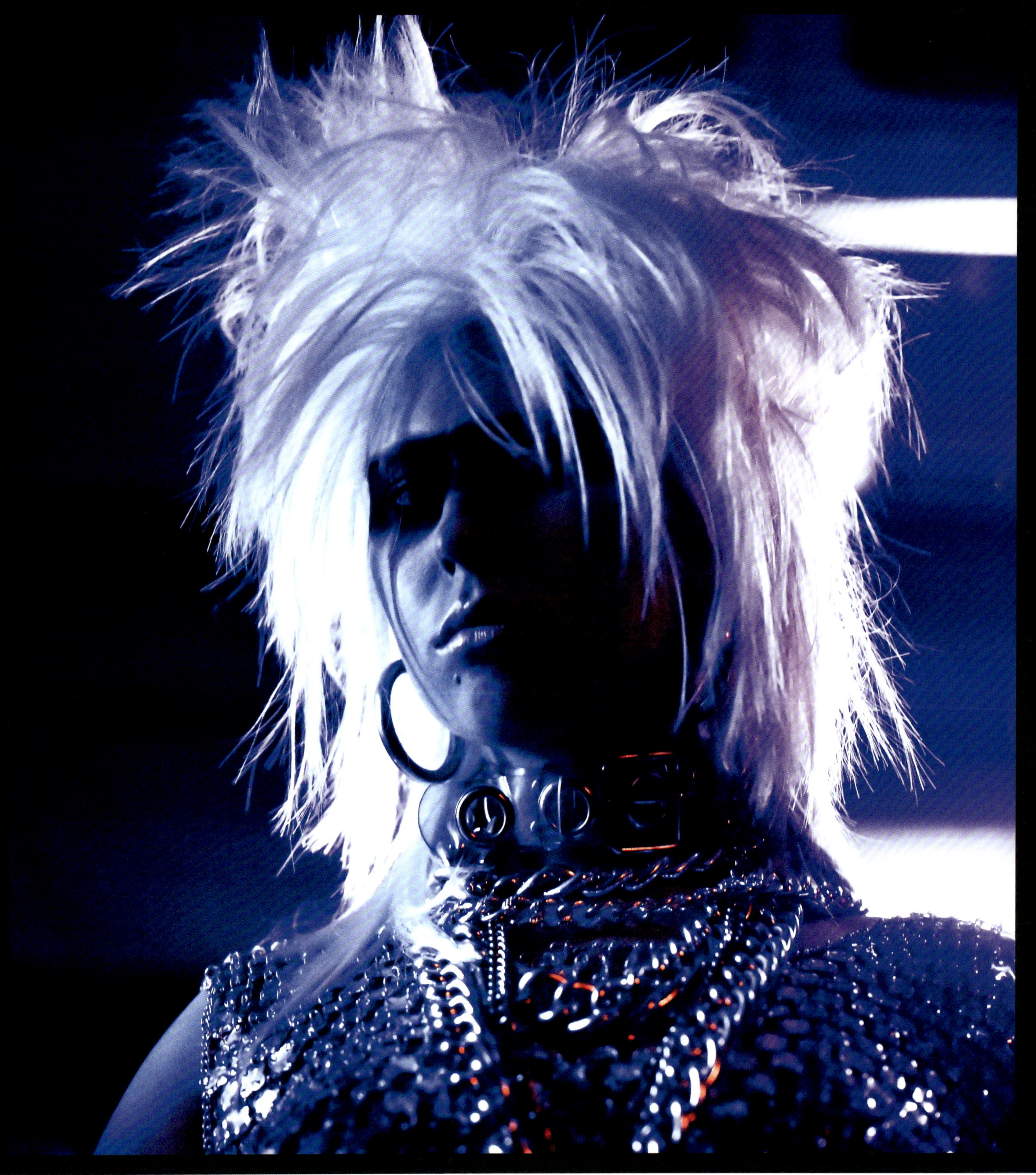

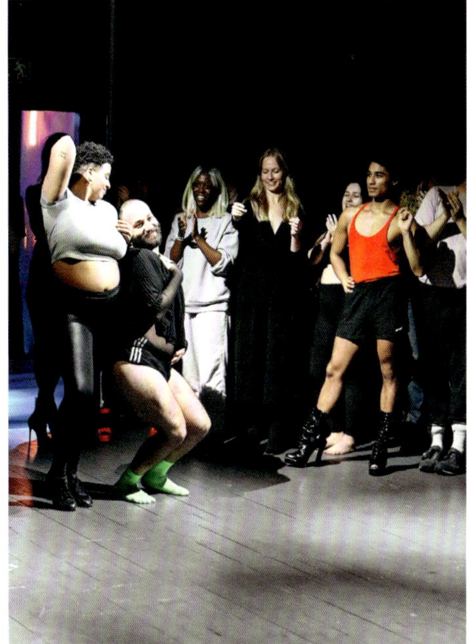

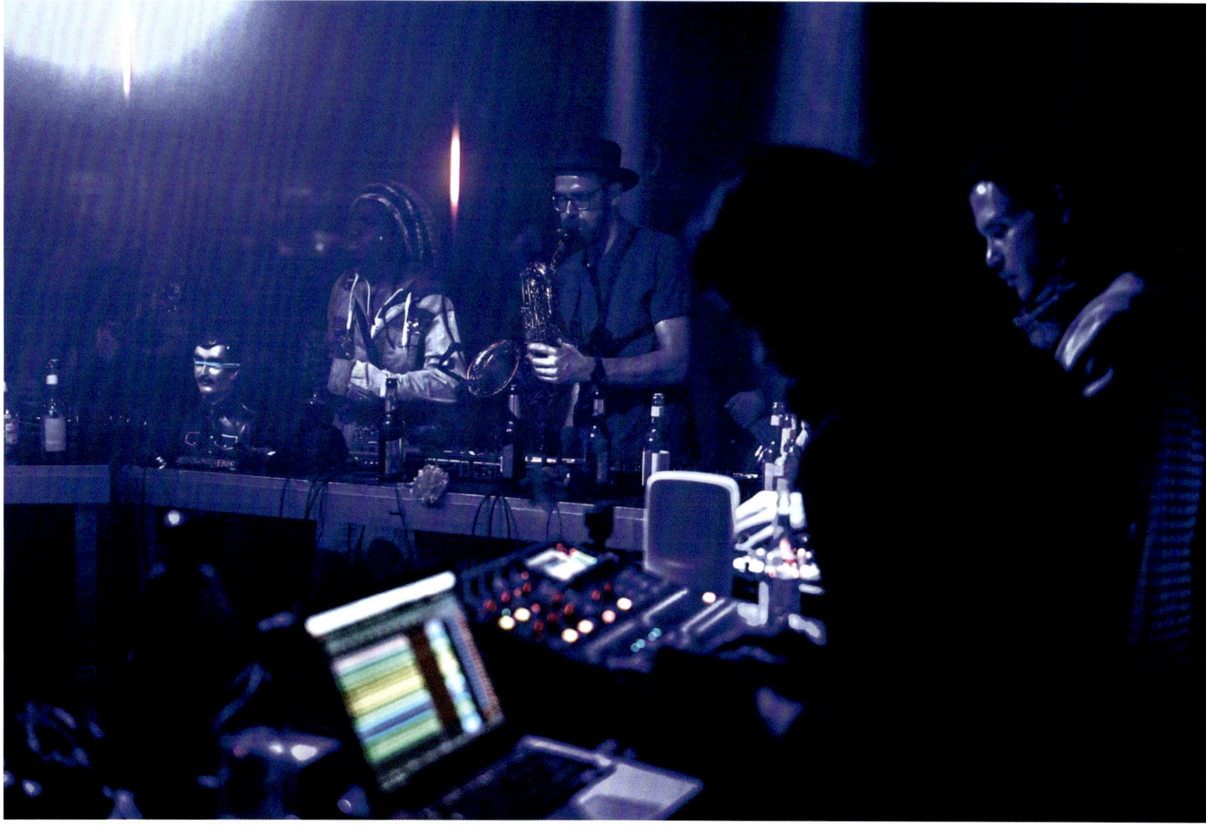

MASTERING THE ART OF PARTY-ING IN BERLIN

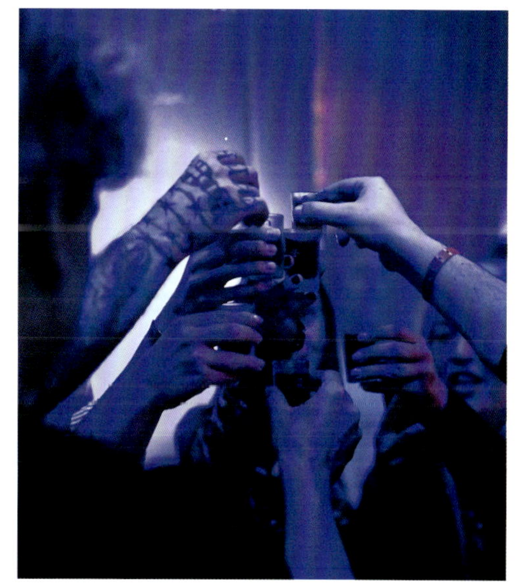

Letting the imagination run wild, giving full rein to creativity, living according to one's own rules and being free – this is what virtually every creative mind dreams of. In 2019, Jägermeister established Night Embassy Berlin to help such people get a little closer to achieving these dreams. Youth-focused radio station Fritz provided broadcasts of its shows from the unique venue in the heart of Berlin's Kreuzberg district, where Jägermeister had been able to create a space for the urban creative scene. Over 100 events – club nights, concerts, fashion shows, workshops and exhibitions – delighted thousands of people and turned Berlin's nights into day for three whole months. But, above all, Night Embassy Berlin gave creatives the inspiration, freedom and, equally importantly, the platform to make their ideas a reality – based on the motto of 'be the meister of your life!' Berlin's pulsating artistic heart brought artists and art enthusiasts together through the night, enabling them to inspire each other over an ice-cold shot as the yellow S-Bahn suburban railway sped past the window.

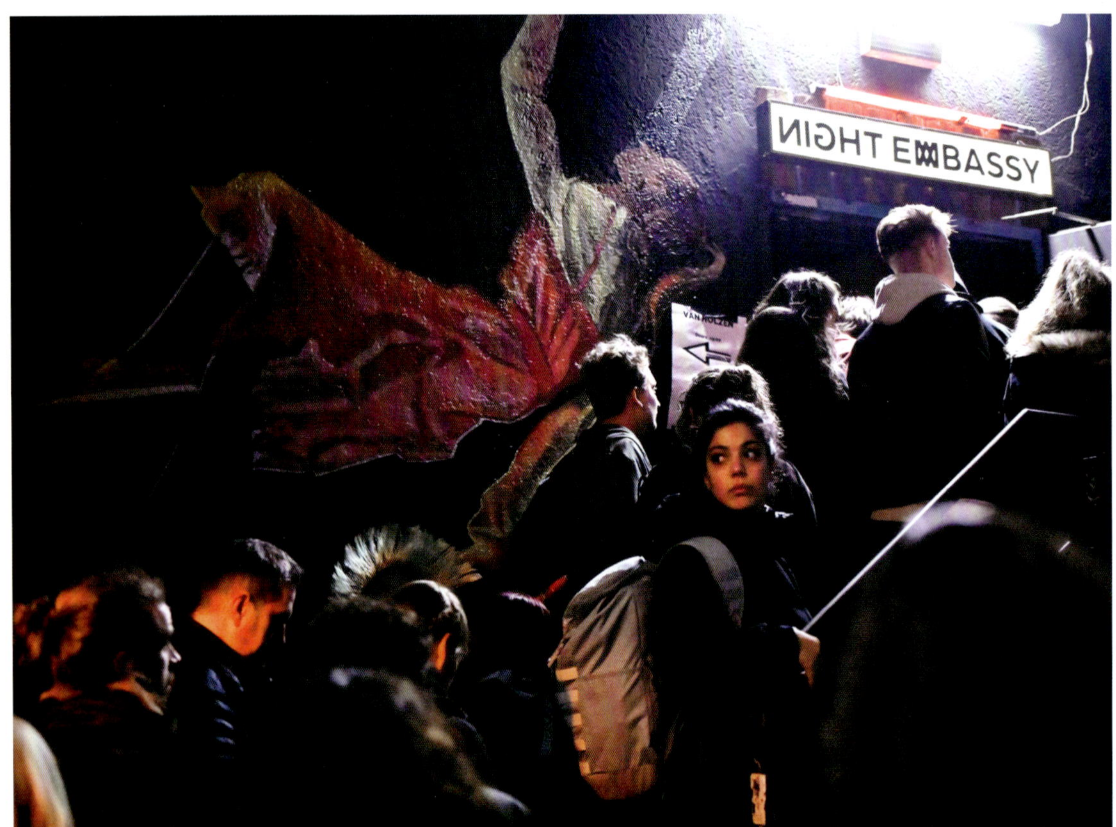

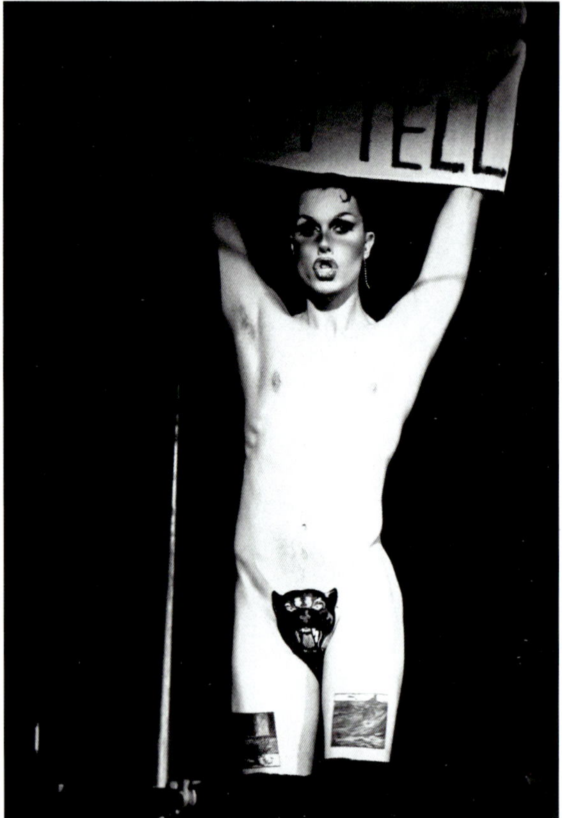

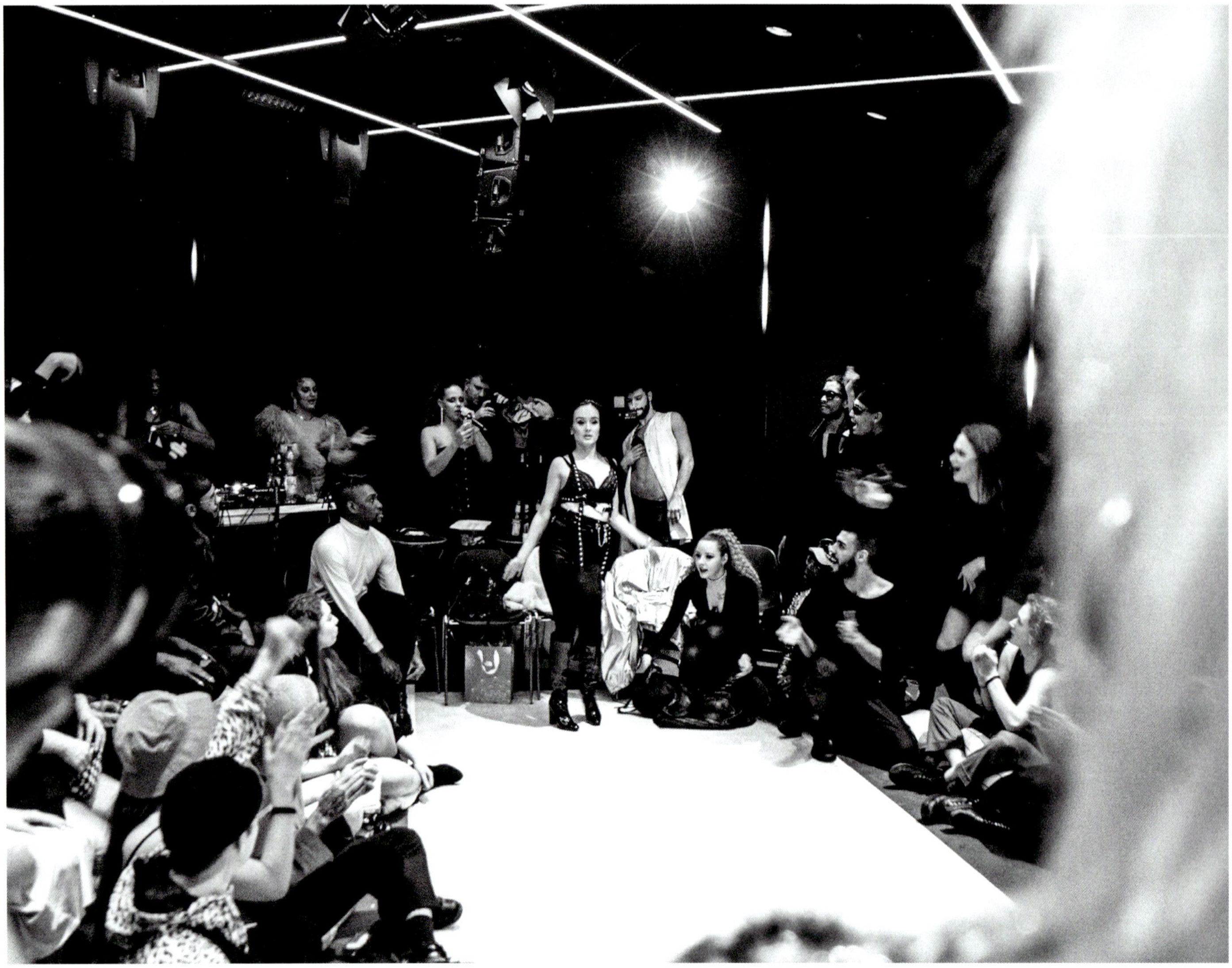

Visitors crowd in front of
the Night Embassy in Berlin
Kreuzberg (top left). Artists and
cultural workers were allowed to
let their creativity run free – and
art fans and interested parties
could take part in it themselves.

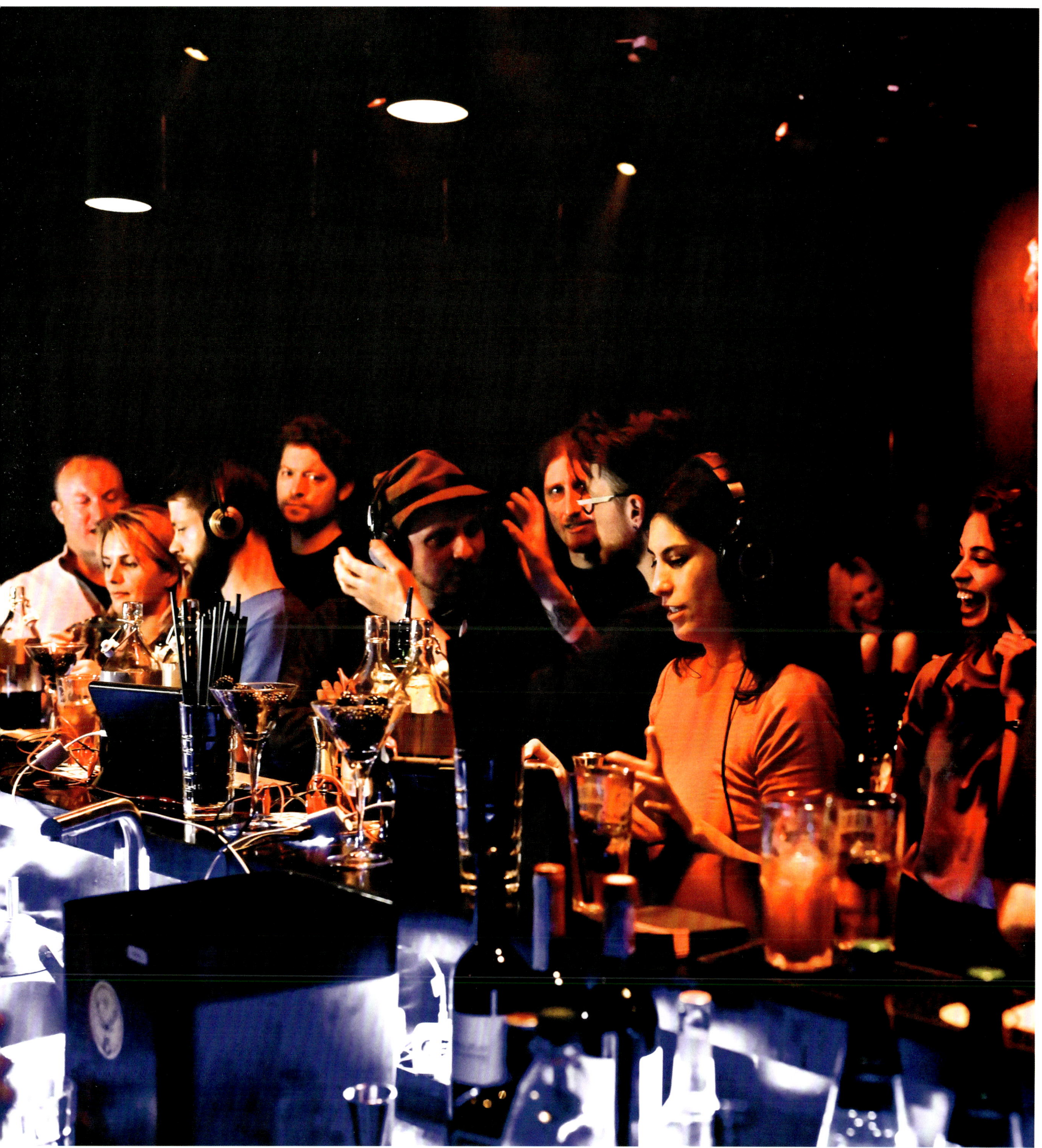

NEW ORLEANS: LIQUEUR SHOTS ON BOURBON STREET

While a legendary brand isn't developed overnight, it is often developed during the night time. This is certainly true for Jägermeister in the USA, a country where shot culture has always been predominant. The early days of Jägermeister's American success story can be clearly traced back to the New Orleans of the late 1960s and early 1970s. It was here, on the world-famous Bourbon Street, that, in 1969, German immigrant Gunter Seutter opened Fritzel's European Jazz Club – a live-music pub that not only served up traditional jazz, but also a mysterious liqueur from 'good old Germany'.

Jägermeister was still unheard of in the United States at the time, but Seutter started serving the aromatic herbal liqueur to his patrons. The unusual flavour, the secret recipe, the unique lettering and stag emblem, and the myths and legends initially surrounding Jägermeister all had an impact. A legend was born, and the rest is history; within a very short space of time, Jägermeister became college students' favourite drink, and remains so to this day. Everyone in America has now heard of the German herbal liqueur – and that's partly thanks to Fritzel's, where you can still sit at long oak bars, listen to the finest Dixieland jazz and enjoy a shot of Jägermeister instead of a bourbon. But it will probably be a while before Bourbon Street is renamed as Jägermeister Street – in the meantime, Jägermeisterstrasse in Wolfenbüttel remains the only one of its kind anywhere.

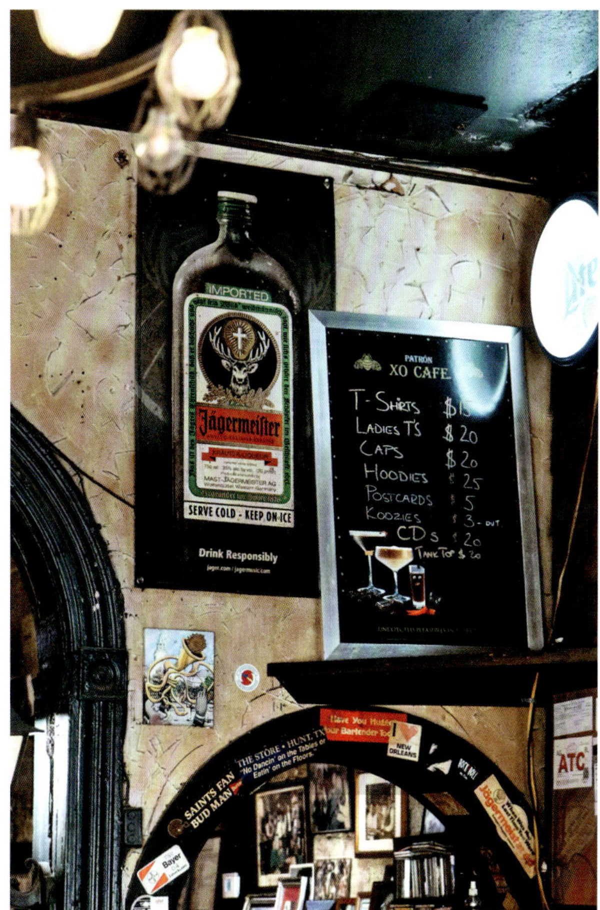

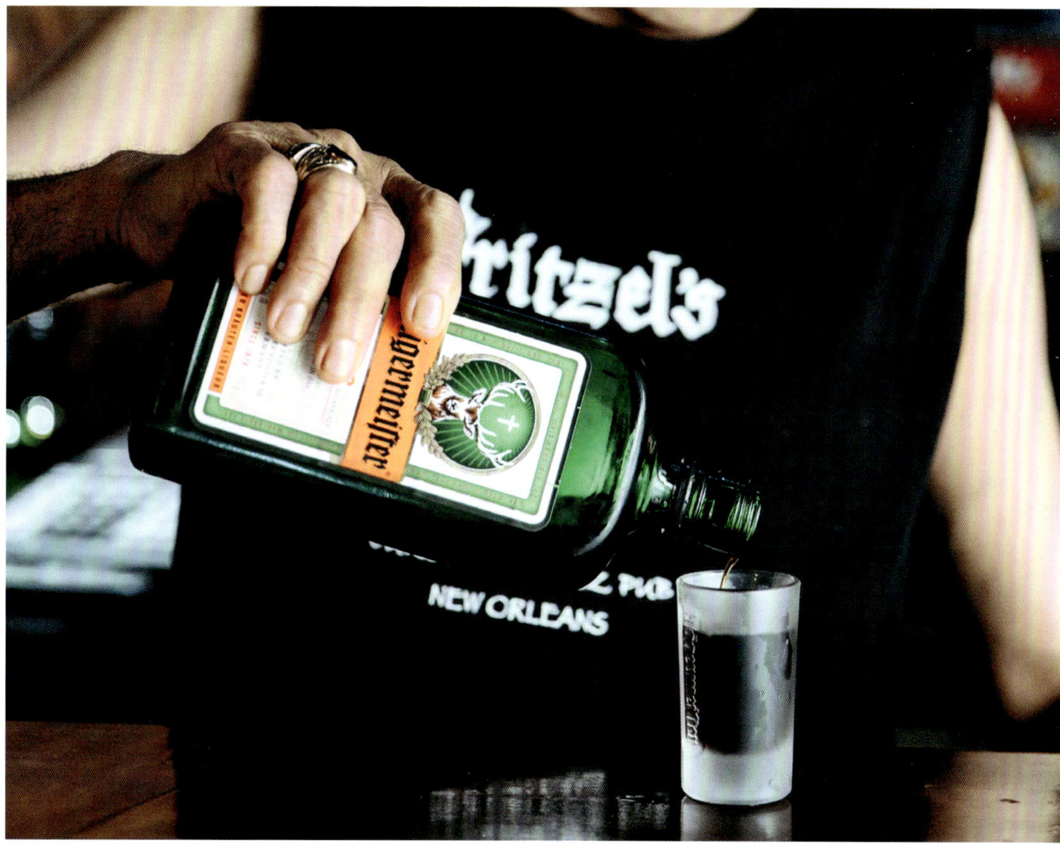

Fritzel's European Jazz Club in New Orleans is considered the birthplace of the Jägermeister cult in the USA, and the Jäger spirit is still omnipresent here today.

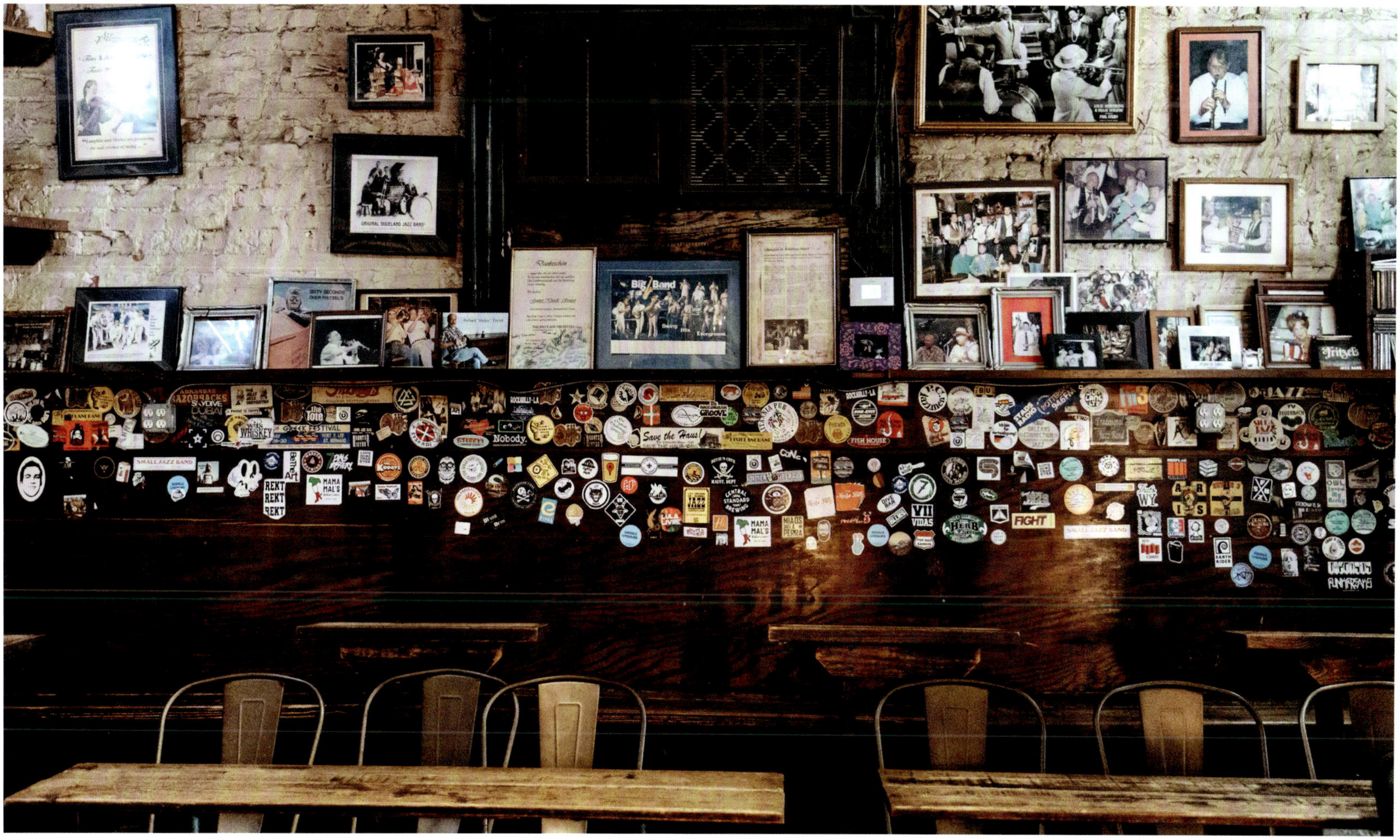

DUBAI:
THE HEIGHT OF ICE-COLD PLEASURE

Whether it be in historic pubs, hip clubs or swanky bars, Jägermeister can be found virtually all over the world. Shot moments don't need to be extreme or break records in order to be special, but there is one particular place on earth where Jägermeister fans can really enjoy this exquisite shot. Anyone ordering a Jägermeister at Dubai's At.mosphere is rewarded with an indulgence experience at the highest level – literally. The exclusive bar is located on the 122nd floor of the legendary Burj Khalifa, currently the world's tallest skyscraper – 422 metres above the ground, to be exact. These unique (and ice-cold) goose-bump-inducing moments will create lifelong memories for Jägermeister fans.

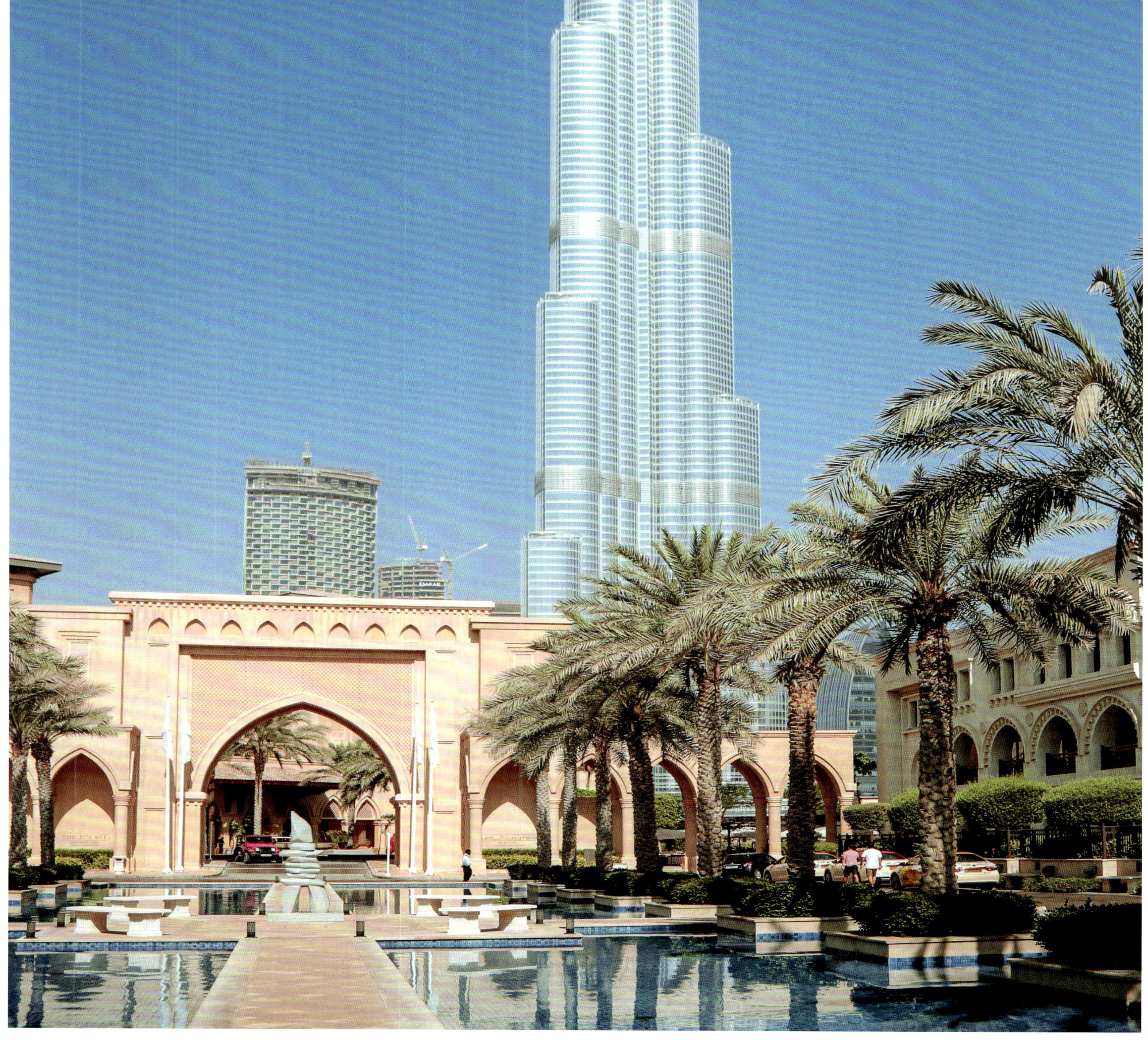

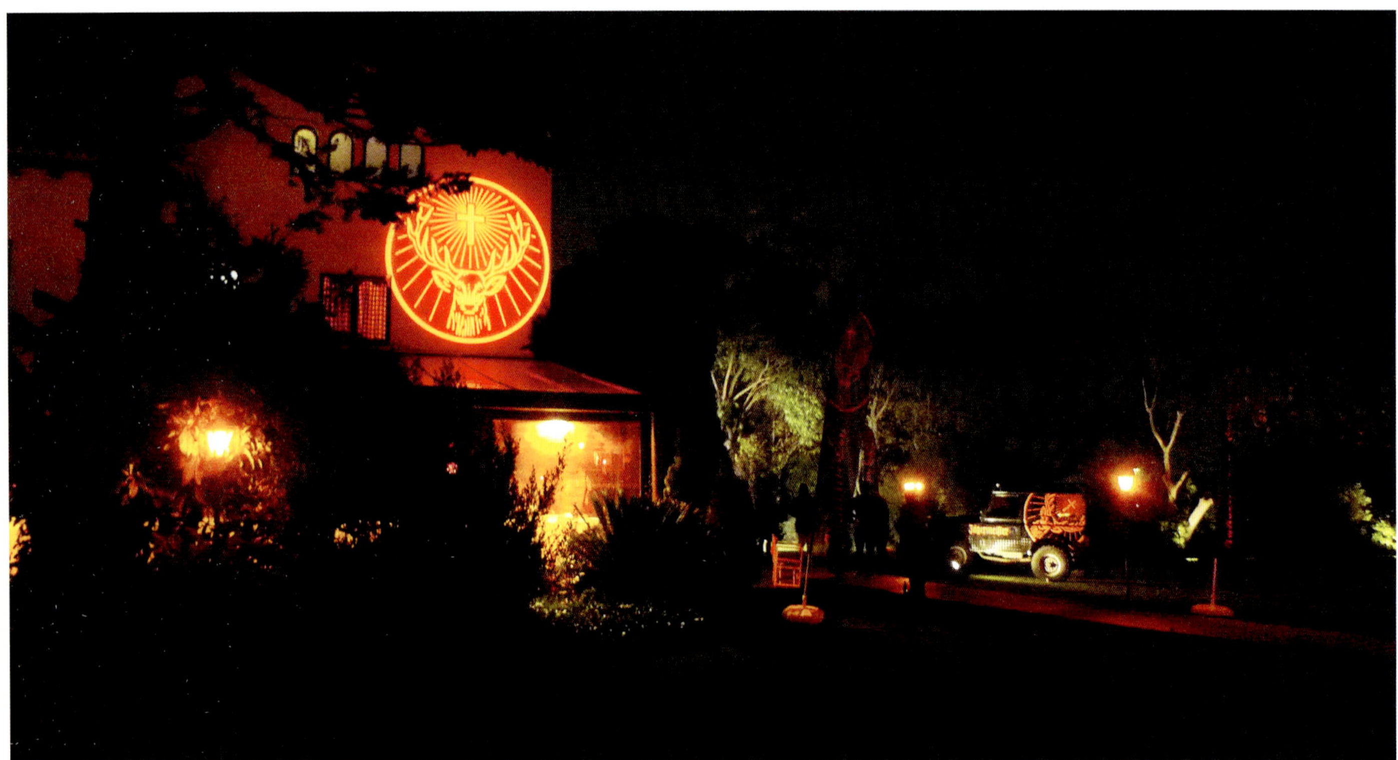

SPAIN:
ULTIMATE HOUSE PARTY

House parties are unpredictable and often wild. Guests go a bit crazy, and you never know what the night is going to bring, but something mad always happens. And just as the vibe hits fever pitch, the drinks suddenly run out. But of course this doesn't happen when Jägermeister is involved – that's a matter of honour, and it's what makes the parties even more legendary. Casa Jäger is the best proof of this. Every year from 2010 to 2014, Jägermeister would invite around 500 guests to villas on the outskirts of Barcelona, Madrid, Bilbao, Valencia and Gijón, which served as extraordinary locations for a crazy party concept that remains unrivalled to this day.

As soon as they arrived, guests would be met by hot beats and Jägerettes serving up ice-cold shots. In the living room, Spain's best-known DJs would belt out beats, transforming the carpet into a dance floor and the sofas into a springboard for stage-divers. But the walls would also be shaking in the other rooms, which were decorated in a Jägermeister theme. Jägermeister Tap Machines were positioned throughout the entire house while, in the gaming room, guests indulged in childhood nostalgia playing Super Mario Kart and GTA on retro consoles. The pillow fights were as wild and spectacular as the house parties themselves. Oh yeah, there would be a pool too, of course. Some details are probably better left out here. Let's just say there were a few creatures of the night who found themselves cooling off in it – sometimes earned, sometimes by chance. No wonder fans of Casa Jäger still rave about their experience being the best house party of all time even years later.

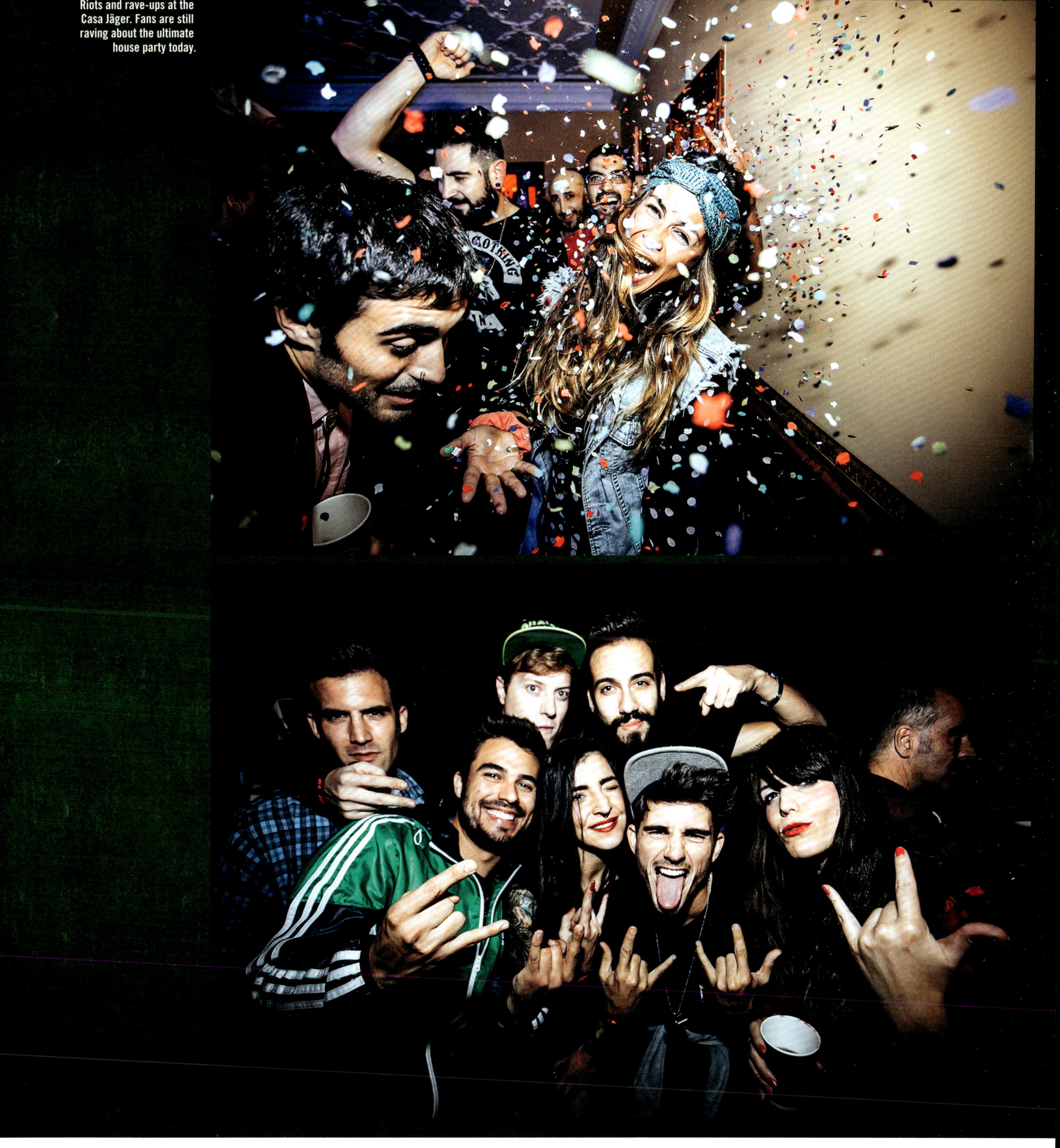

Riots and rave-ups at the
Casa Jäger. Fans are still
raving about the ultimate
house party today.

DEER CROSSING

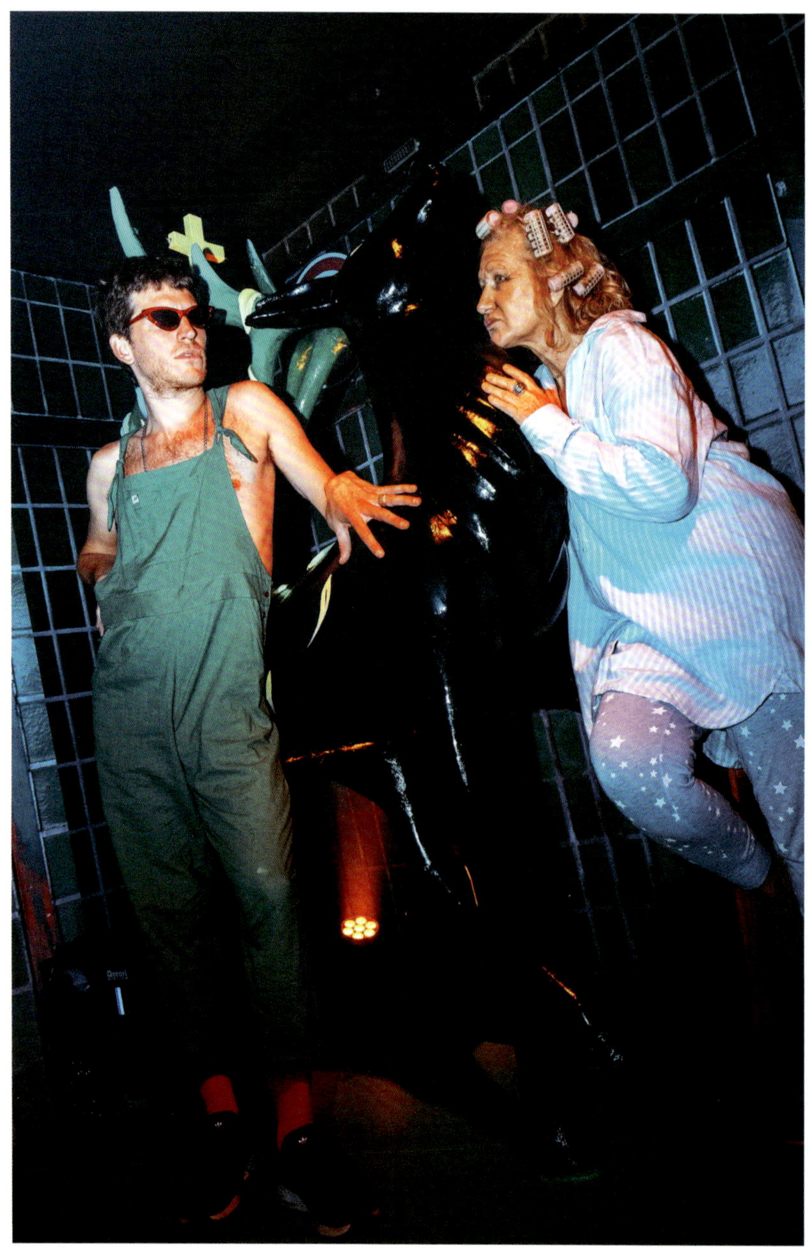

IN BUENOS AIRES

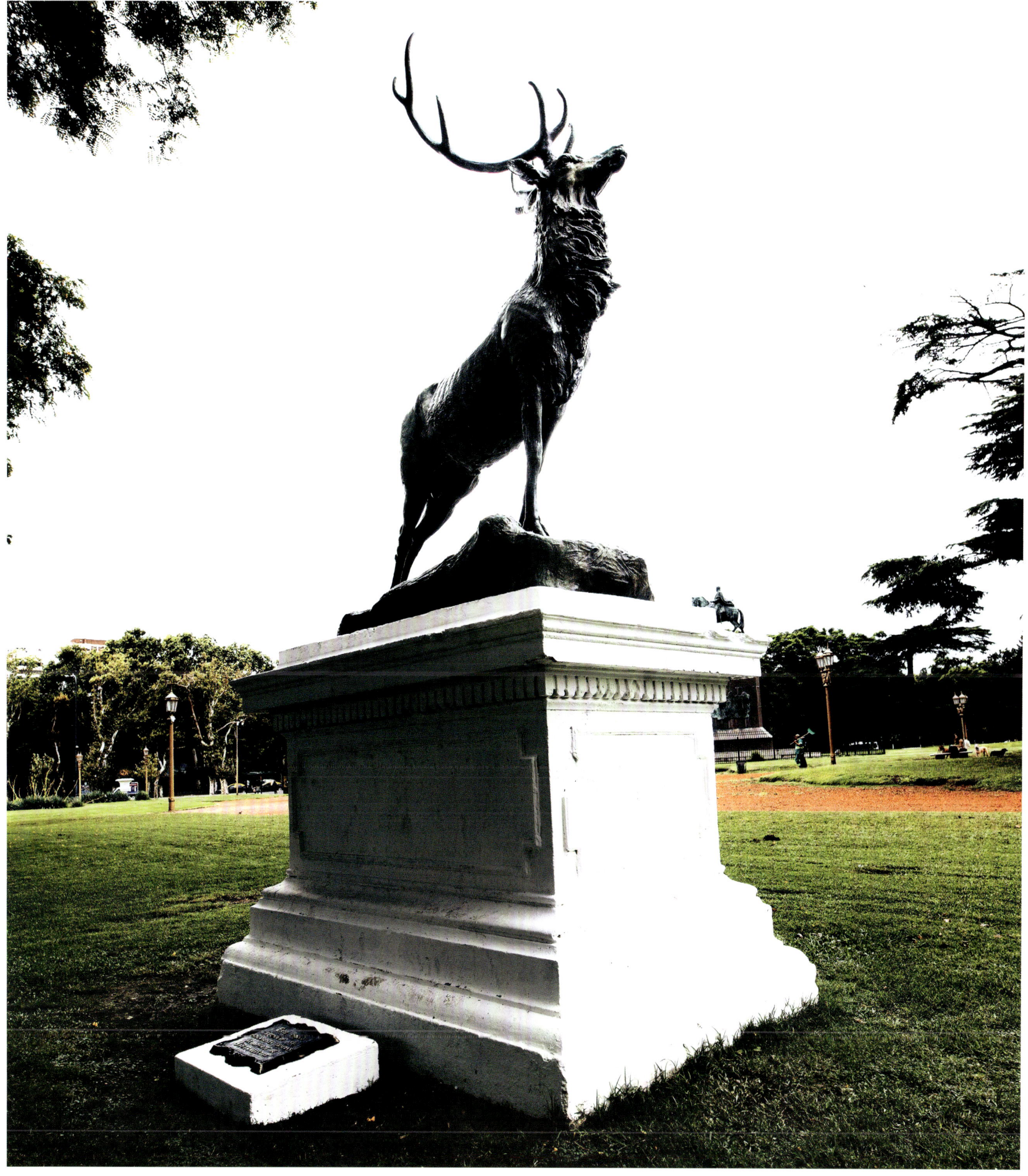

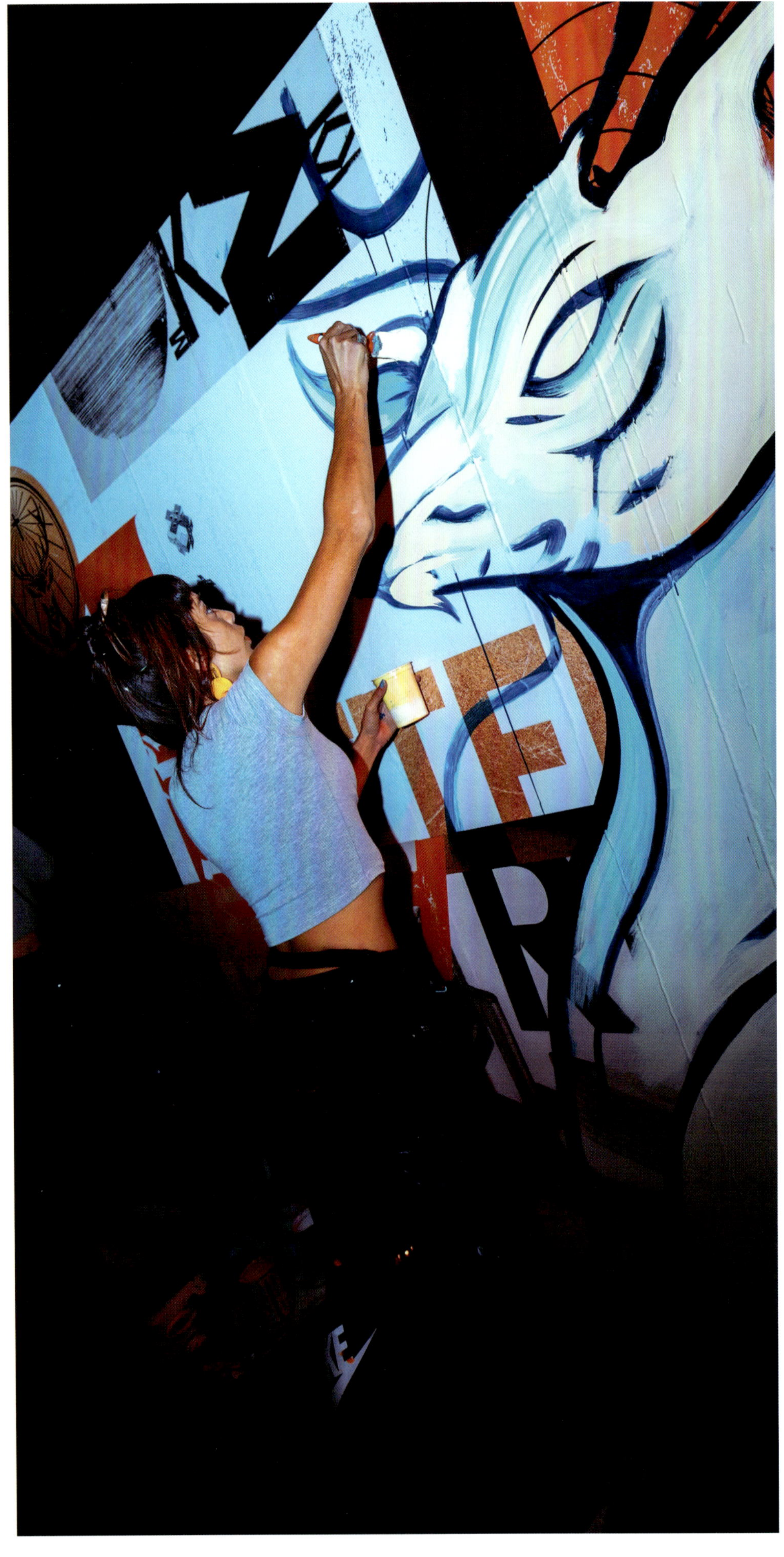

The legend of St Hubertus, the patron saint of hunters, and his encounter with a mighty stag between whose antlers a dazzling crown lit up the night, is the basis for the iconography of the Jägermeister brand. To celebrate this, the Jägermeister Art Project started up a few years ago, holding its first legendary campaign in the Argentinian capital, Buenos Aires. Its Seeber Square, a famous park in the district of Palermo, is home to a sculpture triptych featuring work by French sculptor Arthur Jacques Leduc – a 100-year-old bronze statue of a royal stag whose imposing antlers extend skyward.

But, one morning in December 2014, the antlers suddenly disappeared. Someone had sawn them off in the night and made off with them, so the park managers promptly removed the vandalized sculpture.

Motivated by a desire to give back to the city and its people some of the inspiration that artists draw on every day, the Jägermeister Art Project team got together and came up with the great idea of getting sculptor Juan Del Prado to transform an illustration of the Hubertus stag, originally designed by well-known Argentine artist Fio Silva, into a sculpture.

In March 2018, the new stag statue was officially inaugurated by both artists and doused in ice-cold shots of Jägermeister at a fitting

The sculpture of the Jägermeister deer was created by the sculptor Juan Del Prado based on an illustration by the Argentine artist Fio Silva (left).

location – the former stone base of Leduc's sculpture in Seeber Square. But joy was only short-lived, for the stag was seemingly swallowed up by the earth a few days later. The Jägermeister team asked the Instagram community to help with the hunt for the stag. 'If you know anything or have any idea as to where it could be, please let us know. We're looking for it!'

The wildest theories ran rampant among Jägermeister fans, until someone called Martin Ferrero who claimed to know the truth suddenly got in touch: 'It's at my place of work. I have photos with it, seriously.' And it turned out that the Jägermeister stag was indeed at the city restoration museum where Ferrero worked. It had had to be taken out because the old base had also required restoration.

With help from Jägermeister, the magnificent old statue was beautifully restored and returned to its original location. The statue's base still bears a bronze plaque commemorating the quite extraordinary collaboration with the inscription 'This stag has got its antlers back thanks to Jägermeister's contribution.'

And what happened to the stag from the Jägermeister Art Project? It also found a fitting stomping ground, precisely where it feels most at home: in the lively Argentinian nightlife scene in Buenos Aires!

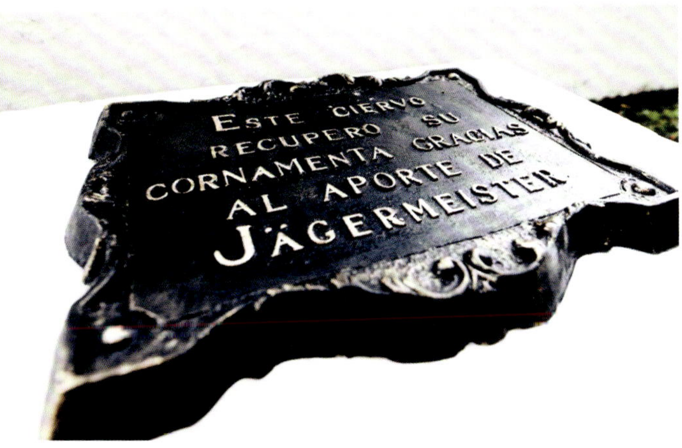

A bronze plaque on the base of the Leduc sculpture in Seeber Square commemorates the collaboration with Jägermeister to this day.

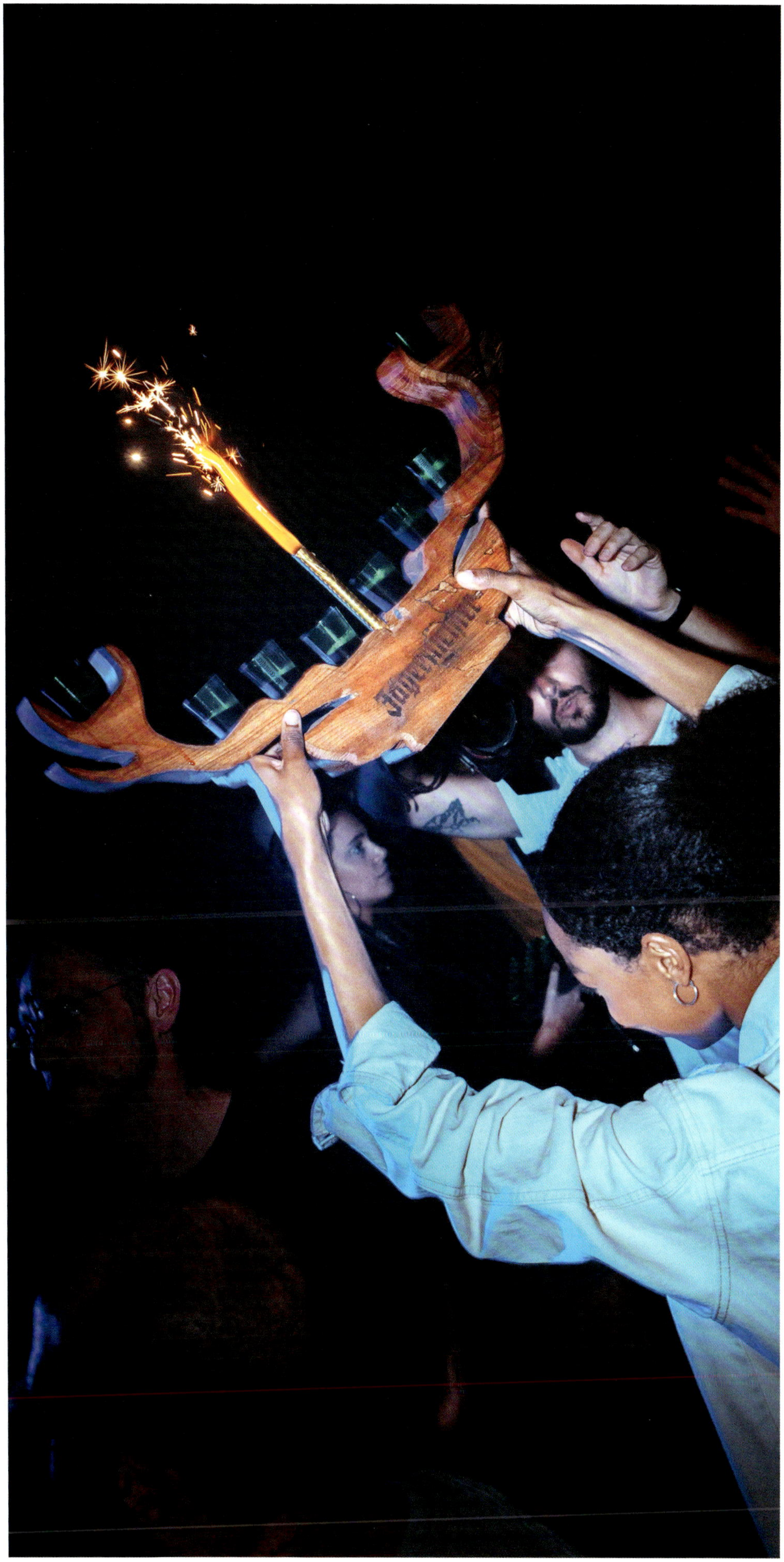

'YE GE' IN THE UNDERGROUND OF SHANGHAI

Shanghai, one of the world's largest cities, is home to a number of good pubs, but none are as legendary and fabled as C's. C's is considered the best-known and most enduring dive bar in the Chinese metropolis, and an institution in the underground music scene. The drinks are cheap and the patrons eclectic. The gloriously shabby, catacomb-like rooms attract musicians, students, spivs, artists, the wealthy, the poor, non-conformists, the bourgeoisie, pimps and night-owls. Weekends in particular see the bar filled with partygoers from all corners of Shanghai, seeking to rock out to pumping electro or hip-hop, or just hang out. And of course there is no way a melting pot like this could be devoid of Jägermeister, both as an ice-cold drink and as a partner in crime. C's regularly hosts Jägermeister music parties, and one of its colourful walls is the result of a graffiti collaboration with (and tribute to) Jägermeister. Incidentally, anyone ordering a Jägermeister at the bar here had better ask for 'Ye Ge' – that's what the herbal liqueur is called in China. And it's no coincidence that the name is reminiscent of *Jäger* (which means 'hunter' in German) both in sound and meaning, with 'Ye Ge' translating to 'wild character'. So, *gānbēi!*

野 格

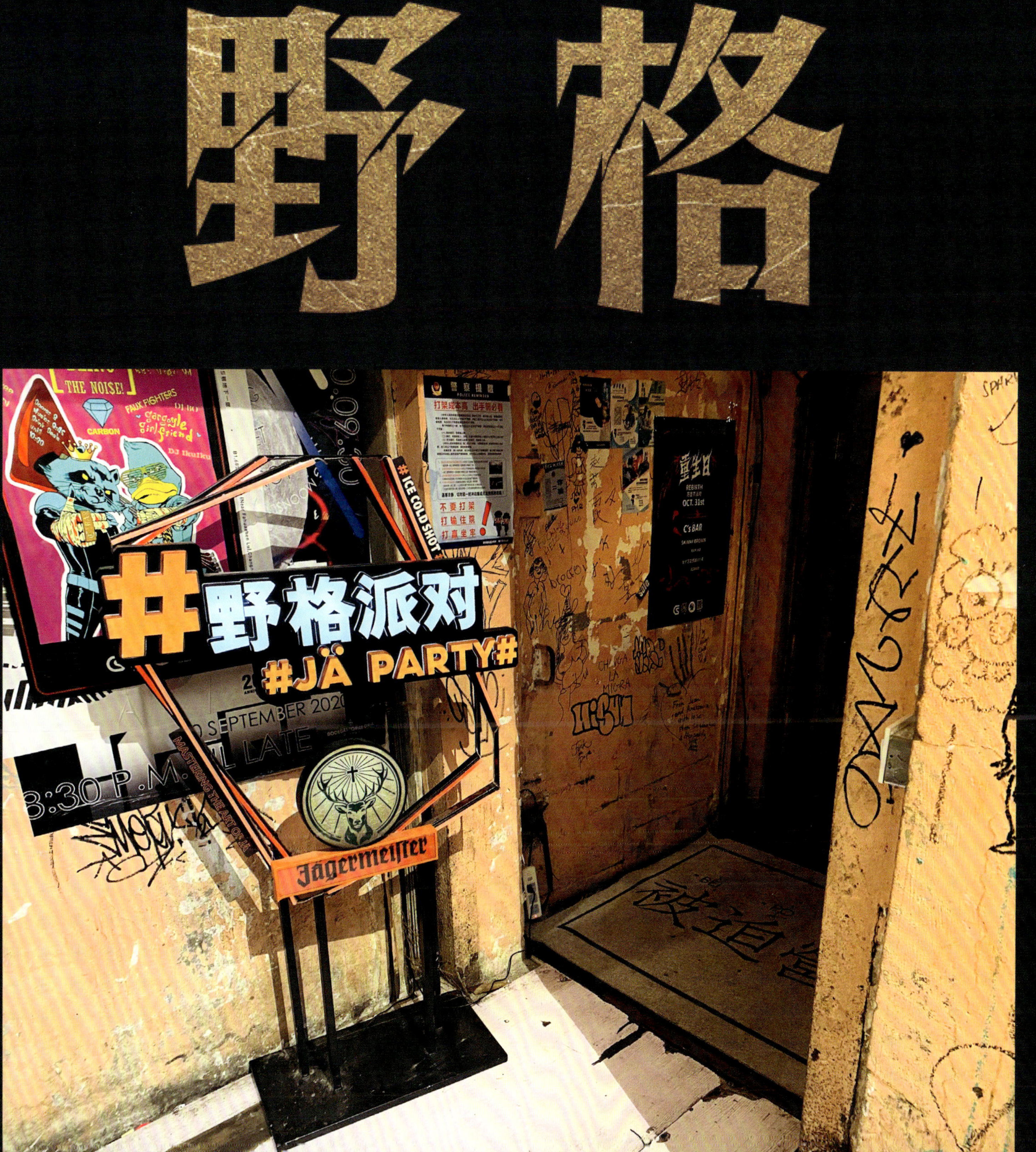

Slightly hidden, slightly wicked: C's, one of the coolest underground bars in all of Shanghai, regularly hosts
Jägermeister parties – or Jä parties, as they are called here.

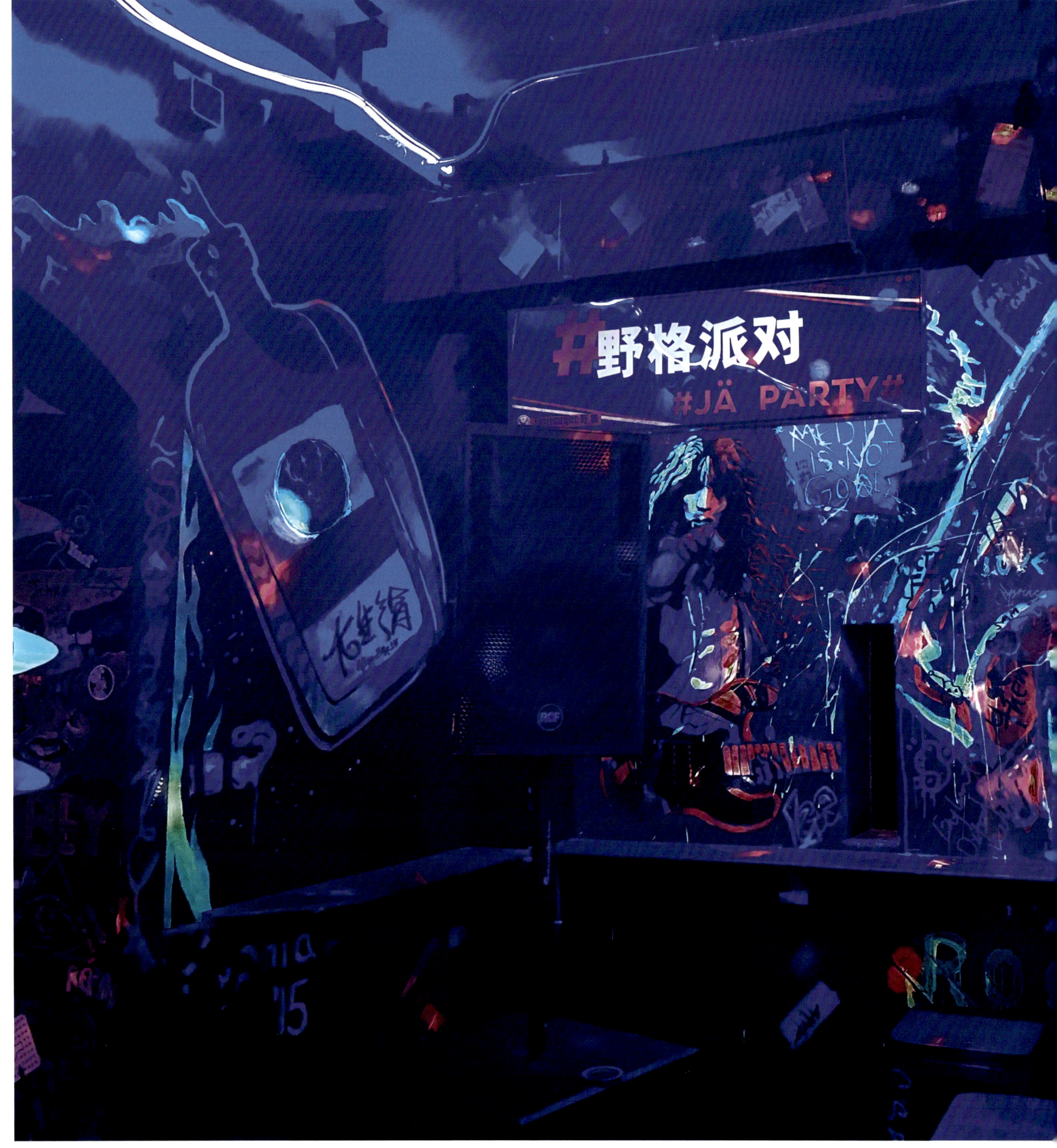
#野格派对
#JÄ PARTY#
MEDIA IS NOT GOD

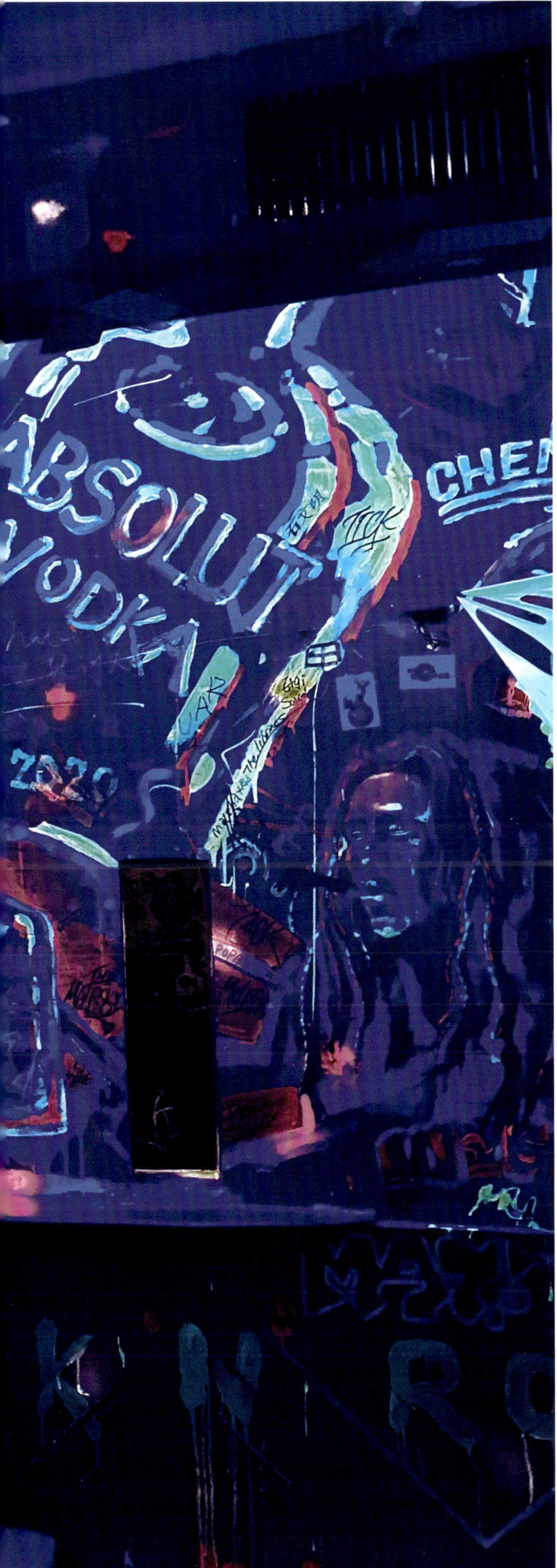

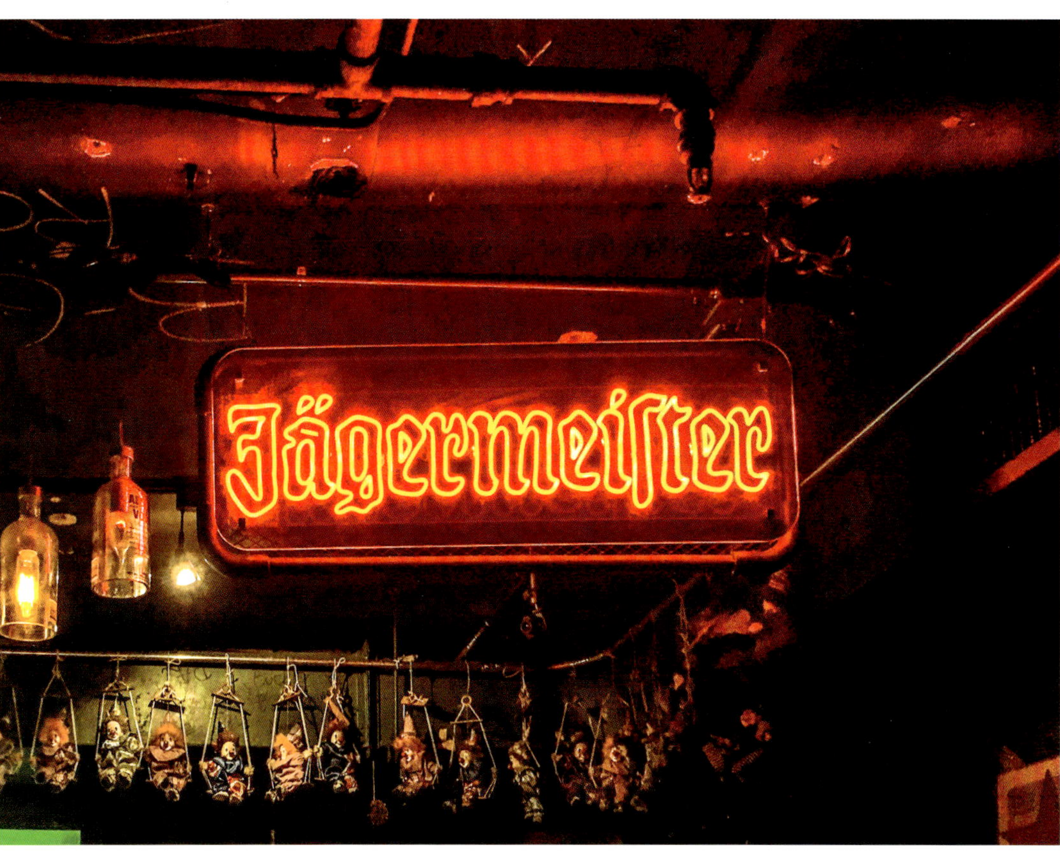

It's always lively in the catacombs of the underground bar at the weekend. And it's not just because the walls are covered with graffiti. The audience at C's is also high-spirited and diverse.

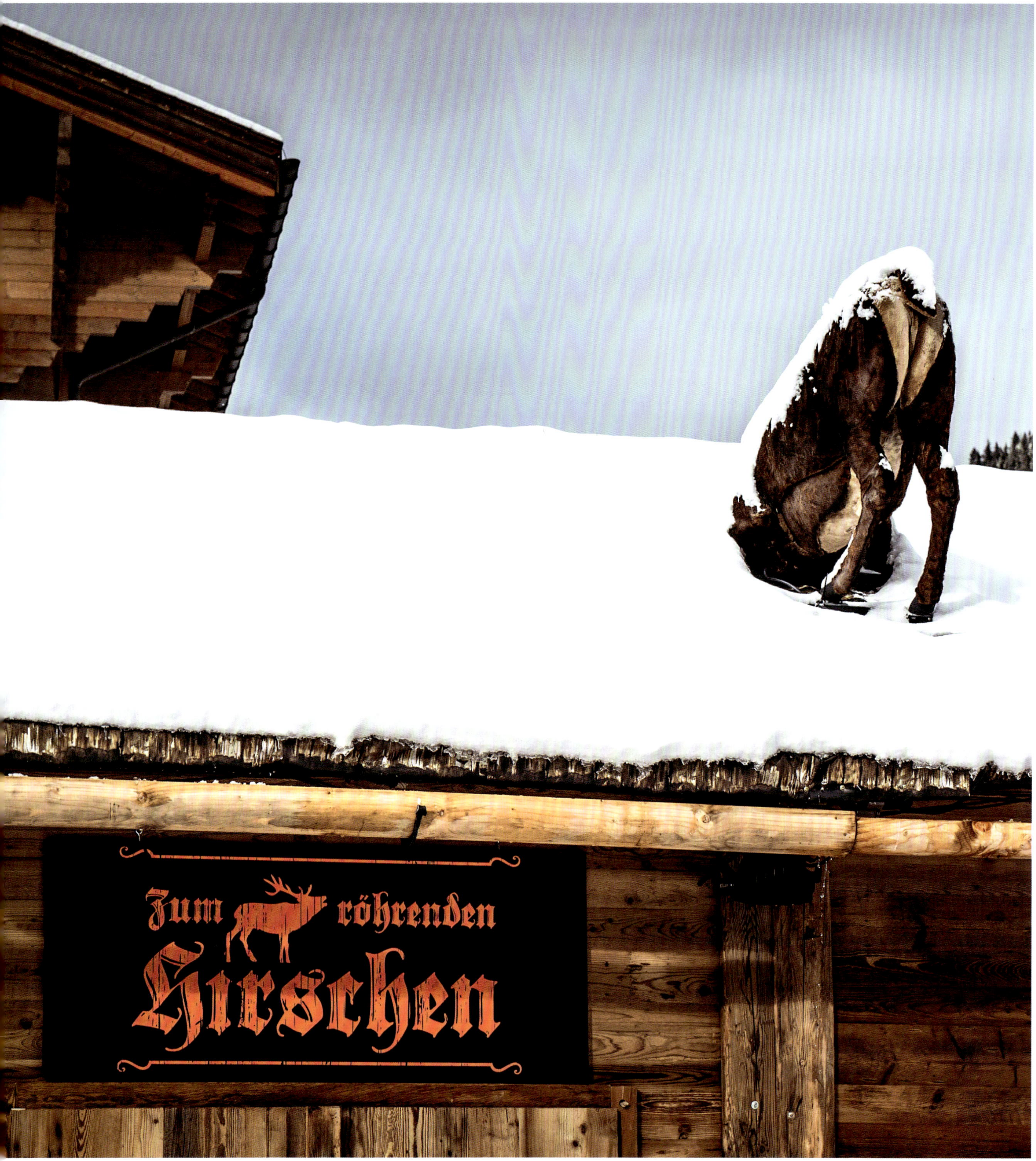

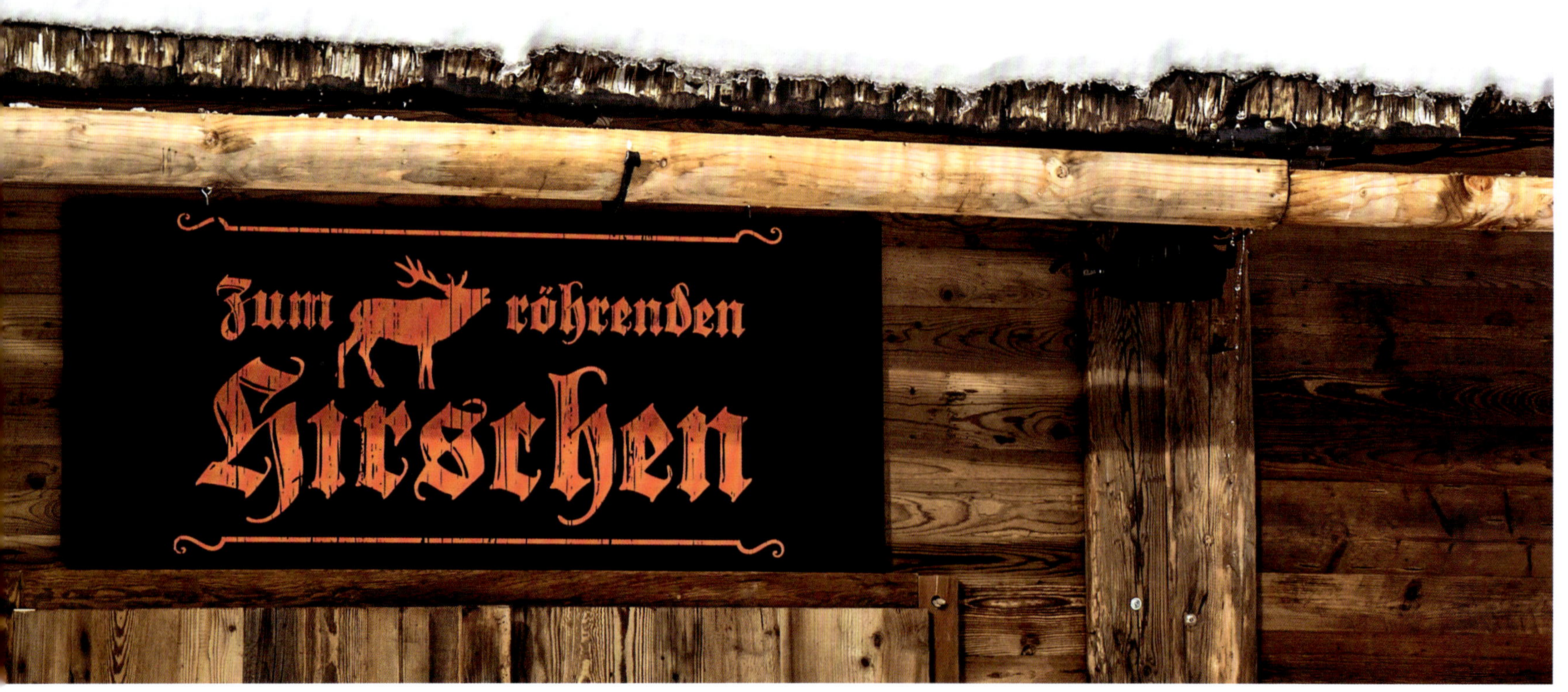

Zum röhrenden
Hirschen

ICE-COLD HEAT IN SWITZERLAND

Idyllic alpine scenery, adrenalin rushes, après-ski – the Lauberhornrennen ski races in the picturesque Bernese Oberland have all of this in spades. The annual ski event is part of the World Cup competition, and is an institution for winter-sport enthusiasts and party people alike. And as the athletes hurl themselves down the longest descent in alpine ski-racing, the Tap Machines at the Jägermeister pub run hot so that fans and curious onlookers can warm up both inside and out. The ice-cold Jägermeister shots, chilled to exactly –18°C, are the main source of warmth here. 'Jägermeister, not Jagertee' (the traditional après-ski tea) has now become the tradition at the Lauberhornrennen.

And speaking of tradition: it was at an alpine ski race, the famous Hahnenkamm Race in Kitzbühel, that Jägermeister's sports advertising began. In 1961, Jägermeister set up advertising boards along the descent, and as the race was broadcast on television, the herbal liqueur's signage made it into German living rooms for the first time.

The Lauberhorn race has been taking place since 1930. Over the years legendary figures such as Toni Sailer,
Willy Bogner, Franz Klammer, Markus Wasmeier and Marc Girardelli raced to victory here.

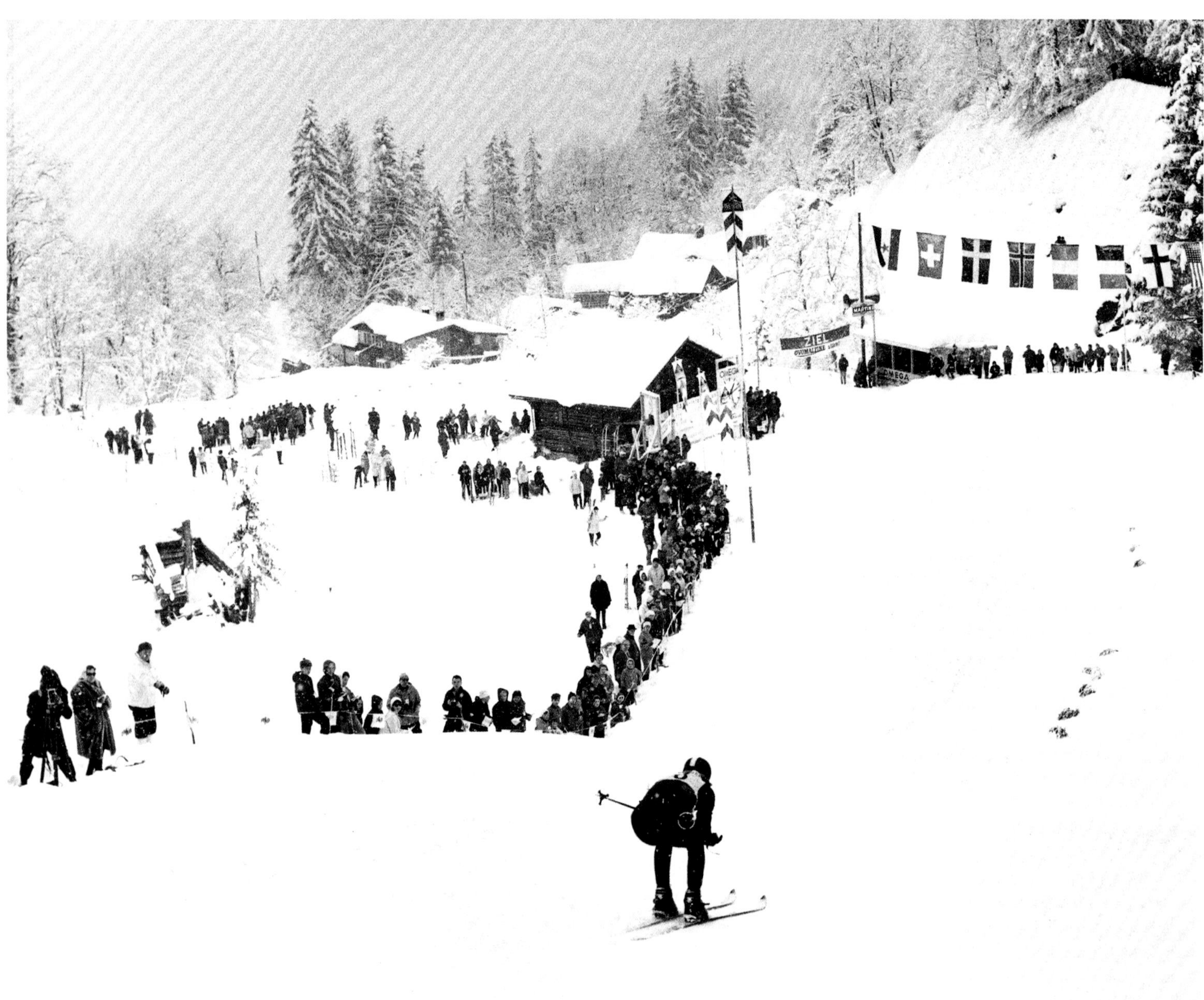

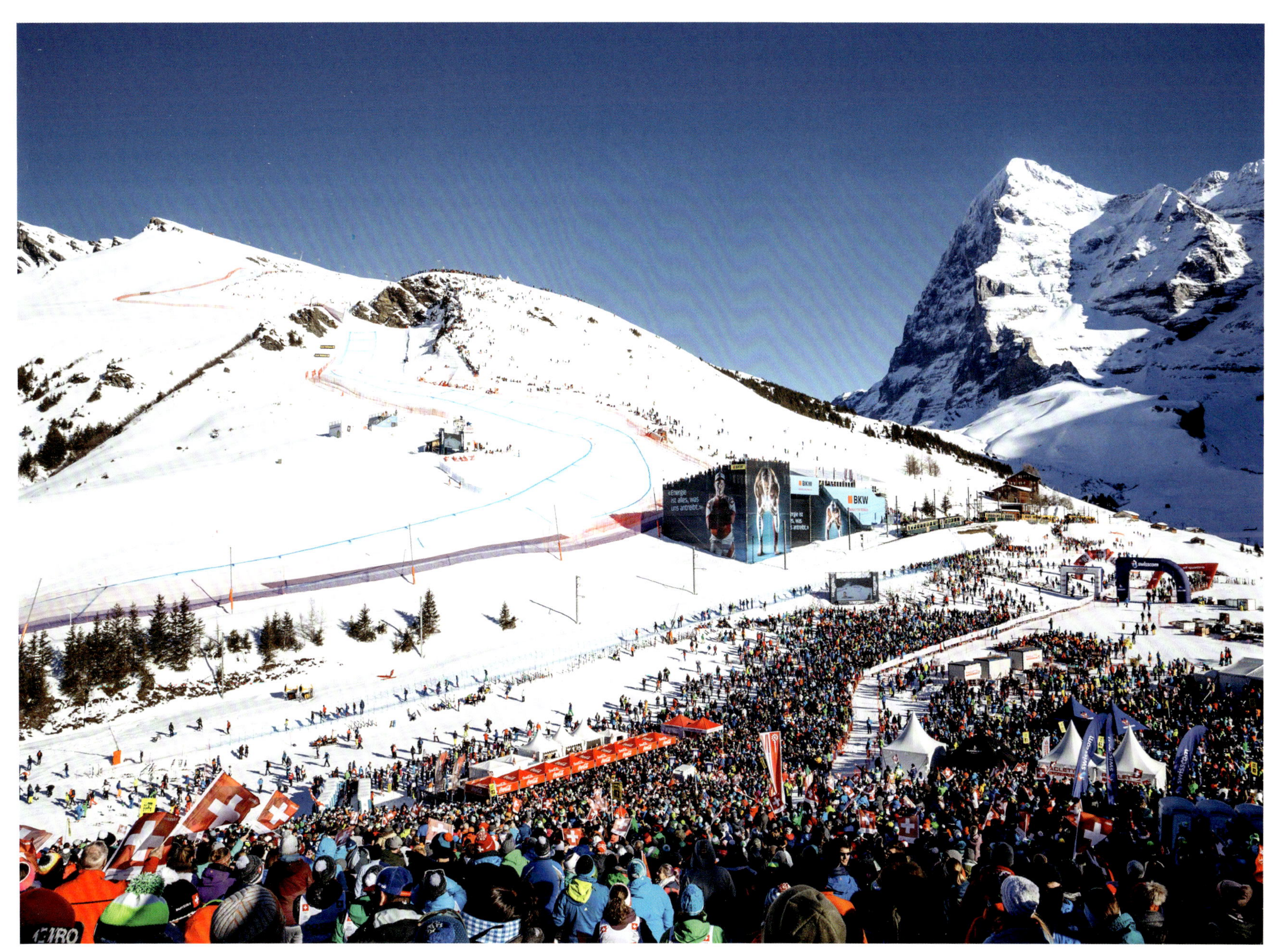

The classic event of the Alpine Ski World Cup is traditionally held in January, a week before the Hahnenkamm race in Kitzbühel. It is now a major event attracting thousands of spectators.

ICE-COLD APPARITION IN THE CZECH REPUBLIC

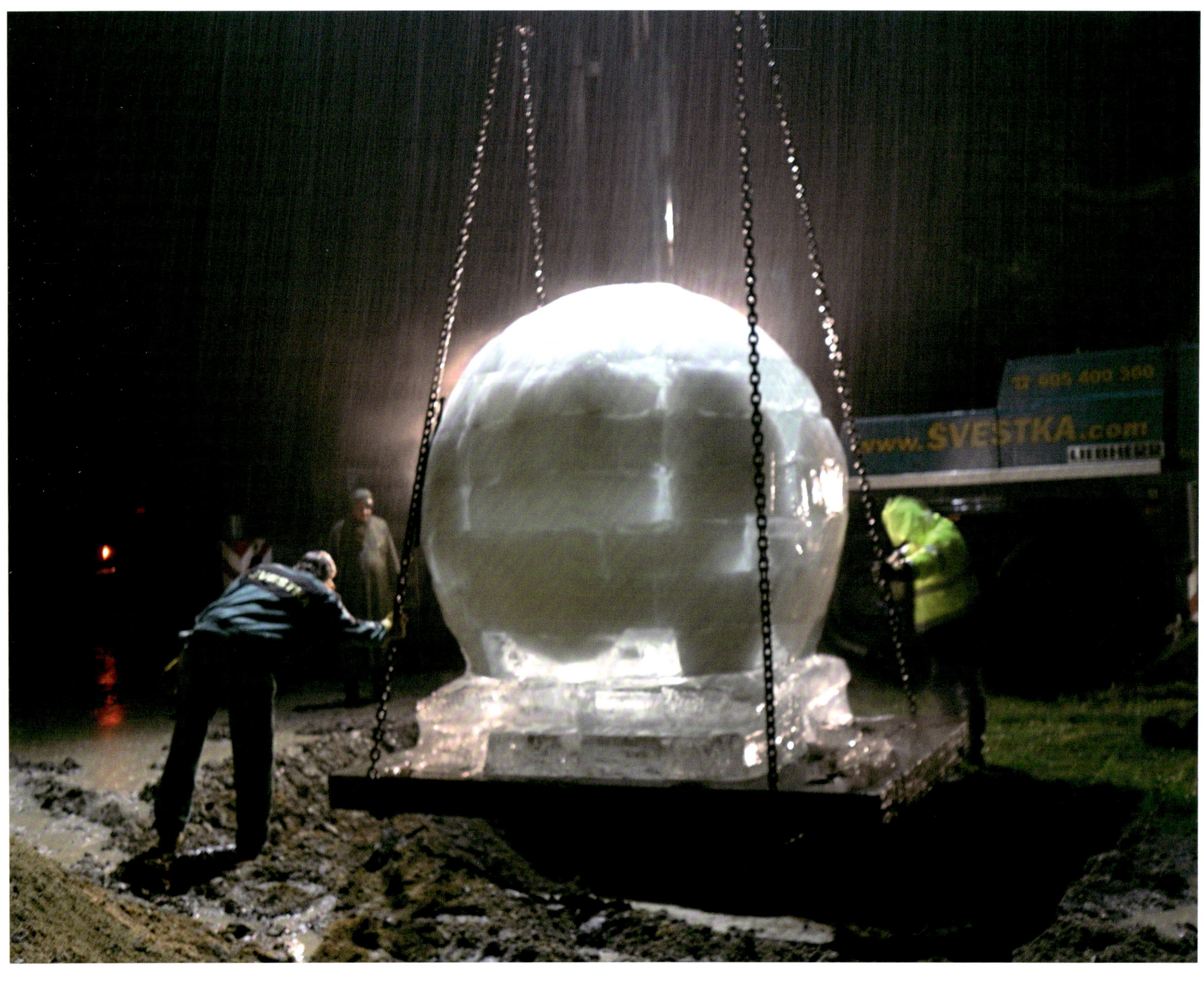

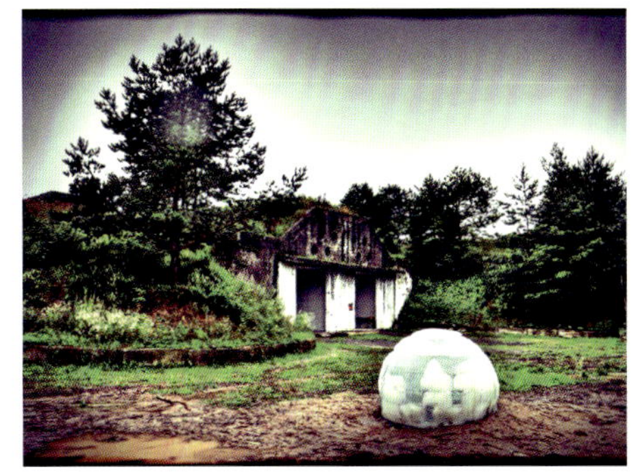

Whether a volcanic eruption or the northern lights, natural phenomena have always wowed people – and a particularly mysterious apparition caused a stir in the Czech Republic a few years ago. In 2011, an object of inexplicable origin appeared near a former Soviet military base some 50 kilometres from Prague: a ball of ice,

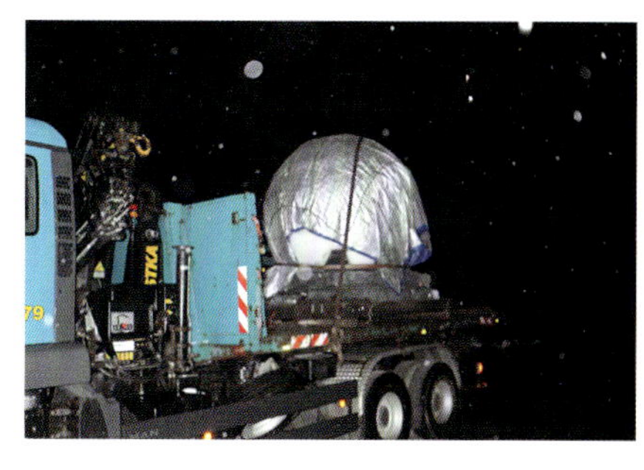

measuring around 2.5 metres in diameter, was glinting in the sunlight in a crater.

People were soon on their phones, posting pictures and making YouTube videos. And, within a few days, the ball became a top headline in nationwide news, even attracting international media attention. Was it a giant hailstone? A military experiment? Or maybe even an extra-terrestrial phenomenon? Only once the curious ice object was ultimately blasted open did the mystery come closer to being solved. A bottle of Jägermeister was hidden inside the ball. Destroyed by the explosion, its shards now lay among the chunks of ice.

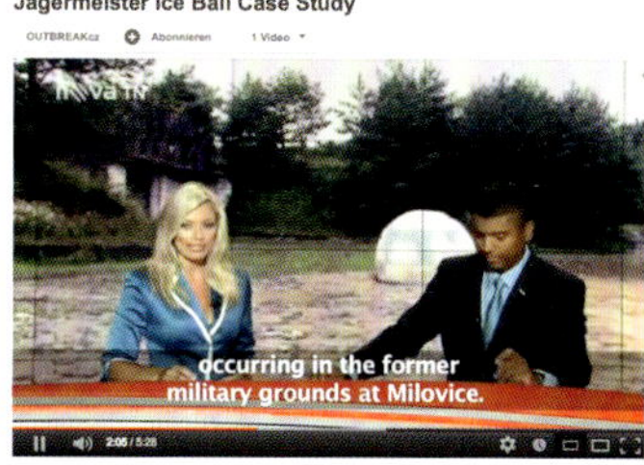

A few days later, Jägermeister finally made public what some had already begun to suspect: the brand itself was behind this phenomenal promotion, which was part of a promotional campaign to raise Czech consumers' awareness of the pleasures of ice-cold Jägermeister. The brand had previously discovered that most Czechs did not drink Jägermeister chilled. In the weeks that followed, ten giant ice balls, which would thereafter be associated with the brand, appeared at various events and festivals. The balls, which had Jägermeister bottles frozen into them, helped to refresh festival-goers, proved a popular subject for photos, and were a huge success.

Since then, there has ceased to be any mystery surrounding how Jägermeister is best served: cold as ice, of course!

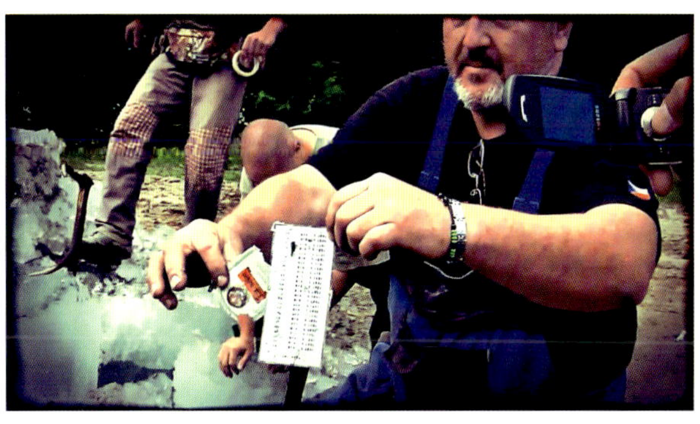

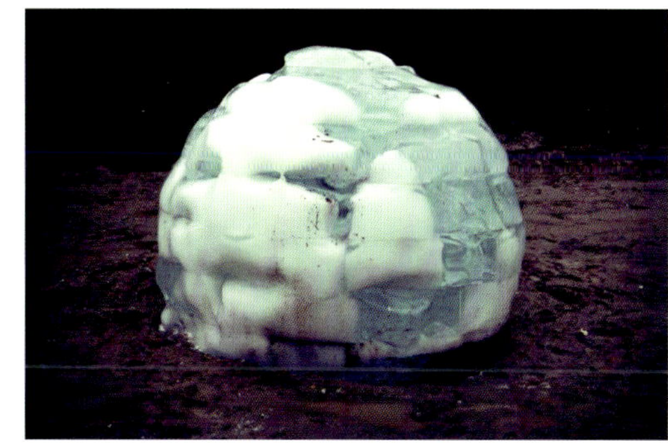

An ice-cold surprise: when the mysterious ice ball was finally blown up, a broken Jägermeister bottle (left) came to light!

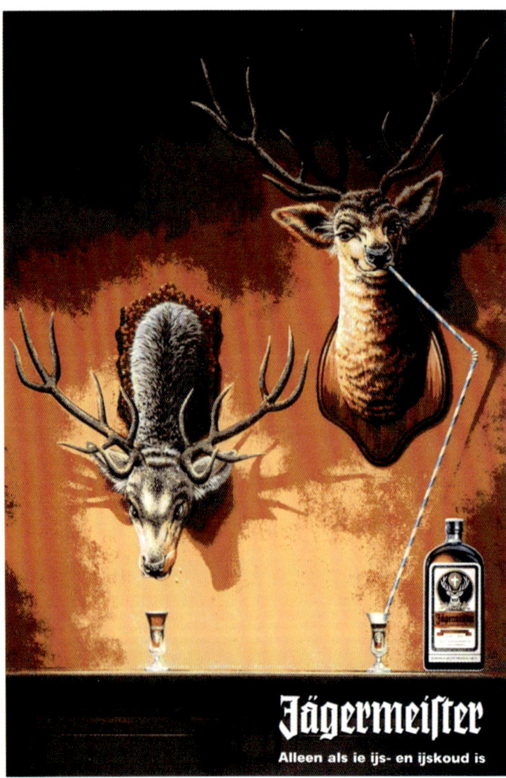
Jägermeister
Alleen als ie ijs- en ijskoud is

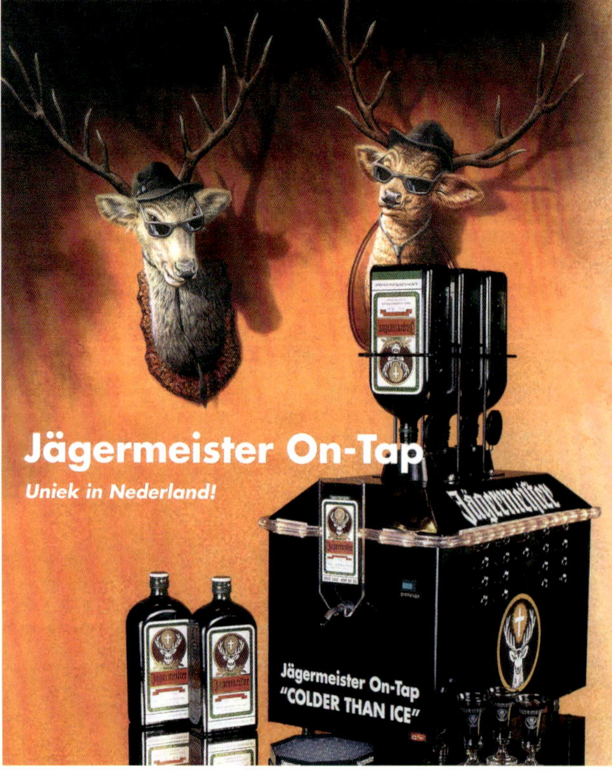
Jägermeister On-Tap
Uniek in Nederland!
Jägermeister On-Tap
"COLDER THAN ICE"

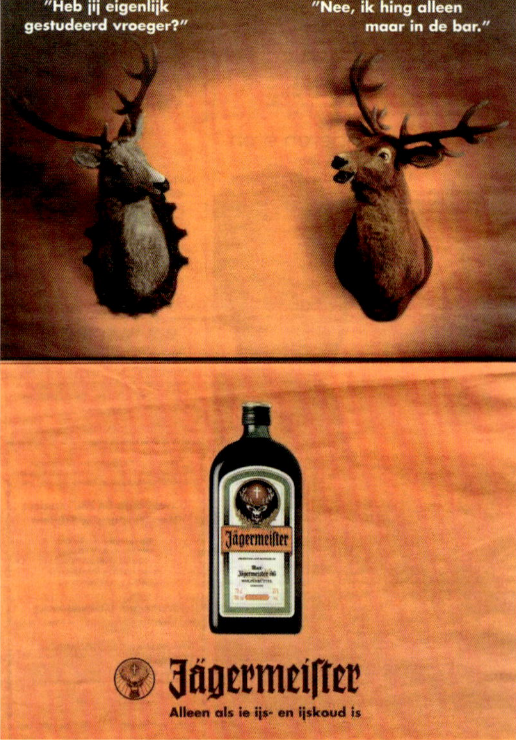
"Heb jij eigenlijk gestudeerd vroeger?"
"Nee, ik hing alleen maar in de bar."
Jägermeister
Alleen als ie ijs- en ijskoud is

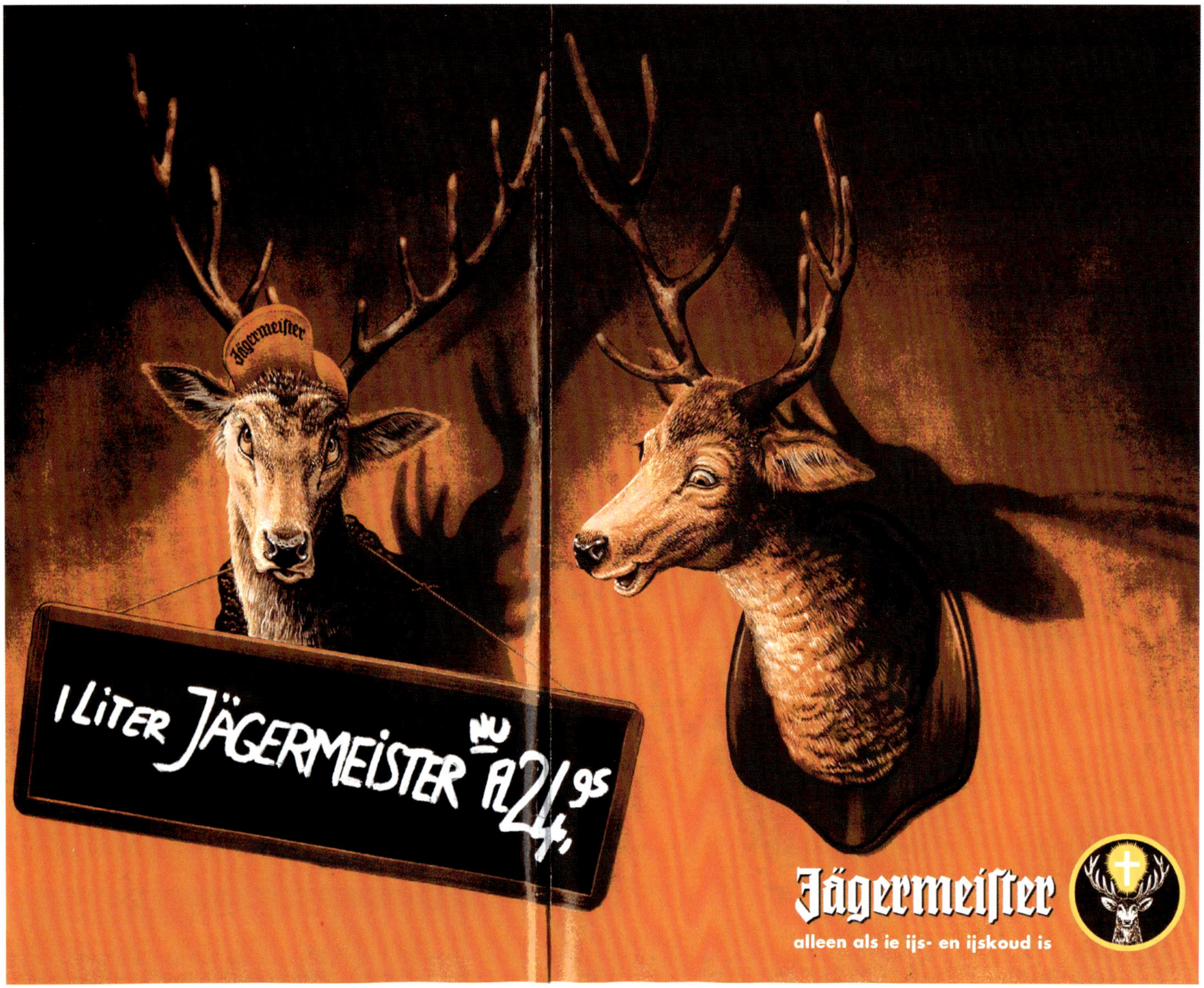

RUDI AND RALPH: THE PERT DUTCH STAGS

Achtung wild! In the early 2000s, this slogan for what remains one of the most popular Jägermeister advertising campaigns ever was intended less as a warning and more as a promise – of wild parties, ice-cold enjoyment and cheeky sayings. Its pin-ups on pub walls were Rudi and Ralph, the two talking stags who, from their vantage point, would comment on all the happenings with a hefty dose of pub humour and self-deprecation. In doing so, they soon found their way into the hearts of Jägermeister fans both young and old.

What many don't know, however, is that Rudi and Ralph actually made their debut in 1999 – and not in Germany but rather in the Netherlands. They were created by the world-famous Henson Company, which had also given the world *The Muppet Show*. As animation technology in the late 1990s still wasn't sophisticated enough, Rudi and Ralph were initially operated by professional puppeteers, with additional movements and facial expressions facilitated by remote control, giving the stags their own personal character – some might even call it charm.

After the ad's very first run, it became clear that Rudi and Ralph were the new advertising stars of Dutch television, promptly becoming recognizable to a whopping 61 per cent of the population. A crazy success story! And so it was that the campaign soon also started running in Germany, where the two thirsty friends spent many years creating countless mad moments, guaranteeing ice-cold drinking pleasure and making the bar the focal point of pub life.

SIBERIA
THE COLDEST GIG OF ALL TIME

A temperature of −18°C is the average for a freezer. It's also the temperature at which Jägermeister's full flavour is unleashed. Briton Charlie Simpson can only smile benignly at this. The singer and frontman of British band Fightstar set what is likely to be the coolest record of all time in the Siberian town of Oymyakon, the coldest consistently inhabited place on earth – with a Jägermeister 'Ice Cold Gig' that lived up to its name.

While very few people indeed can withstand the Siberian cold in Oymyakon for longer than five minutes, the then-27-year-old played a 15-minute unplugged set of songs from his *Young Pilgrim* album there on 24 November 2012 – at −30°C. 'It was unbearably cold, and playing the guitar with gloves was not an option,' he recalls. 'We had to stick heat pads in my sleeves to keep my blood warm and protect my fingers from freezing.' It was an intense experience, but a unique one. 'It was the trip of a lifetime, and a gig unlike any other,' he says. 'It feels great to have set a world record.' Charlie fittingly celebrated his entry into the *Guinness Book of Records* for the coldest concert 'with a few shots of Jägermeister, of course', and this time he didn't need a freezer or a tap machine for his perfect shot moment.

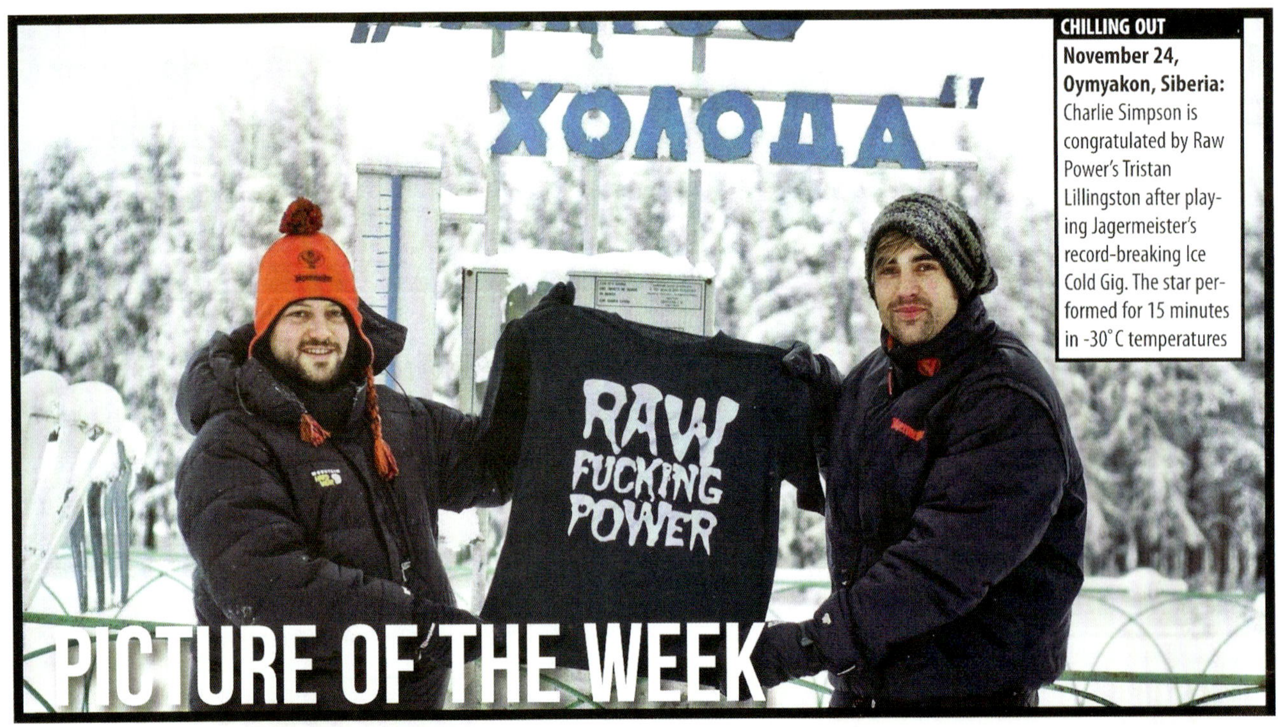

CHILLING OUT
November 24, Oymyakon, Siberia: Charlie Simpson is congratulated by Raw Power's Tristan Lillingston after playing Jagermeister's record-breaking Ice Cold Gig. The star performed for 15 minutes in -30˚C temperatures

Charlie Simpson (right, here with music manager Tristan Lillingston) and his ice-cold record were also celebrated in the British press.

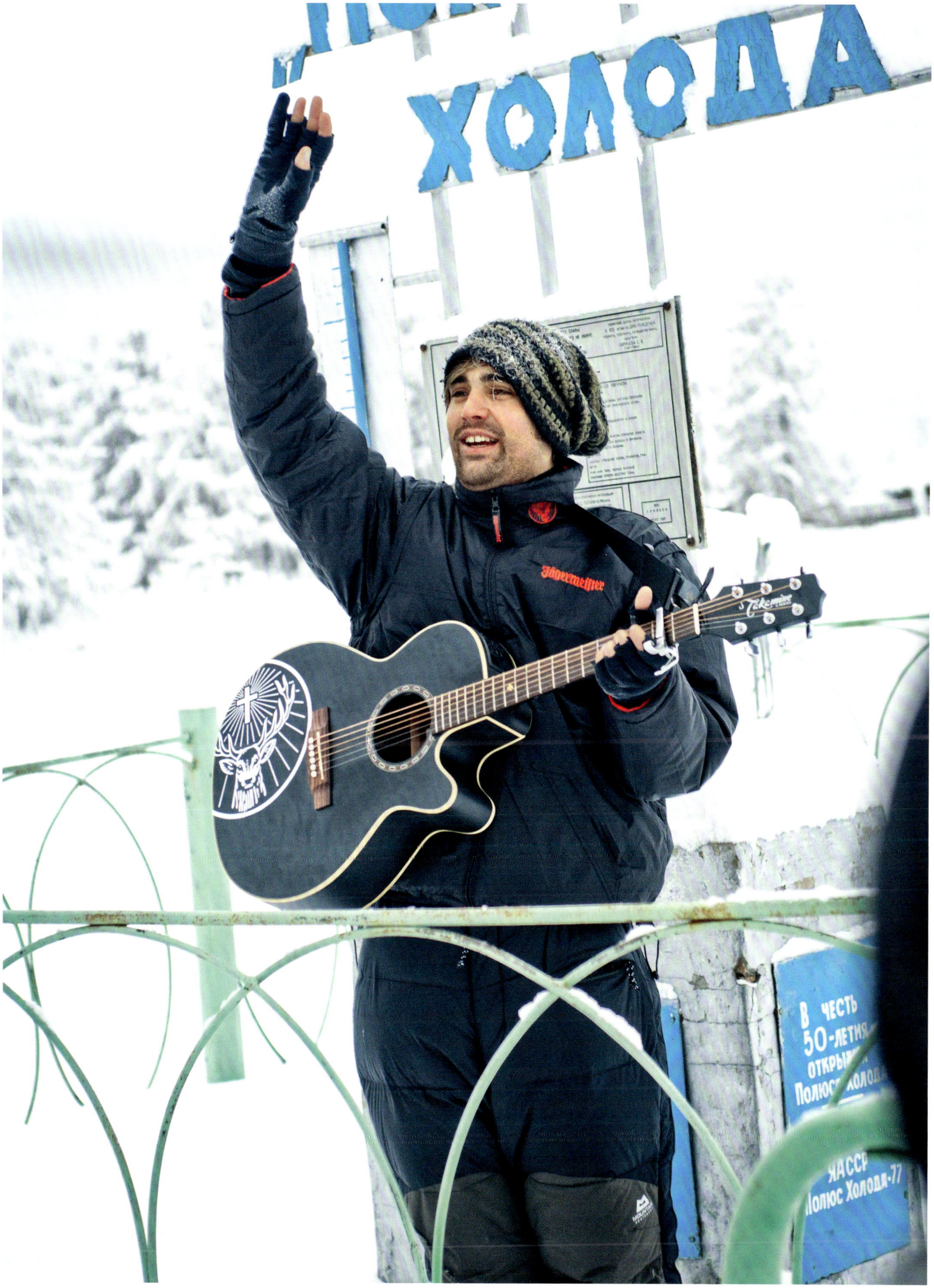

Full-on physical effort: to play his guitar with his bare fingers without catching frostbite, musician Charlie Simpson had to put hand warmers up his sleeves.

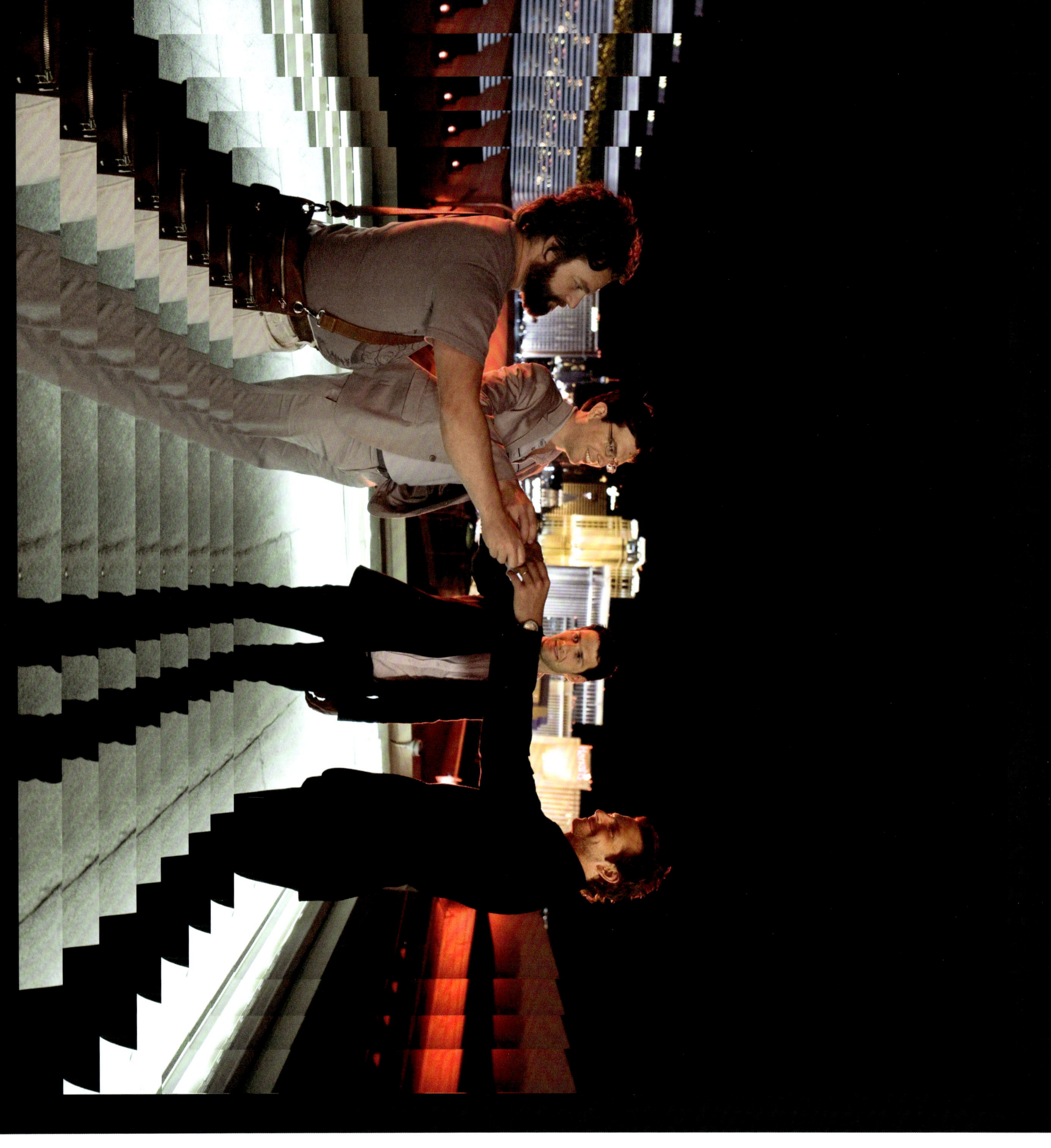

SHOT MOMENTS ON THE
BIG SCREEN

While Jägermeister is world-famous, there are some moments that really are the absolute pinnacle. Or, more accurately, shot moments. One of these is the famous film scene where four friends are sitting on the roof of a hotel, toasting the night ahead with a Jägermeister. The scene from the comedy blockbuster *The Hangover* (2009) – a film which has now become synonymous with out-of-control stag nights – was indeed shot on the roof of the legendary Caesars Palace in Las Vegas, and of course many wolf packs since have tried to recreate the notorious scene on that very same rooftop. Their attempts have been in vain, because the hotel's roof is closed to guests for safety reasons, but fortunately, Sin City has plenty of other places to fittingly propose a toast with Jägermeister – as long as it is ice-cold and shared with others.

Hollywood screenwriters quickly recognized the wild, unique nature of Jägermeister and have repeatedly used it as a symbol of close bonds, waywardness and untamed moments. Over the years, *Sons of Anarchy*, the cult television series about the seedy but friendly motorcycle club SAMCRO, has featured legendary guest stars such as Stephen King, Courtney Love, Henry Rollins, Dave Navarro – and Jägermeister. After a brief guest appearance at the start of the 'Authority Vested' episode (season 5, episode 2), Jägermeister reappears at minute 39 – spoiler alert! – as a 'guest' at the wedding of senior biker and SAMCRO president Jax Teller (Charlie Hunnan).

But Jägermeister wasn't just cast alongside real actors – a role was also made for the German herbal liqueur in the animated cult series *Family Guy*. And the creators of comedy cartoon series *Futurama* similarly like things a little wild: in the 'A Flight To Remember' episode, the Planet Express team takes a holiday flight through space. But Bender falls in love with a rich robot, the spaceship hurtles towards a black hole, and the situation seems hopeless. Only a stiff drink can help – and its ingredients? Pennzoil motor oil and Jägermeister.

Less absurd but all the more momentous was Jägermeister's role in the thriller *Killing Season* (2013) alongside two Hollywood heavyweights – Robert de Niro and John Travolta. To forget the traumas of his past, retired soldier Benjamin (de Niro) lives a reclusive life in a small hut in the woods. One day, he comes across European tourist Emil (Travolta). The pair become friends over a shared shot – until Benjamin finds out Emil's true identity, and a cat and mouse game begins . . .

Not Hollywood, but equally cult-like, is the TV series *Stromberg*, Germany's answer to *The Office*, featuring Christoph Maria Herbst as Bernd Stromberg. After five successful seasons, the head of the Claims Settlement department celebrated his comeback on the big screen in 2014 – as egocentric and snarky as ever. In *Stromberg – Der Film*, he and his staff embark on a unique kind of work outing – Jägermeister included. Cheers then, boss!

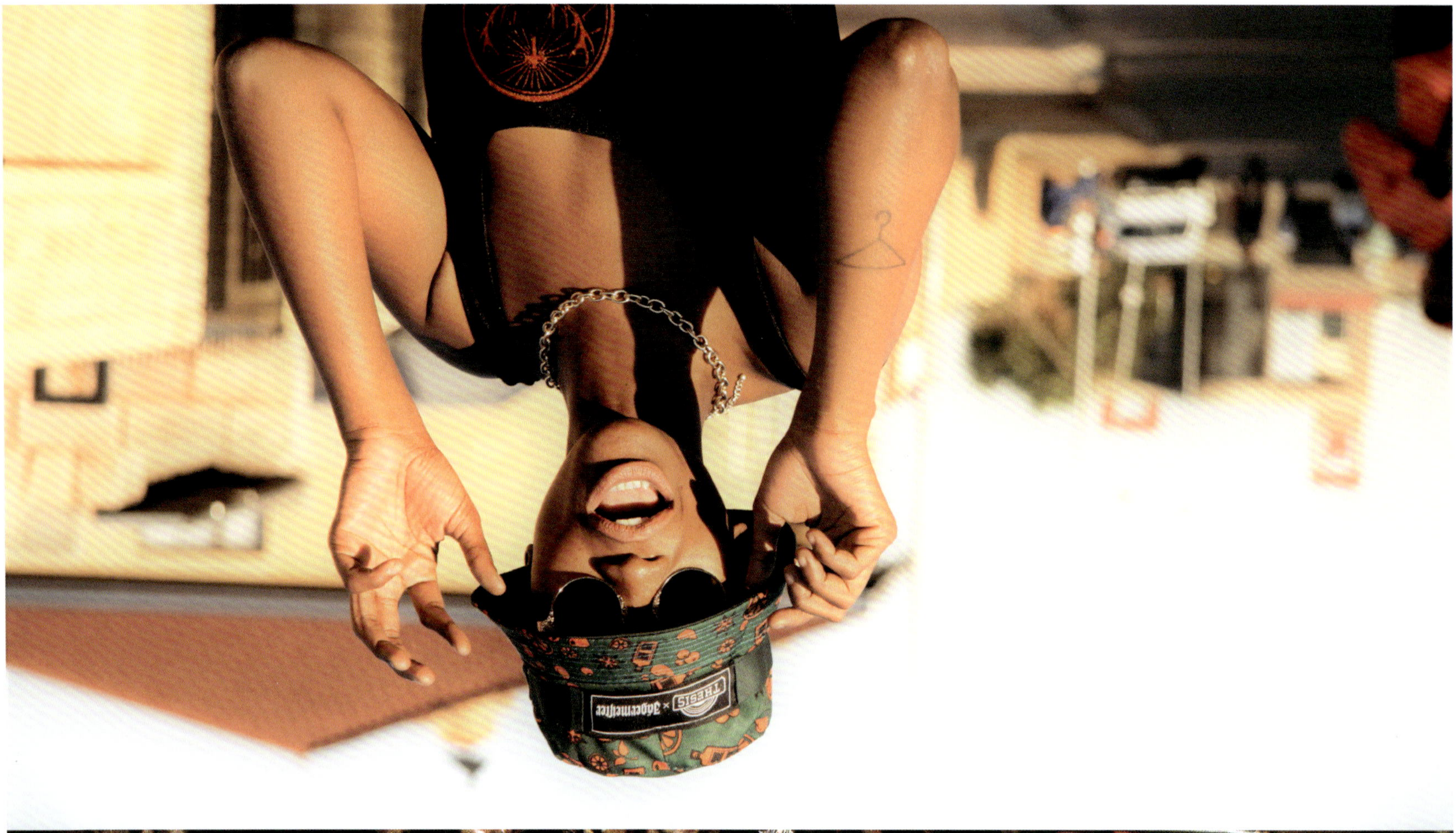

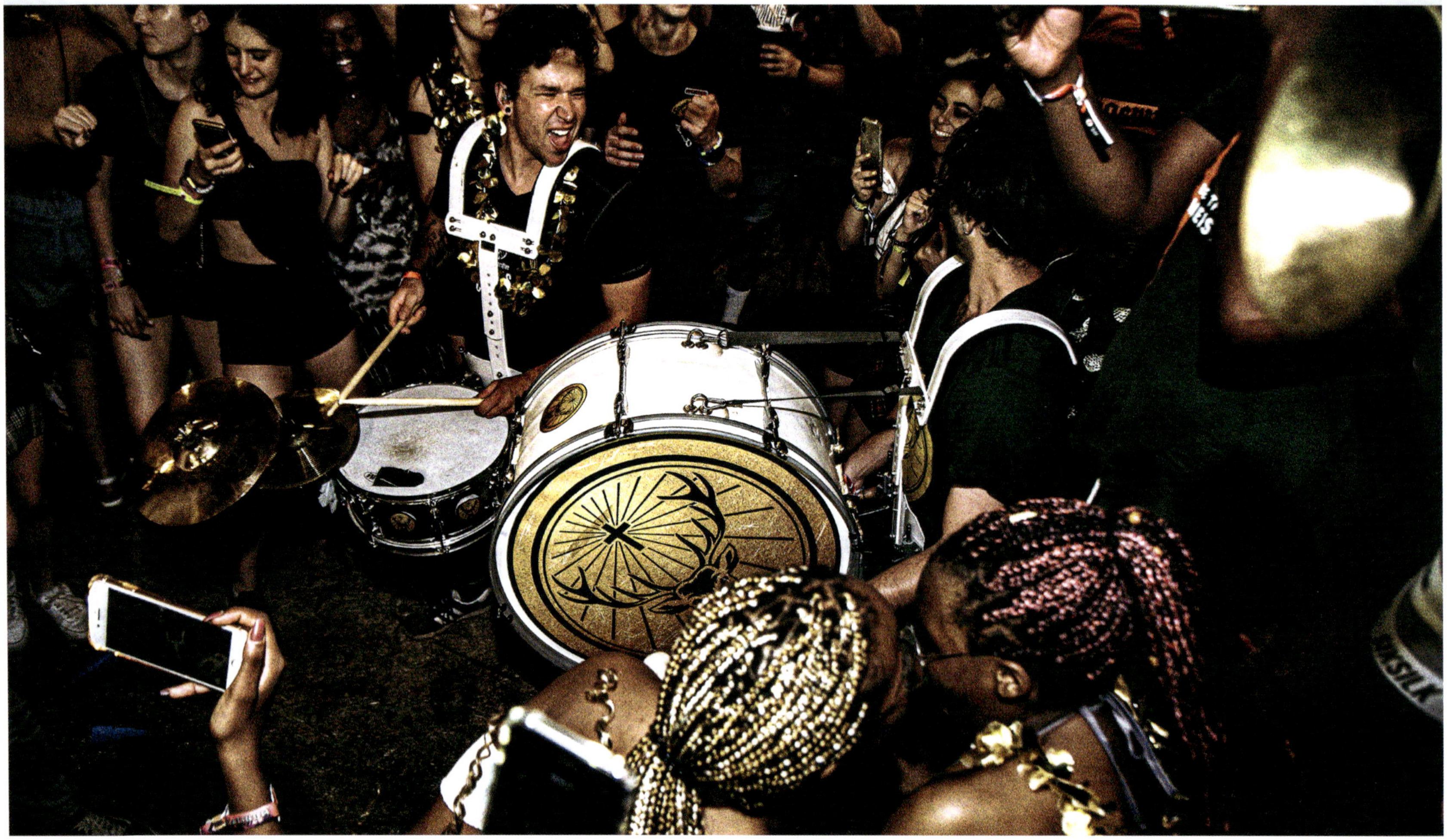

LOUD AND VIBRANT IN SOUTH AFRICA

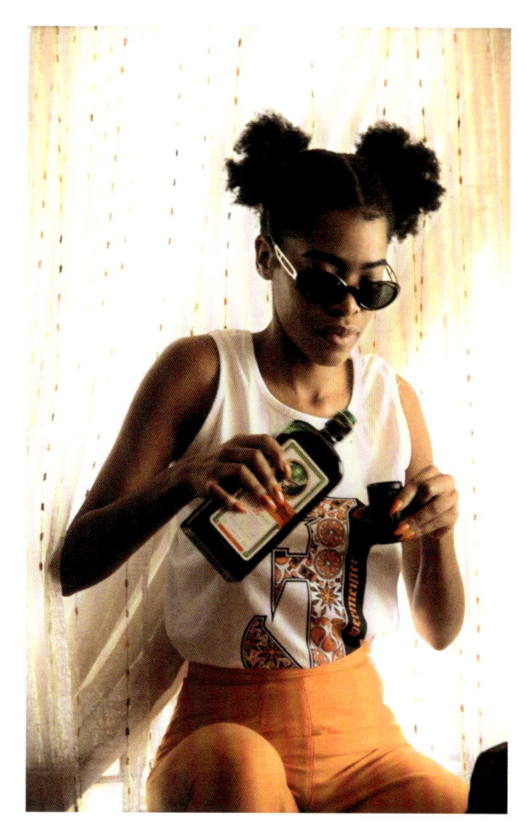

Africa is diverse, fascinatingly modern and wild – just like Jägermeister, so it's no wonder that the German herbal liqueur also has many fans on that continent. The South African version of the Jägermeister Blaskapelle plays a part here too. Since its founding in 2016, the Jägermeister Brass Cartel has twice toured through Swaziland, Namibia and the Democratic Republic of the Congo. In collaboration with top-class artists such as Rouge, Reason, Kenzhero and Grassy Spark, it reinterprets the latest chart hits and popular classics – just like its German role model – and creates a hot vibe for ice-cold shots in the process. But Jägermeister also has a strong presence in Africa when it comes to fashion and lifestyle – once again under the motto of being unique and unusual. Launched in 2018, the Thesis Streetwear collection is Jägermeister's answer to South Africa's exciting culture, which is particularly reflected in the colours and designs of the individual pieces. They are unconventional, unique, South African – and orange! Not only is this colour one of Jägermeister's brand colours, it is also a key part of traditional South African fashion.

Even more fashion in Africa: Jägermeister provided various fabrics for the Fashion without Borders initiative in a bid to support and bring together young and emerging African fashion designers. Five designers each prepared different Jägermeister collections, which were then presented at the Fashion without Borders show in Botswana.

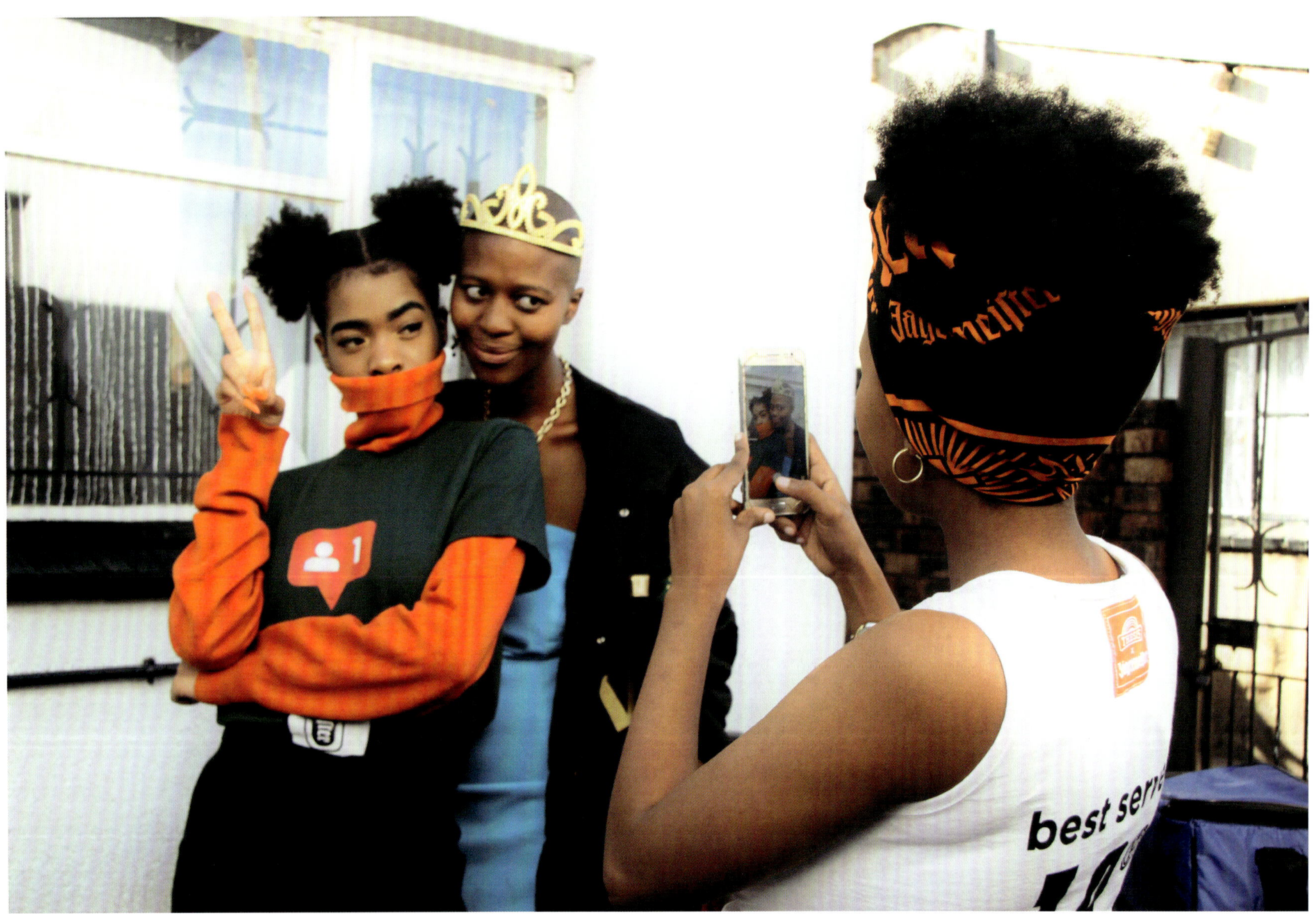

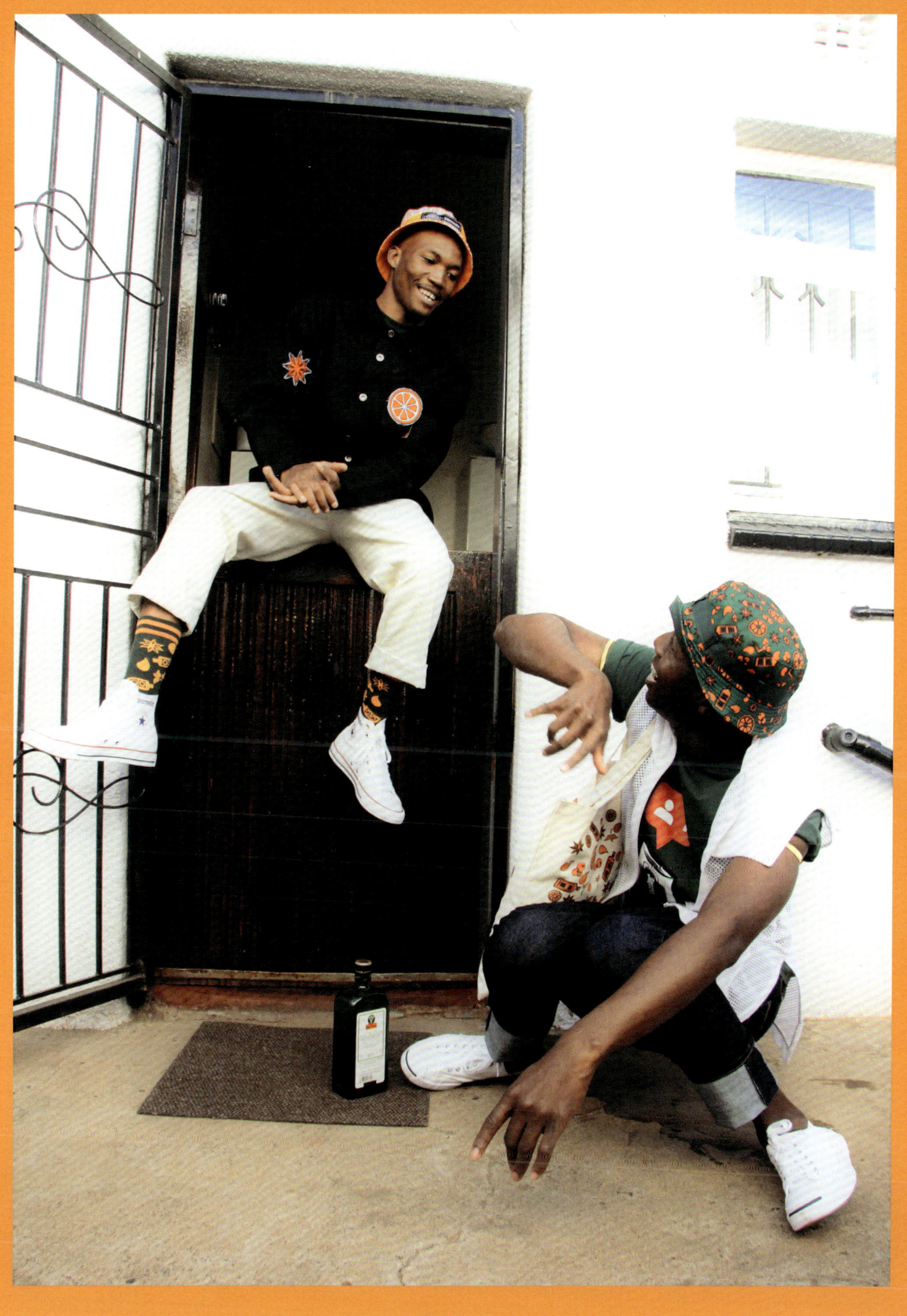

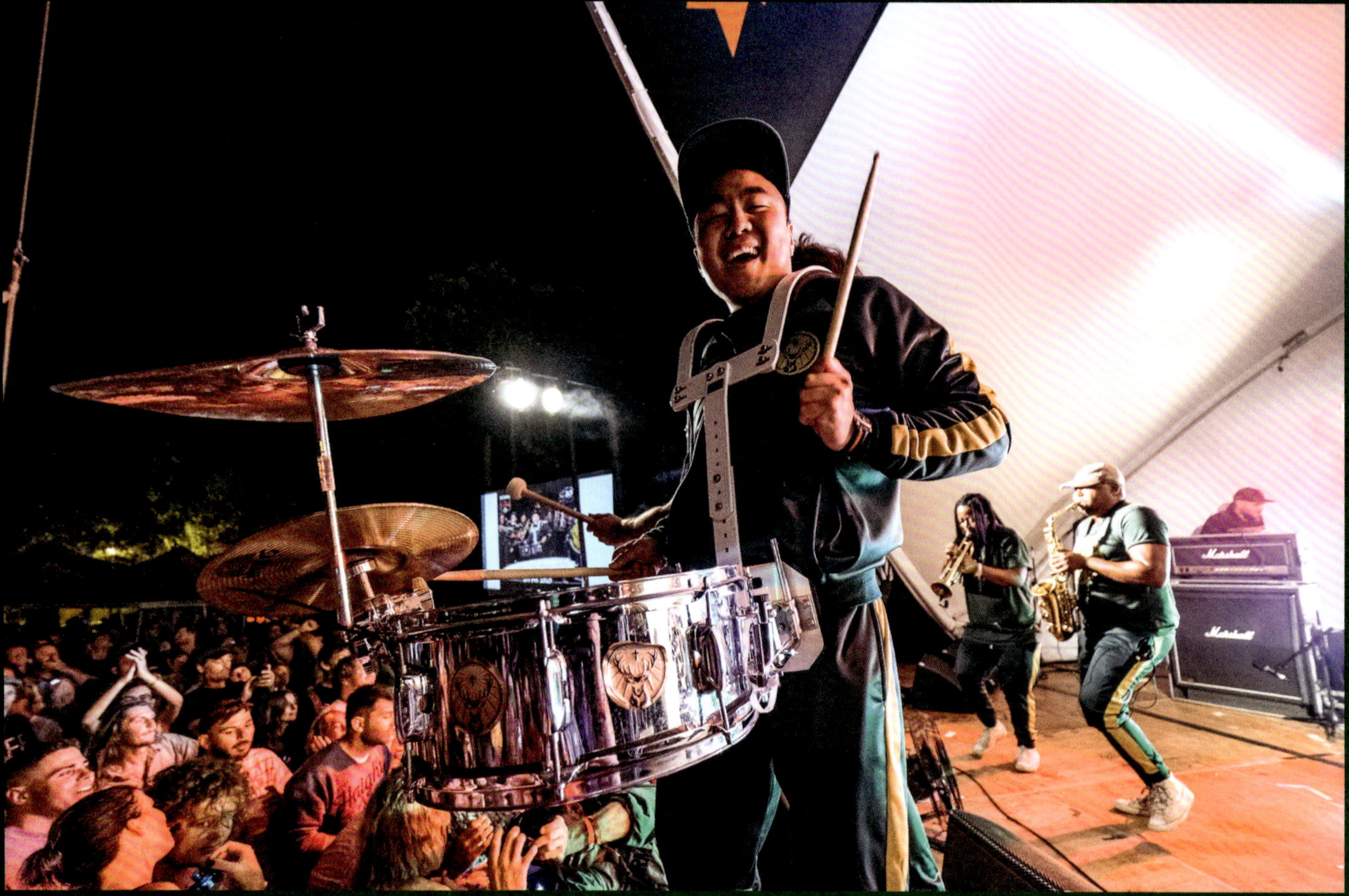

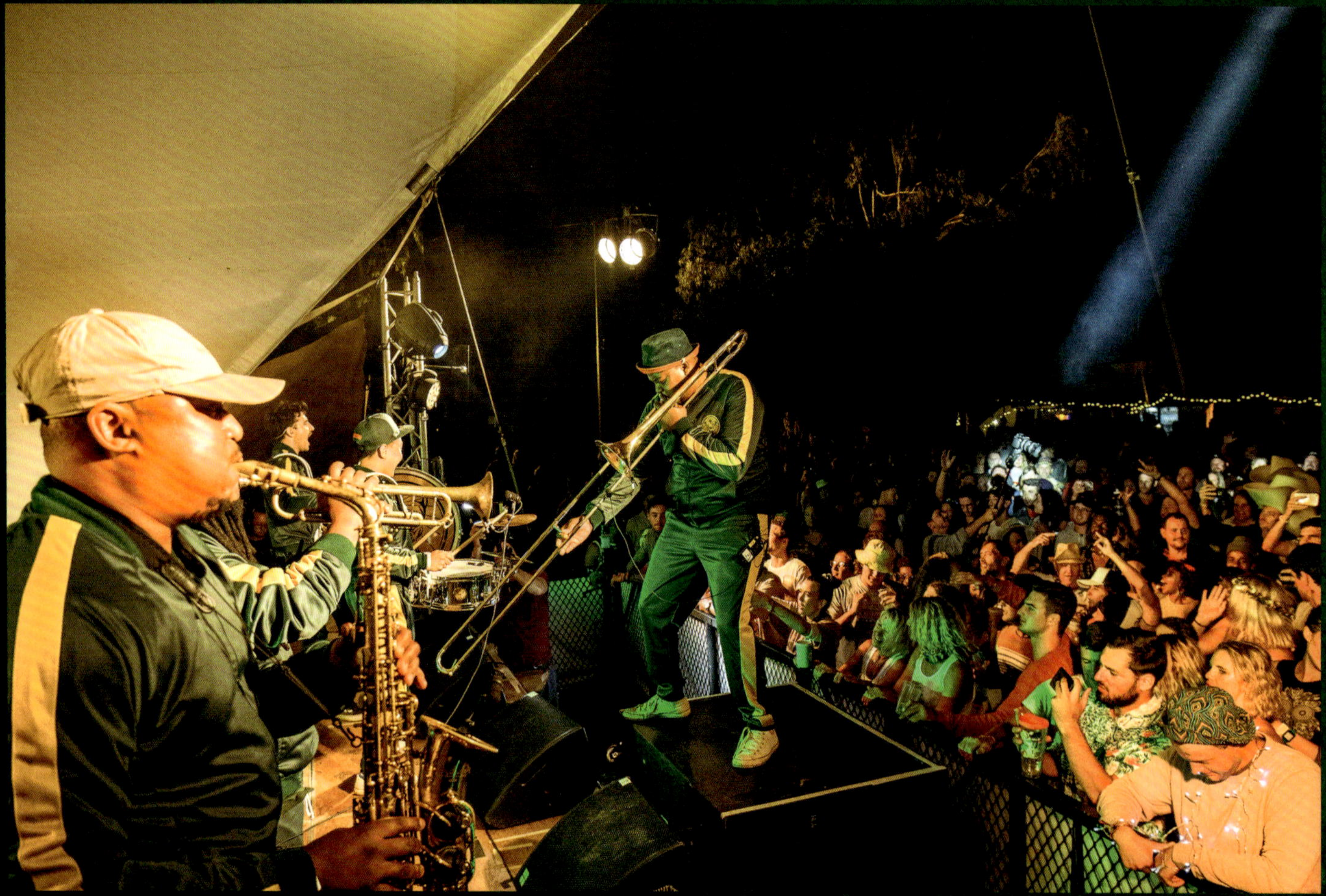

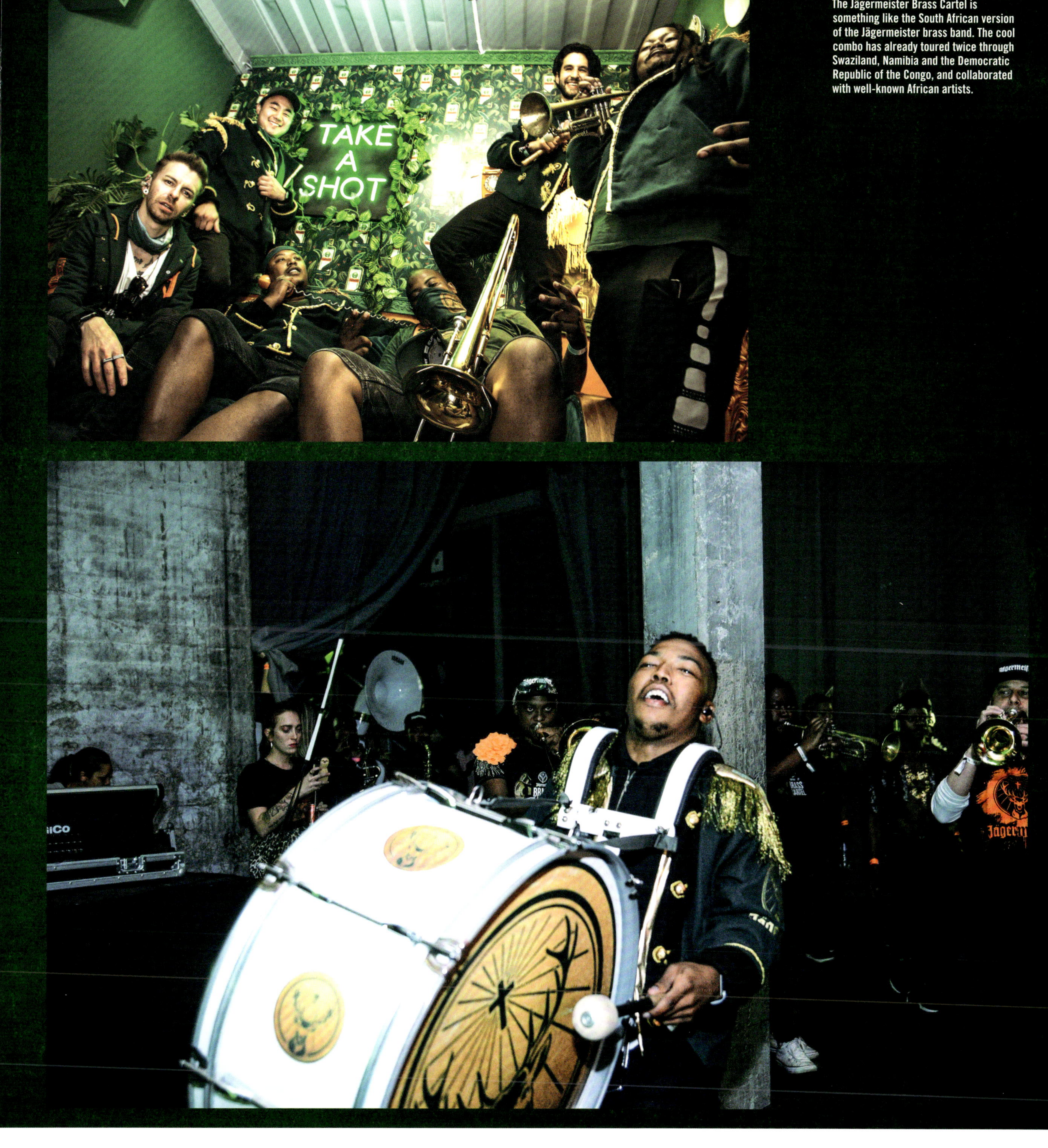

The Jägermeister Brass Cartel is something like the South African version of the Jägermeister brass band. The cool combo has already toured twice through Swaziland, Namibia and the Democratic Republic of the Congo, and collaborated with well-known African artists.

THE 'ICE COLD TOUR' OF INDIA

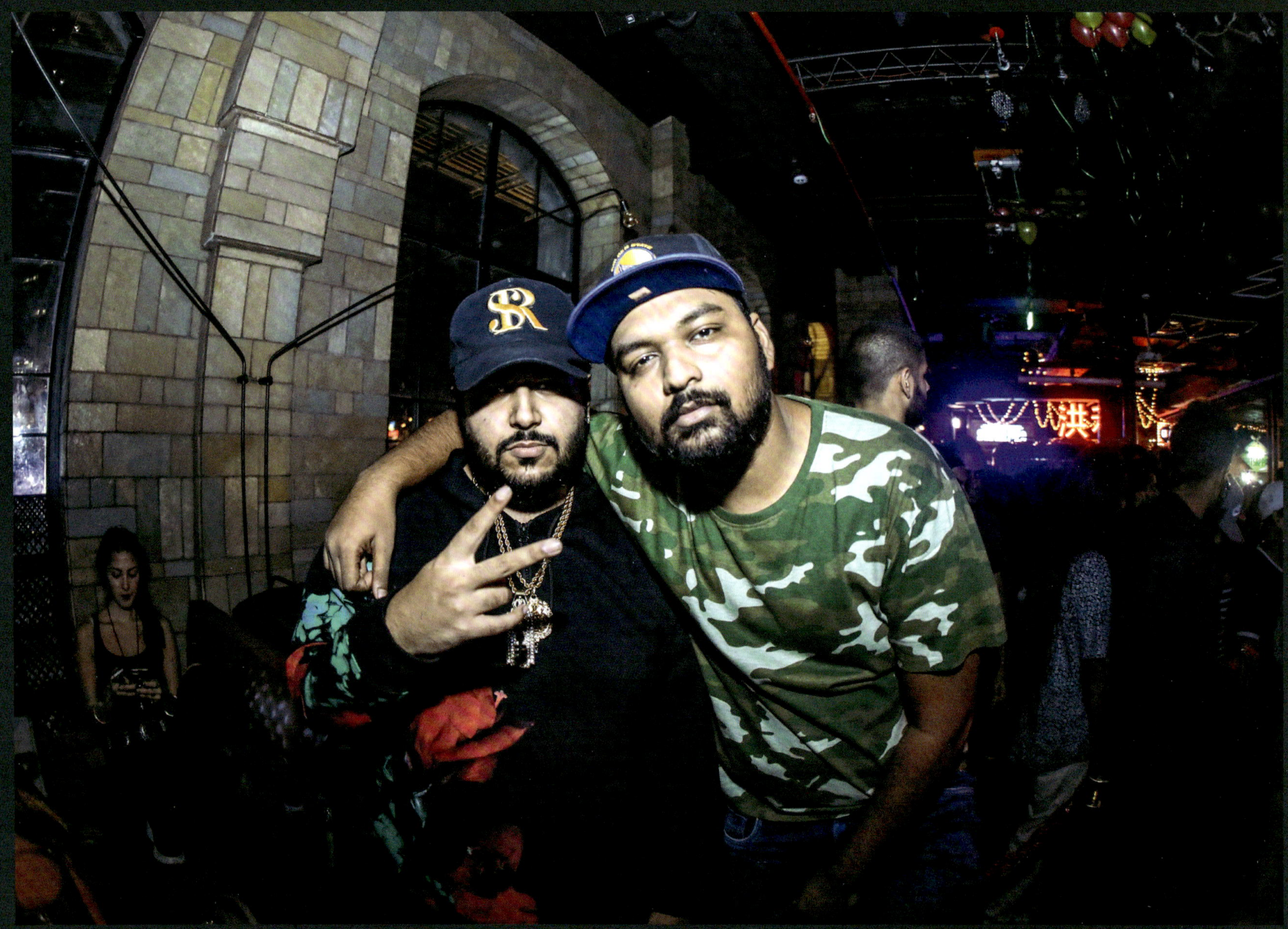

One song, one rapper and thousands of Jägermeister shots: with his song *Ice Cold*, a homage to the internationally loved herbal liqueur, rapper Shah Rule toured eight Indian cities in 2019, getting dance floors throbbing everywhere from New Delhi to Goa, with help from DJ Proof and Jäger Music. 'Ice Cold' in this heat? That's not a problem when you've got shots aplenty!

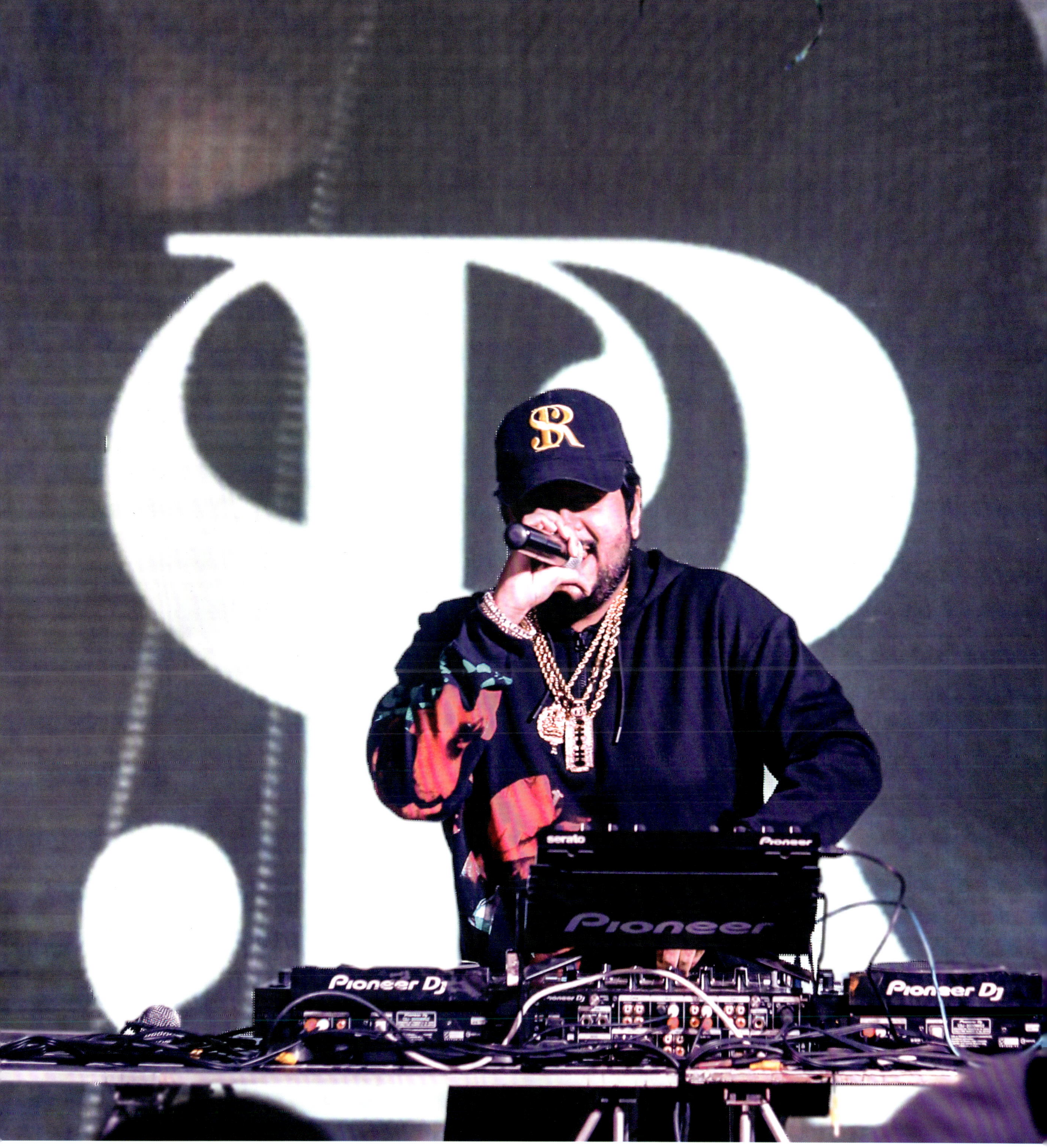

serato
Pioneer
Pioneer
Pioneer DJ
Pioneer DJ

ISRAEL: MEMORIES FOR ETERNITY

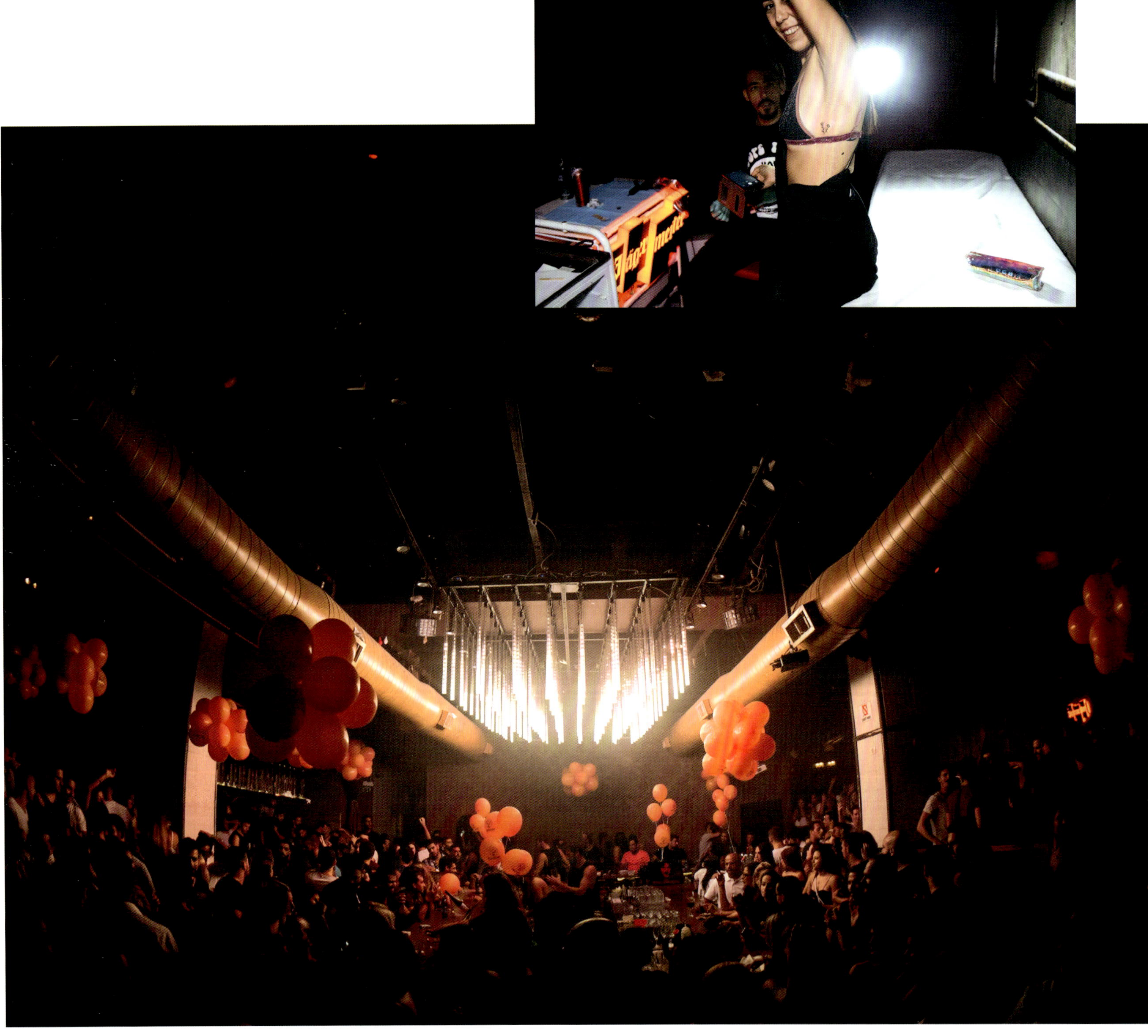

Not all of the visitors to the Jägermeister Tattoo Tour were able to take home a free tattoo, but they all had impressively clear memories of a rather wild evening!

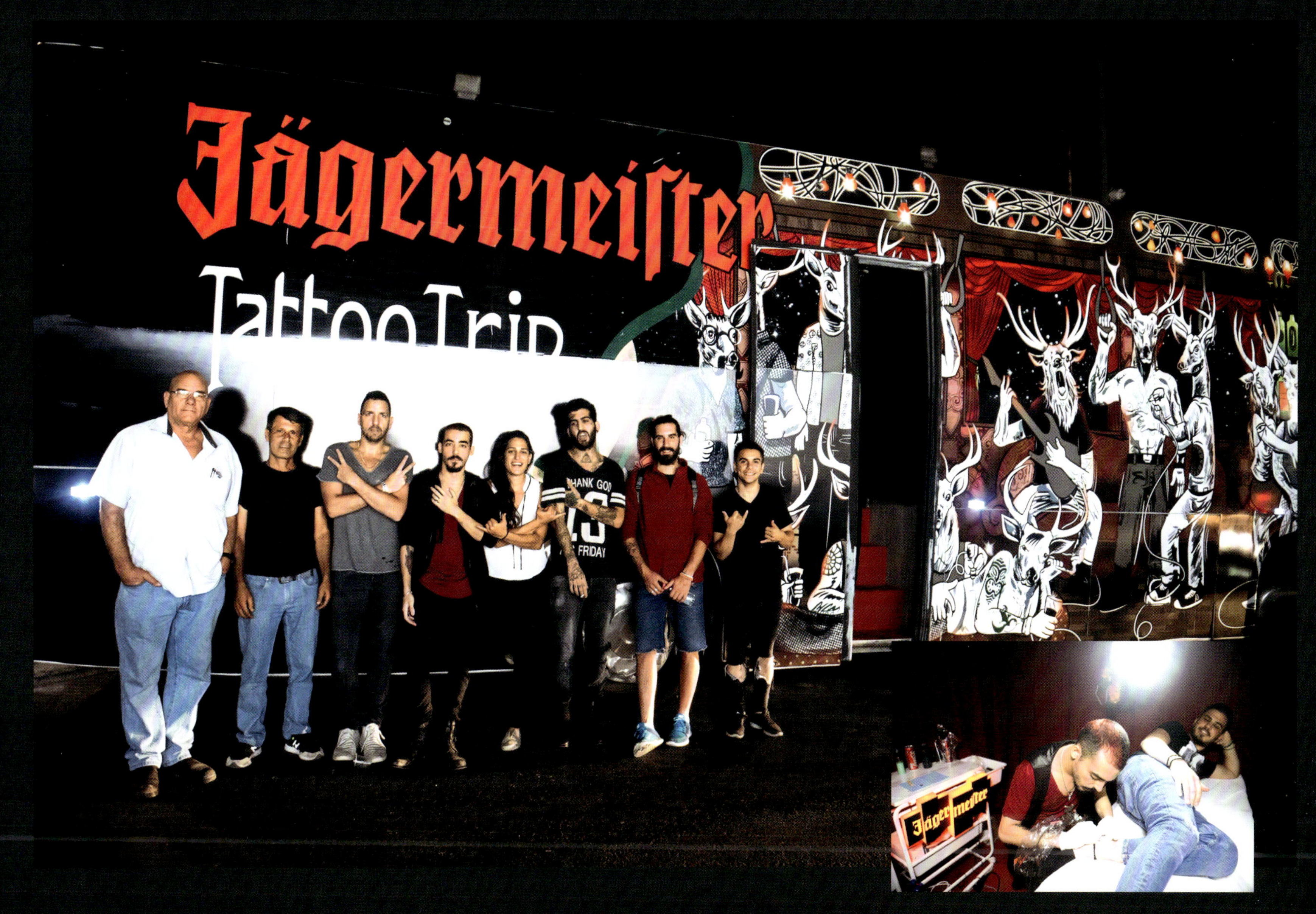

Jägermeister is popular all over the world, and sometimes this love for Jägermeister gets more than skin deep, as proven by the Jägermeister Tattoo Tour in Israel. The 'injection' of originality that was this campaign saw a bus converted into a mobile tattoo studio. At 12 different bars, patrons were given the opportunity to score a free tattoo when buying Jägermeister shots, receiving lifelong Jägermeister souvenirs courtesy of two artists in the mobile tattoo parlour. The promotion was naturally broadcast live on Facebook. By the end of the ten-day tour of Israel, 100 lucky Jäger fans had not only experienced an unforgettably wild night, but were also able to take a memory home with them forever.

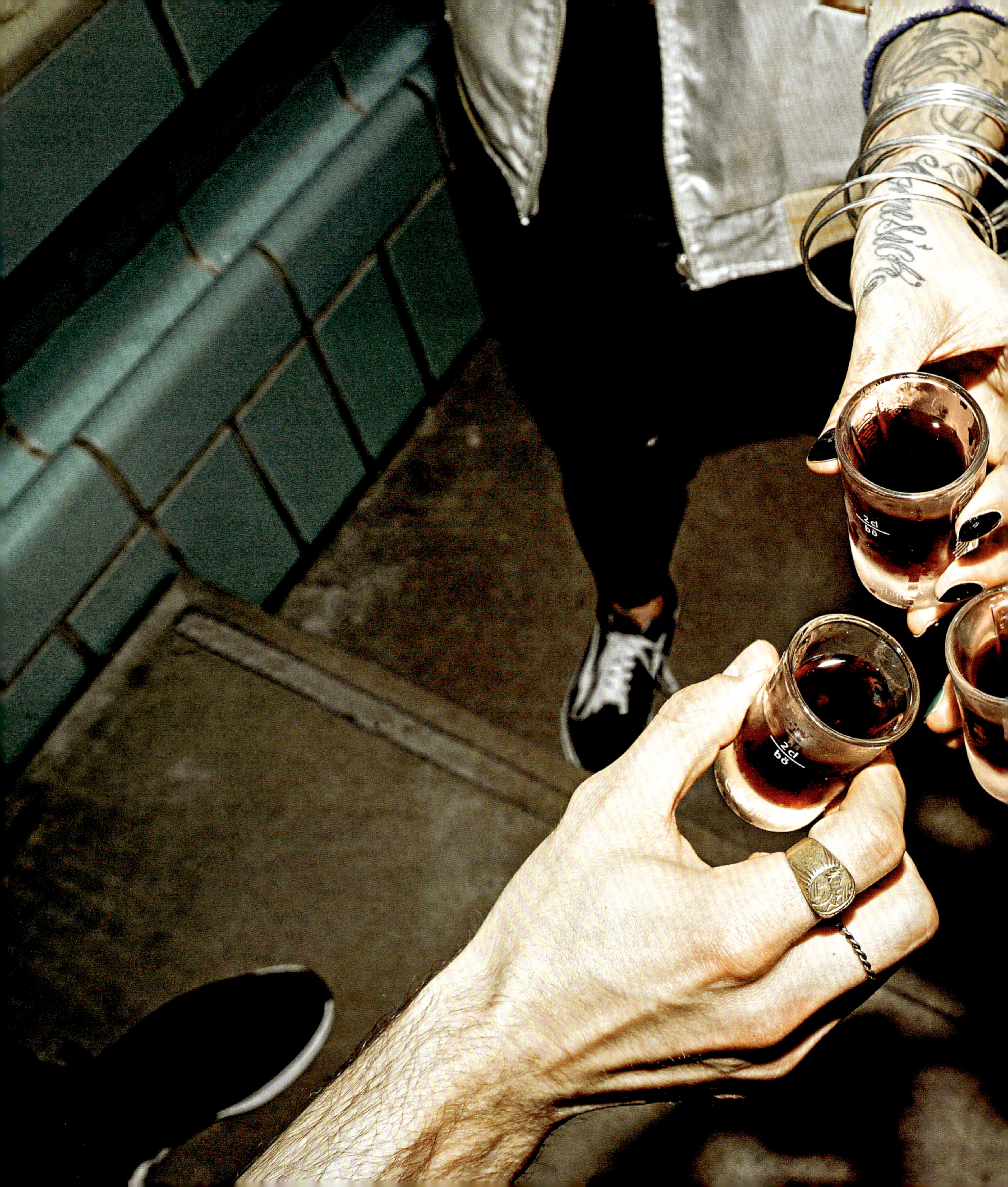

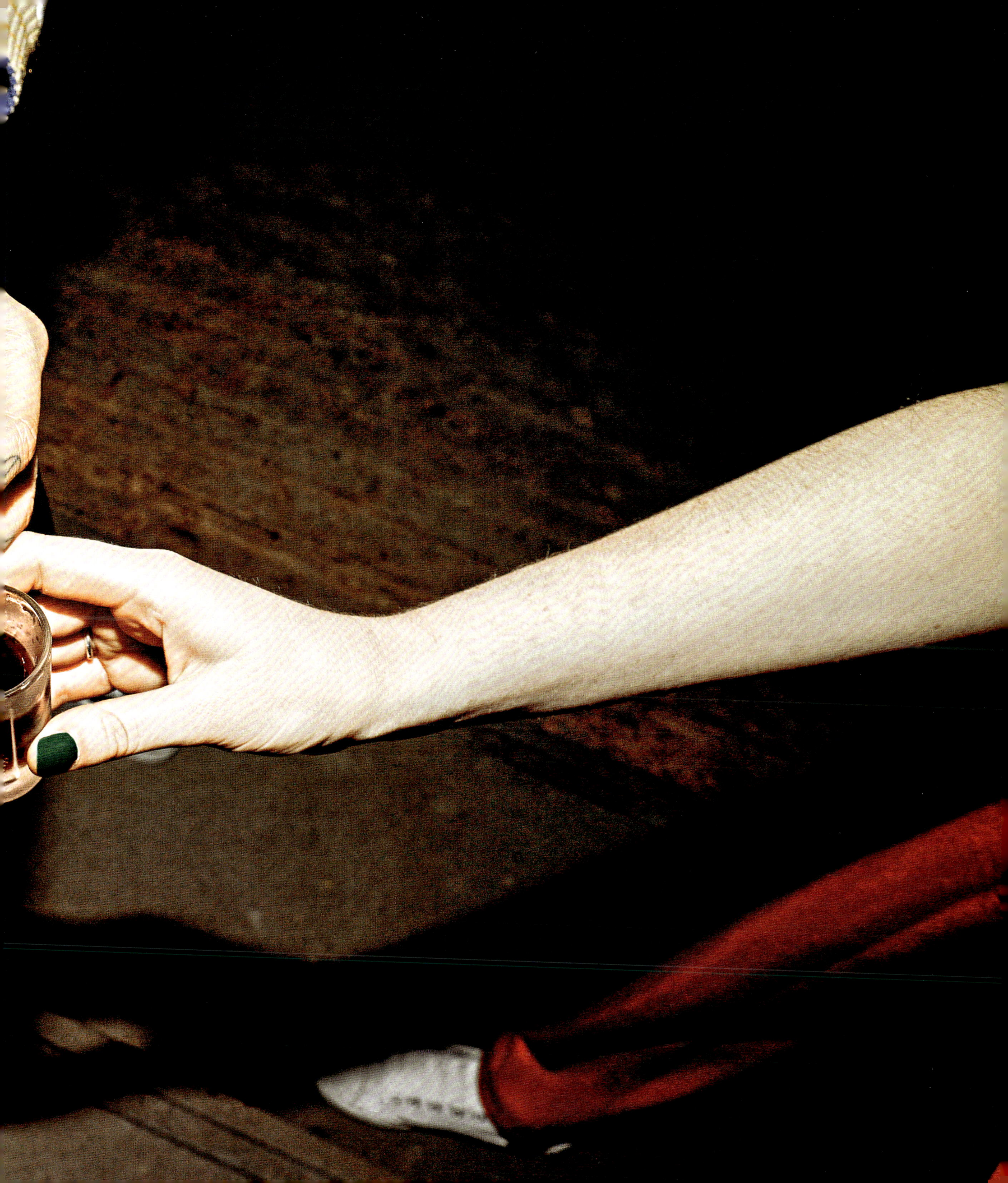

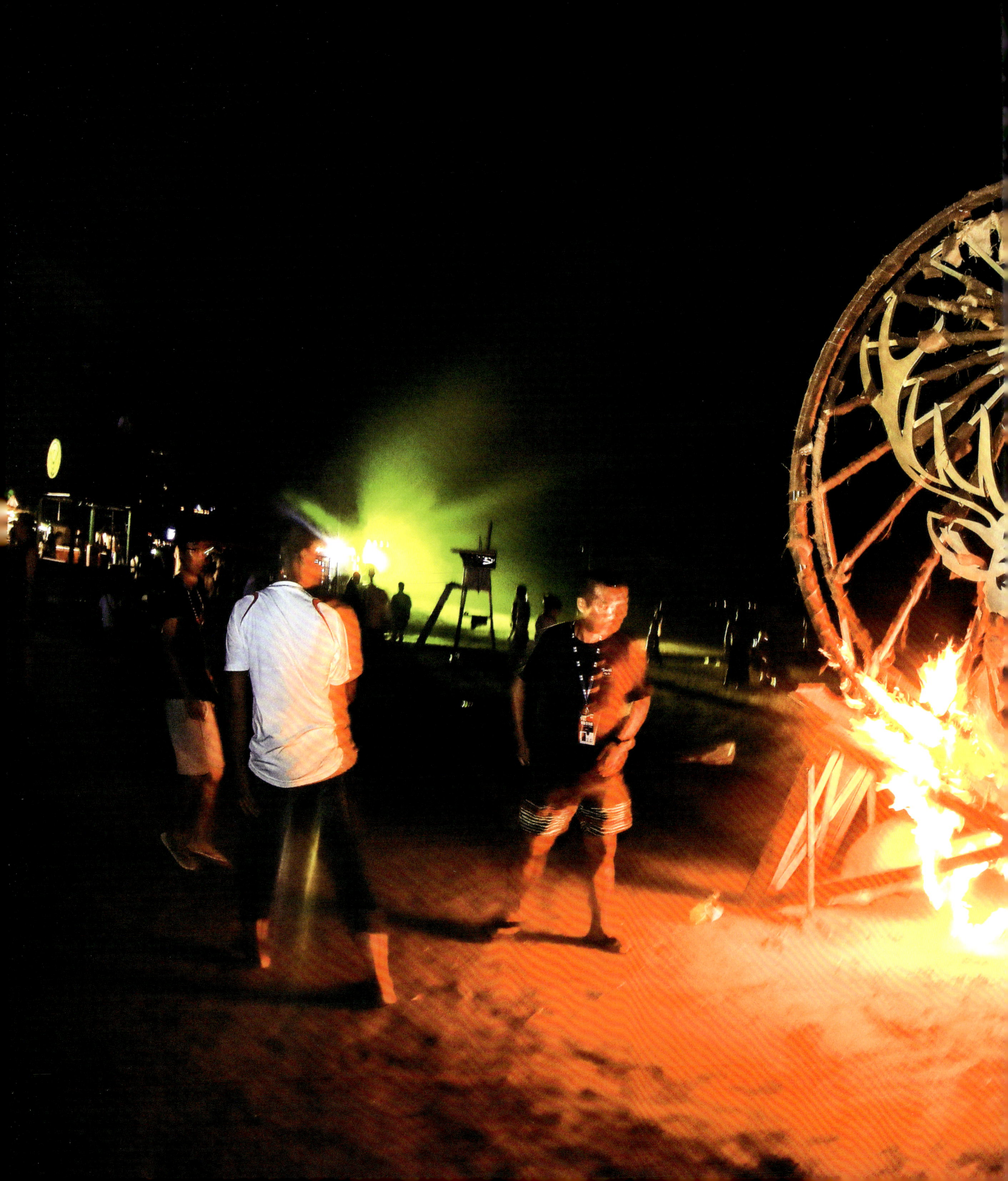

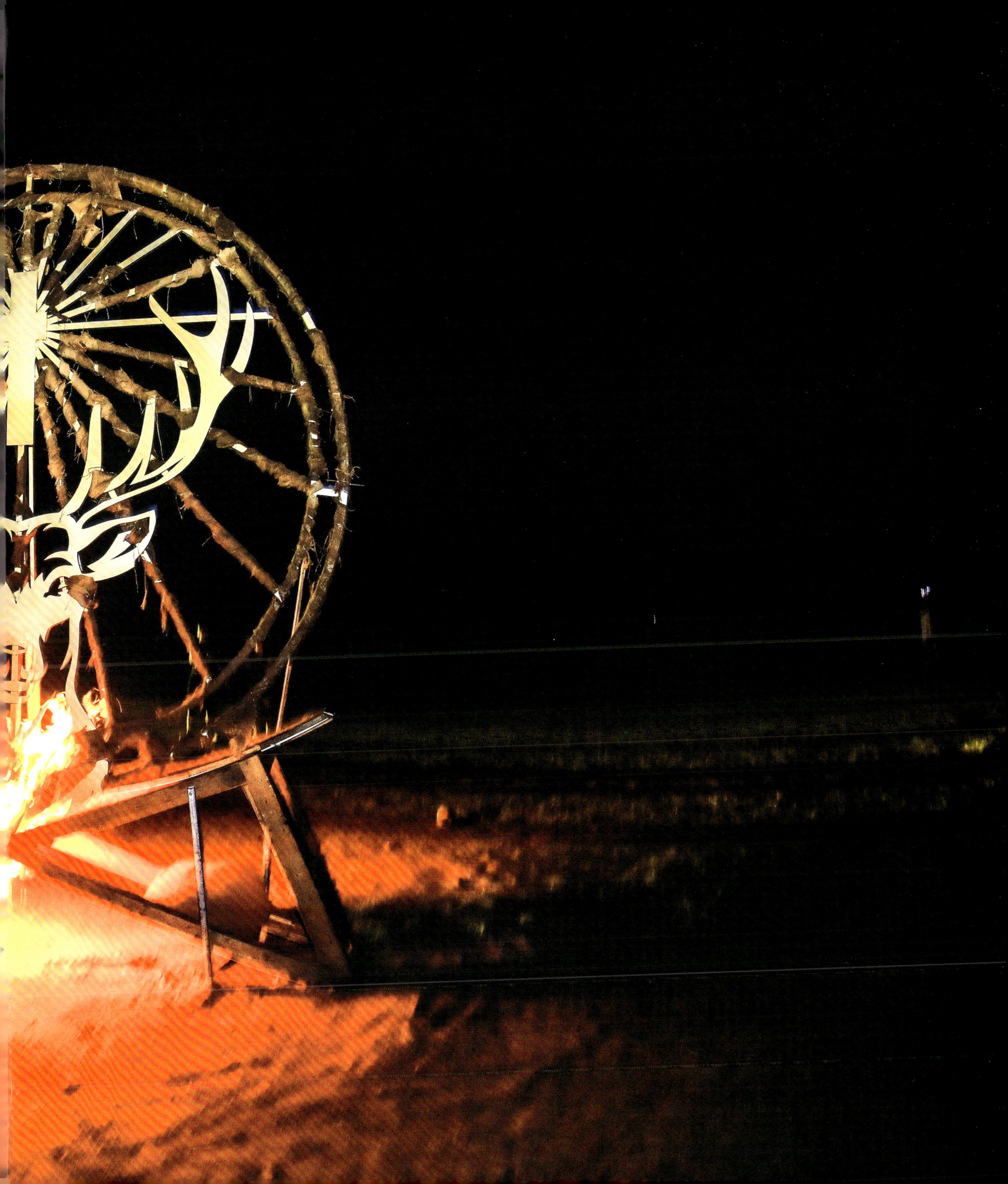

Imprint

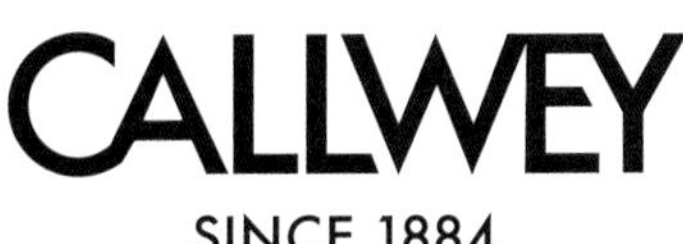

CALLWEY
SINCE 1884

2021 Callwey GmbH
Klenzestrasse 36
D-80469 Munich, Germany
buch@callwey.de
Tel.: +49 89 8905080-0
www.callwey.de
On Instagram: @callwey

ISBN 978-3-7667-2506-6

1st edition 2021

Bibliographic information of
the German National Library

The German National Library lists this publication in
the German National Bibliography; detailed bibliographic data are available online at http://dnb.d-nb.de

This book was produced in Callwey quality:
For the pages, we chose 150 g/sq m Magno
Satin – a picture-printing paper whose silk matt sur-
face gives the content a noble and high-quality char-
acter. This book is printed and bound in Germany by
Optimal Media GmbH in Röbel/Müritz.

Author:
As a creative consultant, author and journalist Anja
Delastik has already created numerous publications.
Yet she is so much more than a talented media professional. As a person with a natural instinct for
trends and people, and with her passion for pop culture, music and art, she got to the heart of the Jägermeister cult in all its diversity in *The best Nights of
your Life*.

Picture Editor:
The best shot, the best moment and the best night of
your life deserve one thing above all else: the best
images that move and touch you. This is exactly what
Heide Christiansen has created in this book with passion and dedication. As a photographer, picture editor and creative director, she works for renowned
magazines and brands and has already illustrated several successful books at Callwey.

Acknowledgements:
The Jägermeister team would like to thank all col-
leagues, partners, the entire Callwey team, friends,
fans and all the supporters, without whom this book
would not have been possible.

Head of Public Relations:
Andreas Lehmann
Archivist:
Florian Eisenblätter
Manager Engagement & Events:
Silke Hulzer
Manager Corporate Communications:
Sonja Wilkens

We hope you enjoy this book:
Project Manager:
Johanna Böshans
Copyeditor & Contributing Editor:
Victoria Wegner
Contributing Editor:
Max Marquardt
Final correction:
Andreas Leinweber
Art Direction & Layout:
Rose Pistola GmbH, Munich
Typesetting: (English edition):
bookwise medienproduktion GmbH
Translation:
Emily Plank, Sylvia Goulding
Digital features:
The Cookie Labs GmbH
Comic:
AR by EyeJack,
Illustrated and animated by Alexander Hare
Production:
Dominique Scherzer / Oliver F. Meier

Picture credits:
Cover: Mast-Jägermeister SE, Wolfenbüttel
Endpapers: Chris Noltekuhlmann
p. 8/9 Tobias Ebert, p. 12-14 David Bonvillain Design
Studio, p. 15 Jack Carson, p. 16/17 Bernd Frank (20),
Thomas Gottfried (1), p. 18/19 Bernd Frank (12),
Thomas Gottfried (1), p. 20 Jaewon Chung, p. 22/23
Jaewon Chung, p. 24 Fritz Rust, Hannover/Mast-Jäger-
meister SE, Wolfenbüttel, p. 25 Rust/imago, p. 26/27
Werner Baum/picture-alliance/dpa, p. 29 Alfred
Kogler privat, p. 29 Renan Kamikoga/unsplash, p. 30
Katarina Benzova, p. 31 Personal Collection, p. 32
Katarina Benzova, p. 33 Katarina Benzova, p. 35 Bild-
agentur Kräling, p. 36 Bildagentur Kräling (4), p. 37
Bildagentur Kräling, p. 38/39 Tattooclique (10), p. 40
Nela König, p. 41 Holmes Hiebert, p. 42 Andreas Röhl,
p. 44 Andreas Röhl, p. 45-47 Historisches Fotoarchiv
Wolfgang Lange, Wolfenbüttel, p. 48/49 Martin Huch,
Produktion: Heide Christiansen, p. 50/51 Holger Ston-
jek, p. 54 Luca Massaro, p. 56/57 Frank Engel, p. 58/59
CHEFBOSS, p. 60 Christoph Neumann, p. 62 PR Jäger-
meister, p. 63 Stone Brewing, p. 64-75 Illustration:
Alexander Hare, p. 76/77 Chris Noltekuhlmann, p. 80
Chris Noltekuhlmann, p. 84/85 Chris Noltekuhlmann,
p. 86-97 Bildagentur Kräling, p. 98-102 Fritz Rust,
Hannover/Mast-Jägermeister SE, Wolfenbüttel, p. 103
Sven Simon/imago, p. 104/105 Werek/Süddeutsche
Zeitung Photo, p. 105 Eintracht Braunschweig, p. 110-
113 Fabrikx Media UG Köln, p. 114-116 Mark Latham,
p. 118/119 Matty Adame/unsplash, p. 120/121
INSOMNIAC, p. 122/123 Hans Just A/S, Denmark, p.
124_1 Jesper Bjarke Andersen, p. 124_2-125 Hans
Just A/S, Denmark, p. 126/127 Daniel Dvorsky/
unsplash, p. 136/137 BRIX & MAAS, p. 139 Fritzel's,
p. 144/145 Chris Noltekuhlmann, p. 148_2 Mark
Latham, p. 150 Milos Potuzak, p. 156 Chris Noltekuhlmann, p. 159 Chris Noltekuhlmann (5), p. 174/175
André Kirsch (23), p. 186/187 Chris Noltekuhlmann,
p. 192/193 Peter Kaaden, p. 194 Chris Noltekuhlmann
(2), p. 196 Chris Noltekuhlmann (2), p. 198 Chris
Noltekuhlmann (2), Produktion: Heide Christiansen,
p. 209-211 André Kirsch, Produktion: Heide Christiansen, p. 212-215 Camille Blake, p. 217 Fritzel's,
p. 218/219 frederick-tadeo/unsplash, p. 230-233
Verein Int. Lauberhornrennen, p. 240 imago,
p. 252/253 Peter Kaaden
Mast-Jägermeister SE, Wolfenbüttel:
p. 34, p. 36 (1), p. 37, p. 43, p. 44 (3), p. 52, p. 53. p. 55,
p. 78/79, p. 81-83, p. 106/107, p. 109, p. 128-135,
p. 138, p. 140-143, p. 146-147_1, p. 148_1, p. 149,
p. 151, p. 152/153, p. 154/155, p. 157, p. 159 (2), p. 160-
164, p. 165, p. 169-173, p. 174/175 (3), p. 176/177,
p. 178-185, p. 189-191, p. 195, p. 197, p. 199-203,
p. 206/207, p. 220-229, p. 234-239, p. 242-251,
p. 254/255

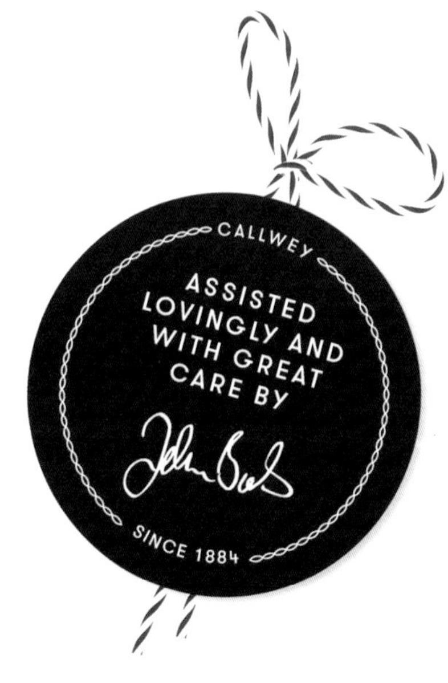

">

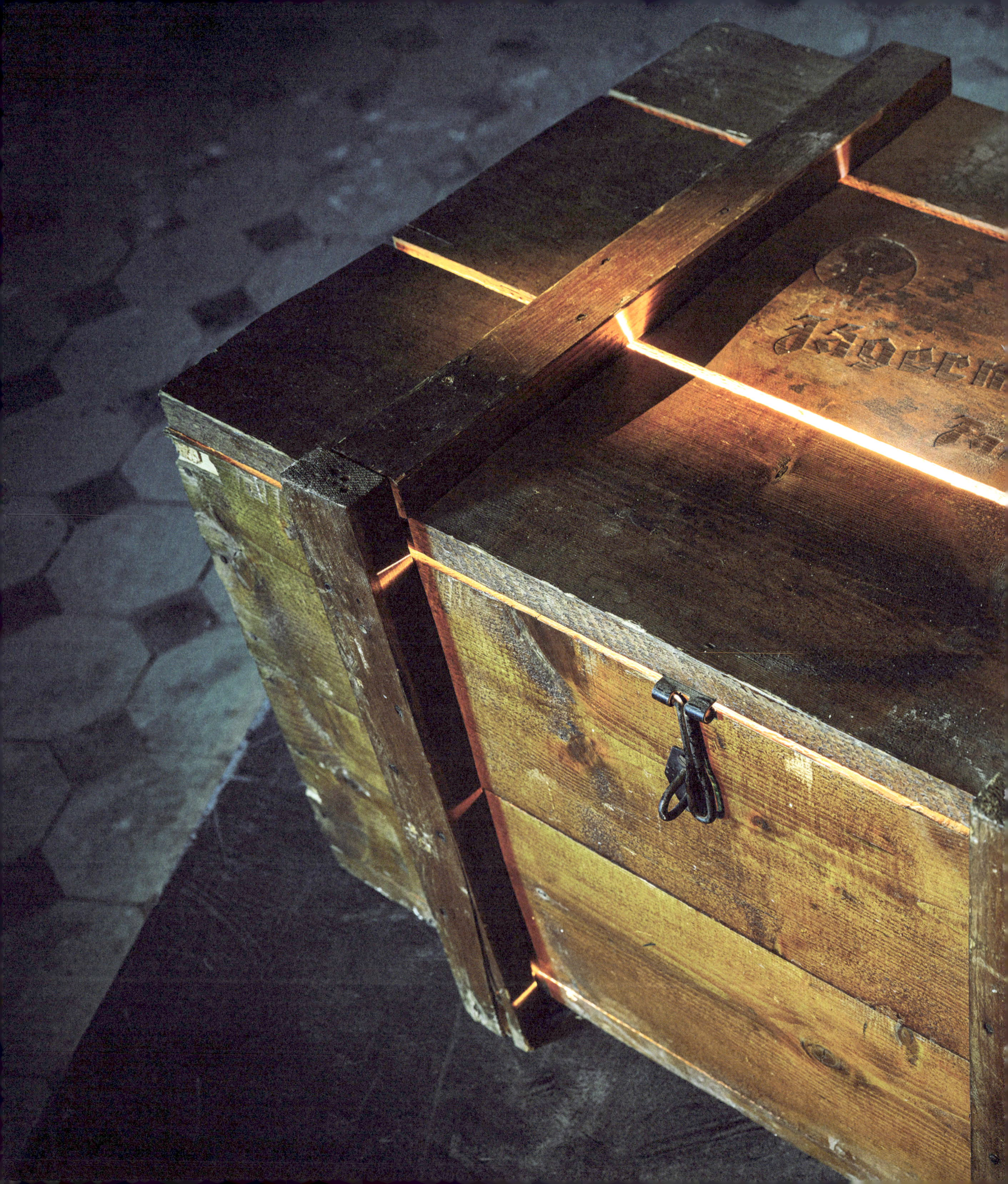